T0192452

Fundamentals of Fluid Mechanics

Patrick Chassaing

Fundamentals of Fluid Mechanics

For Scientists and Engineers

 Springer

Patrick Chassaing
National Polytechnic Institute of Toulouse
Toulouse, France

ISBN 978-3-031-10088-8   ISBN 978-3-031-10086-4   (eBook)
https://doi.org/10.1007/978-3-031-10086-4

To Colette,
 Jean-Camille,
 Benoît,
 Pierre-Marie.

Preface

Discipline among the oldest in physics, Fluid Mechanics was first recognized by its applications whether addressing air flows (mills, marine sailing), water flows (irrigation, hydraulic machines) or geo-fluids motions (meteorology, atmospheric and oceanic circulation).

Over time, the initial high level of empiricism gradually gave way to an increasingly rigorous approach, without eradicating any recourse to experience so far.

Thus, Fluid Mechanics is presented nowadays as a close association of physical concepts and mathematical theory, which describes the originality of this specialist area but also constitutes a source of learning difficulties for the student and teaching complexity for the lecturer.

It is with the aim to overcome such challenges that this book was written, drawing on several years of pedagogical experience in various French Universities and national institutions of higher education in science and technology known as 'Grandes Écoles'. For this purpose, it is organised into three parts.

- The Fluid Mechanics physical concepts according to the continuous medium approach, in addition to the mathematical tools required for the theoretical analysis, are explained at the beginning of the first part. This is followed by the presentation of fluid kinematic concepts and the derivation of the general expressions of mass, momentum and energy balances. It is concluded by identifying two major structuring concepts which are fluid motion models and flow classes.
- The second part is devoted to ideal (inviscid) fluid flows. It includes the potential flow class in the incompressible regime and the quasi one-dimensional motions in the compressible regime.
- Newtonian viscous fluid motions, mainly in the incompressible regime, are the subject of the last part. The corresponding flow class is broached according to increasing Reynolds number values, from creeping motions to laminar boundary layer flows, taking into consideration, in this case, both dynamic and thermal issues.

In order to allow the student to improve his or her comprehensive level throughout personal reading, numerous exercises are available at specific points, with a view to pedagogical efficiency and for self-evaluation purposes. Finally, various additional information items and historical notes are provided for the reader wishing to go beyond simple access to consolidated contemporary knowledge.

Although this work is intended to provide a substantial introduction to a comprehensive study of Fluid Mechanics, it is far from giving a complete overview of the thematically rich content of this specialist field.

However it is the author's expectation to whet the reader's appetite and to stimulate the desire to learn more and to delve deeper into the fascinating field of Fluid Mechanics.

Toulouse, France   Patrick Chassaing
March 2022

Acknowledgements

This fourth edition and the first in the English language, revised and updated, could not have come into being without various sources of assistance, first and foremost that of the National Institute of Electrical engineering, Electronics, Computer Science, Fluid Mechanics and Telecommunications & Networks[1]. We would hereby like to convey our deep gratitude to the director, Professor, Jean-Franois Rouchon.

Mrs. Marianna Braza, Director of Research at the CNRS, the French National Centre for Scientific Research, initiated the editorial project and gave her unfailing support during various stages of the work. We cannot sufficiently express our deepest gratitude for her determination, commitment and tireless dedication throughout this editorial project.

Long postponed, the translation of the book into English could not have been completed without the precious and consistently wise assistance of Mrs. Anne Brittain, English teacher at ENSEEIHT. With our sincere thanks, we would like here to express to her our grateful appreciation for her comments and advice.

We cannot complete these acknowledgements without a special thought for all the students at ENSEEIHT, ENSICA[2], SUPAERO[3] and Toulouse III-Paul Sabatier University who, by their justified demands for pedagogical rigour and clarity, have been a constant source of motivation for the writing of this work.

---

[1] École Nationale Supérieure d'Électrotechnique, d'Électronique, d'Informatique, d'Hydraulique et des Télécommunications (ENSEEIHT) of the Institut National Polytechnique de Toulouse (INPT).

[2] École Nationale Supérieure de Constructions Aéronautiques, currently merged into the Institut Supérieur de l'Aéronautique et de l'Espace (ISAE).

[3] At present ISAE-SUPAERO.

About This Book

The book brings out a unitary and structured understanding journey of fluid mechanics, a discipline present in many natural phenomena and at the very heart of the most diversied industrial applications and human activities. The balance between phenomenological analysis, physical conceptualization and mathematical formulation serves both as a unifying educational marker and as a methodological guide to the three parts of the work.

In the first one, the thermo-mechanical motion equations of a homogeneous single-phase fluid are established, from which flow models (perfect fluid, viscous) and motion classes (isovolume, barotropic, irrotational, etc.) are derived. Incompressible, potential flows and compressible flows, both in an isentropic evolution and shock, of an ideal inviscid fluid are addressed in the second part. The viscous fluid is the subject of the last one, with the creeping motion regime and the laminar, dynamic and thermal boundary layer.

Historical perspectives are included whenever they enrich the understanding of modern concepts. Many examples, chosen for their pedagogical relevance, are dealt with in exercises.

This book is intended as a textbook for undergraduate and graduate students, wishing to acquire a first mastery of fluid mechanics, as well as graduates in specialisation courses and ingeneers of other fields, concerned with structuring or completing what is sometimes a scattered body of knowledge.

Contents

Part I
Fluid Properties and Flow Models

Chapter 1
Physical Concepts and Mathematical Tools

Abstract The description of the physical properties of the fluid in the macroscopic sense of the continuous medium is the subject of the first part of the chapter. The second part explains the mathematical tools required for describing its motion.

1.1 The Different Description Levels of a Fluid Motion

In the absence of atomic reaction, the basic fundamental entity of matter is the molecule. It is thus conceivable to study the movement of a homogeneous fluid[1] from that of each of its molecules. This approach sets a first level of description which is called the «microscopic» level.

However, when the medium is sufficiently dense, it is not necessary for the vast majority of practical applications, to use such a refined description. We then adopt a supra-molecular level, which is the so-called *«macroscopic»* scale. The elementary space entity, at this scale, is therefore no longer the molecule, but another physical entity which actually covers a very high number of molecules. The development of a physical model adapted to this macroscopic approach, can be obtained in two separate ways:

- Indirectly, *i.e.*, starting from an analysis of microscopic order. The transition to the macroscopic scale is then reached by using statistical tools to reconstruct, from the evolutions of a set of molecules, an *"averaged pattern"* of the motion. This is the path of the *kinetic theory of gases* which, for species of low molecular weight, supplies satisfactory physical interpretations and estimates of macroscopic proporties ;
- Directly, which means *"ignoring"* the discrete element that is the molecule and considering that matter is continuously distributed throughout the flow field. This

---

[1] By definition, *strictly speaking*, a fluid is said to be *homogeneous* if it consists of a single and even chemical species. Throughout this book, we also assume that the constitutive species is present at any point in the flow domain and at any time under the same phase (*single-phase or monophasic fluid*). By *extension*, any gas mixture *in fixed composition*, such the air, will be treated as a homogeneous medium.

© The Author(s), under exclusive license to Springer Nature Switzerland AG 2022
P. Chassaing, *Fundamentals of Fluid Mechanics*,
https://doi.org/10.1007/978-3-031-10086-4_1

is the approach of *continuum mechanics*, a concept in which *continuity* refers to a physical meaning as ensuring the **definition** of the so-called *local* values of the macroscopic functions of the flow field, but not necessarily the **continuity**, in the mathematical sense of the term, as we shall see later in Chap. 7.

ADDITIONAL INFORMATION—The Kinetic Theory of Gases—The idea of the kinetic theory of gases takes root in the atomistic design of the philosophers of antiquity. It took scientific shape in 1738 with the work of Daniel Bernoulli [10]. This theory is not concerned with determining the dynamics (position, velocity) of each molecule, but that of the *mean values* relative to an ensemble (very high number) of molecules *in statistical equilibrium*. This fundamental assumption of statistical equilibrium states, for example, that the velocity modules of the different molecules are constantly changing and can take any value between zero and the speed of light (asymptotically), but are distributed according to a *probability law* which is *time independent*. Similarly, under the same assumption, the movement directions are necessarily distributed uniformly along all orientations (isotropy). The kinetic theory is thus concerned with situations of fluid in *dense medium*, where exists a molecular agitation generating a significant number of collisions, thus justifying the assumption of statistical equilibrium. Indirect evidence for the existence of this agitation was provided in 1827 by Brown, in relation to liquids, and in 1908 by Broglie, to gases.

• **Pressure** : According to the kinetic theory of gases, the pressure on a solid wall is defined as the averaged force applied to a surface unit of the wall, by the molecules colliding with it because of their erratic agitation. Its calculation leads to the expression :

$$P = n\frac{mv^{*2}}{3},\tag{1.1}$$

where $n = N/V$ is the number of molecules per unit volume (N, the number of molecules in the volume V), m the common mass of each molecule and $v^* = \sqrt{\overline{v^2}}$ the mean quadratic velocity of the molecular agitation motion.

• **Temperature** : this function, in terms of the kinetic theory, accounts for the energy related to the molecular agitation motion.
The simplest case is where this energy is solely reduced to translational kinetic energy. The temperature is then directly proportional to the *average kinetic energy* of the molecular's agitation motion:

$$\frac{3}{2}k_B T = \frac{1}{2}mv^{*2},\tag{1.2}$$

where k_B is the Boltzmann constant.
This case can be identified with that of the *ideal gas* of classical thermodynamics, as we shall see at a later stage (see Additional Information Sect. 1.2.3.4).

• **The distribution law of the molecular velocities** : the major outcome of the theory expresses the function $F(v_x, v_y, v_z)$, such that $F(v_x, v_y, v_z) \times dv_x dv_y dv_z$

represents the probability that the velocity vector components of the molecular agitation range between v_x and $v_x + dv_x$, v_y and $v_y + dv_y$, v_z and $v_z + dv_z$ respectively.

Assuming that this function does not depend on the directions x, y, z, Maxwell was able to establish that (see for example Bruhat [21], page 466):

$$F(v_x, v_y, v_z) = \left(\frac{m}{2\pi k_B T}\right)^{3/2} e^{-\frac{1}{2}\left\{\frac{m(v_x^2 + v_y^2 + v_z^2)}{k_B T}\right\}}. \tag{1.3}$$

This result was retrieved by Boltzmann in the broader context of the *Statistical Mechanics* and is therefore designated as the distribution law of Maxwell-Boltzmann.

One can easily deduce the probability distribution law relating to *velocity module* values as $F(v) \times dv \equiv Prob\{v \leq ||\boldsymbol{v}|| < v + dv\}$ with :

$$F(v) = 4\pi v^2 \left(\frac{m}{2\pi k_B T}\right)^{3/2} e^{-\frac{1}{2}\left\{\frac{mv^2}{k_B T}\right\}} \tag{1.4}$$

The function $F(v)$ is drawn in Fig. 1.1. From its expression, different characteristic velocities of the molecular agitation can be deduced:

Most probable velocity :

$$v_m = \sqrt{\frac{2k_B T}{m}},$$

Mean velocity :

$$\bar{v} \equiv \int_0^\infty v F(v) dv = \sqrt{\frac{8}{\pi} \frac{k_B T}{m}},$$

Root mean square velocity:

$$v^* \equiv \sqrt{\overline{v^2}} = \sqrt{\frac{3k_B T}{m}}.$$

The maximum relative difference between the two extreme values is less than 22.5%.

1.2   The Macroscopic Concept and Its Consequences

1.2.1   *Validity of the Continuous Assumption*

In the continuity of this lesson, we adopt the perspective of the *continuous medium*. Such a choice has a certain number of consequences which subsequently will be considered. For now, let us simply mention that it is not appropriate for studying all

Fig. 1.1 Characteristic
velocities associated with the
distribution law of
Maxwell-Boltzmann

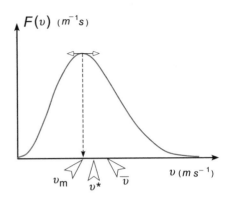

fluid motion problems. It excludes, in particular, flows in rarefied environments where
the weakness of the molecules' space density is incompatible with a localization at
the *macroscopic order* of function values defined by averaging over a very a high
number of molecules (see below Sect. 1.2.2).

To have a better judgment of the relative degree of a fluid flow rarefaction, it is
usual to introduce a dimensionless parameter called the *Knudsen number*, defined
as:

$$K_n = \frac{\ell}{L}, \tag{1.5}$$

where L is a macroscopic length scale of the flow field (a characteristic dimension
of an obstacle for example) and ℓ a length scale representative of the fluid on a
molecular scale.

For common liquids, the molecules move together as far as the repulsive forces
allow. The inter-molecular distance is thus substantially of the same order as the size
δ_m of the molecules[2] so that the characteristic microscopic scale is $\ell \simeq \delta_m$.

As far as gases are concerned, the situation is quite different due to the level
of the molecular agitation in such environments. A realistic estimate of the microscopic scale is then $\ell = \ell_m$, where ℓ_m stands for the average distance travelled by
the molecules between two consecutive collisions, called the *mean free path*.[3] In the
most unfavourable case of gases, the value of the Knudsen number with respect to
a macroscopic scale of about one meter is very low, namely, for air under normal
conditions $K_n \simeq 10^{-7}$.

ADDITIONAL INFORMATION— **Types of Flow Regimes**—According to various
Knudsen number ranges, it is customary to distinguish four types of motion:

---

[2] Typically $\delta_m \simeq \text{Å} = 10^{-10} m$.
[3] For air under normal conditions of temperature and pressure (P = 1 atm, $T = 298.15$ K), the
mean free path is of the order of 0.1 μm.

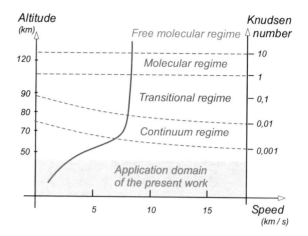

Fig. 1.2 Flow regimes and Knudsen number along the atmospheric reentry trajectory of the space shuttle (red curve)

– discontinuous or free molecular regime for $K_n \geq 10$;
– molecular or Knudsen regime, for $\leq K_n \leq 10$;
– transitional regime for $10^{-2} \leq K_n \leq 10$, including the sliding motion $10^{-2} \leq K_n \leq 10^{-1}$;
– continuum regime for $K_n \leq 10^{-2}$.

The field of applications covered by this course is actually devoted to situations where the Knudsen number is less than 10^{-3} which corresponds to the validity domain of the Navier-Stokes equations.

At high altitude when the pressure decreases, the gas becomes rarefied to the point that, in the Knudsen regime, the number of inter-molecular collisions becomes negligible compared to that of molecules on the walls. The concept of viscosity no longer makes sens and the Boltzmann model should be substituted for the Navier-Stokes equations. These different regimes are illustrated in Fig. 1.2, referring to the situations encountered by the space shuttle during its atmospheric reentry.

EXERCISE 1 The Knudsen number of a hot wire operating in various gases.

The hot-wire thermo-anemometry provides a common means for the measurement of "*local*" properties (temperature, velocity, species concentration) of gas flows. A typical sensor of this measuring device consists of a cylindrical wire, of a few micrometers in diameter, heated by Joule's effect.

QUESTION: Using the values of the mean free path in the following Table 1.1, estimate the value of the Knudsen number for a standard wire of 5 μm in diameter according to the nature of the gas being used.

SOLUTION: The Knudsen number values deduced from its definition and using the previous data are grouped in the following Table 1.2.

Table 1.1 Mean free path of some usual gases under normal conditions

Gas	He	H_2	N_2	O_2	CO_2
Mean free path (nm)	209	176	83	95	39

Table 1.2 Knudsen number of a hot wire operating in some common gases

Gas	He	H_2	N_2	O_2	CO_2
Knudsen number based on the hot wire diameter $\phi = 5\mu$m	0.042	0.035	0.017	0.019	0.008

It appears from reading this table that, with the exception of carbon dioxide, where the regime can be treated under the assumption of a continuous medium, for all other gases, the regime is that of the slip flow ($10^{-2} < K_n < 10^{-1}$) and transition flow ($10^{-1} < K_n < 10$). This observation leads to specific calibrations and to increase the diameter of the wire.

1.2.2 The Fluid Particle Concept

At the macroscopic scale, it is clear that the space location of any function of the flow cannot refer to "*mathematical points*". We are thus led to conceive a *physical domain* which is *infinitesimal on the macroscopic scale*. This is known as the *mesoscopic* order. The volume of such a domain must be both sufficiently *large* to contain a very high number of molecules and sufficiently *small* so that it cannot be possible to distinguish any space heterogeneity of macroscopic order. The physical entity of mesoscopic order associated with such a domain is the *fluid particle*, a concept about which we make three remarks.

(i) Denoting by $\mathcal{D}$ any volume of *macroscopic* order, the passage to the *mesoscopic* limit, mathematically expressed as $\mathcal{D} \to 0$, corresponds to a physical reality symbolized by $\mathcal{D} \to \epsilon$, where ϵ represents the elementary mesoscopic volume of the fluid particle around which the domain $\mathcal{D}$ collapses. Thus the *local* value in the mesoscopic sense of any function of the flow field $f(M, t)$ is always associated with the fluid particle "positioned around" the mathematical point M at time t.

(ii) A sufficient validity condition of the fluid particle concept is that:

$$\frac{1}{n} \ll \epsilon \ll L^3, \tag{1.6}$$

introducing the *fluid particle volume* ϵ.

In the above relation, n is the average number of molecules per unit volume and L the smallest significant scale of macroscopic order. When excluding the rarefaction effects already mentioned, the first inequality in the previous relation (1.6) is generally always satisfied for most flows concerned with common engineering applications. Indeed, under normal conditions, a cube with $2\ \mu m$ edges contains from about 2×10^8 to 2×10^{11} molecules, depending on whether the fluid is a gas or a liquid. As a result, the fluid particle concept will have to be called into question each time the flow characteristics impose very low values of the scale L. This is particularly the case at the crossing of a shock wave (see Chap. 7, addition to Sect. 7.3.7.1).

(iii) When the first inequality in (1.6) is fulfilled, any given fluid particle always contains the same (extremely high) number of molecules. This condition is to be taken in the statistical sense of the microscopic equilibrium according to the kinetic theory of gases. In other words, it does not mean in any way that a given fluid particle always contains the *same molecules* at any time.

Thus, as a mesoscopic entity, the fluid particle is an *elementary macroscopic* domain which is bounded by a surface that is "pervious" to the molecules.

1.2.3 The Functions of the Flow's Macroscopic Description

Considering the fluid particle as the infinitesimal physical system in terms of the continuous medium approach, its state can be defined by the usual thermodynamical and mechanical functions. In practice, restricting our analysis to a homogeneous single-phase medium, these flow field functions consist of *six* real-valued functions:

- the mass per unit volume or density $\rho\,(M,t)$,
- the three components of the velocity vector $V\,(M,t)$,
- the pressure $P\,(M,t)$,
- the temperature $T\,(M,t)$,

where M denotes the position of the fluid particle with respect to a given referential and t is the time.

We will now give some details about how to define these functions in a mesoscopic order, explaining on that occasion the physical meaning of the inequalities (1.6) introduced when defining the *fluid particle volume*.

1.2.3.1 The Density

Firstly, let us consider a *macroscopic* finite volume D containing a mass M of fluid. The ratio of these two quantities defines the mass per unit volume *in the domain D*:

$$\varrho\,(D) = \frac{M}{D}.\qquad(1.7)$$

Now, when adopting a microscopic point of view, one can introduce the mass per unit volume by considering a set $N(V)$ of N molecules included in a given volume V at a given time t. Denoting by m_k the mass of the molecule k, we have:

$$\widetilde{\varrho}(V) = \frac{1}{V} \sum_{k=1}^{k=N(V)} m_k. \tag{1.8}$$

The concept of density, as a mesoscopic order quantity, that is to say *local* at the fluid particle scale, corresponds to the existence of an asymptotic matching between two limits:

(i) *macroscopic*, symbolized by $D \to 0$, such that no space variation can be found over the matching volume ϵ ;
(ii) *microscopic*, symbolized by $V \to \infty$, for which there is no further sensitivity to the number of molecules present in the volume ϵ at this matching scale.

These considerations make it possible to introduce the *mesoscopic* value of the density $\rho(M)$ as :

$$\lim_{D \to 0} \varrho(D) = \lim_{V \to \infty} \widetilde{\varrho}(V) \equiv \rho(M) = \frac{m}{\epsilon(M)},$$

where M denotes the "material point" corresponding to the center of the fluid particle over which the macroscopic domain D "collapses" within the limit of its mesoscopic value ϵ. At this scale limit, we can still speak about density in terms of a ratio between the *elementary* mass m and the *elementary* volume ϵ of a fluid particle, a statement that no longer pertinent below the mesoscopic scale (see Additional Information below).

As a direct consequence of this definition, it is worth noting the absence of any inhomogeneity *inside the fluid particle*, which we can express, in a purely formal way as:

$$\iiint_{\epsilon(M)} f\,dv = f(M,t) \times \left[\iiint_{\epsilon(M)} dv\right] \equiv f(M,t) \times \epsilon(M), \tag{1.9}$$

where M is the "material point" identifying the position of the fluid particle.
The key points of this discussion are outlined in Fig. 1.3 illustrating the double feature in the fluid particle concept as follows:

(i) the removal of microscopic fluctuations, by spatial integration over a sufficiently large "*microscopic*" domain ;
(ii) the smoothing out of any macroscopic order inhomogeneity by suitably reducing the size of the "*macroscopic*" domain.

ADDITIONAL INFORMATION—**Microscopic fluctuations of the density**—
The density $\widetilde{\varrho}$ defined at the microscopic scale, Eq. (1.8) is a *random* function of time, whose value depends, *each time*, on the number of molecules present in

Fig. 1.3 The concept of the mesoscopic limit as the matching of microscopic and macroscopic inhomogeneities

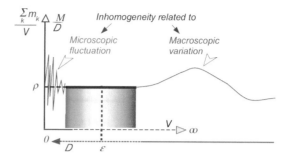

the reference domain V. Indeed, because of the *molecular agitation*, this number varies from one moment to another, so that the density fluctuates in time.

In the framework of the classical statistical mechanics of Boltzmann, it can be shown that the relative density variation, due to fluctuations in molecule numbers actually present in the volume V can be estimated, around the mean value N, as (see Bruhat [21] for example) $\Delta\tilde{\varrho}/\tilde{\varrho} = 1/\sqrt{2N}$.

At the atmospheric pressure level, a cubic millimeter of air contains about $N \simeq 10^{16}$ molecules. The above relation gives, for the fluctuations in density, a relative difference of about 10^{-8}, virtually undetectable. This volume size is therefore still beyond the mesoscopic limit.

On the other hand, a cube of 0.5 μm on each side—the order of the wavelength magnitutude of the blue-coloured light visible by the human eye[4]—one obtains $N \simeq 10^6$ and density variations, by fluctuations in the number of molecules present in the domain, are now of the order of one thousandth, in relative value. This is sufficient to cause some refractive index perturbations which could affect light propagation and induce a significant light scattering on this wavelength. This phenomenon, which is the origin of the sky's *blue* colour, prevents adopting such a space scale in order to give a significant uniform value of the refractive index of the mesoscopic order.

1.2.3.2   The Velocity

As before, the local macroscopic (or mesoscopic) value of the velocity can be defined by an asymptotic matching of the mass-weighted momentum. Thus, successively considering the macroscopic and microscopic levels, one can write:

$$V = \lim_{D \to 0} \frac{Q(D)}{M(D)} = \lim_{V \to \infty} \left(\sum_{k=1}^{k=N(V)} m_k v_k \Big/ \sum_{k=1}^{k=N(V)} m_k \right),$$

---

[4] This colour's wavelengths ranges between 446 and 500 nm.

where $Q\,(D)$ and $M\,(D)$ denote respectively the momentum and the mass of a macroscopic fluid domain D; v_k is the velocity of the molecule k of mass m_k and $N\,(V)$ the number of molecules in a microscopic volume V *at time t*.

1.2.3.3 The Temperature

• In the microscopic sense, the temperature can be expressed by Eq. (1.2), in terms of the kinetic theory of gases of a simple molecular structure (see Sects. 1 and 1.2.3.4).
• In the macroscopic sense, the temperature is defined as a state variable of the system, according to the *classical thermodynamics* of systems in equilibrium (see Appendix A).
Taking, for instance, for the fluid state variables, the *specific internal energy* (e) and the *specific entropy*[5] (s) the *canonical equation of state* is $e = e(\rho, s)$, where ρ denotes the density. According to the laws of classical thermodynamics, any evolution (or transformation) of the system around an equilibrium state is accounted for by variations of the state variables satisfying:

$$de = T ds - P d \left(\frac{1}{\rho} \right).$$
(1.10)

Hence follows the definition of the *thermodynamic* temperature as:

$$T = \left(\frac{\partial e}{\partial s} \right)_{\rho}.$$
(1.11)

1.2.3.4 The Pressure

• In terms of the microscopic kinetic theory, the pressure is made explicit by Eq. (1.1). For a gas of a simple molecular structure, this relation, supplemented by its counterpart for the temperature, leads to a law of state *macroscopically* equivalent to that of an *ideal gas*. (Cf. Additional Information hereafter).
• In the macroscopic sense, the definition of the pressure can be considered in two separate ways, according to a purely *mechanical* sense on the one hand, and a *thermodynamic*, on the other hand.

– In any point of a fluid at rest under the action of gravitational forces alone, there is a force balancing the weight: the clearly identified *hydrostatic* pressure. When the fluid is in motion, d'Alembert's principle can generalize this result, taking into consideration all the *normal stresses*[6] acting on the bounding surface of any fluid particle. We then introduce the *mechanical pressure*, a concept we will have the opportunity to return to in Chap. 3, Sect. 3.2.2.3.

---

[5] A *specific* quantity is defined *per unit mass.*
[6] The stress is the force per unit area, see Sect. 1.3.2.2 of the present Chapter.

– In thermodynamics terms, the pressure can be defined in a way similar to that of the temperature, using, for instance, the canonical law of state. Thus, by direct reference to relation (1.10), one can write:

$$P = \rho^2 \left(\frac{\partial e}{\partial \rho} \right)_s, \qquad (1.12)$$

introducing the concept described as the *thermodynamic* pressure.

ADDITIONAL INFORMATION—The ideal gas and the kinetic theory—For a gas of a simple molecular structure, energies associated with molecule rotation and vibration modes can be neglected compared to that of the translation mode (kinetic energy). This is the kinetic theory of gases in its simple formalism presented previously (see Additional Information in Sect. 1).

Returning then to the expressions of the pressure, Eq. (1.1) and the temperature, Eq. (1.2) and eliminating the term mv^{*2} between these two relations, the result is:

$$P = nk_BT \quad \text{or} \quad PV = Nk_BT.$$

Now, taking for V the volume of one mole of gas ($\mathcal{V}$), n is identified with the Avogadro number $\mathcal{N}$, so that we retrieve the law of an ideal gas $P\mathcal{V} = \mathcal{R}T$, with, for the *Boltzmann constant* $k_B = \mathcal{R}/\mathcal{N} (\simeq 1.38065010^{-23}\ J\ K^{-1})$.

1.2.3.5 The Local Thermodynamic Equilibrium

Classical thermodynamics, from which result some of the above definitions of the temperature and pressure, governs processes involving work and heat exchanges, with reference to a *macroscopic system in equilibrium*. The concept of the macroscopic system, in the sense of a set of a very high number of molecules, applies perfectly to the study of fluid flows, under the assumption of continuous medium, the only case considered here.

In contrast, the equilibrium condition[7] contradicts the very idea of flow, a situation where the properties of the fluid environment (velocity, pressure, temperature ...) obviously vary in space (inhomogeneity) and time (unsteadiness). Thus, any fluid motion falls into the domain of the "thermodynamics of *irreversible processes*". It is a discipline in itself,[8] covering vast fields of physics, whose use however is extremely reduced here by the adoption of assumptions allowing direct transposition of the

---

[7] The "transformations" (changes of system states) in terms of *classical* thermodynamics can only represent the "evolutions" which proceed by equilibrium state sequences, whether real or fictitious.

[8] The thermodynamics of irreversible processes deals with the behaviour of a system whose entropy production is consubstantial to its evolution which is intrinsically irreversible, see Chap. 3 Sect. 3.3.5.1. It borrows methods from classical thermodynamics and implements two major tools: general balance equations and phenomenological relations.

results and concepts of *classical* thermodynamics.

These assumptions are twofold :

• the *weak* departure from equilibrium: deviations from equilibrium are assumed to be sufficiently small to be taken into account by first-order expansions. We are still speaking of *linear approximations*.

• the *local* thermodynamic equilibrium: at any time and *at the scale of the fluid particle volume*, the state of the fluid environment can be characterized by functions defined in terms of the state variables of the thermodynamic equilibrium.

The latter assumption is based on the existence of a clear separation between two time scales characterizing:

(*i*) the variations of *macroscopic order* ($\mathcal{T}$), associated with the departure from equilibrium[9];

(*ii*) the relaxation to equilibrium (τ), by mechanisms inherent in the molecular agitation.

Clearly, the validity of the local thermodynamic equilibrium assumption is reached when the inhomogeneities at the macro level, induce *slow* changes compared with the relaxation time to equilibrium at the microscopic scale, viz $\mathcal{T} \gg \tau$. Under these conditions, the statistical equilibrium conditions apply at the fluid particle scale, despite the fact that, at the overall macroscopic flow scale, the fluid environment is steadily subject to heat and momentum fluxes. This will be the case throughout this lesson, as in most applications of dense fluid mechanics in terms of the continuous medium approach, unless the departure from equilibrium is driven by sustainable causes (optically pumping in laser physics, for example), a situation that exceeds the scope of this book.

EXERCISE 2 The local thermodynamic equilibrium assumption.

A fluid flows over a distance L at the velocity U_0. It is a gas whose molecular agitation, under the considered flow conditions, has the average free path ℓ for characteristic length scale, and for characteristic velocity, $u' = 1.35 \times a$, where a is the speed of sound of the flow under consideration.

QUESTION 1. Estimate, from the preceding parameters, the characteristic time scales of the macroscopic motion, T and the microscopic agitation, τ.

QUESTION 2. Express the ratio T/τ as a function of the Knudsen number K_n and the Mach number $M_a = U_0/a$. Discuss the consequences of this result for an incompressible subsonic flow with $K_n = 10^{-3}$ and $M_a = 0.1$.

---

[9] Regarding the non-equilibrium situation of a flow, we should keep in mind that the departure from equilibrium is to be considered both in space and time. The characteristic time scale T for any scalar function f, should therefore be estimated with respect to the temporal variation, $\left[\frac{1}{f}\frac{\partial f}{\partial t}\right]^{-1}$ and/or the advective/convective variation $\left[\frac{U_j}{f}\frac{\partial f}{\partial x_j}\right]^{-1}$, according to the material derivative (see later Sects. 1.6.1 and 1.6.10).

SOLUTION: **1.** From simple dimensional considerations, we have $T = L/U_0$ and $\tau = \ell/u'$.

2. We easily obtain that $\dfrac{T}{\tau} = \dfrac{L}{U_0} \times \dfrac{u'}{\ell} = \dfrac{L}{\ell} \times \dfrac{1.35 \times a}{U_0} = \dfrac{1.35}{K_n \times M_a}$.

With the given values, we see that $T/\tau = 1.35 \times 10^4 \gg 1$. Thus, this flow corresponds to a situation of a macroscopic motion, whose characteristic time scale is much greater than that of the equilibrium relaxation due to microscopic mechanisms. In this flow, macroscopic changes will be applied to fluid particles whose molecules will always be in a state of statistical equilibrium: this is the very essence of the local thermodynamic equilibrium assumption.

1.3   The Physical Macroscopic Properties of a Fluid

1.3.1   The Notion of Fluid

The concept of fluid can be seen from different points of view, especially that of the physicist and the mechanical engineer.

To a physicist, this concept combines two states of matter, gases and liquids, where one meets neither a periodic spatial organization of atoms, characteristic of a solid crystal, nor a disordered but frozen composition of atoms, as in an amorphous solid. The *intermediate* state corresponding to a liquid can result from increasing the cohesion of the gaseous state, by densification of the collisions between particles or, conversely, decreasing the solid state by disorganization of the microscopic structure. However, such a characterization, with respect to pure species physics, makes it difficult to account for complex situations where the organization degree of the (heterogeneous) medium can vary depending on the motion conditions, as is the case with liquid solutions of macromolecules and polymers, for example.

For the mechanical engineer, the notion of solid is very closely associated with the shape considered as a separate entity. However, there are many deformable solids, so that, here again, the distinction is far from being simple. We will see that many "fluid media" may exhibit various *plastic* behaviours that make quite as many intermediaries between a pure fluid and a deformable solid. Nevertheless, within the limit of the elastic behaviour of solids, a solid material exhibits a fixed deformation as long as the stress field remains unchanged. This feature distinguishes it from the fluid medium which, in the usual sense, is able to deform (or flow) permanently, under the action of shear forces which are *constant over time*.

The mathematical schematization of the fluid behaviour, in the manner addressed in this course, will be given later in Chap. 4, Sect. 4.3.1. We limit ourselves here to introducing some of the physical properties which are characteristic of the fluid medium at the macroscopic scale.

Fig. 1.4 Schematic for the plane Couette flow

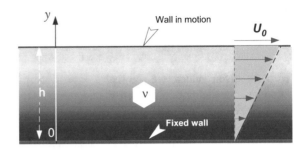

1.3.2 Viscosity

1.3.2.1 The Elementary Evidence

In a fairly intuitive way, we can say that the viscosity concept is associated with any fluid's resistance to being set in motion. To specify this property in a more quantitative way, we consider the experiment of a plane Couette flow.

In this experiment, see Fig. 1.4, a viscous fluid disposed between two parallel planes separated by a thickness h, is set in motion by the relative movement of one of the walls in a rectilinear translation in its own plane.

For simplicity, we assume that the lower wall is fixed and upper wall is in translation at a constant velocity U_o. In the absence of any external force (gravity, pressure), the fluid motion results from the mere displacement of the moving plane wall. When a steady regime has been reached, it is found that, for some fluids, the velocity is linearly distributed across the fluid. One may also speak of a *linear* velocity profile between the two plates. To maintain the uniform motion of the upper wall, it is necessary to apply a tangential force, whose module F, relative to a surface A, is such that:

$$F/A \propto U_o/h.$$

Denoting by μ the proportionality coefficient, called the *dynamic viscosity*, which dimension[10] is $[\mu] = [M] \times [L]^{-1} \times [T]^{-1}$, the previous relation can be rewritten as:

$$\boxed{\frac{F}{A} = \mu \frac{U_o}{h}} \tag{1.13}$$

---

[10] The SI unit is the Pascal-second ($Pa.s$). One also meets the former name of *Poiseuille* (Pl), with $1\,Pl = 1\,Pa.s$. Regarding the choice of the symbol, the notation μ has become customary in the community of hydro-aerodynamicists who adopted it from English scientists. However, as noted by Prandtl [125], the physicists community use instead the symbol η to denote the dynamic viscosity.

Fig. 1.5 The local viscous
shear stress for a non-linear
velocity distribution

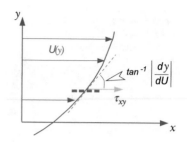

This relation applies well to common fluids such as air and water and corresponds
to a *rheological behaviour* named *Newtonian*, after Isaac Newton. However, as we
shall see later, there are other fluid media where it is not appropriate.

1.3.2.2 The Generalization of the Newtonian Scheme

Relation (1.13) explicitly introduces the notion of stress, which is a force per unit
area F/A. For a force whose direction is parallel to the surface to which it is applied,
we refer to shear stress and we write:

$$\tau_{xy} = \frac{F}{A},$$

where, by convention, the first index (x) identifies the direction of the force and the
second (y) that of the vector normal to the surface.

A first extension of relation (1.13) is to be considered when the velocity field,
reduced to the component $U(y)$ does not linearly vary in the field, see Fig. 1.5. In
this case, the shear stress in x direction acting on a facet of y normal can be simply
written[11]:

$$\tau_{xy} = \mu \frac{dU}{dy}. \tag{1.14}$$

In the general case, the velocity vector has three components. Considering a small
tetrahedral fluid element based on three orthogonal directions (see Fig. 1.6), every
facet has a stress vector, which leads to the existence of nine elementary scalar components. As discussed in Chap. 3, Sect. 3.2.2.1, these nine components denoted τ_{ij},
$(i = 1, 2, 3, \ j = 1, 2, 3)$ define a second-order symmetric tensor, called the *stress
tensor* $\overline{\overline{\tau}}$. The generalization of the linear model (1.14) will thus result in one of the
following relations, depending on whether one uses tensor or index-linked notation:

---

[11] Equation (1.14) is *described* by Newton in his *Principia*. It is for this reason that we refer to it
as *Newton's scheme*. Accordingly, the fluids which obey this relation are called *Newtonian fluids*.

Fig. 1.6 Exploded view of
the viscous stress
components on the three
facets of a tetrahedral fluid
element

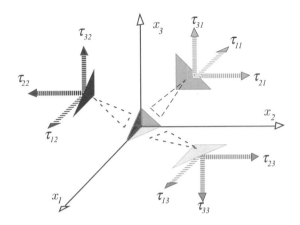

$$\overline{\overline{\tau}} = 2\mu\overline{\overline{S}} + \mu'\vartheta\overline{\overline{I}} \qquad (1.15a)$$

$$\tau_{ij} = 2\mu S_{ij} + \mu'\vartheta\delta_{ij} \qquad (1.15b)$$

In relations (1.15a) and (1.15b), to be derived in Chap. 4-Sect. 4.3.1, $\overline{\overline{S}}$ (S_{ij}) is the
strain rate tensor, whose components are

$$S_{ij} = \frac{1}{2}\left(\frac{\partial U_i}{\partial x_j} + \frac{\partial U_j}{\partial x_i}\right). \qquad (1.16)$$

The trace of this tensor is $\vartheta = S_{ii} \equiv div\,\boldsymbol{V}$. The second order unit tensor is denoted
by $\overline{\overline{I}}$ (δ_{ij}). Finally μ and μ' are two viscosity coefficients similar to those of Lamé's
linear elasticity.

1.3.2.3 Other Rheological Behaviours

The branch of science which studies how deformable materials react when subjected
to stresses is called *rheology*. Among the various mechanical behaviours that are
revealed in this discipline,[12] a first distinction appears depending on whether the
material exhibits a response which is or not time dependent.

  (a) When the behaviour is *time independent*, one can observe, see Fig. 1.7, various
relations between the shear stress τ_{xy} and the component dU/dy of the shear rate
which is applied.
Thus, we can distinguish:

---

[12] The reader interested in an exhaustive presentation of the different types of material behaviours
may usefully refer to the Dictionary of Rheology [60].

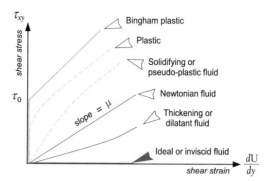

Fig. 1.7 Examples of time-independent rheological behaviours

1. *ideal* or *inviscid* fluids, in the sense that there is no internal friction and therefore no shear stress, whatever the strain rate should be;
2. *Newtonian* fluids, for which a linear relation exists between the stress and strain rate;
3. *nonlinear behaviour* fluids, to which belongs, for example, *Ostwald's scheme*:

$$\tau_{xy} = \eta \left| \frac{dU}{dy} \right|^n \quad for \; n \neq 1.$$

Compared to the Newtonian case, everything happens for these materials as if there were an "*apparent viscosity*" which increases with the velocity gradient, for thickening (or dilatant) fluids and decreases in the case of pseudo-plastic or shear-thinning fluids;
4. *plastic* materials, and in particular the *Bingham* plastic, whose behaviour can be summarized by:

$$\begin{cases} \frac{dU}{dy} = 0, & if \; \tau_{xy} \leq \tau_0, \\ \tau_{xy} = \tau_0 + \mu \frac{dU}{dy}, & if \; \tau_{xy} > \tau_0. \end{cases}$$

For these materials, there is a stress threshold, τ_0, below which no motion occurs and above which they flow, like a Newtonian fluid in the case of the Bingham plastic.

Example The Newtonian scheme is adequate for natural common fluids such as water and air. Gelatin, milk, blood,[13] liquid cement exhibit a pseudo-plastic behaviour, while concentrated aqueous solutions of sugar, starch behave as dilatant fluids. Drilling muds provide a representative example of Bingham plastic.

(b) There are also materials whose response to mechanical stresses involves time. We then find behaviours which may depend on *(i)* the time period of application of

---

[13] This inhomogeneous fluid, regrouping in particular plasma and blood cells, is an extremely complex "*mechanical material*", whose behaviour can change depending on the relative size of the vessels compared to that of the red blood cells.

a steady stress, or *(ii)* the rate of change of a time-dependent stress. For the former, it is as if the fluid had a *"memory effect"* of the stress field to which it is submitted, resulting in behaviours that are depicted as:

1. *rheopectic*, when the apparent viscosity of the fluid increases with the duration of the application of a steady stress;
2. *thixotropic*, when the apparent viscosity decreases over time under a steadily applied stress.

The medium may retrieve, if any, the initial value of the viscosity after a more or less long period at rest.

Example High performance hydraulic safety fluids, as phosphate ester type (Skyrol), used in aviation and the navy, have a viscosity which varies (decreases) with the time of use. This phenomenon of aging, due to the rupture of internal molecular chains, can halve the viscosity after 500 h of use.

(c) Another case of time-dependent behaviour is that of *viscoelastic* materials. These materials have the ability to change their molecular structure at a characteristic time scale Θ. If the stress applied to them varies over time, they may thus exhibit a behaviour of either a viscous fluid or an elastic solid. Denoting by T the time scale representative of this change in stress, the dimensionless parameter

$$De = \Theta/T,$$

characterizes the type of behaviour of these materials. This is the *Deborah* number, whose values $De \ll 1$ correspond to the viscous fluid and those $De \gg 1$ to the elastic solid.

Example In respect to thixotropic fluids one finds paints, drilling muds and clay solutions, mayonnaise, ketchup. The viscoelastic behaviour is that of bread dough, jelly, cornstarch, concentrated solutions of polymers or macromolecules, wet sand.

1.3.2.4 Some Usual Data

We will now consider only the case of Newtonian fluids for which we have introduced the *dynamic viscosity* coefficient μ. One also introduces the *kinematic viscosity* coefficient:

$$\nu = \frac{\mu}{\rho} \tag{1.17}$$

which has the dimension $[L]^2 \times [T]^{-1}$. This coefficient is thus expressed in the international system, in m^2/s or myria Stokes (ma St), with $1\,m^2/s = 10^4$St; the Stokes is the CGS system unity.[14] The following table lists the numerical values

---

[14] The unit of the *dynamic* viscosity in that system is the *Poise* (Po) with $1\,Pl = 10\,Po$.

Table 1.3 Coefficients of dynamic and kinematic viscosity of air and water

Fluid	Density [kg/m^3]	Dynamic viscosity [kg/(m × s)]	Kinematic viscosity [m^2/s]
Air (*gas*) (*Stand.cond.*)	1.29	1.85×10^{-5}	1.43×10^{-5}
Water (*liquid*) (*Stand.cond.*)	10^3	10^{-3}	10^{-6}

Table 1.4 Temperature and pressure effects on the dynamic viscosity of liquids and gases

	Liquid	Gas
Temperature	$\mu \searrow$ as $T \nearrow$ Walther's formula: $\mu = \mu_0 \left(\frac{T}{T_0} \right)^m$	$\mu \nearrow$ as $T \nearrow$ Sutherland's formula: $\frac{\mu}{\mu_0} = \sqrt{\frac{T}{T_0}} \frac{1+C/T_0}{1+C/T}$ C approximately constant for a given gas
Pressure	$\mu \nearrow$ as $P \nearrow$ $\frac{\mu}{\mu_0} = a^{\left(\frac{P}{P_0} - 1 \right)}$	μ approximately constant

of the dynamic and kinematic viscosity coefficients for water and air taken under standard temperature and pressure conditions.

ADDITIONAL INFORMATION—**Viscosity sensitivity to the temperature and pressure**—Experimental evidence has shown that the dynamic viscosity is a physical property non *strictly* independent of flow conditions. The following table lists some directions of variation depending on the temperature and pressure. For gases, the dynamic viscosity increases with the temperature, the opposite occurs with liquids.

- In gases, the viscosity can be analyzed at a *microscopic* scale in terms of momentum transfer by *molecular agitation*. We then show (see Eq. (1.22) of the additional information on the transport coefficients in the kinetic theory at the end of Sect. 1.3.3.4 below) that the dynamic viscosity is a function of the temperature,[15] viz $\mu \propto \sqrt{T}$, where T is the absolute temperature. On this basis, Sutherland was able to establish a formula (see Table 1.4) which is remarkably well verified up to temperatures of 2000 K. For air, the value of the Sutherland constant is $C = 110$ K.

- In liquids, the microscopic modeling is different and more complex. For more details on this issue, one may consult Guyon's *et al.* work [60] for instance.

---

[15] It may seem surprising, at first glance, that the viscosity of a gas does not depend on the pressure. In fact, increasing the pressure results in increasing the density, hence the number of molecules per unit volume; However, one cannot induce an increase of the viscosity, because at the same time, the mean free path decreases.

More practical information and data on the viscosity of simple gases and steam are available in tables [71]. For gas mixtures, one can refer to the paper by Liley [90] and for liquids to Katz's work [82].

1.3.3 Thermal Conductivity

1.3.3.1 The Different Modes of Heat Transfer

If we refer to their physical origins, two basic heat transfer modes can be distinguished:

(a) *radiation*, which is based on a mechanism of remote energy propagation by electromagnetic waves (partially from the spectrum of the visible light, mostly in the infra-red range) emitted by hot bodies. This phenomenon is present in applications where temperature variations are sufficiently high (several hundred degrees K, for example) ;
(b) *conduction*, which results from a gradual energy transfer by molecular agitation through a heat exchange surface.

According to the practical use in *thermal engineering*, the thermal *convection* concept is also introduced whenever a heat exchange takes place in a moving fluid. One then distinguishes different convection modes, depending on whether the motion is the result of the heating alone, *natural convection*, or depends, partly or totally, upon forces of a different origin (such as gravity, pressure, etc). We then speak of *mixed convection* and *forced convection* respectively.

It is important to note that convection is merely the association of conduction and/or radiation with the motion of the fluid. In this respect, it is not considered here as a *specific* mode of heat transfer.

In the following sections, we will assume that the radiative transfer is negligible and therefore retain the conduction as the single mode of basic heat transfer in a fluid, whether moving or at rest.

1.3.3.2 The Elementary Evidence of Thermal Conduction

We consider (Fig. 1.8) a fluid at rest between two parallel plane walls, separated by a spacing h. For every time $t < 0$, the system *fluid–plates* is in thermal equilibrium at the temperature T_0.

At time $t = 0$, the upper plate is heated to a temperature $T_1 \neq T_0$, a condition which is kept unchanged for every $t > 0$. It is assumed that the temperature difference $(T_1 - T_0)$ is such that it induces no fluid motion at a macroscopic scale, so that the heat exchange mode is pure conduction. After a transient phase during which the temperature difference *diffuses* across the thickness of the fluid, a steady state is

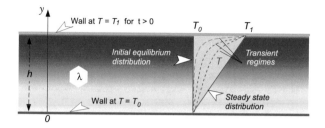

Fig. 1.8 Schematic for pure thermal conduction between parallel planes

reached—see (Fig. 1.8)—where the temperature between the two plates is distributed according to a *linear* profile:

$$\frac{T - T_0}{T_1 - T_0} = \frac{y}{h}.$$

In this situation, a steady heat flux exists between the plates, the intensity of which Q through a surface A parallel to the plates is such that $Q/A \propto |T_{1-T_0}| / h$, a relation which leads to introducing a positive proportionality coefficient λ ($W m^{-1} K^{-1}$), called the *thermal conductivity* of the fluid, such as:

$$\boxed{\frac{Q}{A} = -\lambda \frac{T_1 - T_0}{h}} \tag{1.18}$$

Similarly as for the Newton scheme, Eq. (1.13), relation (1.18) can be generalized to a heat transfer in any direction. This leads to introducing the heat flux density vector q representing the heat flux per unit area, whose expression is given, according to Fourier's law, as:

$$\boxed{q = -\lambda \, grad \, T} \tag{1.19a}$$

$$\boxed{q_i = -\lambda \frac{\partial T}{\partial x_i}} \tag{1.19b}$$

1.3.3.3 Some Usual Data

The thermal conductivity has a dimension of $[M] \times [L] \times [T]^{-3} \times [\Theta]^{-1}$. As we have already done for the dynamic viscosity μ, it is possible to introduce a coefficient whose dimension is $[L]^2 \times [T]^{-1}$, called the *thermal diffusivity*, defined by:

$$\boxed{a = \frac{\lambda}{\rho C_p}} \tag{1.20}$$

Table 1.5 Conductivity and thermal diffusivity of water and air under normal conditions of temperature and pressure

	Thermal conductivity $[kg{\times}m/\left(s^3{\times}K\right)]$	Thermal diffusivity $[m^2/s]$
Air	2.6×10^{-2}	2.24×10^{-5}
Water	0.59	10^{-7}

where C_p stands for the specific heat at constant pressure.

The following table provides the values of these coefficients for air and water.

1.3.3.4 The Prandtl Number

Since the thermal diffusivity and kinematic viscosity have the same dimension, their ratio defines a dimensionless quantity called the *Prandtl number*:

$$P_r = \frac{\nu}{a} = \frac{\mu C_p}{\lambda} \qquad (1.21)$$

The Prandtl number is a *physical property* of the fluid whose value allows selecting the material best suited to an application involving a diffusive heat transfer in a moving fluid (see Additional Information further below). Thus, promoting a high level of heat transfer while minimizing the energy loss by viscosity, leads to choosing fluid materials such as $P_r \ll 1$.

As shown in Table 1.6, these are the liquid metals which are then more appropriate. In contrast, for the lubrication of moving parts of different temperatures, while reducing the heat transfer exchange between some of them, materials such that $P_r \gg 1$ are more favourable. We now have a range of silicone oils, whose Prandtl number can reach values of 10^7.

ADDITIONAL INFORMATION—Selecting the Prandtl number with respect to the applications — As shown in Table 1.6, there is a wide range of fluids with a broad variety of Prandtl number values. It is thus possible to meet various applications needs.

Table 1.6 Prandtl number of some types of fluid materials

Liquid metals	Air (*Standard conditions*)	Organic liquids	Liquid water	Molten glass	Silicone oils
10^{-3} to 10^{-1}	0.71	1 to 10	10	10^3 to 10^4	10 to 10^7

• In the old fast neutron Superphenix French reactor for example, heat was removed by a circulating fluid loop in contact with the reactor core. This fluid, whose motion need not slow the neutrons flux, has to be also an efficient coolant. This is what led to choosing the liquid sodium, which Prandtl number at 450° C is 0.0047.

• In contrast, in the exploitation of oil drilling in deep water, significant temperature differences may exist between the fluid pumped into the Earth's crust and the external environment, at the wellhead in the ocean. One has then to reduce heat losses at this point in order to keep as far as possible the fluidity in the extracted oil and prevent the formation of hydrates which may reduce the production or damage the facility. One injects an insulating fluid into the well around the exhaust duct, whose Prandtl number is as high as possible. For one of such fluids, the MY-T-OIL of the Halliburton Company, the Prandtl number (see Sepulvado [145]) can vary from 1.38×10^{11} for a temperature of 10° C to 1.51×10^8 at 20°C. These values also depend on the shear rate, due to the thixotropic behaviour of this fluid.

ADDITIONAL INFORMATION—**Prandtl number and thermal conductivity sensitivity to the temperature**—Similar to the dynamic viscosity, the thermal conductivity also depends, *a priori*, on the temperature and pressure conditions of the fluid.

• For gases, the kinetic theory provides a framework for interpreting the transfer mechanisms by molecular agitation in a unique way, see further below. It is therefore possible to assess the *transport* coefficients (both viscosity and thermal conductivity) which are involved in the schematization of the agitation effects at a macroscopic scale. This allows demonstrating that for gases, the thermal conductivity tends to increase with the temperature.

• For liquids, such a unitary model for the microscopic interpretation of these coefficients does not exist. One can distinguish basically two mechanisms of heat transfer: one is based on the gradual propagation of the molecule vibrations; The second is found in liquid metals and originates from the electrons that are also responsible for the *electrical conduction*.

• Regarding the Prandtl number of gases, its value remains of the same order of magnitude as that of air, for gases of low molecular weight (0.70 for H_2, 0.73 for N_2, 0.74 for O_2, 0.78 for CO_2 for example). Finally, regarding the influence of the temperature on this number, we can observe a decrease of 9% for air between 100 K ($P_r = 0.77$) and 2000 K ($P_r = 0.70$).

For more quantitative information, the reader can refer to the references already cited in Sect. 1.3.2.4 of this chapter.

ADDITIONAL INFORMATION—**The molecular transport coefficients and the kinetic theory of gases**—The kinetic theory of gases provides a conceptual framework for interpreting the effects of the *microscopic* agitation in gases as the source of *physical* coefficients at the *macroscopic* scale.

• **The mean free path**: This concept plays a crucial role in the analysis of the *momentum* transport by the molecular agitation. It corresponds to the average distance (in a statistical sense) travelled by a molecule between two consecutive collisions over which its momentum is unchanged. From the Maxwell-Bolztmann velocity distribution law, it can be shown that (see [21]) the mean free path ℓ is expressed by:

$$\ell = \frac{\sqrt{2}}{2\pi}\frac{1}{n\sigma^2} \equiv \frac{\sqrt{2}}{2\pi}\frac{m}{\rho\sigma^2},$$

where σ is a length scale (also called a diameter) characteristic of an effective area involved in molecule collisions, independent of the temperature and n the number of molecules per unit volume, with $n \times m = (N/V) \times m = \rho$, where ρ is the density of the gas composed of identical molecules of mass m.

• **The viscosity** : This macroscopic feature reflects the microscopic exchange of *momentum* by collisions between the molecules. It is expressed as a function of the mean free path, *viz* $\mu = 0.499\,n\ell m\bar{v}$, where $\bar{v}$ is the mean velocity of the molecular agitation.[16]

By using the Maxwell-Boltzmann distribution, Eq. (1.4) we obtain

$$\mu = 0.182\frac{\sqrt{mk_BT}}{\sigma^2}. \tag{1.22}$$

• **The thermal conductivity** : the microscopic interpretation of this coefficient is analogous to the previous one, considering now the transport of the molecular agitation *energy*. Its expression is:

$$\lambda = \frac{1}{3}\frac{n}{N}\,c_v\ell\bar{v}, \tag{1.23}$$

where N is the Avogadro number and c_v the heat molar capacity at constant volume.

1.3.4 Compressibility

1.3.4.1   The Compressibility Coefficients and the Speed of Sound in a Fluid at Rest

Another well-known natural property of real fluids is the *compressibility*, that is to say, the volume reduction in response to an increase in pressure. To quantify this

---

[16] This is not exactly the same velocity as that appearing in the definition of the microscopic pressure and temperature in Eqs. (1.1) and (1.2) respectively. It is worth remembering that in this case, the significant velocity is the *mean square velocity* v^*. However, as we have seen previously (see Additional information Sect. 1.1), the values of these two velocities are only slightly different, which validates any order of magnitude reasoning bases on any of them.

effect one can introduce various coefficients measuring the relative decrease $\Delta v / v$
of a unit mass volume of fluid by increasing the pressure by an amount ΔP :

– the *isothermal* compressibility coefficient[17] χ_T:

$$\left. \frac{\delta v}{v} \right|_T \equiv - \left. \frac{\delta \rho}{\rho} \right|_T = -\chi_T \delta P \; ; \qquad (1.24)$$

– the *isentropic* compressibility coefficient χ_S:

$$\left. \frac{\delta v}{v} \right|_S \equiv - \left. \frac{\delta \rho}{\rho} \right|_S = -\chi_S \delta P. \qquad (1.25)$$

The dimension of these two coefficients is the inverse of a pressure. They are respectively equal for an ideal gas to:

$$\chi_T = \frac{1}{P} \quad \text{and} \quad \chi_S = \frac{1}{\gamma P}, \qquad (1.26)$$

where P is the pressure of the gas and γ the polytropic coefficient, or the ratio of the
specific heats at constant pressure and constant volume C_p / C_v.

It follows from the expressions of these compressibility coefficients that for χ_S
or $\chi_T \to 0$, the medium exhibits no change in the volume (or density), irrespective
of the change in the applied pressure, hence the following definition:

> An *incompressible* fluid is a medium whose compressibility coefficient is always
> zero.

It is important to note that, according to this definition, the density of the fluid does not
vary with the pressure, which in no way means that the density is constant anywhere
in the flow field and at any time. In addition, the incompressibility has to be seen as
an "*ideal concept*", since all real material is compressible.

Remark For liquids, the isothermal compressibility coefficient is virtually independent of the pressure. Its value can vary significantly from one liquid to another. Let us
simply mention, for example, that, compared to water, mercury is about 13.3 times
more compressible, while ethyl alcohol is 2.3 times ***less*** compressible.

Compressibility has several consequences, depending on whether or not the fluid is at
rest. Those specifically related to a moving fluid will be discussed in Subsect. 1.3.4.3.
Let us simply mention here that the fluid compressibility is responsible, in any case,
for the propagation of infinitesimal pressure waves, with a velocity a defined by:

---

[17] This coefficient is linked to the elasticity modulus E of the medium by relation $\chi_T = 3 (1 - 2\nu) / E$

Table 1.7 Values of the isothermal compressibility coefficient and the speed of sound for water and air under standard temperature and pressure conditions

	Isothermal compressibility coefficient χ_T (Pa^{-1})	Speed of sound **a** (m/s)
Air (Stand. cond.)	10^{-5}	330
Water	5×10^{-10}	1420

$$a^2 = \frac{\delta P}{\delta \rho}|s = \frac{1}{\rho \chi_S}. \tag{1.27}$$

For periodic waves whose frequencies range between 20 Hz and 20,000 Hz, this speed is that of the propagation of a sound audible to the human ear. Relation (1.27) clearly shows that the more compressible the medium, the lower the propagation speed of the pressure waves. Finally, for an ideal gas, we note that, taking account of (1.26), we still have:

$$\mathbf{a} = \sqrt{\frac{\gamma P}{\rho}}.$$

Table 1.7 lists the values of the isothermal compressibility coefficient and the speed of sound for water and air under standard temperature and pressure conditions.

According to the data in the previous Table one can notice that an increase in pressure of one atmosphere causes a relative decrease in volume of 100% in the case of air,[18] while it is only 1/20,000 (0.005%) for water. One might infer from this data about fluids *at rest* that the same conclusion applies to a moving fluid and thus consider water and air as media respectively incompressible and compressible when they are flowing. We will see later in this course that the situation is rather more complex.[19]

1.3.4.2   The Compressibility of a Moving Fluid. The Isovolume Evolution

When a fluid is moving, it is naturally submitted to pressure forces which can be variable throughout the flow field. The question which then arises, as a consequence of the compressibility "*at rest*", is whether the pressure variations *due to the motion of the fluid* are such that they have or not a significant influence on volume changes of fluid particles when in flow. Consider (Fig. 1.9) a fluid particle centered at time t around a given point M whose elementary volume is denoted by $\epsilon (t)$. Subsequently,

---

[18] This is only an estimate. Indeed, since the pressure variation in this case cannot be taken as infinitesimal, the result, strictly speaking, should be expressed by the integral $\int dP/\chi_T$.

[19] Water is a hundred times more compressible than steel. The incompressible modelling of water is incompatible with some phenomena: the water hammer in penstock pipes, the instability of hydraulic actuators, the connecting of unsteady pressure measuring lines, etc.

Fig. 1.9 Variation of the elementary volume of a fluid particle while in motion

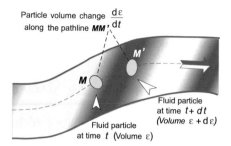

at $t + dt$ this *same* fluid particle will have moved to a point M' and its bounding surface will have changed position and shape.

The problem is then to determine whether such changes occur along with "*dilation/expansion*" or "*compression/contraction*" of the initial particle volume.

We will demonstrate later (Sect. 1.6.7.1) that the change in the particle volume *when following the motion of the particle along its path* is simply expressed by

$$\frac{1}{\epsilon}\frac{d}{dt} = div V. \tag{1.28}$$

Relation[20] (1.28) clearly shows that if the velocity of the flow field is *divergence free*, the initial value of the volume of each fluid particle is unchanged when moving along its respective path. Accordingly, the corresponding evolution will be called an *isovolume* evolution.[21]

We will see in Chap. 3 Sect. 3.1.2 that the mass conservation of a fluid particle during its motion results in the *continuity equation*:

$$\frac{d\rho}{dt} + \rho\, div V = 0.$$

We can therefore draw the following conclusion:

> In every isovolume evolution, the density of each fluid particle remains unchanged during its motion.

It should be noted that the conservation of the density *during the motion of each fluid particle* does not mean in any way that *all* particles which make up the fluid have the *same* density. We will come back to this issue in Chap. 3 Sect. 3.1.3.

Finally, referring to Eqs. (1.25) and (1.26), one might be tempted to give to the previous statement the following alternative formulation:

---

[20] Note that the notation $\frac{d}{dt}$ introduced here will be used in the following of this work with the specific meaning clarified in Sect. 1.6.2.

[21] We are still speaking of this condition as that of a *solenoidal field*.

Any isovolume evolution is that of an incompressible fluid. This reformulation is somewhat reductive, since the compressibility is not the only possible cause for the change in density, as discussed in Sect. 1.3.5 below.

1.3.4.3   The Mach Number and the Compressible Flow Regimes

In addition to the *local* analysis of the previous section, the problem discussed here concerns the influence of the fluid compressibility on the *general* flow pattern. Clearly, the question is to determine whether the natural compressibility of the *fluid at rest* has to be taken into account in *any* mathematical model *representative of a fluid flow*, simply on the ground that *all* real fluids are *actually* compressible at rest.

To discuss this issue, one has to define a parameter which globally accounts for both *static* (fluid at rest) and *dynamic* (fluid in motion) compressibility effects. This parameter is the Mach number, *viz* the ratio between the flow velocity and the speed of sound under the same conditions, namely:

$$\boxed{M = \frac{V}{a}} \tag{1.29}$$

- Firstly, the Mach number allows us to distinguish different flow regimes: subsonic ($M < 1$), transonic ($0.95 < M \lesssim 1.05$), supersonic ($1 < M \lesssim 5$) and hypersonic ($M \gtrsim 5$).
- Secondly, by anticipating the results of the further compressible flows study in Chap. 8, we can use the density expansion ρ, around its incompressible value ρ_i, with respect to the Mach number M:

$$\frac{\rho_i}{\rho} = 1 + \frac{1}{2}M^2 + \mathcal{O}(M^4).$$

Thus, for $M < 0.2$, the change in density[22] is less than 2%. This is the reason which leads us to introduce, within the subsonic flow regime, the class of "*incompressible flows*[23]".

Figure 1.10 schematically summarizes the conclusions of this discussion on the subject of compressibility effects in airflows.

For air under standard conditions of temperature and pressure, the incompressible regime extends up to velocities of about 60 to 70 m/s (200 to 250 km/h). Concerning

---

[22] By a similar technique, it can be shown that the velocity determined by considering that the fluid is incompressible does not differ by more than 1% of that found by taking into account the compressibility, as long as the Mach number remains below 0.2; the difference increases beyond this point at a high (quadratic) rate with the Mach number.

[23] It is noteworthy that the **incompressible flow** notion differs from that of the **incompressible fluid** in that it is an approximation of the fluid motion which remains compatible with $\chi \neq 0$.

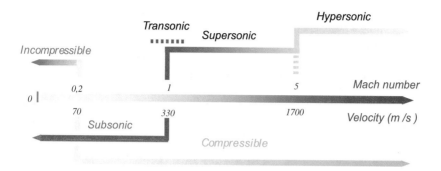

Fig. 1.10 Various air-flow regimes as a function of the Mach number

external aerodynamics problems for example, there is thus a wide flight domain where the air compressibility can be taken as having a negligible influence on the flow properties.

For water, compressibility effects are negligible in most applications, except for such as those mentioned in footnote 19.

1.3.5 Thermal Expansion and the Specific Heat Coefficients

Subjected to a temperature rise of ΔT, the volume of the unit mass of a fluid medium undergoes an increase which can be measured by using the *isobaric thermal expansion* coefficient α, such that:

$$\left.\frac{\delta v}{v}\right|_P \equiv -\left.\frac{\delta \rho}{\rho}\right|_P = \alpha \delta T. \tag{1.30}$$

The coefficient α is homogeneous to the inverse of a temperature. For an ideal gas it is found that $\alpha = 1/T$.

We also have to consider the specific heat coefficients[24] at constant pressure and constant volume, respectively defined as follows for a unit mass:

$$C_p = \left.\frac{\delta Q}{\delta T}\right|_P \quad \text{and} \quad C_v = \left.\frac{\delta Q}{\delta T}\right|_v, \tag{1.31}$$

whose ratio defines the *polytropic coefficient*:

$$\gamma = C_p/C_v. \tag{1.32}$$

---

[24] We still refer to heat capacity.

For air, the value of γ is close to 1.4. The specific heat coefficients are expressed in the international system with the J kg^{-1} K^{-1} unit and their values for air and water are given in Table 1.8.

Finally let us recall that for an ideal gas, the specific heat coefficients at constant pressure and volume are linked by Mayer's relation:

$$C_p - C_v = R, \qquad (1.33)$$

where R (J kg^{-1}K^{-1}) is the universal constant of the ideal gas law (see Reminders in Appendix A).

ADDITIONAL INFORMATION—The undilatable fluid and the Boussinesq approximation—The same problematic as that for compressibility also applies to thermal expansion.

(a) We therefore begin by stating as a definition that:

> An *undilatable fluid* is a medium whose thermal expansion coefficient is zero.

(b) We then aim at expressing the thermal expansion effects *during the motion* by introducing a dimensionless parameter accounting for both *(i) static* (expansibility of the the fluid at rest) and *(ii) dynamic* (flow effects) aspects.

The situation is, in this respect, somewhat more complex than in the case of the compressibility. Indeed, for the natural convection situation, where the expansion of a hot fluid is at the very origin of its motion through *buoyancy* or *Archimedes forces*, the momentum balance equation also involves *inertia* and *viscous* forces. To quantify the relation between these three forces, it is relevant to introduce a dimensionless parameter called the Grashof number defined as:

$$Gr = \frac{g\alpha\Delta T L^3}{\nu^2}, \qquad (1.34)$$

where g denotes the gravity acceleration, ΔT is a difference temperature of reference, L a length scale and ν the kinematic viscosity of fluid.

Now, by introducing a velocity scale U, we can infer, from dimensional considerations, the following expressions:

$$\begin{cases} \text{Buoyant forces} \propto \rho g\alpha\Delta T \\ \text{Inertial forces} \propto \rho U^2/L \\ \text{Viscous forces} \propto \mu U/L^2 \end{cases}$$

We can then verify that $\dfrac{\text{Buoyant forces} \times \text{Inertial forces}}{[\text{Viscous forces}]^2} \propto Gr.$

c) Under specific conditions, a fluid motion model can be implemented where the change in density due to the thermal expansion manifests itself only in the

Table 1.8 Isobaric thermal expansion coefficient and specific heat at constant pressure of air and water

	Air (20° C, 1 Atm)	Water (20° C, 1 Atm)
Isobaric thermal expansion coefficient α (K^{-1})	3.4×10^{-3}	0.18×10^{-3}
Specific heat at constant pressure C_p (J kg^{-1} K^{-1})	1000	4200

buoyancy term of the momentum equation. It is referred to as as the "*Boussinesq approximation*" which allows us to introduce the buoyancy effects in the isovolume and incompressible flows class, in the terms defined in Sect. 1.3.4.3. An example of such a flow will be discussed in Chap. 12 Sect. 12.3.1, *viz* the *natural convection* over a vertical flat plate.

HISTORICAL NOTE—**The Boussinesq-Oberbeck approximation**—In 1902 and 1903, Boussinesq published in two large volumes [16] and [17] (363 pages and 657 pages respectively) a synthesis of the the knowledge and theories on 'Heat' common at the time.

In the second volume, however, he addressed a relatively new issue, that of hydrodynamic convection, and wrote on page vii of the Warning of this work: «...*it should also be pointed out that in most movements caused by heat on our heavy fluids, volumes or densities can be kept almost constant, although the corresponding change in the unit volume weight is precisely the cause of the phenomena to be analyzed. Hence arises the possibility of neglecting density variations, where they are not multiplied by the gravity g, only retaining in the calculations, their product by the gravity.*» Accordingly, Navier's hydrodynamics equations ($\rho = C^t$) can be extended to the thermal convection regime under the *Boussinesq approximation*, as named by Rayleigh in 1916 [132].

In fact, a similar result was independently proposed by Oberbeck [109], in 1879, hence the qualification of the *Oberbeck-Boussinesq approximation*, more in line with the historical evidence.

1.4 The Mathematical Description of the Fluid Motion

After having defined the functions characterizing the fluid "*motion*", viz taken in terms of a *thermo-mechanical evolution*, we will now specify the arguments (or variables) upon which depend the local values of each of these flow field functions.

1.4.1 Lagrangian Variables

The space location method leading to the *Lagrangian variables* is directly inspired from that of the material points in rational mechanics.

Let us suppose that, at a time $t = \tau$, called the tagging instant, one is able to "individualize" each fluid particle with a label or mark which will not affect its thermo-mechanical properties. At any later instant $t > \tau$, it thus becomes possible to *follow* each fluid particle from its position at the tagging instant and to determine its coordinates, in a given referential, from those at the tagging instant (ξ, η, ζ) and of the current time t.

> The four arguments (ξ, η, ζ, t) are the Lagrangian variables.

Thus, as a direct property of this definition, any function $f(\xi, \eta, \zeta, t)$ of the *Lagrangian variables* always refers to the *same* fluid particle. Assuming that at the tagging instant the particle is at a point A, its position P at any later instant t is given by:

$$AP = \int_{\tau}^{t} V(\xi, \eta, \zeta, \alpha) \, d\alpha. \tag{1.35}$$

where V stands for the velocity vector of the marked particle.

Denoting by (X, Y, Z) the coordinates of P, there is, for each fluid particle at any time—and therefore at *every point of the flow field* (Ω)—a one to one relation between these coordinates and those of the particle position at its tagging instant. We symbolically note:

$$\forall P(X_i) \in \Omega, \quad (X, Y, Z) \overset{X_i(\xi, \eta, \zeta, t)}{\rightleftarrows} (\xi, \eta, \zeta). \tag{1.36}$$

1.4.2 Eulerian Variables

When adopting the prospect of an experimenter *external to the flow*, one may decide to describe the flow field in a different way. It consists in observing, at any point M arbitrarily set in the flow field Ω, the value of any property of *the* fluid particle which is positioned at *this* point[25] at a *given* time. This value is thus a function of the four arguments (x, y, z, t), where the first three are none other than the coordinates, in a given referential, of the location M where the observation is taking place and t is the instant of the observation.

---

[25] This notion of "*point*" is, of course, to be taken in terms of the *mesoscopic* continuum analysis.

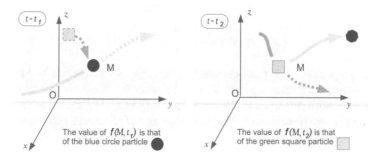

Fig. 1.11 Physical interpretation of a function's value depending on Eulerian variables

> The four arguments (x, y, z, t) are the Eulerian variables.

As illustrated in Fig. 1.11, *different* particles are located at the *same* point M (x, y, z) at *different* times.

The value of any flow field function expressed in Euler variables $f(x, y, z, t)$ is therefore *assigned* to *the only* fluid particle passing through the observation point (x, y, z) at the observation instant t. From now on, only the Eulerian variables will be used in this course.

HISTORICAL NOTE— Euler variables—The term *Eulerian variable* was established by custom, in tribute to the remarkable work published by Euler [56] in 1757, where the *inviscid* fluid motion equations are demonstrated for the first time using this formalism. Nevertheless, the idea of that which we now call *Lagrangian variables* is also present in the introductory section of this paper, where Euler wrote, page 275: «*We should also assume that the state of the fluid at a certain time is known, & that I name the primitive state of the fluid: this state is arbitrary, it is necessary to firstly be known the arrangement of the particles, of which the fluid is composed & the motion which has been transmitted to them, unless the primitive state is that of a fluid at rest*».

ADDITIONAL INFORMATION—Selection of the appropriate variables—With two types of fluid motion description, a matter of choice may arise in practice. Euler variables are perfectly adapted to the intrusive probe measure delivering a signal associated with the flow properties at the sensor location. If the objective is to follow the displacement of a marker from an injection point, as is the case of contaminant dispersion problems *eg*, Lagrangian variables are the most appropriate.

EXERCISE 3 Lagrangian and Eulerian coordinates.

Let us consider a flow related to an orthonormal cartesian coordinate system, in which the particles move according to the following time-dependent law :

$$\begin{cases} X(t) = a\cos(\omega t + \varphi) \\ Y(t) = a\sin(\omega t + \varphi) \\ Z(t) = Z_0 + a\omega t, \end{cases}$$

where a is a positive real homogeneous to a length, ω a positive real homogeneous to a frequency, φ a phase angle and Z_0 a real coordinate. Time $t = 0$ is taken as the *tagging time*, with reference to the definition of the Lagrangian variables. We are interested in three distinct particles $\{L_1\}$, $\{L_2\}$ and $\{L_3\}$, whose coordinates are at $t = 0$:

- for $\{L_1\} : 0, -a, -a\dfrac{\pi}{2}$;

- for $\{L_2\} : a, 0, 0$;

- for $\{L_3\} : 0, +a, +a\dfrac{\pi}{2}$;

Finally, we denote by E an observation point in terms of the Eulerian analysis, whose coordinates in the same frame are $(a, 0, 2a\pi)$.

QUESTION 1. Explicit the time-dependent motion laws for each of the three particles $\{L_1\}$, $\{L_2\}$ and $\{L_3\}$.

QUESTION 2. Show that all these particles will pass through point E *at a given time*.

QUESTION 3. To which particle correspond the Eulerian coordinates $(a, 0, 2a\pi, \frac{2\pi}{\omega})$?

SOLUTION: **1.** The whole question is a matter to determining the phase origin φ as well as the value of the Z_0 coordinate of each particle motion. This is directly obtained by imposing the condition, for each time-dependent motion law, to be verified for the coordinates at the tagging time, *viz*:

$$\{L_1\} \begin{cases} X_1(t) = a\cos(\omega t - \frac{\pi}{2}) \\ Y_1(t) = a\sin(\omega t - \frac{\pi}{2}) \\ Z_1(t) = -a\frac{\pi}{2} + a\omega t, \end{cases} \{L_2\} \begin{cases} X_2(t) = a\cos(\omega t) \\ Y_2(t) = a\sin(\omega t) \\ Z_2(t) = a\omega t, \end{cases} \{L_3\} \begin{cases} X_3(t) = a\cos(\omega t + \frac{\pi}{2}) \\ Y_3(t) = a\sin(\omega t + \frac{\pi}{2}) \\ Z_3(t) = a\frac{\pi}{2} + a\omega t, \end{cases}$$

2. Considering the Z coordinate, it is easy to see that it takes the value $2a\pi$, which corresponds to point E coordinates, for $t = \frac{\pi}{2\omega}, t = \frac{2\pi}{\omega}, t = -\frac{\pi}{2\omega}$ [modulo 2π], with respect to each of the particles under consideration. We can then easily verify that the two other coordinates are well identified with those of point E for each particle, which demonstrates the proposition.

3. It follows directly from the previous question that the Eulerian coordinates $(a, 0, 2a\pi, \frac{2\pi}{\omega})$ correspond to those of the particle $\{L_2\}$ *at time $t = \frac{2\pi}{\omega}$*.

1.5 General Principles for the Study of a Fluid Motion

1.5.1 The Eulerian Viewpoint and the Control Volume

As we have seen, the guiding idea of the Eulerian description originates in the observation, at any fixed point of the flow field and at any time, of the properties of the fluid particles which pass through this point over time. By extension to a finite volume, we are led to looking into the properties of a set of particles which, at any time, are included in a fixed domain Δ of the flow field.

> Any fixed observation domain Δ, in terms of the Eulerian analysis, will be called a *control volume*. It can be chosen arbitrarily within the flow, solely provided that its bounding surface is entirely composed of fluid particles.

As illustrated in Fig. 1.12, the control volume Δ is lawful, its bounding surface Σ consists only of fluid particles, even if some of them are in contact with a solid limit. In contrast, the domain Δ' is unlawful, some portions of the surface Σ' are not made of fluid particles.

More generally, regarding the control volume concept the following three points should be mentioned:

1 Over time, the domain remains *unchanged*, both in position and magnitude, hence the control volume terminology;

2 At *two different times*, the domain does *not* contain *the same* fluid particles. The bounding surface Σ of the domain is "*pervious*" to the fluid and thus subjected to a *flux* of any property which is transported with the fluid particles passing through;

3 The balance equation of any property over Δ can be derived by using either the Eulerian or Lagrangian variables for expressing the local value of the representative function of that property. There is indeed no reason to make a linkage

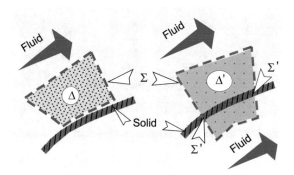

Fig. 1.12 Illustration of a lawful and unlawful control volume

between the adopted type of reasoning—in this case Eulerian—and the variables used to express the arguments of the functions.

1.5.2 The Lagrangian Viewpoint and the Material Domain

A similar extension can also be considered for the Lagrangian case. It leads, according to Lagrange's description, to considering at the tagging instant rather than *a single* particle, *a set* of particles located in a domain $\mathcal{D}$ and to *following the displacement of that domain* by moving at the same velocity as that of each of the fluid particles which composes this domain.

Any moving domain $\mathcal{D}$ in terms of the Lagrangian analysis will be termed a *material domain*. It can be chosen arbitrarily within the flow, solely provided that it is composed of fluid particles.

Figure 1.13 schematically illustrates the main distinctions between a control volume (Euler's analysis) and a material domain (Lagrange's analysis).
Concerning the balance equations over a material domain, it is worth noting that:

- At any time, such a domain always contains the same particles. Accordingly, there is no macroscopic flux of any fluid property through the bounding surface S of the domain, hence the term "*material*" to recall that it is a closed volume which does not exchange mass with the surrounding fluid, *at a macroscopic level*;
- In principle, over time, the domain undergoes changes in position, shape and size;
- Any given function related to a material domain can be expressed with Lagrangian or Eulererian variables.

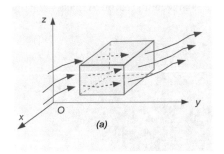
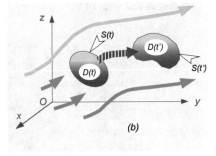

Fig. 1.13 The control volume and the Eulerian viewpoint **a**; The material domain and the Lagrangian viewpoint **b**

Fig. 1.14 Schematic for the
material derivative

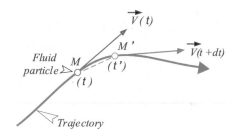

Fig. 1.14 Schematic for the
material derivative

1.6 The Material Derivation

1.6.1 Definition

Since Eulerian variables are not linked to a single fluid particle over time, the problem arises with regard to expressing, with this formulation, any change when *following the motion* of a single particle or a set of the same particles. By definition, such changes will be called *material* and we will refer to a *material derivative*[26] whenever this occurs.

1.6.2 The Material Derivative of a Scalar Function

Let $f(x, y, z, t)$ a real function[27] of the Eulerian variables (x, y, z, t). Its differential is expressed by:

$$df = \frac{\partial f}{\partial t}dt + \frac{\partial f}{\partial x}dx + \frac{\partial f}{\partial y}dy + \frac{\partial f}{\partial z}dz \quad \text{or equally} \quad \frac{df}{dt} = \frac{\partial f}{\partial t} + \boldsymbol{grad}\,f.\frac{d\boldsymbol{M}}{dt},$$

(1.37)

when introducing the vectors $\boldsymbol{grad}\,f \left[\dfrac{\partial f}{\partial x}, \dfrac{\partial f}{\partial y}, \dfrac{\partial f}{\partial z}\right]$ and $\boldsymbol{dM}\,[dx, dy, dz]$ in a Cartesian orthonormal frame.

As illustrated in Fig. 1.14 this expression can be considered as a *material derivative* if and only if the displacement vector $\boldsymbol{dM}$ identifies with that of the change in positions, over the time dt, of the particle located at point M at time t.
Accordingly, we can set:

$$\boldsymbol{dM} = \boldsymbol{MM'} \equiv \boldsymbol{V}dt,$$

where $\boldsymbol{V}$ denotes the velocity of the particle located at point M at time t.

---

[26] Also known as a substantial derivative.

[27] This function is supposed to satisfy all the mathematical requirements regarding the definiton, continuity and differentiability to validate the operations mentioned above.

By introducing this condition into relation (1.37) we obtain the general expression of the *material derivative* of a *scalar* function:

$$\boxed{\frac{\mathrm{d}f}{\mathrm{d}t} = \frac{\partial f}{\partial t} + V.grad\ f}$$ (1.38a)

or equally[28]:

$$\boxed{\frac{\mathrm{d}f}{\mathrm{d}t} = \frac{\partial f}{\partial t} + U_j \frac{\partial f}{\partial x_j} \equiv \frac{\partial f}{\partial t} + U_1 \frac{\partial f}{\partial x_1} + U_2 \frac{\partial f}{\partial x_2} + U_3 \frac{\partial f}{\partial x_3}}$$ (1.38b)

Relation (1.38a) shows that the material derivative expression of a scalar function of the Eulerian variables consists of two additive contributions:

- a *time* variation associated with the *unsteady* character of the flow (see Sect. 1.6.10 below);
- a *transport* variation resulting from the fluid displacement with the velocity V of a non-homogeneous space function in any direction non-orthogonal to the displacement ($grad\ f$). It is called[29] the *convective variation* or *convective transport* or simply the *convection*.

1.6.3 The Material Derivative of a Vector Function

1.6.3.1 General Vector Function

Let $A\ (x, y, z, t)$ be a vector function of the Eulerian variables whose components are $A_i\ (x, y, z, t)$, $i = 1, 2, 3$. By applying relation (1.38b) to each component, we directly obtain the result, in indicial notation:

$$\frac{\mathrm{d}A_i}{\mathrm{d}t} = \frac{\partial A_i}{\partial t} + U_j \frac{\partial A_i}{\partial x_j} \qquad \text{with } i = 1, 2, 3.$$ (1.39)

We can verify that it is indeed a vector equality since, as the index j is a dummy index by summation convention, the solely free index is i taking the three values 1, 2, 3. Relation (1.39) introduces the gradient of a vector A which is a second rank tensor noted $\overline{\overline{grad}}\ A$ whose components are defined in the following matrix:

---

[28] In relations (1.38a) and (1.38b) and systematically in the remaining sections, the use of the romanized font style $\frac{\mathrm{d}}{\mathrm{d}t}$ will systematically denote a material derivation.

[29] It is equally called the *advective variation*, *advective transport* or simply the *advection*.

$$
\overline{\overline{grad}}\; \boldsymbol{A} \; = \;
\begin{pmatrix}
\dfrac{\partial A_1}{\partial x_1} & \dfrac{\partial A_1}{\partial x_2} & \dfrac{\partial A_1}{\partial x_3} \\[2ex]
\dfrac{\partial A_2}{\partial x_1} & \dfrac{\partial A_2}{\partial x_2} & \dfrac{\partial A_2}{\partial x_3} \\[2ex]
\dfrac{\partial A_3}{\partial x_1} & \dfrac{\partial A_3}{\partial x_2} & \dfrac{\partial A_3}{\partial x_3}
\end{pmatrix}
\; = \; \dfrac{\partial A_i}{\partial x_j}
$$

line

column

Symbolizing by $\odot$ the contraction product with respect to the right index, that is to say the derivation or column index of the tensor $\overline{\overline{grad}}\;A$, relation (1.39) can be rewritten in a tenso-vectorial formulation as follows:

$$
\frac{\mathrm{d}\boldsymbol{A}}{\mathrm{d}t} = \frac{\partial \boldsymbol{A}}{\partial t} + \overline{\overline{grad}}\,\boldsymbol{A} \odot \boldsymbol{V}. \tag{1.40}
$$

1.6.3.2 Expression of the Acceleration

The expression of the acceleration using the Eulerian variables can be easily deduced from the previous relations by simply identifying $\boldsymbol{A} \equiv \boldsymbol{V}$, the fluid particle velocity at the observation point and time t.
Thus we have:

$$
\boxed{\frac{\mathrm{d}U_i}{\mathrm{d}t} = \frac{\partial U_i}{\partial t} + U_j \frac{\partial U_i}{\partial x_j}} \tag{1.41a}
$$

$$
\boxed{\frac{\mathrm{d}\boldsymbol{V}}{\mathrm{d}t} = \frac{\partial \boldsymbol{V}}{\partial t} + \boldsymbol{V} \odot \overline{\overline{grad}}\,\boldsymbol{V}} \tag{1.41b}
$$

By specifying the relation in detail between the components, we can verify that the acceleration can also be expressed in a purely vectorial form:

$$
\boxed{\frac{\mathrm{d}\boldsymbol{V}}{\mathrm{d}t} = \frac{\partial \boldsymbol{V}}{\partial t} + \boldsymbol{grad}\left(\frac{\|V\|^2}{2}\right) + curl\,\boldsymbol{V} \wedge \boldsymbol{V}} \tag{1.42}
$$

where the symbol $\wedge$ denotes the cross product and $\|V\|$ is the velocity vector module.

TERMINOLOGY The two additive contributions of any material derivative are of course present in the expression of the Eulerian acceleration. According to the etymology, we will reserve here the term of "*convection*" to situations where the transported quantity is certainly different from the transporting agent which is the velocity. Conversely, when the velocity acts both as the transported and the transporting agent, we will prefer using the "*advection*" qualification.

EXERCISE 4 The velocity and acceleration in Euler's/Lagrange's variables.

Let X, Y, Z be the coordinates, in an orthonormal cartesian coordinate system, of the position *at any time* $t > 0$ of the fluid particle which was located at the tagging time $t = 0$, at the coordinate point ξ, η, ζ.
We give : $X = (1 + t^2) \times \xi$, $\quad Y = \eta$, $\quad Z = \zeta$.

    QUESTION 1 Express the velocity and acceleration with the Lagrangian variables.
    QUESTION 2 Express these same functions with the Eulerian variables.

SOLUTION: **1.** When using the Lagrangian variables, the velocity and acceleration are directly obtained by taking the time derivative of the position vector, in the form:

$$
\text{Velocity :} \quad
\begin{cases}
\dfrac{dX}{dt} = 2\xi t \\[2mm]
\dfrac{dY}{dt} = 0 \\[2mm]
\dfrac{dZ}{dt} = 0
\end{cases}
\qquad
\text{Acceleration :} \quad
\begin{cases}
\dfrac{d^2 X}{dt^2} = 2\xi \\[2mm]
\dfrac{d^2 Y}{dt^2} = 0 \\[2mm]
\dfrac{d^2 Z}{dt^2} = 0
\end{cases}
$$

2. Now, if we adopt the Eulerian variables (x, y, z) of a given point where we observe *at time* t this particle, which means that at that time, the observation point (x, y, z) coincides with the position of the particle (X, Y, Z). So we have :

$$x = X \quad\quad y = Y \quad\quad z = Z,$$

which accounts for all the exploitable information, since the position at the tagging time has no relevance in terms of the Eulerian analysis. To express the velocity and acceleration using the Eulerian variables, we simply substitute the coordinates ξ, η, ζ with the above values in the previous Lagrangian expressions of these vectors. We thus obtain :

$$
\text{Velocity :} \quad
\begin{cases}
U = \dfrac{2xt}{1 + t^2} \\[2mm]
V = 0 \\[2mm]
W = 0
\end{cases}
\qquad
\text{Acceleration :} \quad
\begin{cases}
dU/dt = \dfrac{2x}{1 + t^2} \\[2mm]
dV/dt = 0 \\[2mm]
dW/dt = 0
\end{cases}
$$

The expression of the acceleration can also be derived by solely using the Eulerian variables and by immediately applying relation (1.41a) for instance. Thus one obtains for the x component

$$\frac{dU}{dt} = \frac{\partial U}{\partial t} + U \frac{\partial U}{\partial x},$$

Fig. 1.15 Correspondence
between the positions of a
fluid line element at τ and t
times

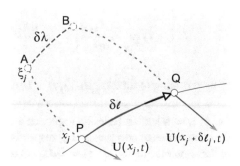

which gives $\dfrac{dU}{dt} = 2x\dfrac{\left(1+t^2\right) - 2t^2}{\left(1+t^2\right)^2} + \dfrac{4xt^2}{\left(1+t^2\right)^2} \equiv \dfrac{2x}{\left(1+t^2\right)}$, which is of course
identical to the expression derived with the Lagrangian calculus. The previous results
can easily be generalized to any trajectory defined as $X = f(t) \times \xi,\ Y = \eta,\ Z = \zeta$,
where $f(t)$ is any time dependent function.

1.6.4 The Material Derivative of a Fluid Line Element

Let us consider now an elementary vector $PQ = \delta\ell$ associated with an elementary oriented segment of a *fluid line*, a curve where every point, by definition—see
Chap. 2 Sect. 2.1.4—consists of a fluid particle moving at the local fluid velocity
(see Fig. 1.15).

The question is to express the material derivative $\dfrac{d(\delta\ell)}{dt}$, where the variation
is taken according to the elementary displacement vector moving with the fluid.
According to the Lagrangian framework, the velocity at point P and time t can be
expressed as:

$$U_i(P,t) = \left.\frac{dX_i^A(\xi_j, \alpha)}{d\alpha}\right|_{\alpha=t}, \tag{1.43}$$

where $X_i^A = X_i^A(\xi, \eta, \zeta, t)$ is the time-dependent displacement law of the fluid particle located at point $A\,(\xi, \eta, \zeta)$ at the tagging time τ and which is located at point
P at time t.

Denoting by $AB = \delta\lambda(\xi, \eta, \zeta)$ the line element at the tagging time, we obtain

$$X_i^B(t) = X_i^A(\xi, \eta, \zeta, t) + \delta\lambda_i(\xi, \eta, \zeta, t),$$

where $X_i^B(t)$ is the time-dependent displacement law of the particle located at point
B at the tagging time and passing to point Q at time t.
We deduce the Lagrangian expression of the velocity at time t and point Q as:

$$U_i(Q,t) \equiv \left.\frac{\mathrm{d}X_i^B(\alpha)}{\mathrm{d}\alpha}\right|_{\alpha=t} = \left.\frac{\mathrm{d}X_i^A(\xi_j,\alpha)}{\mathrm{d}\alpha}\right|_{\alpha=t} + \left.\frac{\mathrm{d}\,\delta\lambda_i(\xi_j,\alpha)}{\mathrm{d}\alpha}\right|_{\alpha=t},$$

or, by returning to time t, $\quad U_i(Q,t) - U_i(P,t) = \dfrac{\mathrm{d}\,\delta\ell_i}{\mathrm{d}t}.$ (1.44)

Let us now adopt an Eulerian approach and accordingly denote by x_j and $x_j + \delta\ell_j$ the coordinates of points P and Q respectively. The corresponding velocities in each of these points are : $U_i(P,t) = U_i(x_j,t)$ and $U_i(Q,t) = U_i(x_j + \delta\ell_j, t)$.
By a first order expansion, we can write

$$U_i(x_j + \delta\ell_j, t) = U_i(x_j,t) + \frac{\partial U_i}{\partial x_j}\delta\ell_j,$$

which yields the expression of the velocity difference as:

$$U_i(Q,t) - U_i(P,t) = \frac{\partial U_i}{\partial x_j}\delta\ell_j.$$ (1.45)

Comparing relations (1.44) and (1.45) provides the expected result:

$$\boxed{\frac{\mathrm{d}(\delta\ell_i)}{\mathrm{d}t} = \frac{\partial U_i}{\partial x_j}\delta\ell_j}$$ (1.46)

1.6.5 The Material Derivation of a Fluid Line Integral

Let $f(x_j,t)$ be a scalar function of the Eulerian variables. We consider the line integral G along an oriented arc $\widehat{PQ}(t)$ of a *fluid line* at time t, which is the vector whose components are

$$G_i = \int_{\widehat{PQ}(t)} f(x_j,t)\,dx_i,$$ (1.47)

where dx_i, $i = 1,2,3$ denotes the components of the vectorial line element, *viz* $d\boldsymbol{\ell} = dx_i\boldsymbol{e}_i$.

The question is to express $\dfrac{\mathrm{d}G_i}{\mathrm{d}t}$, where the derivative is taken in terms of the material derivation, *i.e.*, with a displacement which follows that of the moving fluid curve. As shown in the definition (1.47), the time dependency now results in both the integrand and the limit of integration. By applying the transformation $t \to \tau$ (tagging time), the fluid line $\widehat{PQ}(t)$ retrieves its initial location $\widehat{AB}(\tau)$, similarly to what was carried out in the previous Section, see Fig. 1.15. By the same transformation, we obtain :

$$f(x_j, t) \rightarrow f[x_j(\xi, \eta, \zeta, t), t] \equiv \widetilde{f}(\xi_j, t) \text{ and } dx_i \rightarrow \frac{\partial x_i}{\partial \xi_j} d\xi_j.$$

The line integral G_i can then be re-expressed along the contour $\overset{\frown}{AB}(\tau)$ to obtain:

$$G_i = \int_{\overset{\frown}{AB}(\tau)} \widetilde{f}(\xi_j, t) \frac{\partial x_i}{\partial \xi_j} d\xi_j,$$

which can be easily differentiated with respect to time, since the limit of integration is no longer time dependent. We thus obtain:

$$\frac{dG_i}{dt} = \int_{\overset{\frown}{AB}(\tau)} \left[\frac{d\widetilde{f}}{dt} \frac{\partial x_i}{\partial \xi_j} + \widetilde{f} \frac{d}{dt} \left(\frac{\partial x_i}{\partial \xi_j} \right) \right] d\xi_j = \int_{\overset{\frown}{AB}(\tau)} \left[\frac{d\widetilde{f}}{dt} \frac{\partial x_i}{\partial \xi_j} + \widetilde{f} \frac{\partial}{\partial \xi_j} \left(\frac{dx_i}{dt} \right) \right] d\xi_j$$

or equally

$$\frac{dG_i}{dt} = \int_{\overset{\frown}{AB}(\tau)} \frac{d\widetilde{f}}{dt} \frac{\partial x_i}{\partial \xi_j} d\xi_j + \int_{\overset{\frown}{AB}(\tau)} \widetilde{f} \frac{\partial U_i}{\partial \xi_j} d\xi_j.$$

To revert to an expression along the contour $\overset{\frown}{PQ}(t)$, we simply apply the opposite transformation $\tau \rightarrow t$, which yields the final result:

$$\boxed{\frac{dG_i}{dt} = \int_{\overset{\frown}{PQ}} \frac{df}{dt} dx_i + \int_{\overset{\frown}{PQ}} f \, dU_i} \tag{1.48}$$

1.6.6 The Material Derivative of the Circulation

1.6.6.1 The Circulation Around a General Contour

For any given oriented element $\delta\boldsymbol{\ell} = \boldsymbol{t} ds$ of a curve $(\mathcal{C})$, the elementary circulation of the velocity vector $\boldsymbol{V}$ is classically defined as the scalar product:

$$\delta\Gamma = \boldsymbol{V} . \delta\boldsymbol{\ell} = \boldsymbol{V} . \boldsymbol{t} \, ds.$$

We apply here this definition to a *fluid curve*, which, as previously mentioned (see Sect. 1.6.4), is a *material* line which moves as the fluid particles of which it is made up. The circulation along such a ' frozen' curve *at time t* is expressed as:

$$\Gamma(x, y, z, t) = \int_{\mathcal{C}(t)} \delta\Gamma = \int_{\mathcal{C}(t)} U_i(x, y, z, t) dx_i, \tag{1.49}$$

where the Eulerian coordinates (x, y, z, t) of the particle at the current point of the curve make it possible to express the components dx_i ($i = 1, 2, 3$), of the oriented curve element $d\ell = t\,ds$.

The question is to express $\dfrac{d\Gamma}{dt} \equiv \dfrac{d}{dt} \displaystyle\int_{\mathcal{C}(t)} U_i(x, y, z, t)\,dx_i$, the derivative being taken in the material sense, *i.e.*, following the curve during its displacement. The dependence on time again results from both the element to be integrated and the integration limit. We will therefore follow a procedure similar to that of the previous section.

Thus, we apply the transformation $t \rightarrow \tau$, the tagging time, so that the coordinates (x, y, z) of the particle at time t become (ξ, η, ζ) those at the tagging time τ; The fluid curve $\mathcal{C}(t)$ becomes $\mathcal{C}(\tau)$ and the differential element takes the form:

$$dx_i = \frac{\partial x_i}{\partial \xi}\,d\xi + \frac{\partial x_i}{\partial \eta}\,d\eta + \frac{\partial x_i}{\partial \zeta}\,d\zeta.$$

The expression of the circulation, Eq. (1.49) becomes :

$$\Gamma(\xi, \eta, \zeta, t) = \int_{\mathcal{C}(\tau)} U_i[x(\xi, \eta, \zeta), y(\xi, \eta, \zeta), z(\xi, \eta, \zeta), t)]\left(\frac{\partial x_i}{\partial \xi}\,d\xi + \frac{\partial x_i}{\partial \eta}\,d\eta + \frac{\partial x_i}{\partial \zeta}\,d\zeta\right).$$

Since the integration is carried on a current time-independent curve the derivation can be directly taken, which gives, by reducing the notations:

$$\frac{d\Gamma}{dt} = \int_{\mathcal{C}(\tau)} \frac{dU_i}{dt}\left(\frac{\partial x_i}{\partial \xi}\,d\xi + \frac{\partial x_i}{\partial \eta}\,d\eta + \frac{\partial x_i}{\partial \zeta}\,d\zeta\right) +$$

$$\int_{\mathcal{C}(\tau)} U_i\,\frac{d}{dt}\left(\frac{\partial x_i}{\partial \xi}\,d\xi + \frac{\partial x_i}{\partial \eta}\,d\eta + \frac{\partial x_i}{\partial \zeta}\,d\zeta\right) \qquad (1.50)$$

Now, $\dfrac{d}{dt}\left(\dfrac{\partial x_i}{\partial \xi}\,d\xi + \dfrac{\partial x_i}{\partial \eta}\,d\eta + \dfrac{\partial x_i}{\partial \zeta}\,d\zeta\right) = \dfrac{\partial(\frac{dx_i}{dt})}{\partial \xi}\,d\xi + \dfrac{\partial(\frac{dx_i}{dt})}{\partial \eta}\,d\eta + \dfrac{\partial(\frac{dx_i}{dt})}{\partial \zeta}\,d\zeta$

$$= \frac{\partial U_i}{\partial \xi}\,d\xi + \frac{\partial U_i}{\partial \xi}\,d\eta + \frac{\partial U_i}{\partial \xi}\,d\zeta,$$

the last equality resulting from the fact that the current point along a fluid curve moves with the velocity of the coinciding fluid particle.

Substituting in Eq. (1.50) one finally obtains

$$\frac{d\Gamma}{dt} = \int_{\mathcal{C}(\tau)} \frac{dU_i}{dt}\left(\frac{\partial x_i}{\partial \xi}\,d\xi + \frac{\partial x_i}{\partial \eta}\,d\eta + \frac{\partial x_i}{\partial \zeta}\,d\zeta\right) + \int_{\mathcal{C}(\tau)} U_i\left(\frac{\partial U_i}{\partial \xi}\,d\xi + \frac{\partial U_i}{\partial \xi}\,d\eta + \frac{\partial U_i}{\partial \xi}\,d\zeta\right)$$

where Γ and U_i are to be considered as functions of ξ, η, ζ, t.

To revert to the space coordinates at the current time t, it is just necessary to make the opposite transformation $\tau \to t$, which brings the curve $\mathcal{C}(\tau)$ back on the curve $\mathcal{C}(t)$ and gives:

$$\frac{\mathrm{d}\Gamma}{\mathrm{d}t} = \int_{\mathcal{C}(t)} \frac{\mathrm{d}U_i}{\mathrm{d}t} \mathrm{d}x_i + \int_{\mathcal{C}(t)} U_i \mathrm{d}U_i \equiv \int_{\mathcal{C}(t)} \frac{\mathrm{d}U_i}{\mathrm{d}t} \mathrm{d}x_i + \int_{\mathcal{C}(t)} \mathrm{d}\left(\frac{U_i U_i}{2}\right) \qquad (1.51)$$

1.6.6.2 The Material Derivative of the Elementary Circulation

Returning to the circulation $\delta\Gamma$ along an arc element $\delta\ell$, the previous relation is reduced to $d\Gamma = \dfrac{\mathrm{d}U_i}{\mathrm{d}t} dx_i + U_i dU_i$, or again, in vector form,

$$\frac{\mathrm{d}(\delta\Gamma)}{\mathrm{d}t} = \frac{\mathrm{d}V}{\mathrm{d}t} \cdot \delta l + V.\mathrm{d}V \qquad (1.52)$$

1.6.6.3 The Material Derivative of a Closed Path Circulation

In the case where the curve $\mathcal{C}(t)$ is a closed path and in the absence of discontinuity in the velocity square around such a contour, the expression (1.51) reduces to the following formula:

$$\frac{\mathrm{d}\Gamma_O}{\mathrm{d}t} = \oint \frac{\mathrm{d}U_i}{\mathrm{d}t} \mathrm{d}x_i \qquad (1.53)$$

which makes it possible to state that:

> Whithout any discontinuity in the velocity square, the material derivative of the velocity circulation along a closed fluid path, is equal to the acceleration circulation along this same path.

1.6.6.4 Kelvin's Circulation Theorem

The result we have just obtained leads to particularly important consequences for flows whose *acceleration derives from a potential*. We will examine in more detail in Chap. 5 the assumptions which make it possible for such a potential to exist. It suffices here to remember that, by definition, the acceleration potential is a function $\varphi(x_j, t)$ such that at each time t we have:

$$\frac{\mathrm{d}V}{\mathrm{d}t} = \mathbf{grad}\ \varphi. \tag{1.54}$$

In this case, the differential element of relation (1.53) can be written:

$$\frac{dU_i}{dt}dx_i = \frac{\partial\varphi}{\partial x_i}dx_i \equiv d\varphi.$$

Hence, the acceleration is a point function whose curvilinear integral is identically zero along any closed path. We can then state the following result, known as *Kelvin's theorem on the circulation*:

> In any acceleration potential flow, the velocity circulation around a closed curve moving with the fluid remains constant over time.

HISTORICAL NOTE—Kelvin's circulation theorem—In an article published in 1869, W. Thomson (Lord Kelvin) [133] resumes and extends the previous results due to Lagrange and Helmholtz, to which he contributes new demonstrations, in a form more similar to the modern approach. One of the key points of these demonstrations is the velocity circulation conservation along a fluid line driven by the fluid particle motion in a flow with an acceleration potential, a result to which its name remained attached, see Lamb [86], Loitsyanskii [91] or Batchelor [5], for example.

1.6.7 The Material Derivative of a Volume Integral

1.6.7.1 The Volume of a Fluid Domain

By adopting *Lagrange's reasoning*, we consider, at time t, a material fluid volume $\mathcal{V}(t)$ occupying the finite domain $\mathcal{D}(t)$. By expressing the volume element in *Euler variables*, we can write:

$$\mathcal{V}(t) = \iiint\limits_{\mathcal{D}(t)} dxdydz. \tag{1.55}$$

The question is now to express the variation of this volume amount *when following the motion of the material domain* $\mathcal{D}(t)$ assumed to be subject to that of its component fluid particles. It is therefore necessary to express:

$$\frac{\mathrm{d}}{\mathrm{d}t}\mathcal{V}(t) = \frac{\mathrm{d}}{\mathrm{d}t}\left[\iiint\limits_{\mathcal{D}(t)} dxdydz\right],$$

Fig. 1.16 The correspondence between the domain positions at τ and t times

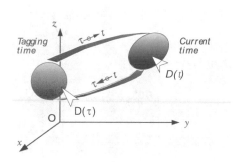

knowing that at the tagging time we have $\mathcal{D}(t) = \mathcal{D}(\tau)$, a domain whose volume is expressed by:

$$V(\tau) = \iiint_{\mathcal{D}(\tau)} d\xi d\eta d\zeta.$$

By changing the variables $(x, y, z) \rightarrow (\xi, \eta, \zeta)$ into Eq. (1.55) we note that (see Fig. 1.16) :

1. the domain $\mathcal{D}(t)$ becomes the domain $\mathcal{D}(\tau)$;
2. the volume element $dx\,dy\,dz$ becomes $J(t)\,d\xi d\eta d\zeta$, $J(t)$ being the Jacobian determinant of the transformation defined as:

$$J(t) = \begin{vmatrix} \dfrac{\partial x}{\partial \xi} & \dfrac{\partial x}{\partial \eta} & \dfrac{\partial x}{\partial \zeta} \\ \dfrac{\partial y}{\partial \xi} & \dfrac{\partial y}{\partial \eta} & \dfrac{\partial y}{\partial \zeta} \\ \dfrac{\partial z}{\partial \xi} & \dfrac{\partial z}{\partial \eta} & \dfrac{\partial z}{\partial \zeta} \end{vmatrix} \equiv \dfrac{D(x, y, z)}{D(\xi, \eta, \zeta)}.$$

So we can write $V(t) = \displaystyle\iiint_{\mathcal{D}(\tau)} J(t)\,d\xi d\eta d\zeta$.

Since the domain $\mathcal{D}(\tau)$ is independent of the current time t, the material derivation of the integral is thus easily obtained as:

$$\frac{\mathrm{d}}{\mathrm{d}(t)} V(t)\,(t) = \frac{\mathrm{d}}{\mathrm{d}t} \iiint_{\mathcal{D}(\tau)} J(t) d\xi d\eta d\zeta = \iiint_{\mathcal{D}(\tau)} \frac{\mathrm{d}J(t)}{\mathrm{d}t} d\xi d\eta d\zeta$$

Then applying the opposite transformation $(\xi, \eta, \zeta) \rightarrow (x, y, z)$ — whose Jacobian determinant is none other than $J^{-1}(t)$ — to the last integral above, we find:

$$\frac{d}{dt}\mathcal{V}(t) = \iiint\limits_{\mathcal{D}(t)} \frac{1}{J(t)} \frac{dJ(t)}{dt} dxdydz$$

Now, we have :

$$\frac{dJ(t)}{dt} = \begin{vmatrix} \frac{\partial \frac{dx}{dt}}{\partial \xi} & \frac{\partial \frac{dx}{dt}}{\partial \eta} & \frac{\partial \frac{dx}{dt}}{\partial \zeta} \\ \frac{\partial y}{\partial \xi} & \frac{\partial y}{\partial \eta} & \frac{\partial y}{\partial \zeta} \\ \frac{\partial z}{\partial \xi} & \frac{\partial z}{\partial \eta} & \frac{\partial z}{\partial \zeta} \end{vmatrix} + \begin{vmatrix} \frac{\partial x}{\partial \xi} & \frac{\partial x}{\partial \eta} & \frac{\partial x}{\partial \zeta} \\ \frac{\partial \frac{dy}{dt}}{\partial \xi} & \frac{\partial \frac{dy}{dt}}{\partial \eta} & \frac{\partial \frac{dy}{dt}}{\partial \zeta} \\ \frac{\partial z}{\partial \xi} & \frac{\partial z}{\partial \eta} & \frac{\partial z}{\partial \zeta} \end{vmatrix} + \begin{vmatrix} \frac{\partial x}{\partial \xi} & \frac{\partial x}{\partial \eta} & \frac{\partial x}{\partial \zeta} \\ \frac{\partial y}{\partial \xi} & \frac{\partial y}{\partial \eta} & \frac{\partial y}{\partial \zeta} \\ \frac{\partial \frac{dz}{dt}}{\partial \xi} & \frac{\partial \frac{dz}{dt}}{\partial \eta} & \frac{\partial \frac{dz}{dt}}{\partial \zeta} \end{vmatrix},$$

or again with more compact notations:

$$\frac{dJ(t)}{dt} = \frac{D(U, y, z)}{D(\xi, \eta, \zeta)} + \frac{D(x, V, z)}{D(\xi, \eta, \zeta)} + \frac{D(x, y, W)}{D(\xi, \eta, \zeta)}.$$

We can deduce

$$\frac{1}{J(t)} \frac{dJ}{dt} = \frac{D(U, y, z)}{D(x, y, z)} + \frac{D(x, V, z)}{D(x, y, z)} + \frac{D(x, y, W)}{D(x, y, z)}$$

$$= \frac{\partial U}{\partial x} + \frac{\partial V}{\partial y} + \frac{\partial W}{\partial z} \equiv div V. \qquad (1.56)$$

Thanks to this last result the particle volume variation becomes:

$$\frac{d}{dt}\mathcal{V}(t) = \iiint\limits_{\mathcal{D}(t)} div V \, dxdydz. \qquad (1.57)$$

By taking the limit $\mathcal{D}(t) \to 0$, the volume $\mathcal{V}(t)$ becomes the fluid particle volume $\epsilon(t)$, so that relation (1.55) leads to:

$$\frac{d\epsilon(t)}{dt} = div V \times \epsilon(t).$$

We will retain this relation in the previously stated form (1.28):

$$\boxed{\frac{1}{\epsilon(t)} \frac{d\epsilon(t)}{dt} = div V} \qquad (1.58)$$

This result, probably going back to Euler [56] in 1757, expresses the following:

> The material rate of change of the cubic expansion velocity is equal to the flow field velocity divergence. Accordingly, any *solenoidal* field is that of an *isovolume* motion.[30]

EXERCISE 5 The velocity field of an isovolume flow.

We consider a flow whose velocity field in Eulerian variables has the following components:

$$\begin{cases} U = x^2 + y + z \\ V = x - xy + z \\ W = x + y - xz \end{cases}$$

QUESTION Is this flow an *isovolume* flow?

SOLUTION: As we have just seen, the question is basically a matter of knowing whether the velocity field is divergence free. Now :

$$\frac{\partial U}{\partial x} = 2x \qquad \frac{\partial V}{\partial y} = -x \qquad \frac{\partial W}{\partial z} = -x.$$

The flow is indeed isovolume.

EXERCISE 6 The isovolume flow with zero acceleration.

Let us consider the flow defined by the following Eulerian velocity field:

$$\begin{cases} U = x + y \\ V = -x - y \\ W = 0 \end{cases}$$

QUESTION 1. Show that the acceleration of the fluid particles is zero.
QUESTION 2. Is the evolution *isovolume* ?

SOLUTION: **1.** Let us firstly express the acceleration components:

$$\begin{cases} \frac{dU}{dt} = U\frac{\partial U}{\partial x} + V\frac{\partial U}{\partial y} = (x + y)(1) - (x + y)(1) \equiv 0 \\ \frac{dV}{dt} = U\frac{\partial V}{\partial x} + V\frac{\partial V}{\partial y} = (x + y)(-1) - (x + y)(-1) \equiv 0 \\ \frac{dW}{dt} = 0 \end{cases}$$

This proves the first statement.

---

[30] By definition, in such an evolution, the fluid particle volume does not change its value in the moving fluid.

2. We immediately obtain $\dfrac{\partial U}{\partial x} = 1,\ \dfrac{\partial V}{\partial y} = -1,\ \dfrac{\partial W}{\partial z} = 0$ so well that $div\,V = 0$.

1.6.7.2 The Volume Integral

Let us now consider a finite fluid domain $\mathcal{D}(t)$ containing, at time t, the amount $F\,(t)$ of a given scalar function $f\,(t)$:

$$F\,(t) = \iiint\limits_{\mathcal{D}(t)} f\,(x, y, z, t)\ dx dy dz. \tag{1.59}$$

As before, the differential element is taken in Eulerian variables, and the problem is to express the material derivative $\dfrac{dF}{dt}$. The same mode of reasoning will therefore be used and we will return to the domain at the tagging time $\mathcal{D}\,(\tau)$ by making the change of variables:

$$(x, y, z) \rightarrow (\xi, \eta, \zeta)\,.$$

We will then have the following transformations:

$$f\,(x, y, z, t) \rightarrow f\,[x\,(\xi, \eta, \zeta)\,,\,y\,(\xi, \eta, \zeta)\,,\,z\,(\xi, \eta, \zeta)\,,\,t] \equiv g\,(\xi, \eta, \zeta, t)\,,$$
$$dx dy dz \rightarrow J d\xi d\eta d\zeta,$$
$$\mathcal{D}\,(t) \rightarrow D\,(\tau)\,,$$

so that the integral (1.57) becomes :

$$F\,(t) = \iiint\limits_{\mathcal{D}(\tau)} g\,(\xi, \eta, \zeta, t)\ J\ d\xi d\eta d\zeta.$$

Since the integration is carried out over a domain which is no longer dependent upon the current time, it is easy to derive this expression with respect to t:

$$\frac{dF\,(t)}{dt} = \iiint\limits_{\mathcal{D}(\tau)} \frac{d}{dt}\,[g\,(\xi, \eta, \zeta, t)\ J]\ d\xi d\eta d\zeta.$$

All that remains now is to take the opposite transformation $(\xi, \eta, \zeta) \rightarrow (x, y, z)$ in order to express the result with respect to the domain at the current time:

$$\frac{dF\,(t)}{dt} = \iiint\limits_{\mathcal{D}(t)} \frac{1}{J}\frac{d}{dt}\,[f\,(x, y, z, t)\ J]\ dx dy dz \equiv \iiint\limits_{\mathcal{D}(t)} \left[\frac{df}{dt} + f\frac{1}{J}\frac{dJ}{dt}\right]\ dx dy dz.$$

Introducing relation (1.56) we finally find:

$$\frac{d}{dt} \iiint_{\mathcal{D}} f \, dv = \iiint_{\mathcal{D}} \left(\frac{df}{dt} + f \, div \, V \right) dv \qquad (1.60)$$

We can give an equivalent expression of (1.60) by introducing relation (1.38a). The result is indeed:

$$\frac{d}{dt} \iiint_{\mathcal{D}} f \, dv = \iiint_{\mathcal{D}} \left(\frac{\partial f}{\partial t} + V.grad \, f + f \, div \, V \right) dv,$$

which directly leads to:

$$\frac{d}{dt} \iiint_{\mathcal{D}} f \, dv \equiv \iiint_{\mathcal{D}} \left(\frac{\partial f}{\partial t} + div \, (f \, V) \right) dv \qquad (1.61)$$

1.6.7.3 The Physical Interpretation

In the application of Ostrogradski's theorem to the divergence term in the volume integral of relation (1.61) we have identically:

$$\frac{d}{dt} \iiint_{\mathcal{D}} f \, dv \equiv \iiint_{\mathcal{D}} \frac{\partial f}{\partial t} dv + \iint_{\mathcal{S}} f V.n \, d\sigma. \qquad (1.62)$$

The physical interpretation of this result can be explained in terms of the Lagrangian analysis by considering the material domain $\mathcal{D}$, consisting of the same particles over time, whose mobile boundary $\mathcal{S}$ necessarily moves with the local velocity of the fluid particles which compose the domain. At a macroscopic scale, there is therefore no mass flux through the surface bounding the domain. Hence intuitively, the change in the volume integral can be decomposed into two contributions:

– a *time variation*, at a fixed volume

$$\iiint_{\mathcal{D}} \frac{\partial f}{\partial t} dv;$$

– a *space variation*, at a fixed time :

$$\iiint_{d\mathcal{D}} f dv;$$

where $d\mathcal{D} = \delta \mathcal{D}^+ - \delta \mathcal{D}^-$ accounts for the variation between the domains $\mathcal{D}(t)$ and $\mathcal{D}(t + dt)$, as outlined in Fig. 1.17.

With reference to relation (1.62) it appears that:

Fig. 1.17 Space variation
due to the material volume
displacement according to
the Lagrangian approach

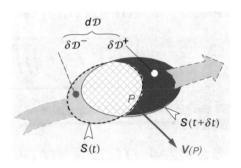

$$\iiint\limits_{d\mathcal{D}} f\, dv = \iint\limits_{S} f\, \boldsymbol{V}.\boldsymbol{n}\, d\sigma.$$

This identity shows that the purely *space* contribution to the change in the balance of f over the *mobile* domain is solely due to the motion, at the velocity $\boldsymbol{V}(P, t)$, of the boundary $S(P, t)$ of the domain. It is this motion which causes the change in the space location (volume denoted by $d\mathcal{D}$), of the fluid particles formerly included in the volume $\mathcal{V}(t)$ of the domain $\mathcal{D}(t)$.

1.6.8 The Material Derivative Related to Mass-Weighted Balances

The previous formulae are directly concerned with *volume* balances, where the function f stands for the amount of any fluid motion property *per unit volume*. Subsequently, we will have to use *mass* balances, which amounts to considering mass weighted functions and thus substituting ρf to f in the previous relations.
New results can be put forward in this case, taking into account that:

$$\frac{\partial \rho}{\partial t} + div\,(\rho \boldsymbol{V}) \equiv \frac{d\rho}{dt} + \rho\, div\boldsymbol{V} = 0. \tag{1.63}$$

a relation, known as the continuity equation, which will be demonstrated in Chap. 3 Sect. 3.1. We will simply examine here the resulting simplifications of the material derivative expressions of mass integrals, due to Eq. (1.63).

1.6.8.1 The General Expression

The following result, sometimes referred to as *Reynolds' formula*,[31] provides the material derivative expression of a mass-weighted volume integral, namely:

---

[31] We also find the *Reynolds' theorem* denomination, see for example, Candel [25]. This name will not be adopted here to avoid any ambiguity with Reynolds-Rayleigh transport *theorems* of the following Sect. 6.9.

$$\boxed{\frac{\mathrm{d}}{\mathrm{d}t} \iiint_{\mathcal{D}} (\rho f)\, dv = \iiint_{\mathcal{D}} \rho \frac{\mathrm{d}f}{\mathrm{d}t}\, dv}$$ (1.64)

To demonstrate this result, let us apply relation (1.61) to the left hand side term of the above equality. Thus it follows that:

$$\frac{\mathrm{d}}{\mathrm{d}t} \iiint_{\mathcal{D}} (\rho f)\, dv = \iiint_{\mathcal{D}} \left[\frac{\partial(\rho f)}{\partial t} + div\,(\rho f \boldsymbol{V}) \right] dv$$

$$= \iiint_{\mathcal{D}} \rho \left[\frac{\partial f}{\partial t} + \boldsymbol{V}.\boldsymbol{grad}\, f \right] dv + \iiint_{\mathcal{D}} f \left[\frac{\partial \rho}{\partial t} + div\,(\rho \boldsymbol{V}) \right] dv\,.$$

The last right hand side integral is identically zero, due to the continuity Eq. (1.63). It is then sufficient to use relation (1.38a) in the remaining integral to obtain the expected result.

1.6.8.2   The Conservative and Transport Formulations

It is a straightforward result of the previous demonstration that

$$\iiint_{\mathcal{D}} \left[\frac{\partial(\rho f)}{\partial t} + div\,(\rho f \boldsymbol{V}) \right] dv = \iiint_{\mathcal{D}} \rho \left[\frac{\partial f}{\partial t} + \boldsymbol{V}.\boldsymbol{grad}\, f \right] dv,$$

from which is deduced by taking the limit to the particle volume:

$$\boxed{\frac{\partial (\rho f)}{\partial t} + div\,(\rho f \boldsymbol{V}) = \rho \left(\frac{\partial f}{\partial t} + \boldsymbol{V}.\boldsymbol{grad}\, f \right) \equiv \rho \frac{\mathrm{d}f}{\mathrm{d}t}}$$ (1.65a)

or

$$\boxed{\frac{\partial (\rho f)}{\partial t} + \frac{\partial \left(\rho f U_j \right)}{\partial x_j} = \rho \left(\frac{\partial f}{\partial t} + U_j \frac{\partial f}{\partial x_j} \right) \equiv \rho \frac{\mathrm{d}f}{\mathrm{d}t}}$$ (1.65b)

Thus, we are facing two alternate material derivative formulations of a mass weighted function :

– The *conservative* formulation, corresponding to the operator

$$\frac{\partial}{\partial t} (\rho \bullet) + div\,(\rho \bullet \boldsymbol{V})\,,$$ (1.66)

This denomination refers to the conservation of the quantity ($\rho \bullet$) by space integration over any volume bounded by stream surfaces, in direct application of Ostrogradski's theorem.

– The *transport* type formulation, which refers to the operator

$$\frac{\partial}{\partial t}(\bullet) + V.grad(\bullet). \tag{1.67}$$

directly associated with the advection/convection term.

1.6.9 The Transport Theorems of Reynolds-Rayleigh

The duality of reasoning on a material domain (Lagrange) and a control volume (Euler) raises the issue of the equivalence between the expressions resulting from both formulations. These equivalences are determined by what is conventionally called, in a generic manner, the *transport theorems*[32] assigned either to Reynolds, see Candel [25], or Rayleigh, see Panton [111], for instance.

We will state:

The material derivative of the integral $\iiint\limits_{\mathcal{D}(t)} f(x, y, z, t)dv$ over a *material*

domain *moving* at the *local velocity of the fluid* can be expressed by

$$\frac{\mathrm{d}}{\mathrm{d}t}\iiint\limits_{\mathcal{D}(t)} f(x, y, z, t)dv = \iiint\limits_{\Delta} \frac{\partial f}{\partial t}dv + \iint\limits_{\Sigma} f\,V.n\,d\sigma, \tag{1.68}$$

where Δ is the *fixed* volume, bounded by the surface Σ, which *coincides* with the domain $\mathcal{D}(t)$ *at time* t.

Proof We denote, see Fig. 1.18, by $\mathcal{D}(t)$ the *moving* material domain and by Δ the *fixed* control volume, coinciding with the domain $\mathcal{D}(t)$ *at time* t.

By definition, the differential element δF of a function f with respect to the volume element δv can be expressed as

$$\delta F(x, y, z, t) = f(x, y, z, t) \times \delta v(x, y, z, t).$$

---

[32] In the most general formulation of these theorems, see for example Candel [25], the surface of the *moving* domain is supposed to move at *any* velocity. We limit ourselves here to the only case of a motion *at the velocity of the fluid at the coinciding point of the surface*.

Fig. 1.18 Mobile material
domain $\mathcal{D}$, at times t and
$t + dt$, and fixed control
volume Δ, coinciding with
$\mathcal{D}$ at time t

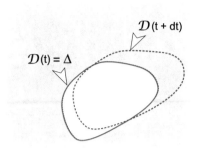

By integrating the left hand side over the material domain $\mathcal{D}(t)$ and the right hand side over the control volume Δ *coinciding* with $\mathcal{D}(t)$ at time t, we obtain

$$\iiint_{\mathcal{D}(t)} \delta F = \iiint_{\Delta} f\,\delta v, \text{ or simply } F(t) = \iiint_{\Delta} f\,\delta v.$$

Now, by taking the material derivative of both sides, it is deduced:

$$\frac{dF}{dt} = \frac{d}{dt}\left(\iiint_{\Delta} f\,\delta v\right) = \iiint_{\Delta} \frac{d}{dt}(f\,\delta v),$$

the last equality is allowed since the integration operates over a volume which is fixed over time.

Now, $\dfrac{d(f\,\delta v)}{dt} = \dfrac{df}{dt}\cdot \delta v + f\cdot \dfrac{d(\delta v)}{dt}$ and since δv is the fluid particle volume $\dfrac{d(\delta v)}{dt} = div\,V \times \delta v$.

We can therefore expand the integral on Δ as follows:

$$\iiint_{\Delta} \frac{d}{dt}(f\,\delta v) = \iiint_{\Delta} \left(\frac{Df}{Dt} + f\,div\,V\right)\delta v$$

$$= \iiint_{\Delta} \left(\frac{\partial f}{\partial t} + V.\mathbf{grad} + f\,div\,V\right)\delta v$$

$$= \iiint_{\Delta} \frac{\partial f}{\partial t} + div(f\,V)\,\delta v.$$

The expected result is deduced by applying Ostrogradski's relation to the second integral of the last equality.

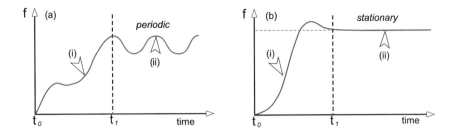

Fig. 1.19 The Eulerian function evolution towards a periodic $(a-ii)$ and stationary regime $(b-ii)$ after a transient phase (i)

1.6.10   The Case of a Stationary Motion

1.6.10.1   Definition

By definition, the motion is said to be *stationary in time* or *steady* if any function of the flow field *in Eulerian variables* is *time independent*.

$$
\text{Stationary motion} \Longleftrightarrow \quad \frac{\partial f}{\partial t} \equiv 0, \forall f \text{ a function of the } \textbf{Eulerian variables.}
$$

This notion, as it has been defined and will be used in the following sections, should not be confused with that of a *steady flow regime*. To have a better understanding of the distinction, we take the case of a fluid which is set in motion from rest in a given frame of reference.

The change in initial or boundary conditions of the state of rest, which drives the motion, is supposed to occur at time t_0, the new situation remaining henceforth unchanged at every later instant. For $t > t_0$, we generally distinguish (i) a "*transient*" regime, during which the flow adjusts itself to the new conditions and (ii) a "*long-lasting*" flow regime, for which any evolution resulting from the initial condition change has disappeared. In the latter case, the flow regime establishment does not necessarily result in a stationary (or invariant in time) flow character.

Indeed, by translating a solid into a fluid initially at rest, one can generate, under certain conditions, a long-lasting flow regime of *periodic* characater,[33] as shown in Fig. 1.19a. The steady motion thus corresponds to the particular case where the flow regime is time independent, see Fig. 1.19b.

---

[33] This is the case, for example, of the velocity behind a circular cylinder, whose translation in a fluid at rest at infinity, in a direction normal to its axis corresponds to a Reynolds number between 50 and 100.

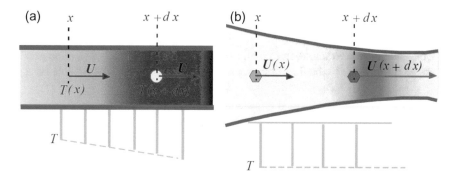

Fig. 1.20 Material variation origins in a steady motion: **a** pure convection in a constant velocity field; **b** pure advection with zero property gradient in the flow direction

1.6.10.2 Properties

In any stationary flow, the set of material derivation formulas is simplified by only retaining the advective-convective variations. Thus, in a stationary motion, the material derivative expression of an Eulerian variable function $f(x, y, z)$ reduces to:

$$\frac{df}{dt} = U_j \frac{\partial f}{\partial x_j}.$$

This relation clearly shows that in a stationary motion, the fluid particle acceleration is not *necessarily* zero.

Example To illustrate the physical meaning of this variation, we consider the case of a one-dimensional situation where the material derivative is reduced to:

$$\frac{df}{dt} = U \frac{\partial f}{\partial x}.$$

We will apply this formula on the one hand to the velocity component $U(x)$ and on the other hand to a scalar, such as the temperature for instance, $T(x)$ in the two situations illustrated in Fig. 1.20.

1. In the case of a constant velocity flow, Fig. 1.20a, the material variation can solely originate from the existence of a function's gradient component along the flow direction. We then have:

$$\frac{dT}{dt} = U \frac{\partial T}{\partial x} \neq 0 \quad \text{and} \quad \frac{dU}{dt} = U \frac{\partial U}{\partial x} = 0.$$

2. In the isothermal situation, the material variation is of a purely advective type. It is necessarily linked to the velocity variation in the flow direction:

$$\frac{\mathrm{d}T}{\mathrm{d}t} = U\frac{\partial T}{\partial x} = 0 \text{ and } \frac{\mathrm{d}U}{\mathrm{d}t} = U\frac{\partial U}{\partial x} \neq 0.$$

Consequently for example, the situation as outlined in 1.20 (b), is that of an accelerated motion due to the flow cross section reduction of an incompressible fluid.

Chapter 2
Flow Kinematics

Abstract This chapter is devoted to the presentation of concepts resulting from the mere presence of a vector field representative of a fluid motion. We will consider, first of all, that this is the flow velocity field $\mathbf{V}\,(M, t)$ in a given domain of space Ω. We will then consider the case of the pseudo-vector curl (rotational) of the velocity field. The study of the motion generation under the action of the forces involved in various situations will be the subject of the next chapter.

2.1 Specific Lines and Surfaces of a Flow Field

2.1.1 The Streamline and Streamtube

2.1.1.1 Definitions

Any curve whose tangent in each of its points is, at a *given time*, colinear with the *instantaneous flow* velocity at this point and that time is called a *streamline*.
A set of streamlines leaning at a *given time* on any closed contour is called a *streamtube*.

As outlined in Fig. 2.1, we can note that, at the same fixed point of the flow field, the streamline representative curve varies over time, unless the motion is stationary.

Remark This last formulation proposal is that which is usually used by the physicist. However, and from a purely mathematical point of view, it is sufficient, in order to ensure the invariance over time of the streamline geometry, that the velocity field can be put in the form $\mathbf{V}\,(x, y, z, t) = g\,(t)\,\mathbf{U}\,(x, y, z)$ where the time variable (t) is decoupled from the other space variables (x, y, z) according to both functions g and $\mathbf{U}$. The steady regime is then only a particular case for which $g(t) = C^t$.

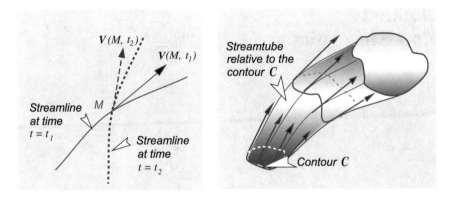

Fig. 2.1 Schematic for a streamline and streamtube

2.1.1.2   The Streamline Equation

In an orthonormal reference frame, the differential equation of any streamline is
written as follows:

$$\frac{dx}{U\,(x, y, z, t)} = \frac{dy}{V\,(x, y, z, t)} = \frac{dz}{W\,(x, y, z, t)}\,, \qquad \textbf{where } t \textbf{ is fixed.} \qquad (2.1)$$

Setting the ratio's common value equal to $d\alpha$, α denoting a given real, the parametric
equations of the streamlines passing, at *any time*, through point $M_0\,(x_0, y_0, z_0)$ for
$\alpha = 0$ are thus :

$$\begin{cases} x = x\,(x_0, y_0, z_0, t, \alpha) \\ y = y\,(x_0, y_0, z_0, t, \alpha) \\ z = z\,(x_0, y_0, z_0, t, \alpha) \end{cases} \qquad (2.2)$$

The streamlines thus constitute a family of curves depending on *two* parameters: they
vary in space—through the *geometric* parameter α—and in time—with the variable t.

2.1.1.3   The Streamline and Stream Function

In the case of the two-dimensional steady plane flow[1] of a constant density fluid, we
will see (Chap. 5 Sect. 5.3.3) that there is a function $\psi\,(x, y)$ such that:

$$U\,(x, y) = \frac{\partial \psi\,(x, y)}{\partial y} \quad \text{and} \quad V\,(x, y) = -\frac{\partial \psi\,(x, y)}{\partial x}\,. \qquad (2.3)$$

---

[1] A stream function can also be introduced in three-dimensional flows under a space coordinate
invariance by axial or spherical symmetry, see the next paragraph under Additional Information.

Equation (2.1) gives $\dfrac{dx}{\partial\psi/\partial y}=-\dfrac{dy}{\partial\psi/\partial x}$, or again $\dfrac{\partial\psi}{\partial x}dx+\dfrac{\partial\psi}{\partial y}dy\equiv d\psi=0$.

Thus, the lines $\psi\,(x,\,y)=C^{t}$ are nothing more than the streamlines of such a flow. The function $\psi\,(x,\,y)$ is therefore called the *stream function*.

ADDITIONAL INFORMATION—Geometric invariance—Two-dimensionality— Some flow fields can be described by vectors involving a reduced number of components and space coordinates. Such particular situations can result from the combination of the following two properties:

- *A* Two-component field, for which one of the three components of any vector is equal to zero;
- *A* Two-dimensional field, in which the velocity vector only depends on two space coordinates.

Let us consider, for example, in an orthonormal Cartesian frame, the velocity vector of general components $U(x,\,y,\,z)$, $V(x,\,y,\,z)$, $W(x,\,y,\,z)$ at a time that is unnecessary to specify for the present discussion.

- The special case $U = U(x,\,y,\,z)$, $V = V(x,\,y,\,z)$, $W = 0$ is that of a *two-component* but *three-dimensional* field.

- In the case where $U = U(x,\,y)$, $V = V(x,\,y)$ and $W = W(x,\,y)$, we are dealing with a *two-dimensional* flow with *three* components.

By associating the *two-component* condition with that of *bidimensionality*, we can, depending on the geometry of the reference frame in question, identify various flow invariance situations, some of which are described in Table 2.1.

In the following for brevity sake, we will use the following terms:

- a « *two-dimensional plane* » $\partial/\partial z = 0$ and two-component field in orthonormal Cartesian coordinates;

- a « *two-dimensional with axi-symmetry* » $\partial/\partial\theta = 0$ and two-component field in polar coordinates.

2.1.2 The Pathline

2.1.2.1 Definition

A *pathline* or *trajectory* is the curve along which a fluid particle of fixed identity moves in time.

Table 2.1 Examples of two-component and two-dimensional fields

Invariance	Coordinates/Components	Geometry	Bidimensionality type
$\dfrac{\partial}{\partial z} \equiv 0$	Cartesian (x, y)		Plane (invariance by translation in the direction which is normal to the flow plane)
	Polar (r, θ)		
$\dfrac{\partial}{\partial \theta} \equiv 0$	Polar (r, z)		Revolution or axi-symmetric (invariance by rotation around an axis.)
	Spherical (r, φ)		

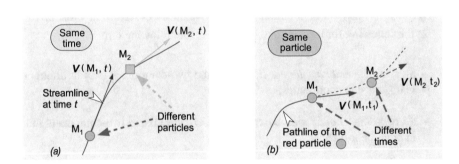

Fig. 2.2 Comparison between **a** streamlines and **b** path lines

In comparison with the previous case and as outlined in Fig. 2.2, it appears that the ***streamline*** refers to the *same **time*** but is composed of different *particles*, whereas the ***pathline*** is a curve *parameterized in time* which refers to the **same** (fixed identity) ***particle***.

2.1.2.2 Equation

The differential parametric equations of the trajectories are, by definition,

$$
\begin{cases}
\dfrac{dx}{dt} = U\,(x, y, z, t) \\[2mm]
\dfrac{dy}{dt} = V\,(x, y, z, t) \\[2mm]
\dfrac{dz}{dt} = W\,(x, y, z, t)
\end{cases}
\tag{2.4}
$$

Mathematically speaking, the trajectories constitute a curve family with a *single* parameter whose equations are derived from those of the streamlines by merging the geometric and temporal parameters in relation (2.2).

From the physics point of view, this means that *time* has become the trajectory path variable. Consequently, for a steady motion, any streamline and pathline which have a common point are identical, since the differential streamline equation becomes in this case:

$$
\frac{dx_i}{d\alpha} = U_i\,(x, y, z, \alpha) \quad i = 1, 2, 3.
$$

We thus retrieve the trajectory equation, by merely identifying the parameter (α with t), which has no longer any temporal connotation in a steady state.

2.1.3 The Streakline

2.1.3.1 Definition

A streakline is the line traced out, *at a given time t*, by all the particles which have passed through a particular point at some earlier time.

As shown in Fig. 2.3 and as a direct consequence of the definition, the streakline is a time-varying curve composed of different particles. The streakline as a curve related to a given point can be generalized in *streaksurface* and *streakvolume*, by considering, at time t, the sets of particles which have all passed through the same line or surface element respectively, at any earlier time.

2.1.3.2 Equation

By integrating equations (2.4) with the condition that the particle whose streakline is related to point $M_0\,(x_0, y_0, z_0)$ is located at this point at $t = \tau$ we obtain:

Fig. 2.3 Pathlines and
streakline related to point A

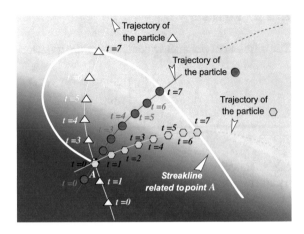

$$\begin{cases} x = x\,(x_0, y_0, z_0, t, \tau) \\ y = y\,(x_0, y_0, z_0, t, \tau) \\ z = z\,(x_0, y_0, z_0, t, \tau) \end{cases} \qquad (2.5)$$

Thus, when t *is fixed* and τ varies over the interval $[\,0, +\infty)$, the previous equations
are those of the streakline related to point M_0 at time t. On the other hand, for τ *fixed*
and when t varies, these same equations identify with those of the particle trajectory
which, at time τ, is located at the starting point M_0 of the streakline.

2.1.4 *The Fluid-line*

A fluid-line (surface) is any material curve (surface) subjected to the same set of fluid
particle motion which compose it at a given time.

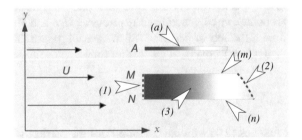

Fig. 2.4 Elements for comparison between a streakline (surface) and a fluid-line (surface).
(a), (m) (n), Streaklines related to points A, M and N, referring to the same time t' ;
(1), (2) Fluid-lines referring to dif- ferent times t and t' ;
(3) A streaksurface related to the segment MN referring to time t'

All the elements we have just introduced are closely related to flow field visualization by a tracer, as we will see at the end of this chapter. In order to clarify the distinction between streaklines (surfaces) on the one hand, and fluid lines (surfaces) on the other, we present, in Fig. 2.4, some schematics of the characteristic elements for discussing this issue in a steady flow.

2.2 Fluxes and Flow Rates

2.2.1 The General Flux Notion

Let $q\,(x, y, z, z, t)$ be any flow field function and S a *fixed control* surface. The fluid particles when passing through S with the velocity $\mathbf{V}\,(M, t)$ are producing an advection flux whose *density vector* $q\mathbf{V}$ is associated to the *elementary advection flux* dQ through the surface element $d\sigma$, *viz*:

$$dQ \equiv q\mathbf{V}.\mathbf{n}\,d\sigma . \tag{2.6}$$

By convention, the direction of the vector normal to the surface element $\mathbf{n}$ will always be positively oriented outwards, so that the **outgoing** flux of any positive scalar property q is positive.

The entire flux through S is expressed as $\quad Q = \iint\limits_{S} dQ \equiv \iint\limits_{S} q\mathbf{V}.\mathbf{n}\,d\sigma . \tag{2.7}$

The integral value generally depends on the function under consideration (q), the boundary area (S) and time (t).

Fig. 2.5 Schematic for the advective flux through an elementary surface

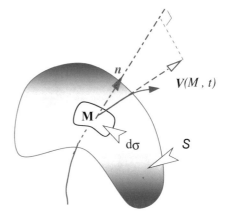

Fig. 2.6 Flow rate and
stream function in a
two-dimensional plane flow

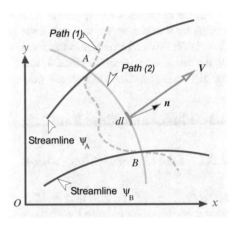

2.2.2 Volume and Mass Flow Rates

For $q \equiv 1$, Eq. (2.7) is that of the volume flow rate[2] (m^3/s). $Q_v = \iint_S \mathbf{V}.\mathbf{n}\, d\sigma$.

For $q \equiv \rho$, Eq. (2.7) defines the mass flow rate (kg/s) $Q_m = \iint_S \rho \mathbf{V}.\mathbf{n}\, d\sigma$.

2.2.3 The Volume Flow Rate and Stream Function

Under the same assumptions as in Sect. 2.1.1.3, it is possible to express the scalar product $\mathbf{V}.\mathbf{n}$ in terms of each vector component in an orthonormal reference frame. Indeed, for a two-dimensional plane flow, the surface element is simply:

$$d\sigma = 1 \times dl\,,$$

by choosing the span unit as the cross dimension and denoting as dl the length element along the curved arc AB (Fig. 2.6). We then obtain

$$\mathbf{V}.\mathbf{n} = U\frac{dy}{dl} - V\frac{dx}{dl}\,.$$

Now, introducing the stream function $\psi(x, y)$, the result is $\mathbf{V}.\mathbf{n}\, d\sigma = \dfrac{\partial \psi}{\partial y} dy + \dfrac{\partial \psi}{\partial x} dx \equiv d\psi$, hence

---

[2] From a mathematical point of view, we can generalize the expression to any vector as $\iint_S \mathbf{A}.\mathbf{n}\, d\sigma$.
We then speak of the vector $\mathbf{A}$ *flux integral* through the surface S. We will avoid this vocabulary in order to preserve the notion of flux density by advection.

$$Q_v = \int_A^B d\psi \equiv \psi_B - \psi_A \tag{2.8}$$

The simple difference between the constant values of the two stream functions there-fore gives the volume flow rate amount of the fluid passing between these two lines. It is worth noticing that the result is independent of the path followed between the points taken on each line.

Remark For most practical applications in hydraulics, where water can be regarded as an incompressible medium, the flow rate is considered in the volume sense. On the other hand, in the aeraulic field or when the gas density varies, it is the mass flow rate which becomes significant.

2.3   Small Displacement Kinematics

We intend here to present the kinematic approach associated with a change in posi-tion *on the particle volume scale* and within the limit of a displacement sufficiently small to be considered as that of an elementary motion.
To define the problem, let us consider, see Fig. 2.6, two fluid particles located respec-tively at points M and N at time t. Their positions at time $t + dt$ are labelled as M' and N' respectively.
    The study assumption concerning the "infinitesimal" space variation can be con-sidered in the following two ways:
    (i) at a **fixed** time t, with

$$\mathbf{dM} = \mathbf{MN}, \tag{2.9}$$

where MN is the distance between two *distinct* particles;
    (ii) at a **variable** time, with

$$\mathbf{dM} = \mathbf{MM'} \equiv \mathbf{V}\,dt, \tag{2.10}$$

where MM' is the displacement of the *same* particle during the time interval dt.
The small displacement assumption allows approximating *any space variation by a first-order development in the displacement*, regardless of the analysis type, (2.9) or (2.10), under consideration. We will firstly carry out a pure space study at a *fixed time* and then apply the results to the fluid particle displacement along its trajectory (*variable time*).

Fig. 2.7 Schematic for the
infinitesimal displacement
analysis

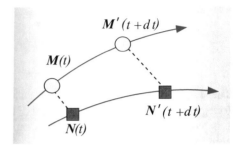

2.3.1 The Space Variation Analysis of a Vector Field

Since the points are infinitely close, the difference in velocities between them can be
expressed by the differential:

$$\mathbf{V}(N) - \mathbf{V}(M) \equiv \mathbf{dV} = \overline{\overline{grad}}\,\mathbf{V} \odot \mathbf{dM}, \qquad (2.11a)$$

where the vector $\mathbf{dM}$ is defined by relation (2.9). The operator $\overline{\overline{grad}}\,\mathbf{V}$ denotes the
(second-order) velocity gradient tensor, as defined in Chap. 1, Sect. 1.6.3.1. The
symbol $\odot$ is that of the tensor contraction product on the (right) derivation index of
the velocity gradient tensor. By using index notations, the previous relation becomes:

$$U_i(N) - U_i(M) = \frac{\partial U_i}{\partial x_j}\,dx_j\,. \qquad (2.11b)$$

Then let us introduce the unique decomposition of the velocity gradient tensor in a
symmetric part $(\overline{\overline{S}})$ and an antisymmetric part $(\overline{\overline{R}})$. We obtain:

$\overline{\overline{grad}}\,\mathbf{V} = \overline{\overline{S}} + \overline{\overline{R}}$ with,	(2.12)
$\overline{\overline{S}}$, whose components are $S_{ij} = \dfrac{1}{2}\left(\dfrac{\partial U_i}{\partial x_j} + \dfrac{\partial U_j}{\partial x_i}\right)$	(2.13)
$\overline{\overline{R}}$, whose components are $R_{ij} = \dfrac{1}{2}\left(\dfrac{\partial U_i}{\partial x_j} - \dfrac{\partial U_j}{\partial x_i}\right)$	(2.14)

Equivalent forms can be given to expressions (2.13) and (2.14) according to the
following considerations:

- The *antisymmetric* tensor $\overline{\overline{R}}$ can be derived from the pseudo-vector $\mathbf{\Omega}$ called the
  curl (rotational) of the velocity field, with

$$\mathbf{\Omega} = curl\,\mathbf{V}, \qquad (2.15)$$

and whose Cartesian components are

$$\Omega_1 = \frac{\partial U_3}{\partial x_2} - \frac{\partial U_2}{\partial x_3} \; ; \quad \Omega_2 = \frac{\partial U_1}{\partial x_3} - \frac{\partial U_3}{\partial x_1} \; ; \quad \Omega_3 = \frac{\partial U_2}{\partial x_1} - \frac{\partial U_1}{\partial x_2} \, .$$

It may be noted that $\Omega_k = -\epsilon_{ijk} R_{ij}$, where ϵ_{ijk} is the alternating symbol[3] (see Appendix B). We can verifiy that:

$$\overline{\overline{R}} \odot \mathbf{dM} = \mathbf{\Omega} \wedge \mathbf{dM}, \quad \text{where } \wedge \text{ is the cross (vector) product symbol.}$$

- The *symmetric* tensor $\overline{\overline{S}}$ is called the *strain-rate tensor*. Its trace is:

$$\vartheta \equiv S_{ll} = \frac{\partial U_l}{\partial x_l} \equiv div \, \mathbf{V}. \tag{2.16}$$

It can be split again into a *spherical* part and a *deviatoric* or *traceless* component according to:

$$S_{ij} = \frac{1}{3}\vartheta\delta_{ij} + D_{ij} \tag{2.17}$$

$$\text{with } D_{ii} \equiv 0 \tag{2.18}$$

Given all these elements, relation (2.11a) can be expressed as:

$$\mathbf{V}(N) = \mathbf{V}(M) + \mathbf{\Omega} \wedge \mathbf{dM} + \frac{1}{3}\vartheta\overline{\overline{I}} \odot \mathbf{dM} + \overline{\overline{D}} \odot \mathbf{dM} \tag{2.19a}$$

or, in index notation

$$V_i(N) = V_i(M) + \epsilon_{ijk}\Omega j \, dx_k + \frac{1}{3}\vartheta\delta_{ik} \, dx_k + D_{ik} \, dx_k \tag{2.19b}$$

We will now provide the detailed geometric meaning of each right hand side term of these relations. To do so we will consider an elementary fluid cuboid or parallelepipedic shape fluid piece.[4] This domain is bounded by rectangular facets whose length edges, on a *supra-particle* scale, will be noted as dx, dy and dz respectively.

---

[3] Levi-Civita symbol in two-dimensions.

[4] By such an *elementary* fluid portion, is meant a volume of sufficiently small dimensions so that the analysis can be restricted to the first order in any space expansion. However, the dimensions in question are higher than those of the *fluid particle* whose mesoscopic scale excludes, by definition, any space variation at its scale.

Fig. 2.8 Schematic for the
elementary translation
motion

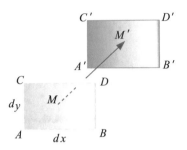

2.3.1.1   Translation

If relation (2.19a) reduces solely to its first right hand side term, it implies, taking
$M \equiv A$ for example, that the velocities at the other three points B, C and D are
identical to $\mathbf{V}(A)$.

As illustrated in Fig. 2.7, this corresponds to a change in the location of the fluid
parcel by a mere *translation* of vector $\mathbf{MM}' = \mathbf{V}dt$, M being any point in the domain.

2.3.1.2   Rotation

Let us now consider a facet such as $ABDC$ undergoing a *solid body rotation* around
an axis Az normal to its plane (Fig. 2.8). Within the *small displacement* limit, the
choice of the rotation axis location at point A rather than at the $ABDC$ surface center
is without consequence on the demonstration.

Now, in any axial solid body rotation, whose angular velocity is ω, we know that
the velocity vector of a point located at the distance $\mathbf{r}$ from the rotation axis is given
by:

$$\mathbf{V}(r) = \omega \wedge \mathbf{r}.$$

Hence, these velocity vector components in cylindrical coordinates are $V_r = 0$, $V_\theta = \omega r$ and $V_z = 0$. By now using the curl of the velocity vector, we deduce that

$$\mathbf{\Omega} \equiv \Omega\, \mathbf{e}_z = \left(\frac{\partial V_\theta}{\partial r} + \frac{V_\theta}{r} \right) \mathbf{e}_z = 2\omega \mathbf{e}_z.$$

The second term in the right hand side of relation (2.19a) is thus directly associated
with the *local* fluid motion rotation at an angular velocity equal to half the curl of
the velocity. In fluid mechanics, it is common practice to refer to the vorticity vector
ω defined as:

$$\omega \equiv \frac{1}{2}\mathbf{\Omega} = \frac{1}{2}\, curl\, \mathbf{V}. \tag{2.20}$$

Fig. 2.9 Schematic for the
elementary solid-body
rotation

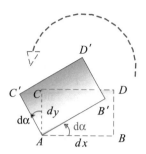

Such a result can also be derived directly from a first-order analysis of the rotation
(see Proof below). In the end, we can conclude:

> The antisymmetric part of the velocity gradient tensor accounts for a *purely solid*
> (no-deformation) local rotation. The local angular velocity vector is given as half
> the local vorticity.

A particularly important class of fluid motions is that which the velocity vector
satisfies, *at any point in the field*, the condition *curl* $\mathbf{V} = \mathbf{0}$. By definition, the corresponding flows are referred to as *irrotational*. Their properties will be studied in
more detail in the following Chaps. 5 and 6.

Proof By rotating the edge AB of the domain $ABDC$ with an elementary angle $d\alpha$
(see Fig. 2.9), we have $BB' \approx dx\, d\alpha$, and similarly for the edge AC, with $CC' \approx$
$dy\, d\alpha$. Within the limit $d\alpha \to 0$, the distance BB' is still equal to the difference in
the displacements along the axis y of point B relative to point A. If we then denote
by U and V the velocity vector components at these points, we have :

$$BB' \simeq [V(B) - V(A)]\, dt = \left[V + \frac{\partial V}{\partial x}dx - V \right] dt = \frac{\partial V}{\partial x}dx\, dt.$$

Similarly, it can be demonstrated that:

$$CC' \simeq -[U(C) - U(A)]\, dt = -\frac{\partial U}{\partial y}dy\, dt.$$

By equaling the two expressions of BB' (resp. CC'), we deduce that:

$$\frac{d\alpha}{dt} \simeq \frac{\partial V}{\partial x} \simeq -\frac{\partial U}{\partial y} \equiv \frac{1}{2}\left(\frac{\partial V}{\partial x} - \frac{\partial U}{\partial y} \right),$$

which shows that the angular rotation velocity in the plane (x, y) is nothing more
than half the vorticity component along z.

2.3.1.3 Elongational Deformation

We now come to the analysis of the last two terms on the right hand side of Eq. (2.19a) which, let us recall, are related to the symmetrical part $\overline{\overline{S}}$ of the velocity gradient tensor. Let us firstly observe that, for a *rigid, non-deformable body*, this side reduces solely to the first two terms, so that the contributions in question are necessarily associated with the fluid parcel *deformations*. This is what justifies the name of the *strain rate tensor* for $\overline{\overline{S}}$.

We begin by addressing the diagonal term effects, assuming for example that the tensor $\overline{\overline{S}}$ reduces to :

$$\overline{\overline{S}} = \begin{pmatrix} \dfrac{\partial U}{\partial x} & 0 & 0 \\ 0 & \dfrac{\partial V}{\partial y} & 0 \\ 0 & 0 & 0 \end{pmatrix}$$

The evolution of the $ABDC$ facet subjected to these velocity gradient components is shown in Fig. 2.10 and it should be noted that (see Proof below):

> The diagonal terms of the symmetrical part of the velocity gradient tensor account for the elongational velocities in each direction x, y, z.
>
> Their sum is equal to the velocity vector divergence and represents the *cubic expansion rate* of an elementary fluid volume.

Depending on whether the cubic expansion rate is positive or negative, there is an expansion or compression of the fluid parcel volume. An important specific case is where this cubic expansion rate is zero. It corresponds to an evolution termed *isovolume* which will be studied in more detail later in this course.

Proof Since there is no velocity gradient component U along the y axis, we have $U\,(A) = U\,(C)$, see Fig. 2.10. Thus, the AC edge moves by a translation along the x axis at the velocity U. The same is also true for the BD edge, the velocity being equal now to:

$$U + \frac{\partial U}{\partial x}dx.$$

Fig. 2.10 Elongational deformation

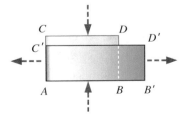

The distance AB between the two edges AC and BD therefore does not remain constant and is subject to the variation:

$$dAB \simeq [U(B) - U(A)]\,dt = \left[U + \frac{\partial U}{\partial x}dx - U\right]dt = \frac{\partial U}{\partial x}dx\,dt.$$

Since we have $AB = dx$, the variation rate of the length AB is therefore:

$$\frac{1}{AB}dAB = \frac{\partial U}{\partial x}dt.$$

It thus appears that the component S_{11} accounts for the elongational velocity in the x direction:

$$S_{11} \equiv \frac{\partial U}{\partial x} = \frac{1}{AB}\frac{dAB}{dt}.$$

It could be likewise demonstrated that S_{22} and S_{33} are the elongational velocities along y and z. This variation in the edge length obviously leads to a variation in the volume which, for a parallelepiped parcel of sides AB, AC, AD is equal to

$$\frac{dV}{V} = \frac{dAB}{AB} + \frac{dAC}{AC} + \frac{dAD}{AD}.$$

Based on the previous expressions, the result obtained is still:

$$\frac{1}{V}\frac{dV}{dt} = \frac{\partial U}{\partial x} + \frac{\partial V}{\partial y} + \frac{\partial W}{\partial z} = S_{ll} \equiv \vartheta. \tag{2.21}$$

We thus retrieve, for *the trace* of the $\overline{\overline{S}}$ tensor, the expression which was mentioned in Chap. 1, Sect. 1.6.7.1.

2.3.1.4 Angular Deformation

Finally, we discuss the case where the $\overline{\overline{S}}$ tensor diagonal terms are zero. Subjected solely to the non-diagonal term effects, the change in the $ABCD$ facet of the fluid parcel is such as that shown in Fig. 2.11.
We can express that:

> The non-diagonal terms of the symmetrical part of the velocity gradient tensor account for the angular deformation rates of an elementary body of fluid.

Proof Let us consider the AB edge rotation by an angle $d\alpha$ which we can approximate to its tangent. Then we have $d\alpha \simeq tg\alpha \simeq BB'/AB$. Now, the displacement

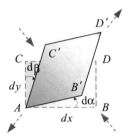

Fig. 2.11 Angular deformation

BB' relative to point A can be approximated by $BB' \approx [V(B) - V(A)]\, dt = \dfrac{\partial V}{\partial x} dx dt$. We deduce $\dfrac{d\alpha}{dt} \simeq \dfrac{\partial V}{\partial x}$. Similarly, the AC edge rotates by an angle $d\beta$ such that $\dfrac{d\beta}{dt} \simeq -\dfrac{\partial U}{\partial y}$. Thus the angle $(\mathbf{AB}, \mathbf{AC})$ varies by the amount $d\,(\mathbf{AB}, \mathbf{AC}) = d\beta - d\alpha$ and we have :

$$\frac{d\,(\mathbf{AB}, \mathbf{AC})}{dt} = -\left(\frac{\partial V}{\partial x} + \frac{\partial U}{\partial y}\right) \equiv -2S_{12}.$$

2.3.2   The Infinitesimal Motion Components of an Elementary Fluid Body

We reinterpret the previous results here according to a variable time analysis, where the space difference results from the same elementary fluid body motion along its trajectory within the limit $dt \to 0$. We then have:

$$\mathbf{V}\left(M'\right) = \mathbf{V}(M) + \mathbf{\Omega} \wedge \mathbf{dM} + \frac{1}{3}\vartheta\overline{\overline{I}} \odot \mathbf{dM} + \overline{\overline{D}} \odot \mathbf{dM}.$$

Thus, as outlined in Fig. 2.12, the change in velocity of an elementary fluid portion between two "infinitely" close positions, consists of three additive motions:

- a translation;
- a rotation;
- a cubic and angular deformation.

Remark The familiar reader of text books dealing with elasticity or more generally continuum mechanics will have noticed the very strong analogy between this small space variation analysis of the ***fluid motion velocity*** field, and that of the small ***deformable solid displacements***. This can be ascertained, for example, in the treatises on continuous media mechanics, such as those of Roy [141], Coirier [39] or Thual [158].

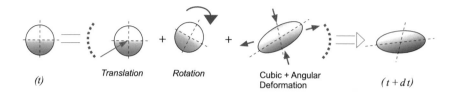

(t)   Translation   Rotation   Cubic + Angular Deformation   (t + dt)

Fig. 2.12 Schematic for the infinitesimal motion components of an *elementary* fluid portion

HISTORICAL NOTE—**Dilatation, condensation and rotation according to Cauchy**—In the «*Memoir on the dilations, condensations and rotations produced by a change of shape in a material point system* » which he published in 1841, Cauchy [28] puts forward the relation expressing the changes in the deformable volume geometry, for the general case of any displacement and then according to the small space variation analysis. His objective is clearly stated in the introduction: «*To be able to easily apply Geometry to Mechanics, it is not enough to know the various shapes that lines or surfaces can exhibit, and the properties of these lines or surfaces, but it is still important to know what shape changes bodies considered as material point systems can undergo, and to which general laws these shape changes can be attributed.*»

HISTORICAL NOTE—**Stokes' elementary fluid motion analysis**—It is to Stokes [149] in 1845, that the first full infinitesimal variation analysis of the fluid velocity field as well as the physical interpretation of the result returns. Apart from the use of some tensor analysis, its presentation is very similar to that adopted here. In this regard, let us quote his conclusion: «*Hence the most general instantaneous motion of an elementary portion of a fluid is compounded of a motion of translation, a motion of rotation, a motion of uniform dilatation, and two motions of shifting of the kind just mentioned.*» These last two motions correspond very precisely to the angular offset of two initially orthogonal edges of the present analysis.

However, we owe it to historical truth to mention the luminous anticipation of the approach found in Euler's publication [56] in 1757.

EXERCISE 7 The pure elongational distortion motion.

We consider the steady flow whose fluid particle position with respect to an orthonormal reference frame is governed by the following parametric equations:

$$\begin{cases} x = U_0 \alpha \\ y = y_0\, e^{\,a U_0 \alpha} \\ z = z_0\, e^{-a U_0 \alpha} \end{cases}$$

where α denotes a parameter whose dimension is a time, U_0 an arbitrary velocity, y_0, z_0 two reference coordinates and a a non-zero real.

QUESTION 1. Express the velocity vector components U, V, V, W as a function of the coordinates (x, y, z) and solely constants a and U_0.

QUESTION 2. Is the motion isovolume?

QUESTION 3. Derive the expressions of the vorticity components and of the strain-rate tensor. Give a physical interpretation of the results.

$SOLUTION$: 1. Since the dimension of α is that of a time, we immediately have

$$U\,(x, y, z) = \frac{dx}{d\alpha} = U_0\,,$$

$$V\,(x, y, z) = \frac{dy}{d\alpha} = aU_0y_0e^{aU_0\alpha} \equiv aU_0y\,,$$

$$W\,(x, y, z) = \frac{dz}{d\alpha} = -aU_0y_0e^{-aU_0\alpha} \equiv -aU_0z\,,$$

which shows that the motion is indeed stationary, since the velocity components when specified using the Eulerian variables are time independent.

2. The evolution is in fact isovolume since :

$$div\,\mathbf{V} = \frac{\partial U}{\partial x} + \frac{\partial V}{\partial y} + \frac{\partial W}{\partial z} = aU_0 - aU_0 = 0\,.$$

3. All components of the velocity gradient tensor are zero, with the exception of

$$S_{22} = aU_0 \quad \text{and} \quad S_{33} = -aU_0\,.$$

Accordingly, the vorticity is zero. The motion is *irrotational*, with pure *elongational deformation*, as shown in the following figure. It is a constant velocity flow in the x direction, exponentially expanding in the y direction and contracting, equally exponentially, along the second transverse coordinate z, as illustrated in the (a) part of Fig. 2.13.

Any fluid element, moving by translation along x, is stretched along y and squeezed along z of an equal amount in order to retain unchanged its elementary volume, Fig. 2.13**b**.

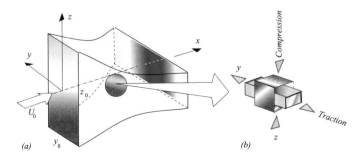

Fig. 2.13 Distortion duct **a** and schematic of the pure elongational deformation **b**

EXERCISE 8 Rotational and irrotational circular motions.

Let us consider a steady flow whose velocity field components are, in cylindrical coordinates and for $r \neq 0$:

$$V_r = 0 ; \quad V_\theta = Ar + \frac{B}{r} ; \quad V_z = 0 ,$$

where A and B are two non-dimensional constants.

QUESTION 1. Is the motion isovolume?

QUESTION 2. Derive the expressions of the vorticity and strain-rate tensor components.

QUESTION 3. Discuss the special cases $A = 0$ and $B = 0$ and physically interpret the results.

SOLUTION: **1.** In cylindrical coordinates, we have $div\, \mathbf{V} \equiv \dfrac{1}{r}\dfrac{\partial (r V_r)}{\partial r} + \dfrac{1}{r}\dfrac{\partial V_\theta}{\partial \theta} +$
$\dfrac{\partial V_z}{\partial z}$ which clearly shows that the considered motion is isovolume.

2. From the expressions of the curl operator components in cylindrical coordinates (see Appendix), it follows that:

$$\Omega_r \equiv \frac{1}{r}\frac{\partial V_z}{\partial \theta} - \frac{\partial V_\theta}{\partial z} = 0 ,$$

$$\Omega_\theta \equiv \frac{\partial V_r}{\partial z} - \frac{\partial V_z}{\partial r} = 0 ,$$

$$\Omega_z \equiv \frac{1}{r}\left[\frac{\partial (r V_\theta)}{\partial r} - \frac{\partial V_r}{\partial \theta}\right] = 2A .$$

Similarly, we obtain for the strain-rate tensor components :

$$S_{rr} \equiv \frac{\partial V_r}{\partial r} = 0 , \quad S_{\theta\theta} \equiv \frac{1}{r}\frac{\partial V_\theta}{\partial \theta} + \frac{V_r}{r} = 0 , \quad S_{zz} \equiv \frac{\partial V_z}{\partial z} = 0 ,$$

$$S_{r\theta} \equiv \frac{1}{2}\left(\frac{\partial V_\theta}{\partial r} - \frac{V_\theta}{r} + \frac{1}{r}\frac{\partial V_r}{\partial \theta}\right) = -\frac{B}{r^2} ,$$

$$S_{rz} \equiv \frac{1}{2}\left(\frac{\partial V_r}{\partial z} + \frac{\partial V_z}{\partial r}\right) = 0 ,$$

$$S_{\theta z} \equiv \frac{1}{2}\left(\frac{1}{r}\frac{\partial V_z}{\partial \theta} + \frac{\partial V_\theta}{\partial z}\right) = 0 .$$

3. Whatever the values of A and B, provided they are not simultaneously zero, we have $\mathbf{V}.\mathbf{e}_r = 0$, so that the streamlines are circles centered at the frame origin. In addition, the vorticity is related solely to the A value, while the B value is responsible

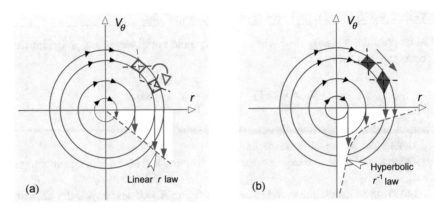

Fig. 2.14 Pure rotational solidifying flow **a** and irrotational flow with pure angular deformation **b**

for the deformation. These two properties are therefore mutually exclusive for special cases where one or other of these constants is zero.

1. For $B = 0$, the motion has a linearly increasing velocity distribution. It is purely rotational. The fluid behaves like a non-deformable, rigid solid and rotates '*in bulk*'. This is the situation outlined in Fig. 2.14a and still referred to as a *solidifying motion*;
2. For $A = 0$, the motion is irrotational. This result is obtained by a pure fluid particle *angular* deformation with no change in its elementary volume. This configuration corresponds to a velocity field whose module hyperbolically decreases with the distance from the rotation axis, as outlined in Fig.2.14b.

EXERCISE 9 The simple shear flow.

Denoting by Γ a constant real, whose dimension is a time inverse, we give the Cartesian components of the velocity field $U(x, y, z, t) = \Gamma y$, $V(x, y, z, t) = 0$, $W(x, y, z, t) = 0$.

QUESTION 1. Derive the expressions of the tensor components $\overline{\overline{grad}}\,\mathbf{V}$, $\overline{\overline{S}}$ and $\overline{\overline{R}}$.

QUESTION 2. What are the eigendirections of the strain-rate tensor?

QUESTION 3. Interpret this pure plane shear motion in terms of rotation and deformation.

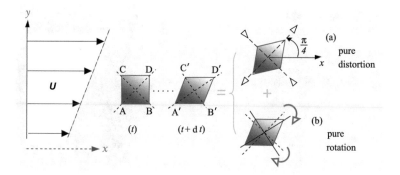

Fig. 2.15 Decomposition of a simple shear into pure distortion **a** and pure rotation **b**

SOLUTION: **1.** By directly applying the general definitions we have:

$$\overline{\overline{grad}}\,\mathbf{V} = \begin{pmatrix} 0 & \Gamma & 0 \\ 0 & 0 & 0 \\ 0 & 0 & 0 \end{pmatrix}, \quad \overline{\overline{S}} = \frac{1}{2}\begin{pmatrix} 0 & \Gamma & 0 \\ \Gamma & 0 & 0 \\ 0 & 0 & 0 \end{pmatrix}, \quad \overline{\overline{R}} = \frac{1}{2}\begin{pmatrix} 0 & \Gamma & 0 \\ -\Gamma & 0 & 0 \\ 0 & 0 & 0 \end{pmatrix}.$$

2. Due to the two-dimensional nature of the flow, the eigenvalues (λ) of $\overline{\overline{S}}$ are simply obtained by solving the equation:

$$\begin{vmatrix} -\lambda & \Gamma \\ \Gamma & -\lambda \end{vmatrix} = 0,$$

which leads to two real eigenvalues $\lambda = \pm\,\Gamma$. The eigendirections, of general equations $-\lambda x + \Gamma y = 0$, are thus the bisectors of the axes $\{x, y\}$.

3. Since the evolution is clearly isovolume, the result of the first question shows that an elementary fluid body undergoes both a deformation and rotation.

From the answer to question 2, it follows that the deformation is purely angular (no cubic or volumetric deformation) and that it is directed along the axes which make the angles $\pi/4$ and $3\pi/4$ with the general flow direction. This deformation corresponds to a stretch in the first bisector direction and compression in the second bisector direction. It is by the concomitant rotation of the fluid element on itself that the AB and CD facet parallelism with the x axis is obtained, as illustrated in Fig. 2.15.

2.4   Vorticity and Vortex

With the exception of irrotational motions, a vorticity vector field is associated, by Eq. (2.15) with the velocity field of any fluid flow.

Fig. 2.16 Vorticity lines and
tube

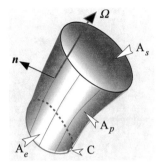

We will focus here, firstly, on the kinematic properties[5] of the vorticity field. We will then discuss some specific aspects of '*vortex structures*', in the sense of vorticity concentrating physical entities (see Sect. 2.4.4).

2.4.1 The Vorticity Vector, the Vorticity Field

As previously defined—see Eq. (2.20)—the vorticity vector is half the curl of the velocity vector[6]

$$\omega = \frac{1}{2}\Omega = \frac{1}{2}\,curl\,\mathbf{V}. \tag{2.22}$$

The modulus of the vorticity vector is simply called the *vorticity*.

2.4.2 The Vorticity Line, Tube

By analogy with the streamline (tube) basic principles (Sect. 2.1.1.1), we introduce those of the vorticity line[7] (tube). Thus the vorticity lines are the curves whose tangent, at any time, is colinear to the vorticity vector at the same point. A *vorticity tube* is the surface constituted by the vorticity lines passing through each point of a same closed curve at a given time (Fig. 2.16).

If we denote $(\omega_x, \omega_y, \omega_z)$ as the vorticity vector components, the vorticity line differential equations are:

$$\frac{dx}{\omega_x} = \frac{dy}{\omega_y} = \frac{dz}{\omega_z} = d\alpha\,,$$

where α stands for any parameter *other than time.*

---

[5] Therefore we are not aiming here at determining the vorticity for a given flow. Such an objective can only be addressed on the basis of the vorticity transport equations of Chap. 9.

[6] In some textbooks, the vorticity is taken as the curl of the velocity (Batchelor [5], Panton [111]). In others, it is the present definition which is adopted (Ziegler [170]).

[7] We are also speaking of a vortex line (tube) without necessarily referring to a physical entity.

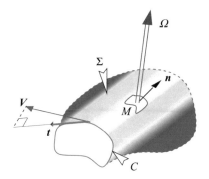

2.4.3   The Vorticity Intensity. Circulation. Stokes' Theorem

Denoting by Σ any control surface in the flow field, the scalar amount I defined by
the following surface integral is called the *vorticity intensity* through Σ:

$$I = \iint_{\Sigma} \boldsymbol{\omega}.\mathbf{n}\,d\sigma \qquad (2.23)$$

where $\mathbf{n}$ the unit normal vector pointing outwards of the surface element $d\sigma$.

Denoting by C any curve and by $d\boldsymbol{\ell}$ the elementary oriented arc on this curve,
the elementary *circulation* of the velocity vector is, see Chap. 1 Sect. 1.6.6.1 $d\Gamma =$
$\mathbf{V}.d\boldsymbol{\ell}$, or, along the curve C:

$$\Gamma = \int_{C} \mathbf{V}.d\boldsymbol{\ell} \qquad (2.24)$$

The velocity circulation along a closed contour and the vorticity intensity (see figure
opposite) are linked by Stokes' theorem which states that:

> The velocity circulation around a closed smooth contour is equal to the surface
> integral over a surface Σ bounded by this contour of the normal component of
> the curl of the velocity field (vorticity flux):
>
> $$\Gamma\,(\text{closed contour}) \equiv \oint \mathbf{V}.d\boldsymbol{\ell} = \iint_{\Sigma} \boldsymbol{\Omega}.\mathbf{n}\,d\sigma = 2 \iint_{\Sigma} \boldsymbol{\omega}.\mathbf{n}\,d\sigma = 2I \qquad (2.25)$$

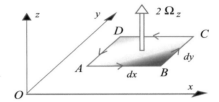

Fig. 2.18 Stokes-Green's formula interpretation around an elementary plane curve

Example In the particular case of a two-dimensional plane motion, it is possible to give a very simple illustration of Stokes' formula.[8] For this purpose, let us consider the elementary contour $ABCD$ (see Fig. 2.18) and adopt as surface Σ the area of the rectangle, $dx\,dy$. From its definition, the circulation of the velocity $d\Gamma$ along this elementary contour is simply equal to:

$$d\Gamma = U dx + \left(V + \frac{\partial V}{\partial x} dx\right) dy - \left(U + \frac{\partial U}{\partial y} dy\right) dx - V dy \equiv \left(\frac{\partial V}{\partial x} - \frac{\partial U}{\partial y}\right) dx\,dy\,,$$

or, $d\,\Gamma = \Omega_z d\sigma$, with $d\sigma = dxdy$.

We recognize Stokes' formula applied to the infinitesimal contour area where it reduces to Green's formula :

$$\boxed{\frac{d\Gamma}{d\sigma}\,(\textbf{closed contour}) = 2\omega.\mathbf{n}} \qquad (2.26)$$

which depicts the curl of a vector field as the elementary surface density of this vector field circulation.

HISTORICAL NOTE—Stokes' theorem—The '*Stokes' formula*' term, quite widespread and familiar nowadays, hides a more complex historical author-ship, as highlighted in the survey published by Katz [83], in 1979. The first *writing* relating to this formula causes it to appear in the statement of one of the questions (no. 8) on the list proposed by Stokes to the candidates for the Cam-bridge Smith Prize in 1854. Earlier, a letter from Thomson to Stokes, dated July 2, 1850, explicitly cites this formula. Finally, the first published demonstration was written by Hankel in 1861, according to Katz [83].

---
[8] The result is then known as Green's formula.

2.4.4 The Vortex—Eddy

As we have already noted, the purely kinematic notions of vorticity line and tube are not necessarily associated with *material entities* which can be identified as *physical structures* within the flow.

Let us consider, for instance, the two-dimensional plane laminar flow of a viscous fluid in the vicinity of a fixed flat wall, as shown in Fig. 2.19a.

In this zone a velocity gradient exists which, for the sake of simplicity, can be assumed, here, to being that of a simple shear flow *viz dU/dy*. There is thus a vorticity vector field, whose axes are orthogonal to the plane of the Figure and regularly distributed at every point in the field. The fluid particles thus move along straight paths while rotating on themselves[9] without inducing the slightest vortex structure, as outlined in Fig. 2.19a.

The situation we have just described is physically quite different from that outlined in Fig. 2.19b. In this case, only the fluid particles on the red line are subject to a rotation on themselves. The circulation of the velocity is therefore zero along any contour not encircling the curve in question. In this flow, the vorticity is therefore *localized* or *concentrated* on a specific space element (line) which constitutes a physical entity in this flow.

With a view to further clarifying the distinction between both situations, we will speak of a *vortex* structure, or following Mathieu's terminology [96], *eddy* structure,[10] whenever the space vorticity distribution is located in a particular geometry characterizing the physical entity.

We will clarify this proposition by considering two *idealized* typical examples in the following sections.

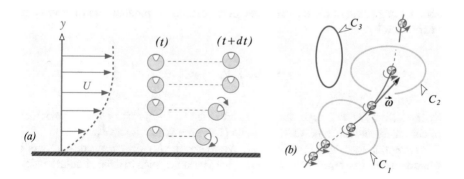

Fig. 2.19 Flow field: **a** regularly distributed vorticity, **b** vortex line singularity

---

[9] There is of course also a particle deformation, as we have seen in Sect. 2.3.2, which has not been included in the figure for the sake of clarity.

[10] In the absence of possible confusion, we will also speak, in short, of a "vortex" or "eddy".

Remark In practice, there are many situations where different types of vortex structures are forming within a flow.

They are present, for example, in the wake of bluff bodies, with in particular the alternating eddies of the Bénard–Von Kármán vortex street downstream of a circular cylinder. The setting in motion of a viscous fluid around a streamlined body is also responsible for the emission of a starting vortex at the trailing edge of the body.

The gravity flow of a liquid in a cylindrical tank through a central orifice generates a vortex structure which can be visualized by the free surface deformation, as also observed in water flowing from a bathtub or wash basin. In geofluid motions, tornado or waterspout phenomena also provide particularly striking examples.

Finally, many other manifestations are associated with secondary vortices which originate in various internal flow geometries, a bent pipe, an abrupt widening, a curvature, a non-circular cross-section, etc.

2.4.4.1 The Eddy Structure with a Vortex Core

In a two-dimensional axisymmetrical situation (r, θ), we consider a fluid medium rotating around an axis (z) with an azimuthal velocity component $V_\theta (r)$ such that:

$$
\begin{cases}
\text{for } 0 < r \leqslant a \quad V_\theta (r) = \omega_0 r , \\[2mm]
\text{for } a < r \quad V_\theta (r) = \dfrac{\omega_0 a^2}{r} ,
\end{cases}
$$

where ω_0 and a are two real constants. In light of the results of exercise 8, such a velocity field can be interpreted as that of a fluid core with a radius a in bulk rotation at the angular velocity ω_0 with, beyond this radius, a strictly irrotational circular motion. On a circle of radius r centered at the origin, the circulation of the velocity is expressed by:

$$
\Gamma (r) = \int_0^{2\pi} V_\theta (r) \, r d\theta ,
$$

which yields $\Gamma (r) = 2\pi\omega_0 r^2$ for $r < a$ and $\Gamma t (r) = 2\pi\omega_0 a^2$ for $r \geqslant a$, from which we can easily deduce, by Stokes' formula (2.25) the vortex intensity.

We are therefore dealing with an axi-symmetrical *eddy structure*, consisting of a cylindrical core of a circular cross-section with radius a, such that the circular motion of the fluid in an infinite space is that of a bulk rotation for $r \leq a$ and an irrotational motion for $r > a$.

We will see in the continuation of this course how, at an initial time, such a flow field in a viscous Newtonian fluid is generated. Such a structure, whose core later diffuses over time, due to the viscosity, will then be called a *viscous core vortex*.

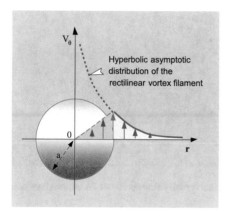

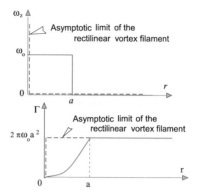

Fig. 2.20 Schematic for the azimuthal velocity field of a vortex core (left), vorticity and circulation distributions of a vortex core and a rectilinear vortex filament (right)

2.4.4.2 The Vortex Filament, Vortex Sheet

A rectilinear *vortex filament* is obtained by passing to the limit of the previous vortex core structure by[11] assuming the three conditions :

1. the radius of the core tends towards zero ($a \rightarrow 0$) ;
2. the angular velocity tends towards infinity ($\omega_0 \rightarrow \infty$) ;
3. the vorticity intensity remains constant ($\omega_0 a^2 = C^t$) .

We then transform the characteristics of an eddy structure with a finite area vortex core to those of the straight vortex filament, as outlined in Fig. 2.20.
By setting $\Gamma = 2\pi\omega_0 a^2$, the velocity field of the rectilinear vortex filament is therefore defined, for $r \neq 0$ by the hyperbolic decreasing law:

$$V_\theta\,(r) = \frac{\Gamma}{2\pi}\,\frac{1}{r}\,. \qquad (2.27)$$

Let us recall that such a field corresponds to an irrotational motion.[12]
Thus, it is important to carefully differentiate between the vorticity, vortex and eddy concepts, both from a physical and mathematical point of view:

– the *vorticity vector* in any point is *physically* associated to the instantaneous rotation of the fluid particle on itself during its motion;
– *vorticity lines, tubes, surfaces*…are kinematic objects which *mathematically* determine the geometry of those regions where the motion occurs with the local rotation of the particles on themselves;

---

[11] In the sense of the continuous medium.
[12] This character applies everywhere to the exclusion, of course, of the origin ($r = 0$).

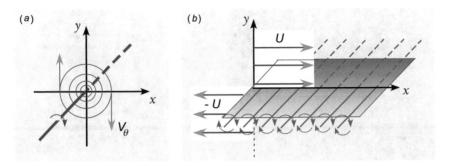

Fig. 2.21 Schematic of a rectilinear vortex **a** and the vorticity straight lines of a vortex sheet **b**

– *vortex, eddy structures* finally correspond to *physical entities* associated with the material vorticity concentration in distinct flow field areas. Thus, a vortex line/surface can always be considered as a vortex line/surface. On the other hand, not every vorticity line/surface is necessarily a vortex or eddy structure, as a physical entity.

This distinction is illustrated by the two examples in Figure 2.21.

In part (*a*) of the Figure is shown the distribution of the azimuthal velocity field component of a rectilinear vortex filament with a z axis, as described above. Outside this axis, the flow is *circular* and irrotational, this line structure concentrating the entire vorticity.

The schematics in part (*b*) is that of a two-dimensional plane *mixing layer*, whose velocity field consists in a step in the $y = 0$ plane. The *vorticity lines* of this flow are also the straight lines of this plane parallel to the z axis. However, it would be inappropriate to speak of them as *vortex lines*, since the structure concentrating the vorticity in this flow is now the $y = 0$ plane, with a parallel and not circular motion on either side of this plane. As such, only the *vorticity sheet* can still be termed as a *vortex sheet*.

Remark In some of the examples just presented, the vortex structures (line, plane) correspond to **mathematical singularities**, since the vorticity becomes infinite on 'infinitely small' geometries in the mathematical sense. This is obviously a matter of theoretical concern, as such singularities should be considered as asymptotically valid limits of a viscous fluid flow.

HISTORICAL NOTE—**Rankine's vortex**—The rectilinear vortex (Sect. 2.4.4.2) is a singular solution, *i.e.* not defined on the axis, of the Euler equations, as we will see in Chap. 6–6.3.1.4. The eddy structure with a vortex core in Sect. 2.4.4.1 can be seen as a *viscous regularization*, as a solution of the Navier-Stokes model. It is customary to attribute the authorship to Rankine and to speak of it in terms of *Rankine's vortex*.

2.4.5 The Vortex Persistence

The persistence of the rotational/irrotational character in a fluid motion is a major property of flows where an *acceleration potential* exists.

It is expressed in practice through two theorems, due to Lagrange and Helmholtz (the first of his four theorems).

Lagrange's theorem—Several equivalent formulations of this theorem can be given. We have chosen the following two:

> In an inviscid barotropic[13] fluid flow whose external forces derive from a potential energy function, if there is, at a given time, a domain where the motion is irrotational, this motion will remain irrotational in the domain at all times.

> Any flow, smoothly generated from rest, of an inviscid barotropic fluid whose external forces derive from a potential energy function, is irrotational.

Proof The most direct demonstration of this theorem comes from the simple application of Kelvin's theorem (see Sect. 1.6.6.4). Denoting by $\mathcal{D}$ the domain over which the motion is irrotational at a given time, the circulation along any closed curve drawn within the domain is null at that time, according to Stokes' theorem. It will remain thus over time, when the curve moves *within the domain* $\mathcal{D}$, according to Kelvin's theorem. Hence, the vorticity will remain zero over any surface in the domain bounded by the curve. The arbitrary choice of the curve within $\mathcal{D}$ ensures the nullity of the vorticity on any surface inside the domain over time.

ADDITIONAL INFORMATION—**The physical interpretation of Lagrange's theorem**—Lagrange's theorem states that the vortical nature of the motion cannot be transferred to a fluid domain which is initially devoid of it. We will see later that this is a property essentially due to the *non-viscous* character of the medium. The viscosity of a real fluid flowing around a solid, for example, is responsible for the vorticity *generation* near the wall and its *diffusion* into a possibly previously irrotational part of the flow field.

In addition to the viscous fluid situation, Lagrange's theorem does not apply either when:

1. external forces do not derive from a potential energy function, as is the case, in particular, in non-Galilean reference frames, when Coriolis forces cannot be neglected (typical geofluid flow situations);
2. in a non-barotropic flow of an ideal fluid, with the presence of heat generation by compression and shock for example.

HISTORICAL NOTE—**Lagrange's theorem**—In the precursory work already mentioned [85], Lagrange set out in 1781, the conditions under which the circulation of the velocity along an elementary path following the edges of the trihedron reference, $d\Gamma_{xyz} \equiv U\,dx + V\,dy + W\,dz$ is a differential, or:

$$d\Gamma_{xyz} = d\phi. \tag{2.28}$$

He then demonstrated, by an original type of reasoning which Stokes later questioned [149], the following three statements:

(i) If the condition (2.28) is verified *at a given time*, then this condition is satisfied *at all times*;
(ii) If there is a time t where the condition (2.28) is not fulfilled, then this condition is never satisfied;
(iii) When the motion occurs from a fluid at rest, the condition (2.28) is always satisfied *at all times*.

HISTORICAL NOTE— **Thomson's theorem**—Lagrange's original demonstration was revisited by William Thomson (Lord Kelvin) [157], in 1869. He put forward a new demonstration, closer to the modern approach, of Lagrange's theorem statements, which he also explains in terms of rotationality. It is for this reason that his name remains associated with results on rotational properties, in addition to Lagrange.

2.4.6 Helmholtz's Theorems

2.4.6.1 Helmholtz's First Theorem

This first theorem directly expresses the persistence of any vortex structure. It can be stated, for example, as follows:

> If, in a flow with an acceleration potential, a fluid surface is a vortex tube at a given time, it will remain so over time as it moves with the fluid.

Proof All necessary information for the demonstration of this first theorem has been given previously. Therefore, the presentation which is given in detail here has only a pedagogical purpose of applying some of the elements introduced previously.

Let us consider, in a flow with an acceleration potential, a vortex tube at time t, on the surface of which we examine the closed contour ABCD, see Fig.2.22. By definition, the vorticity vector at any point on the tube surface is orthogonal to the normal unit vector **n**. The consequence, according to Stokes' theorem, is that:

$$\Gamma_{ABCD} = 0.$$

As a fluid surface, the vortex tube has moved in the flow at time $t+dt$.

At that time, the fluid line ABCD has moved to A'B'C'D'. Now, the material derivative of the circulation, Eq. (1.53), under the acceleration potential assumption yields:

$$\frac{d\Gamma_{ABCD}}{dt} = 0 \quad \text{(Kelvin's theorem)}.$$

Fig. 2.22 Schematic for a
closed contour on a vortex
tube surface

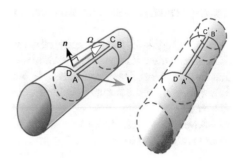

It is deduced that $\Gamma_{A'B'C'D'} = 0$, therefore the vorticity flux through the surface of
the vortex tube at time $t + dt$ is zero.

Since the ABCD contour is arbitrary, the zero flux is due to the orthogonality of
the vorticity vector with the surface normal to the tube at this new time, which clearly
demonstrates that it is a vortex tube.

2.4.6.2 Helmholtz's Second Theorem: The Vortex Intensity Conservation

Helmholtz's *second* theorem states that:

> The circulation is the same in all circuits embracing the same vortex tube.

This theorem plays, with respect to the vorticity flux, the same role as that of the
flow rate conservation (mass flux) for a stream tube.

Proof With reference to Fig. 2.16, let us consider a portion of a vortex tube bounded
by an inlet cross-section A_e, an outlet cross-section A_s, and a lateral surface A_p.
Owing to the characteristic property of the curl operator, we have $div\,\mathbf{\Omega} = 0$. Consequently, according to Ostrogradski's theorem, we can write

$$0 = \iiint_D div\,\mathbf{\Omega}\,dv = \iint_{A_e+A_s+A_p} \mathbf{\Omega}.\mathbf{n}d\sigma,$$

from which we immediately deduce, since the normal to the peripheral surface (A_p)
is by definition orthogonal to the vorticity vector :

$$-\iint_{A_e} \mathbf{\Omega}.\mathbf{n}d\sigma = \iint_{A_s} \mathbf{\Omega}.\mathbf{n}d\sigma.$$

This is the expected result, according to Stokes' theorem.

2.4.6.3   The Cross-Section Variation of a Vortex Tube

To simply highlight the influence of a variation in a vortex tube cross-section, we consider a vortex filament or tube whose cross-section $\delta\sigma$ is small enough to consider that the vorticity (Ω) is constant over the whole area. The elementary circulation $\delta\Gamma_O$ on the closed contour bounding this elementary section is then expressed by $\delta\Gamma_O = \Omega\,\delta\sigma$.

Between two elementary sections of such a vortex filament, Helmholtz's second theorem allows us to write that :

$$\boxed{\Omega_1\,\delta\sigma_1 = \Omega_2\,\delta\sigma_2 \quad \text{or} \quad \frac{\Omega_1}{\Omega_2} = \frac{\delta\sigma_2}{\delta\sigma_1}} \tag{2.29}$$

This relation demonstrates that:

– any decrease (*resp.* increase) in the cross-section of an elementary vortex filament results in an increase (*resp.* decrease) in the vorticity intensity;
– the vortex tube cross-section cannot cancel out inside the fluid domain, since the vorticity would then be infinite. As a result, the vortex tubes either close on themselves or end at the flow boundaries, whether it is a free surface or a solid wall.

Relation (2.29) and its consequences outlined above constitute *Helmholtz's third theorem*.

HISTORICAL NOTE—**Helmholtz and the vortices**—The first major results on vortices in fluid flows were published by Helmholtz [68], in 1858. Some were re-demonstrated, in another way, some years later, by Thomson (Lord Kelvin) [157], in 1869. We also find in this last article the result about the vortex tube's intensity variation a by a change in its cross-section due to a length variation, *i.e.*, the vortex stretching phenomenon which is the subject of the following section.

2.4.7   The Vortex Stretching

We consider a vortex filament or elementary vortex tube whose vorticity vector module is constant over the whole cross-section aera $\delta\sigma$. During the time interval dt, one end of the tube is stretched in a direction colinear with the local vorticity vector, *i.e.*

$$\mathbf{MM'} \equiv \delta\mathbf{M} = \alpha\boldsymbol{\Omega}\,,$$

by denoting α a positive constant, according to the notations in Fig. 2.23.

The rate of length variation at the vortex extremity M is expressed, within the context of the small displacement analysis in Sect. 2.4, by:

Fig. 2.23 Stretching of an
elementary vortex tube

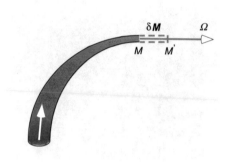

$$\frac{d\,\delta\mathbf{M}}{dt} = \overline{\overline{grad}}\mathbf{V} \odot \delta\mathbf{M}, \text{ hence } \frac{1}{\|\delta\mathbf{M}\|}\frac{d\delta\mathbf{M}}{dt} = \overline{\overline{grad}}\mathbf{V} \odot \frac{\delta\mathbf{M}}{\|\delta\mathbf{M}\|} \equiv \overline{\overline{grad}}\mathbf{V} \odot \frac{\delta\mathbf{\Omega}}{\|\delta\mathbf{\Omega}\|} .$$

In addition, using the vorticity transport equation to be established in Chap. 9
Sect. 9.2.2, the vorticity variation when following point M motion is given by :

$$\frac{d\mathbf{\Omega}}{dt} = \overline{\overline{grad}}\mathbf{V} \odot \mathbf{\Omega} ,$$

neglecting viscosity effects and under the influence of conservative body forces. We
deduce from this:

$$\frac{1}{\|\mathbf{\Omega}\|}\frac{d\mathbf{\Omega}}{dt} = \overline{\overline{grad}}\mathbf{V} \odot \frac{\mathbf{\Omega}}{\|\mathbf{\Omega}\|} .$$

By identifying with the previous expression of the rate of change in the length at
point M, we obtain (*Helmholtz's fourth theorem*):

$$\frac{1}{\|\mathbf{\Omega}\|}\frac{d\mathbf{\Omega}}{dt} = \frac{1}{\|\delta\mathbf{M}\|}\frac{d\delta\mathbf{M}}{dt} .$$

We will group this result with that of the previous section, Eq. (2.29) and write in
abridged notations:

$$\boxed{\frac{\Omega(t)}{|\Omega_0|} = \frac{\delta\ell(t)}{\delta\ell_0} = -\frac{\delta\sigma_0}{\delta\sigma(t)}} \tag{2.30}$$

We can therefore retain the following:

> Without viscosity and under the influence of conservative body forces, any
> stretching (*resp.* compression) of a vortex filament increases (*resp.* reduces)
> the vorticity and decreases (*resp.* increases) the cross-section.

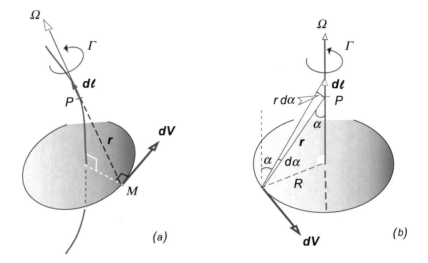

Fig. 2.24 Velocity induced by a vortex filament **a** and a vortex straight-line **b**

2.4.8 The Biot-Savart Law

The Biot-Savart law is derived from the field theory in electromagnetism where it provides the expression of the magnetic field generated in any point in the space by an electric current in a conductor of a given geometry.

The transposition of the Biot-Savart formula to fluid mechanics simply consists in assigning the magnetic field vector to the velocity vector on the one hand, and the intensity of the current to the vorticity intensity on the other hand. With the notations in Fig. 2.24, this results in the relation :

$$ d\mathbf{V} = -\frac{\Gamma}{4\pi} \frac{d\mathbf{l} \wedge \mathbf{r}}{\|\mathbf{r}\|^3} \qquad (2.31) $$

In the above formula, $d\mathbf{V}$ is the velocity element generated at point M in the fluid by a vortex line element $d\mathbf{l}$ located at point P whose distance from M is denoted as $\mathbf{PM} = \mathbf{r}$.

EXERCISE 10 The velocity induced by a vortex straight-line.

We consider, in an infinite fluid medium, a rectilinear vortex with a circulation Γ.

QUESTION What is the velocity induced by this vortex at a point located at a distance r from its axis?

SOLUTION: In direct application of relation (2.31) and denoting by r the distance between point M and the current point of the rectilinear vortex, see Fig. 2.24b, the result of the cross product $d\boldsymbol{\ell} \wedge \mathbf{R}$ at point M is an orthogonal vector to the meridian plane passing through this point. In other words, the velocity vector is reduced solely to its azimuthal component V_θ.

In addition, it is clear that $d\boldsymbol{\ell} \wedge \mathbf{r} = -r \sin \alpha \, d\ell \, \mathbf{e}_\theta$.

Thus, $V_\theta = \dfrac{\Gamma}{4\pi} \displaystyle\int_{-\infty}^{+\infty} \dfrac{\sin \alpha}{r^2} d\ell$, or again with $\sin \alpha \, d\ell = r d\alpha$,

$$V_\theta = \frac{\Gamma}{4\pi} \int_0^\pi \frac{d\alpha}{r} = \frac{\Gamma}{4\pi} \int_0^\pi \frac{\sin \alpha d\alpha}{R}$$

.

After integration, we obtain $V_\theta (R) = \dfrac{\Gamma}{2\pi} \dfrac{1}{R}$ which is certainly again the result Eq. (2.27).

2.5 The Application to Flow Visualization

2.5.1 General Information on Flow Visualization Techniques

Many of the elements which have just been presented have applications in fluid motion visualization. We give here some examples aiming at a better understanding of the differences between the various representations of flow fields. With regard to visualization techniques, two main principles can be very schematically discerned, depending on whether the agent that makes it possible to *see the flow* is *(i)* introduced for this purpose from the outside or *(ii)* derived from an intrinsic characteristic of the motion itself. In the first case, we will speak of visualization by tracer and in the second case, by property.

- In practice, there are many possibilities of *tagging* a flow 'from the outside', by injecting a tracer, the difficulty being that:

  - the injection should *disturb* the flow as little as possible;
  - the tracer can behave throughout the field as a *passive contaminant* whose motion is perfectly matching that of the fluid particles.

  Thus, in hydrodynamics, various dyes can be used such as methylene blue, fluorescein, etc. In aerodynamics, many wind tunnel visualizations are obtained by means of smoke injection. The tracer can also be of a medium whose phase is different from that of the flowing fluid. Thus, hydrogen bubbles generated by electrolysis can be injected as a marker into salt water flows. In the same vein, both in liquid and gas flows, solid particles can also be used.
- The visualization by property is, by its very definition, exempt from the difficulties inherent to the flow tagging from the outside. In contrast, it is not available for any fluid flow. Its two predilection domains are those of some reactive flows, in liquid or gaseous phase, and gas flows with variable density.

Fig. 2.25 Oil-coloured tracer visualization of the attached rollers in the wake of a cylinder in the laminar regime, adapted from Dupin & Teissié-Solier [53], IET 1928

In the first case, a chemical reaction can sign the mixing of fluid streams by the formation of a visible constituent.

In the second case, the variation in the optical index of the medium, resulting from that of the density, can be directly used as a means of visualization, as we will see later. This type of method can be applied to high-speed flows, where it enables highlighting effects related to the Mach number and in particular the presence of shock waves. It also applies to low speed air motions in the presence of temperature differences (*e.g.* vortex structures in natural convection).

In addition, it provides a visual representation of the mixing of gases in air, and therefore be adopted as a tracer visualization process, by injecting a gas of a different nature or temperature than that of the moving fluid.

2.5.2 *Examples of Visualization by a Tracer*

2.5.2.1 Streamlines and Pathlines in a Steady Flow Regime

The wake that develops downstream of an infinite span cylinder normal to the direction of a uniform flow, can exhibit vortex structures of a very different typology, depending on the flow velocity to infinity U_∞, the fluid viscosity ν and the cylinder diameter SD. The dimensionless parameter which differentiates between the various vortex configurations is the Reynolds number, R_e, defined as :

$$R_e = \frac{U_\infty D}{\nu}. \tag{2.32}$$

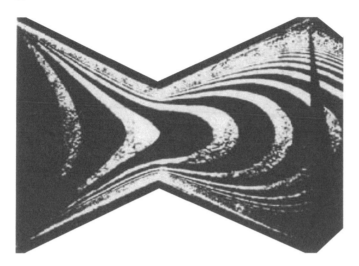

Fig. 2.26 Fluid lines and streaksurfaces of a non-Newtonian fluid flow in a convergent-divergent, adapted from Chassaing & Clet [34] 1969, courtesy of ENSEEIHT

For R_e values ranging between a few units and a few tens, but below a so-called *critical value*, in the order of 45 to 48, the wake is referred to as '*laminar*'. It consists of two attached counter-rotating rollers located immediately downstream of the obstacle, as shown in Fig. 2.25.

Since the flow is stationary, the tracer-coloured strips visualize the streamlines or the trajectories of the flowing fluid, in this case oil. The laminar nature of the motion is clearly visible, both in the 'undisturbed' external flow and within the counter-rotating roll ups of the wake.

2.5.2.2 Fluid Lines and Streaksurfaces in a Steady Flow

To obtain a whole flow field representation, streaklines (surfaces) are appropriate in combination with fluid lines (surfaces). The photograph in Fig. 2.26 provides such an illustration.

The medium is a carboxymethylcellulose aqueous solution at a $6g/\ell$ dilution, a fluid with a *non-Newtonian* character.

Hydrogen bubbles are generated within the moving fluid by electrolysis along a straight wire normal to the main flow direction (here from left-to-right). The electrical current intensity is modulated over time according to a square wave. The result is an intermittent bubble production which appears as clear stripes on the plate, and are streaksurfaces related to the time exposure of the fluid to the electrolysis current. The upstream and downstream edges which separate the bright areas from the dark background are nothing more than fluid lines.

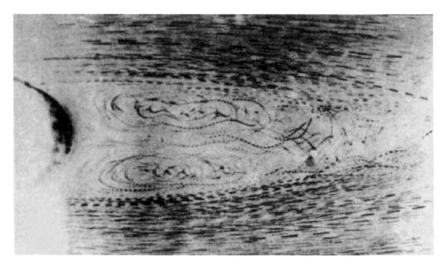

Fig. 2.27 Chronophotography of the streamlines in the wake of a cylinder around the critical Reynolds number $R_e \approx 47.5$, adapted from Dupin & Teissié-Solier [53], IET 1928

In particular, it should be noted that there is no separation of the fluid motion from the wall in the divergent part of the duct, as a result of the non-newtonian behaviour of the medium. By comparison, the flow of a Newtonian fluid would be more *axial* with, *(i)* the presence of a central core with a cross-section close to that of the throat, *(ii)* a reduced lateral expansion in the divergent part and *(iii)* the onset of detachments in the divergent part.

2.5.2.3 Stream Lines in an Unsteady Flow

Returning to the flow around the cylinder, and when the Reynolds number exceeds the critical value of about 47.5 for the experiments reported by Dupin and Teissié-Solier [53], the *steady* roll-up structures in the wake change radically and give rise to an *unsteady* vortex shedding, with an alternating vortex detachment on each side of the cylinder's main cross-section.

In this case, the unsteady wake exhibits a clearly marked periodic character. Slightly below this critical value, the unsteadiness begins to manifest itself by fluctuations within the attached roll-up structures, and oscillations of both the upstream detachment points and the downstream stagnation point.

In the photograph in Fig. 2.27, the fluid is none other than water seeded with a metallic tracer (aluminum powder) which its reflectivity makes visible under suitable lighting. The plate corresponds to a fixed exposure time, so that the tracer motion is evidenced by segments whose length is proportional to the module of the local velocity vector and directed along the local tangent to the streamlines.

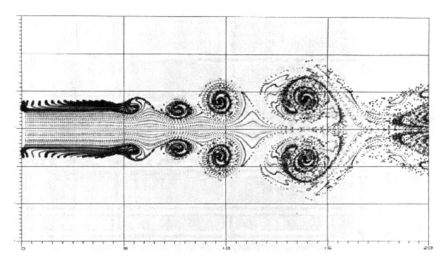

Fig. 2.28 Streakline visualization of the numerical simulation of a plane jet at a Reynolds number of 1000, from Sevrain & Braza [19], 1989 courtesy of IMFT

2.5.2.4 Streaklines in an Unsteady Flow

From the beginning of computational fluid mechanics, scientists have been facing a *representation* problem of the computed flow fields in a form which could be *naturally meaningful* to the non-specialist. In this respect, streaklines have proved to be a very valuable tool for such a visualization goal, as outlined in the following illustration of an unsteady jet flow.

From the experimental point of view, a jet is generated by injecting mass and momentum flow-rates into an infinite atmosphere at rest through a circular orifice or rectangular slot. In the simulation shown in Fig. 2.28, the jet is exiting from a slot whose width d and lengthspan L are such that $L \gg d$, in order to give the flow a two-dimensional plane character. Similarly to the wake flow, it is known that various vortex structures can form in a jet, depending on the Reynolds number value at the exit $R_e = U_0 d/\nu$. In the computed jet by Braza and Sevrain [19] by solving the two-dimensional unsteady Navier-Stokes equations (Cf. Chap. 4), the exit Reynolds number is $R_e = 1000$.

To visualize the results of this simulation, the streakline technique is used. Coloured particle emission points are located within the jet, on either side of the axis and slightly apart from the slot. It is thus possible to observe several characteristic mechanisms of this type of flow *(i)* the external fluid entrainment immediately downstream of the exit section, *(ii)* jet boundary instability (oscillation), *(iii)* primary eddy roll-up and *(iv)* vortex pairing of the downstream moving structures.

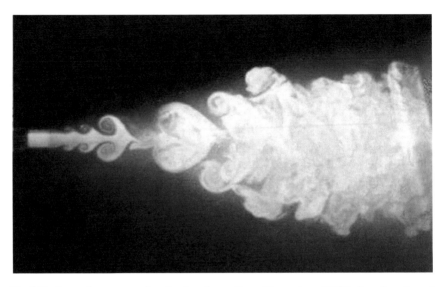

Fig. 2.29 Laser slice tomography of a plane jet at a Reynolds number of 1700, from Sevrain et al. [146], 1990 courtesy of IMFT

2.5.2.5   Fluid Lines and Streak-Surfaces in Unsteady Motions

The numerical simulation results can be directly compared to the experiment as long as the visualization process is based on the same technique (tagging with a passive contaminent). In Fig. 2.29 the global visualization of a plane jet at a Reynolds number $R_e = 1700$ is experimentally obtained by the laser tomography technique.

The provided information is on the fluid-lines and streaksurfaces as previously mentioned (Fig. 2.26). Since the fluid here is a gas, the tagging is achieved by seeding the jet with fine solid particles. The visualization is obtained by taking a photograph of the flow lit by a coherent light sheet.

While confirming the observations from the numerical simulation (Fig. 2.28), this visualization also highlights an additional phenomenon, *i.e.*, the three-dimensionalisation of the flow far downstream, with a gradual coherence loss of the eddy structures and the evolution towards a turbulent agitation state.

2.5.3   *Examples of Visualization by Property*

In all the above examples, the visualization technique is intrusive in the sense that it requires the introduction of a *non-present* property in the *natural* flow. There are situations where this constraint can be avoided by taking advantage of an intrinsic motion characteristic. A widespread situation in aerodynamics is that of gas motion with high enough variations in temperature and pressure to induce a change in density which can modify the fluid refraction index.

Several, more or less related, techniques exist to visualize the flow field of fluids with a density gradient along an optically fixed direction: the shadowgraph method, the Schlieren technique, strioscopy, differential interferometry, etc. They make it possible, for example, to reveal low-speed thermal convection structures, shock waves in supersonic flow or the mixing phenomenology of different gases. An illustration of the latter type is given in Fig. 2.30.

It is a plane jet whose Reynolds number at the exit is $R_e = 4\,000$. The density variation is obtained by heating the effluent. Despite the difference in the value of the Reynolds number with that of the configuration described in Sect. 2.5.2.5, the influence of the visualization mode on the informative content should be noted. Thus, for instance, the methods related to the Schlieren system provide information that cannot be directly interpreted in terms of trajectory, streamline or fluid line. In addition, they are *global* visualizations, in the sense that the light ray diffraction due to the variation in the fluid index is *integrated* across the whole thickness of the fluid slice undergoing the density variation.

Fig. 2.30 Differential interferometry visualization of a plane jet at an exit Reynolds number of 4000 Sevrain et al. [146], 1990 courtesy of IMFT

Chapter 3
The Fundamental Balances of a Fluid Motion

Abstract The focus of this chapter is to deduce the equations which directly result from the application, to a moving fluid, of the laws and general principles of mechanics and thermodynamics. A *Lagrangian approach* is adopted where the flow field functions are described by using the *Eulerian* variables. The *mass conservative* nature of the physical system under consideration is addressed in the first place. Then, the *fundamental equation of Dynamics* and the *first and second laws of Thermodynamics* are applied for deriving the momentum, kinetic energy, internal energy and entropy transport equations.

3.1 Mass Conservation

3.1.1 The Global Formulation

Let us consider, at a reference time τ, a *material fluid domain* of finite expansion $\mathcal{D}(\tau)$ including a given set of fluid particles in any flow. During its motion, this domain will be changing in position, shape and size, but will remain composed, according to the Lagrangian tracking mode, of the same fluid particles. The fluid mass within the domain therefore remains unchanged at any time, which is expressed as:

$$\frac{\mathrm{d}}{\mathrm{d}t}\left[\iiint_{\mathcal{D}(t)} \rho(x,y,z,t)\, dv\right] = 0.$$

In the previous integral, $\rho(x,y,z,t)$ is the local density expressed in Eulerian variables (x,y,z,t) and $dv = dxdydz$ the volume element expressed using the same variables. By applying the material derivative relation of a triple integral—see Eqs. (1.60) and (1.61)—the immediate result is:

$$\iiint_{\mathcal{D}(t)} \left[\frac{\mathrm{d}\rho}{\mathrm{d}t} + \rho\frac{\partial U_i}{\partial x_i}\right] dv \equiv \iiint_{\mathcal{D}(t)} \left[\frac{\partial \rho}{\partial t} + \frac{\partial(\rho U_i)}{\partial x_i}\right] dv = 0. \qquad (3.1)$$

© The Author(s), under exclusive license to Springer Nature Switzerland AG 2022
P. Chassaing, *Fundamentals of Fluid Mechanics*,
https://doi.org/10.1007/978-3-031-10086-4_3

3.1.2 The Local Expression

By taking the limit $\mathcal{D} \to 0$, in terms of the mesoscopic *fluid particule* concept previously defined in Chap. 1, Sect. 1.2.3.1, we deduce from Eq. (3.1) the *local* mass balance equation, called the *continuity equation*:

$$\boxed{\frac{\mathrm{d}\rho}{\mathrm{d}t} + \rho \frac{\partial U_i}{\partial x_i} \equiv \frac{\partial \rho}{\partial t} + \frac{\partial (\rho U_i)}{\partial x_i} = 0} \qquad (3\text{-}2\mathrm{a})$$

or in vector form:

$$\boxed{\frac{\mathrm{d}\rho}{\mathrm{d}t} + \rho \, div\mathbf{V} \equiv \frac{\partial \rho}{\partial t} + div\,(\rho\mathbf{V}) = 0} \qquad (3\text{-}2\mathrm{b})$$

HISTORICAL NOTE—**Euler's continuity equation**—The equation which we have just introduced was formulated almost identically (except for the density notation) by Euler [56], in 1757. Its demonstration remains highly modern. Euler is also the person who inspired its current name: «*This is therefore a very remarkable condition, which already establishes a certain relation between the three speeds* [along] x, y & z, *with regard to the density of the fluid q…Having been provided by the consideration of the fluid continuity, this formula already involves some kind of ratio which should prevail between the quantities u, v, w & q.*»
The particular case of the incompressible fluid, where the relation reduces to the condition of a solenoidal velocity field (see next section), is also addressed by Euler. Similar results are also found in Lagrange [85], in 1781 under the name of «*density equation*» .

3.1.3 The Isovolume Condition and the Incompressible Fluid Assumption

We mentioned in the first Chapter Sect. 1.6.7.1, that the *isovolume* character of the motion[1] is simply stated by the condition $div\mathbf{V} = 0$. We will now investigate if this statement is equivalent to $\rho = C^t$ *anywhere in the fluid* and *at any time*. As the previous relation clearly shows, it is obvious that:

$$\rho(x, y, z, t) = C^t \implies div\mathbf{V} = 0.$$

It is therefore the reciprocal proposal which we should examine, which, to simplify, we will do by considering a stationary flow regime. In this case, relation (3-2b) results

---

[1] It should be recalled that such a motion can equally be qualified as isochoric, isometric or simply as a constant-volume flow.

Fig. 3.1 Density
stratification in an isovolume
flow of an inhomogeneous
fluid

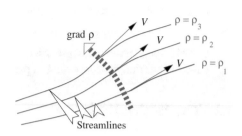

in:

$$div\,(\rho\mathbf{V}) \equiv \rho div\,\mathbf{V} + \mathbf{V}.\mathbf{grad}\,\rho = 0,$$

which, under the isovolume condition reduces to $\mathbf{V}.\mathbf{grad}\,\rho = 0$.
Discarding the trivial case $\mathbf{V} = \mathbf{0}$, this relation leads us to consider two physical
interpretations:

(i) the density ρ is constant throughout the fluid ($\mathbf{grad}\,\rho = 0$);
(ii) the variation in density is, at any point, orthogonal to the velocity vector at that
point.

This second case is that of a density stratified flow where, as outlined in Fig. 3.1, the
density can vary from one streamline to the next.
Geofluid motions provide many examples of stratified flows, owing to salinity and/or
temperature in the ocean, temperature and humidity in the atmosphere.
Thus, in an *isovolume* evolution, the density of *each fluid particle* is unchanged
along its trajectory ($\frac{d\rho}{dt} = 0$), which does not mean that the density is the same
throughout the fluid. The latter situation, which is that of classical hydrodynamics,
is only fulfilled by assuming a *homogeneous* density environment.
In addition, to remove any ambiguity, we will state, by definition, the following
equivalence:

$$\text{Isovolume } evolution \text{ and homogeneous } fluid \iff \begin{cases} div\mathbf{V} = 0 \\ \text{and} \\ \rho = \rho_0 = C^t \end{cases} \quad (3.3)$$

TERMINOLOGY The property $\rho(x, y, z, t) = C^t$ states that the fluid density is the
same throughout the flow field and at any time. It should not be mistaken for
the condition $\frac{d\rho}{dt} = 0$, as we have just demonstrated. A flow for which the
latter is satisfied is sometimes referred to as *incompressible*, a qualifier which
may be somewhat inappropriate, with respect to the various sources of density
variation. Moreover, the *fluid incompressibility*, in terms of any compressibility
coefficient equal to zero (see Chap. 1 Sect. 1.3.4.1), means that variations in

pressure have no effect on the density *at any point in the field*, while relation $\dfrac{\mathrm{d}\rho}{\mathrm{d}t} = 0$ applies only to pressure variations taken when moving with *each* fluid particle. Finally, the incompressibility feature is by no means exclusive of other density variation factors, owing in particular to temperature or composition.

3.2 The Motion Equations

3.2.1 *The Momentum Equation in Integral Form*

The Fundamental Equation of Dynamics applied to a material fluid domain which we follow throughout its motion states that

> The material derivative of the momentum torsor is equal to the external forces torsor applied to the domain.

- The elements of the momentum torsor are:

  – the resultant force $\displaystyle\iiint_{\mathcal{D}(t)} \rho \mathbf{V} dv$,

  – the resultant moment at point O $\displaystyle\iiint_{\mathcal{D}(t)} (\mathbf{OM} \wedge \rho \mathbf{V})\, dv$.

- With regard to the external forces applied to the fluid domain $\mathcal{D}(t)$, we will distinguish two types, as shown in Fig. 3.2, depending on whether they are applied to a point:

  – inside the *volume*, viz at point M;

  – on the *surface*, viz at point P.

  (a) The *volume* or *body forces* are remotely applied to any material element located at point M in the volume $\mathcal{D}(t)$. In this work, they will reduce solely to gravity forces.

Fig. 3.2 Schematic of the external forces in the momentum balance

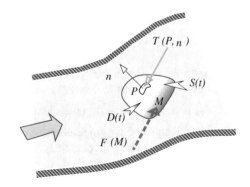

We will denote by $\mathbf{F}(M, t)$ the value, *per unit mass*, of this force at point M and time t.

By referring to $dv(M)$ as the volume element at any point M, the external body force resultant is expressed by :

$$\iiint_{\mathcal{D}(t)} \rho(M, t)\, \mathbf{F}(M, t)\, dv(M) .$$

(b) The *surface forces* are applied from the environment by *contact* through any surface element $\mathcal{S}(t)$ surrounding any point P. Accordingly, it is clear that this force depends, at a given time, on both the point location and the surface element orientation around this point. We will thus represent by $d\mathbf{F}_S(P, \mathbf{n}, t)$ the elementary value of this force acting on a facet centered at point P, of unit normal $\mathbf{n}$ and of elementary area $d\sigma$.

By dividing this force by the area on which it is applied, we define the *local stress vector*:

$$\mathbf{T}(P, \mathbf{n}, t) = \frac{d\mathbf{F}_S(P, \mathbf{n}, t)}{d\sigma} . \qquad (3.4)$$

The resultant external surface force is therefore expressed as:

$$\iint_{\mathcal{S}(t)} d\mathbf{F}_S(P, \mathbf{n}, t) = \iint_{\mathcal{S}(t)} \mathbf{T}(P, \mathbf{n}, t)\, d\sigma .$$

In application of the fundamental dynamic equation, and using Reynolds' formula,— see Chap. 1, Eq. (1.64), we obtain the following two vectorial equations:

$$\text{Resultant Force} \quad \iiint_{\mathcal{D}(t)} \rho \frac{d\mathbf{V}}{dt} dv = \iiint_{\mathcal{D}(t)} \rho \mathbf{F} dv + \iint_{\mathcal{S}(t)} \mathbf{T} d\sigma \qquad (3.5)$$

$$\text{Resultant Moment} \quad \iiint_{\mathcal{D}(t)} \rho \frac{d}{dt}(\mathbf{OM} \wedge \mathbf{V}) dv = \iiint_{\mathcal{D}(t)} (\mathbf{OM} \wedge \rho \mathbf{F}) dv + \iint_{\mathcal{S}(t)} (\mathbf{OM} \wedge \mathbf{T}) d\sigma$$

$$(3.6)$$

3.2.2 The Surface Force Analysis

3.2.2.1 The Stress-Tensor Definition

At any point P of the surface in an orthonormal Cartesian coordinate system, the vectorial surface element $\mathbf{n}\,d\sigma$ has three independent components, along the three base vectors $\mathbf{i}, \mathbf{j}, \mathbf{k}$. To each of these components is associated a normal facet to which an elementary force applies, which we will note in short and respectively $d\mathbf{F}_1, d\mathbf{F}_2, d\mathbf{F}_3$. By definition of the stress vector, Eq. (3.4), we can write :

$$d\mathbf{F}_1 = \mathbf{T}_1\,d\sigma, \qquad d\mathbf{F}_2 = \mathbf{T}_2\,d\sigma, \qquad d\mathbf{F}_3 = \mathbf{T}_3\,d\sigma,$$

by denoting the three stress vectors respectively by $\mathbf{T}_1, \mathbf{T}_2$ and $\mathbf{T}_3$.

By expanding these vectors into components in the reference frame $(\mathbf{i}, \mathbf{j}, \mathbf{k})$, we will write:

$$\mathbf{T}_1 = \sigma_{11}\mathbf{i} + \sigma_{21}\mathbf{j} + \sigma_{31}\mathbf{k},$$
$$\mathbf{T}_2 = \sigma_{12}\mathbf{i} + \sigma_{22}\mathbf{j} + \sigma_{32}\mathbf{k},$$
$$\mathbf{T}_3 = \sigma_{13}\mathbf{i} + \sigma_{23}\mathbf{j} + \sigma_{33}\mathbf{k},$$

or, in index notation $\qquad \mathbf{T}_m = \sigma_{lm}\,\mathbf{e}_l, \ m = 1, 2, 3,$ $\qquad\qquad$ (3.7)

using Einstein's summation convention on the l index and now referring to the unit vectors of the reference frame as $\mathbf{e}_l, \ l = 1, 2, 3$.

Relation (3.7) introduces nine elements $\sigma_{lm}, (l = 1, 2, 3)$ and $(m = 1, 2, 3)$ which, as we will see in the next paragraph, are the components of a tensor, called the *stress tensor*. In the previous index notation, the following naming convention, as shown in Fig. 3.3, is used:

1. the first index, or left index (l), identifies the component of the stress vector in the reference frame $(\mathbf{i}, \mathbf{j}, \mathbf{k}\ )$;
2. the second index, or right index (m), identifies the facet by the direction of its unit normal vector.

The components σ_{11}, σ_{22} and σ_{33} are referred to as *normal* stresses, the other components as *tangential* or *shear* stresses.

3.2.2.2 The Stress-Tensor Properties

For the further development of this work, we will keep in mind the following two major results, which apply at any time t:

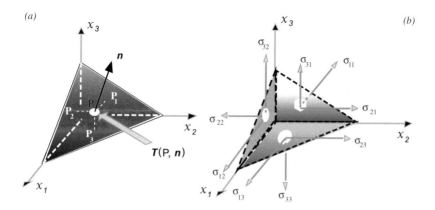

Fig. 3.3 Stress vector (**a**) and stress tensor components on a facet of any orientation (**b**)

(a) It is equivalent to describing the external surface forces by means of (i) a stress *vector* field $\mathbf{T}(P, \mathbf{n})$ or (ii) a second order stress *tensor* field $\overline{\overline{\sigma}}(P)$, on account of the relation:

$$T_i(P, n_j) = \sigma_{ij}(P) . n_j(P) \tag{3.8}$$

where n_j ($j = 1, 2, 3$) are the direction cosines of the local normal vector at the surface.

(b) The stress tensor $\overline{\overline{\sigma}}(P)$ is symmetric, viz $\quad \sigma_{ij} = \sigma_{ji}.$ (3.9)

Proof The stress tensor (i)—Existence — To demonstrate Eq. (3.8) and justify the tensorial nature of $\overline{\overline{\sigma}}$, we apply Eq. (3.4) to an elementary tetrahedral domain, see Fig. 3.3. We have thus:

$$\iiint_{\mathcal{D}(t)} \rho \frac{d\mathbf{V}}{dt} dv = \iiint_{\mathcal{D}(t)} \rho \mathbf{F} dv + \iint_{\Sigma'} \mathbf{T}(P, \mathbf{n}') ds - \sum_{l=1}^{l=3} \iint_{\Sigma_l} \mathbf{T}(P_l, \mathbf{n}_l) ds,$$

where it is recalled that the normal vector is positively oriented when pointing *outwards*. Σ' refers to the slanted facet and Σ_l, $l = 1, 2, 3$ to each of the three facets of respective normals $\mathbf{i}$, $\mathbf{j}$ and $\mathbf{k}$. By taking into account that this is an elementary domain, the previous relation becomes, within the mesoscopic limit $\mathcal{D}(t) \rightarrow 0$:

$$\rho \frac{d\mathbf{V}}{dt} dv = \rho \mathbf{F} dv + \mathbf{T} ds - \mathbf{T}_1 ds_1 - \mathbf{T}_2 ds_2 - \mathbf{T}_3 ds_3.$$

Now, the surface elements are linked by $ds_1 = \mathbf{n}.\mathbf{i}\, ds, ds_2 = \mathbf{n}.\mathbf{j}\, ds, ds_3 = \mathbf{n}.\mathbf{k}\, ds,$ or again in indexed form, by introducing the direction cosines ($n_l = \mathbf{n}.\mathbf{k}_l$, $l = 1, 2, 3$) of the unit normal vector $\mathbf{n}$:

$$ds_l = n_l\,ds, \ \ l = 1, 2, 3.$$

With these area elements, the momentum equation becomes:

$$\mathbf{T}ds = \mathbf{T}_1 n_1 ds + \mathbf{T}_2 n_2 ds + \mathbf{T}_3 n_3 ds + \left(\rho\frac{d\mathbf{V}}{dt} - \rho\mathbf{F}\right)dv,$$

that is $\mathbf{T} = \mathbf{T}_1 n_1 + \mathbf{T}_2 n_2 + \mathbf{T}_3 n_3 + \mathcal{O}\,(L)$, where the last term order is that of a length scale L which is representative of the surface element $(ds \propto L^2)$ and volume element $(dv \propto L^3)$.

By taking the mesoscopic limit $L \to 0$, the previous expression shows that $\mathbf{T} = \mathbf{T}_\alpha n_\alpha$. Then using Eq. (3.7) we deduce that $\mathbf{T} = n_\alpha \sigma_{l\alpha} \mathbf{e}_l$. This last result makes it possible to define the components T_l of the stress vector by Eq. (3.8) and justifies the second order tensorial character of the operator $\overline{\overline{\sigma}}$, linking two vectors $\mathbf{n}$ and $\mathbf{T}$.

Proof The stress tensor (ii)—Symmetry — To ascertain the symmetry property—Eq. (3.9)—we will apply the fundamental dynamic equation to the moments of the forces acting on the infinitesimal domain, Eq. (3.6). Denoting by ϵ_{ijk} the alternating pseudo-tensor,[2] this equation becomes:

$$\iiint_{\mathcal{D}(t)} \rho\epsilon_{ijk}x_j\left(\frac{dU_k}{dt} - F_k\right)dv = \iint_{\mathcal{S}(t)} \epsilon_{ijk}x_j T_k ds,$$

or again by using Eq. (3.8) $\displaystyle\iiint_{\mathcal{D}(t)} \rho\epsilon_{ijk}x_j\left(\frac{dU_k}{dt} - F_k\right)dv = \iint_{\mathcal{S}(t)} \epsilon_{ijk}x_j\sigma_{kl}n_l ds$.

By applying the divergence formula to the surface integral the result is:

$$\iiint_{\mathcal{D}(t)} \rho\epsilon_{ijk}x_j\left(\frac{dU_k}{dt} - F_k\right)dv = \iiint_{\mathcal{D}(t)} \frac{\partial\left(\epsilon_{ijk}x_j\sigma_{kl}\right)}{\partial x_l}dv.$$

By taking the mesoscopic limit $\mathcal{D} \to 0$, we deduce

$$\rho\epsilon_{ijk}x_j\left(\frac{dU_k}{dt} - F_k\right) = \frac{\partial\left(\epsilon_{ijk}x_j\sigma_{kl}\right)}{\partial x_l}.$$

However, as we will see below (Sect. 3.2.2.4), the momentum equation of the fluid motion can be locally expressed as:

$$\rho\frac{dU_k}{dt} = \rho F_k + \frac{\partial\left(\sigma_{kl}\right)}{\partial x_l}.$$

---

[2] Also called the permutation symbol.

After substituting this result in the equation for the moment there remains:

$$\epsilon_{ijk} x_j \frac{\partial (\sigma_{kl})}{\partial x_l} = \frac{\partial \left(\epsilon_{ijk} x_j \sigma_{kl}\right)}{\partial x_l} \,,$$

or in the end $\sigma_{kl}\, \epsilon_{ijk} \dfrac{\partial x_j}{\partial x_l} = 0$. Now, the left hand side of this equality is always identically zero, unless $j = l$, so the condition is still written $\epsilon_{ijk}\sigma_{jk} = 0$ which, given the properties of the alternating pseudo tensor is still expressed by $\sigma_{jk} - \sigma_{kj} = 0$ whatever the value of $k = 1,\ 2,\ 3$, QED.

3.2.2.3 The Mechanical and Thermodynamic Pressure

In the absence of motion, the surface forces existing within a fluid medium are pressure forces normal to any surface element. This pressure is none other than the *hydrostatic* pressure for a heavy fluid at rest, or more generally the *thermodynamic* pressure for any environment in an equilibrium state. The stress tensor is then *spherical* and is therefore written as:

$$\overline{\overline{\sigma}}^{\,0} = -P\overline{\overline{I}} \qquad \text{or} \qquad \sigma_{ij}^0 = -P\delta_{ij} \,.$$

When the fluid is in motion, we can apply:

$$\overline{\overline{\sigma}} = -P\overline{\overline{I}} + \overline{\overline{\tau}}, \qquad \text{or} \qquad \sigma_{ij} = -P\delta_{ij} + \tau_{ij} \,, \tag{3.10}$$

denoting by P the *local* pressure in terms of usual thermodynamics and still referred to as the *static* pressure. The tensor $\overline{\overline{\tau}}\ (\tau_{ij})$ then identifies the stresses specifically due to the fluid motion.

From the tensor trace σ_{ij}, we can also introduce another quantity P', acting as a pressure. It is called the *mechanical pressure* and defined as :

$$P' = -\frac{1}{3}\sigma_{ii} = -\frac{\sigma_{11} + \sigma_{22} + \sigma_{33}}{3} \,.$$

It then appears that $P' = P - \dfrac{\tau_{11} + \tau_{22} + \tau_{33}}{3}$ so that the mechanical and thermodynamic pressures are not *a priori* the same. This will be discussed further in Chap. 4 Sect. 4.3.1.2.

HISTORICAL NOTE— **The hydrostatic pressure according to Barré de Saint-Venant**—The '*static*' qualifier (here for the pressure and later on for the temperature) may be somewhat misleading, as it does not refer to a value in a fluid *at rest*. The static pressure and temperature are also not identified with the *hydrostatic values* measured locally by a probe *moving at the flow velocity*.
One could not be clearer in this regard than Barré de Saint-Venant [3] who, on page 573 of the paper published in 1872 writes:

«It should also be noted that the part of the three normal components $[\frac{1}{3}(\sigma_{11} + \sigma_{22} + \sigma_{33})$, with the present notations]…is not, as some authors believed, their purely hydrostatic part, or what it would be reduced to if the motion suddenly stopped; because gravity, or the vertical weight of water to the free surface, does not solely constitute its intensity, the inertia presently at work also has its part.»

3.2.2.4 The Local Expression of the Momentum Equation

By introducing relation (3.8) in Eq. (3.5) we obtain, in index notations:

$$\iiint\limits_{\mathcal{D}(t)} \rho \frac{\mathrm{d}U_i}{\mathrm{d}t} dv = \iiint\limits_{\mathcal{D}(t)} \rho F_i dv + \iint\limits_{\mathcal{S}(t)} \sigma_{ij}\, n_j\, ds\,.$$

By applying the divergence theorem to the surface integral of the right hand, the result is:

$$\iiint\limits_{\mathcal{D}(t)} \rho \frac{\mathrm{d}U_i}{\mathrm{d}t} dv = \iiint\limits_{\mathcal{D}(t)} \rho F_i dv + \iiint\limits_{\mathcal{D}(t)} \frac{\partial}{\partial x_j}\left(\sigma_{ij}\right) dv\,.$$

After taking the mesoscopic limit, we obtain the *open momentum equation*[3] which governs the local motion of any fluid particle:

$$\boxed{\rho \frac{\mathrm{d}U_i}{\mathrm{d}t} = \rho F_i + \frac{\partial \sigma_{ij}}{\partial x_j}} \tag{3.11}$$

Owing to the continuity equation, Eq. (3-2a), we can rewrite this result in a conservative form:

$$\boxed{\frac{\partial\left(\rho U_i\right)}{\partial t} + \frac{\partial\left(\rho U_i U_j\right)}{\partial x_j} = \rho F_i + \frac{\partial \sigma_{ij}}{\partial x_j}}\,. \tag{3.12}$$

3.3 The Energy Balance Equations

3.3.1 The Kinetic Energy Theorem

3.3.1.1 The Local Formulation

The equation governing the local kinetic energy is obtained by simple scalar multiplication of the momentum Eq. (3.11) by U_i. Thus the result is:

---

[3] This qualifier refers to the fact that, at this stage, the surface forces are not explicitly stated, so that there are excess unknowns in the equation.

$$\rho \frac{d}{dt} \left(\frac{1}{2} U_i U_i \right) = \rho U_i F_i + U_i \frac{\partial \sigma_{ij}}{\partial x_j} , \tag{3.13}$$

where the right hand side includes the power of all forces, both internal and external, applied to the system. We will therefore introduce:

$$\mathcal{P}^v_{ext} + \mathcal{P}^s_{ext} + \mathcal{P}^v_{int} + \mathcal{P}^s_{ext} = \rho U_i F_i + U_i \frac{\partial \sigma_{ij}}{\partial x_j} , \tag{3.14}$$

where the terms of the left hand side are the power *per unit volume* of the external and internal body and surface forces respectively.

To identify such various contributions, we will begin by providing details related to the power of the *external* forces alone. To do so, we return to the finite domain $\mathcal{D}(t)$ introduced in the momentum balance analysis (see Fig. 3.3).

We then have, for the respective powers of these forces on the domain:

- the power of the **external** *body forces* $\mathbb{P}^v = \displaystyle\iiint\limits_{\mathcal{D}(t)} \rho U_i F_i dv$,

- the power of the **external** *surface forces* $\mathbb{P}^s = \displaystyle\iint\limits_{S(t)} U_i T_i ds$.

Owing to relation (3.8) and by using the divergence theorem, the power of the external surface forces is still written as:

$$\mathbb{P}^s = \iint\limits_{S(t)} U_i \sigma_{ij} n_j ds \equiv \iiint\limits_{\mathcal{D}(t)} \frac{\partial}{\partial x_j} \left(U_i \sigma_{ij} \right) dv .$$

After taking the mesoscopic limit $\mathcal{D}(t) \to 0$, we obtain the expressions of the *local* powers, per unit volume, of the *external* body and surface forces, namely:

$$\boxed{\mathcal{P}^v_{ext} = \rho U_i F_i} \tag{3.15a}$$

$$\boxed{\mathcal{P}^s_{ext} = \frac{\partial \left(\sigma_{ij} U_i \right)}{\partial x_j}} \tag{3.16a}$$

Using these latter relations in Eq. (3.14) we deduce that:

$$\boxed{\mathcal{P}^v_{int} = 0} \tag{3.15b}$$

$$\boxed{\mathcal{P}^s_{int} = -\sigma_{ij}\frac{\partial U_i}{\partial x_j}}$$ (3.16b)

Now, by introducing relation (3.10) in the expressions of the power of the surface forces, we can explicit the role of the pressure. We have indeed:

$$\mathcal{P}^s_{ext} = -\frac{\partial\left(P\delta_{ij}U_i\right)}{\partial x_j} + \frac{\partial\left(\tau_{ij}U_i\right)}{\partial x_j} \equiv -\frac{\partial\left(PU_i\right)}{\partial x_i} + \frac{\partial\left(\tau_{ij}U_i\right)}{\partial x_j},$$

$$\mathcal{P}^s_{int} = +P\delta_{ij}\frac{\partial\left(U_i\right)}{\partial x_j} - \tau_{ij}\frac{\partial U_i}{\partial x_j} \equiv +P\frac{\partial\left(U_i\right)}{\partial x_i} - \tau_{ij}\frac{\partial U_i}{\partial x_j}.$$

Finally, the kinetic energy balance can be put in the following form:

$$
\begin{aligned}
\rho\frac{\mathrm{d}\left(\frac{1}{2}U_iU_i\right)}{\mathrm{d}t} &= \rho U_i F_i \dots\dots\dots\text{.external body force power} \\
&- \frac{\partial}{\partial x_i}\left(PU_i\right)\dots\dots\text{ external pressure force power} \\
&+ \frac{\partial}{\partial x_j}\left(\tau_{ij}U_i\right)\dots\dots\text{ external viscous force power} \\
&+ P\frac{\partial U_i}{\partial x_i}\dots\dots\dots\text{internal pressure force power} \\
&- \tau_{ij}\frac{\partial U_i}{\partial x_j}\dots\dots\dots\text{internal viscous force power}
\end{aligned}
$$ (3.17)

3.3.1.2 The Physical Interpretation

First of all, we can notice that for an *isovolume* situation, the power of the *internal* pressure forces is identically zero since:

$$-P\frac{\partial U_i}{\partial x_i} \equiv -P\,div\mathbf{V} = 0.$$

We can then note that, as a scalar quantity, the expression of the power of the internal surface forces is independent from the index's notation choice. We can therefore also write, by permuting the indices i and j :

$$\mathcal{P}^s_{int} = -\sigma_{ji}\frac{\partial U_j}{\partial x_i} \equiv -\sigma_{ij}\frac{\partial U_j}{\partial x_i},$$

the last identity resulting directly from the symmetry of the stress tensor Eq. (3.9). Substituting this expression in Eq. (3.16b) we finally obtain

$$\mathcal{P}^s_{int} = -\sigma_{ij}\frac{\partial U_i}{\partial x_j} \equiv -\sigma_{ij}\frac{\partial U_j}{\partial x_i} = -\frac{1}{2}\sigma_{ij}\left(\frac{\partial U_i}{\partial x_j} + \frac{\partial U_j}{\partial x_i}\right) \equiv -\sigma_{ij}S_{ij}. \qquad (3.18)$$

We can therefore state that:

> With no fluid particle deformation during the motion, the internal surface force power is zero.

ADDITIONAL INFORMATION—**A mechanical analogy of the internal friction forces involved in the energy balance of a moving fluid**—In an attempt to illustrate the dual characteristic of (*i*) the *internal* surface forces acting by (*ii*) the *deformation* within the domain, we will take an analogy inspired by deformable solid mechanics. For this purpose, we consider the system consisting of an assembly of two equal masses, connected together by a spring of negligible mass.

This assembly is entirely surrounded by a perfectly elastic envelope, without stiffness nor mass, playing no other role than that of distinguishing the *interior* of the system from its *environment*, see Fig. 3.4. The entire system is subjected to a constant gravity field. At the origin of time, the system is supposed to possess a double potential energy, (*i*) from the gravity, by an elevation above the ground level and (*ii*) from a tension-compression of the spring, whose deformation is set in a direction orthogonal to that of gravity. For an *external observer*, the release of this double energy causes, at a later stage, a falling action along the vertical with a periodical variation of the envelope's extension along the horizontal, which reflects the oscillations of the *internal* masses connected to the spring.

It is assumed that, when in contact with the ground, the system perfectly rebounds elastically and that the friction of the envelope with the environment can be disregarded.

If the spring connecting the two internal masses is perfectly elastic, the initial potential energy of the system will be preserved during its evolution (conservative external forces and absence of internal damping). On the other hand, if the spring motion is damped, the system will evolve towards a fixed elongation (no further oscillations) while continuing its rebounds indefinitely. This evolution results from the orthogonality of the motion directions along which the system's two energy modes are distributed.

If the system, with a damped spring, is now arranged so that its elongation fluctuates vertically, the whole motion (the center of gravity movement and the oscillations around this centre) will eventually stop completely after a sufficient time. Indeed, since the internal forces are no longer conservative, the interchanges between the two energy modes at each rebound cannot take place with the same total amount. In other words, a part of the *total energy* is irreversibly dissipated *within the system*, even though the *external forces* remain conservative. For this analogy, in the evolution of the vertically positioned mechanical system, the damping coefficient of the spring plays a role similar to that of the *viscosity* (internal friction) in a real fluid flow.

Fig. 3.4 The deformable
system for the mechanical
analogy of the internal force
power

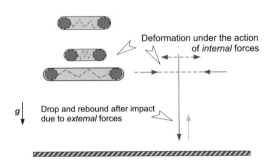

An important result for practical applications can be deduced from the local kinetic
energy balance when the following assumptions are fulfilled:

3.3.1.3   Bernoulli's Theorem

- a stationary motion;
- an inviscid fluid;
- a constant density[4];
- conservative body forces (existence of a body force potential).

Under these conditions and using the conservative form of relation (3.17) we obtain:

$$\frac{\partial}{\partial x_j}\left[\left(\rho\frac{U_iU_i}{2}\right)U_j\right] = -\rho U_i\frac{\partial \mathcal{F}}{\partial x_i} - \frac{\partial (PU_i)}{\partial x_i}\,,$$

where we have put $F_i = -\frac{\partial \mathcal{F}}{\partial x_i}$, denoting by $-\mathcal{F}$ the body force potential function.
Returning to vector notations, the previous relation is still written as:

$$div\left(\rho\frac{V^2}{2}\mathbf{V}\right) = -\rho\mathbf{V}.\boldsymbol{grad}\,\mathcal{F} - div\,(P\mathbf{V})\,,$$

that is $\mathbf{V}.\boldsymbol{grad}\left(\rho\frac{V^2}{2}\right) + \rho\frac{V^2}{2}div\,\mathbf{V} = -\rho\mathbf{V}.\boldsymbol{grad}\,\mathcal{F} - \mathbf{V}.\boldsymbol{grad}\,P - Pdiv\,\mathbf{V}.$

Since the evolution is isovolume, the final result is:

$$\mathbf{V}.\boldsymbol{grad}\left[\rho\frac{V^2}{2} + P + \rho\mathcal{F}\right] = 0.$$

Thus $\mathcal{H} \equiv P + \rho\frac{V^2}{2} + \rho\mathcal{F} = C'$ **along a streamline** .   (3.19)

We will therefore state:

---

[4] We will see later that this result applies more generally to a *barotropic evolution*, for which the
density varies only in relation to the pressure.

> In a stationary motion of an inviscid incompressible fluid under the influence of conservative body forces with $(-\mathcal{F})$ as potential, the following quantity
>
> $$\mathcal{H} \equiv P + \rho\frac{V^2}{2} + \rho\mathcal{F}$$
>
> remains constant on any streamline.

This result is known as Bernoulli's theorem. We will specify **weak** Bernoulli's theorem, because we will have the occasion to demonstrate that another theorem exists, referred to as **strong**, where the conservation of the $\mathcal{H}$ quantity applies to the entire flow field (see Chap. 5 Sect. 5.6).

According to this result, a fluid particle energy consists of three additive contributions (pressure, kinetics and potential). During the motion, any variation in the amount of one of these elements is necessarily counterbalanced by the other two. Hence there is an "ideal energy conversion" along a streamline. We will see in Chap. 5 several examples of how this theorem can be applied.

ADDITIONAL INFORMATION—Practical considerations — In many *aerodynamic* applications, variations in the potential (gravity) energy of the gaseous fluid are generally negligible compared to those in kinetic energy and pressure. Thus Eq. (3.19) reduces to:

$$P + \frac{1}{2}\rho V^2 = C^t \text{ along any streamline}. \tag{3.20a}$$

Now, since all the terms of the previous equation have the dimension of a pressure, the following names are usually used

- P the *static* pressure ;
- $\frac{1}{2}\rho V^2$ the *dynamic* pressure ;
- $P + \frac{1}{2}\rho V^2$ the *total* or *stagnation* pressure, as we will explain at a later stage.

On the other hand, as far as **hydrodynamics** is concerned, gravity forces often play a dominant role. The potential function of the body force is $\mathcal{F} = -gz$, where g is the module of the gravity acceleration and z the vertical coordinate positively oriented upwards. Relation (3.19) is then expressed in the form:

$$\frac{V^2}{2g} + \frac{P}{\rho g} + z = C^t \text{ along a streamline}, \tag{3.20b}$$

where all quantities now have the dimension of a length (*liquid height*).
Therefore, the left hand side of this relation is the total energy of the fluid particles per unit weight which is known as the total head.
We can thus distinguish:

- $z + \frac{P}{\rho g}$ the *piezometric* head;
- $\frac{V^2}{2g} + \frac{P}{\rho g} + z$ the *total* head.

Fig. 3.5 Total head and
piezometric lines of the
pressure flow of an inviscid
heavy fluid

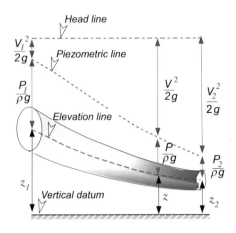

Bernoulli's theorem can then be graphically represented by the variations of these different heights along the circuit, in terms of head and piezometric lines for example. As shown in Fig. 3.5, the result is demonstrated, for a pressure flow, by a horizontal total head line, as a consequence of the absence of dissipation by viscosity, as we will subsequently explain.

HISTORICAL NOTE—**Bernoulli's theorem**—This theorem, which is central to the derivation of many practical results, has its origin in Daniel Bernoulli's treatise on hydrodynamics [10], which was published in 1738. This scientist solved the problem concerning the determination of the flow velocity issued by gravity from an orifice at a vessel bottom. In this regard, he considered horizontal fluid slices, of infinitesimal thickness and applies the principle of 'vis viva' (energy conservation according to the kinetic energy theorem). He reduced this theorem to the conservation of the total mechanical energy: E_c (kinetic) + E_p (potential). He obtained the result by writing the balance between (*i*) '*ascensus potentialis*' the ascent potential, which actually corresponds to the kinetic energy (E_c) and (*ii*) '*decenscus actualis*' *i.e.*, the actual potential energy decrease with height ($-E_p$).

However, the expression linking the velocity and pressure in a moving fluid, as set out by Bernoulli, is not that of the relation which is currently referred to by his name. According to Tokaty [161], the present formulation would have been obtained by Lagrange from the integration of Euler's equations.

Fig. 3.6 Schematic for the application of the first law of thermodynamics to a moving fluid

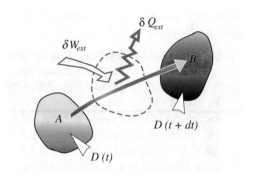

3.3.2 The First Law of Thermodynamics in a Fluid Motion

3.3.2.1 The Material Derivative Formulation

The first law of thermodynamics[5] states that, for a *closed system*, the total energy, *i.e.*, the sum of the internal energy (E_i) and kinetic energy (E_c), is a **state function**, in the sense that its value depends only on the state of the system. Its **variation** when the system changes from a state A to a state B therefore depends only on the difference in the values taken by this function in states A and B. The first law then requires that this variation algebraically accounts for the total **energy exchange**, as the sum total of work (W_{ext}) and heat (Q_{ext}) transferred to or from the system with the environment during the transformation.

This can be expressed in terms of the equivalence:

$$d\,(E_i + E_c)|_A^B = \delta W_{ext}|_A^B + \delta Q_{ext}|_A^B$$

As such, this principle assumes that the initial and final states are both states of *thermodynamic equilibrium*.

To apply this result to a fluid domain being tracked as it moves (Fig. 3.6) we will identify :

$A \rightarrowtail \mathcal{D}\,(t)$ and $B \rightarrowtail \mathcal{D}\,(t + dt)$, and make use of the *local thermodynamic equilibrium* assumption between these two states, see Chap. 1 Sect. 1.2.3.5.

Under this assumption, the first law can be expressed in terms of a *material derivative*, in the form:

$$\frac{d}{dt}\,(E_i + E_c) = \mathcal{P}_{ext}^{\mathcal{W}} + \mathcal{P}_{ext}^{\mathcal{Q}} , \tag{3.21}$$

---

[5] A summary of thermodynamic concepts focusing on the needs of the present work is given in Appendix *A*.

where $\mathcal{P}_{ext}^{W}$ and $\mathcal{P}_{ext}^{Q}$ refer respectively to the mechanical and thermal powers exchanged with the environment.

*T*ERMINOLOGY—**Open system, closed system, isolated system**—Let us recall that, by definition, a system which does not exchange matter with the environment is called a *closed* system. We still sometimes use the term *conservative* system instead of the more accurate *conservative mass* system. Thus, the only possible exchanges with the environment are of an energetic nature, in the form of work and/ or heat. This situation should therefore not be mistaken for that of an *isolated* system for which all exchanges with the environment (mass + energy) are zero. The bounding surface of such a system is therefore necessarily *adiabatic*. When the system freely exchanges energy and matter with the environment, it is referred to as an *open* system.

HISTORICAL NOTE—**The First Law of Thermodynamics**—It is historically extremely difficult, if not impossible, to answer the question "Who actually first discovered the first Law of Thermodynamics?"
There are several reasons for this observation:
(1) When the issue entered the field of scientific current events of the time (the first half of the 19th century), it was approached in different ways by different authors, more or less independently;
(2) The work of each author was probably not generally known or identified (ignorance of anteriorities which is difficult to detemine as being accidental or deliberate);
(3) The issues which are now clearly identified as the first and second laws were not originally identified, since the opinion was still dominated by the interpretation of heat in terms of "caloric";
(4) As far as the first law is solely concerned, it can be seen either as a principle of work-heat equivalence or that of an energy conservation, which gives rise to several possibilities for experimental assessment.

Heat production by mechanical action undoubtedly dates back to the dawn of time, if this phenomenon was the origin of fire by the friction of various solid materials in prehistoric times. To arrive at a conclusion on the quantitative equivalence, the road was long and sometimes controversial, before the "reciprocal" nature of the exchanges between heat and "vis viva" was ascertained and the "universal" value of the conversion factor "quantified", *i.e.*, independentely of the phenomenon/medium used to measure it (the heating by a metal machining, the stirring of a fluid, the compression of a gas...)
In 1847, James Prescott Joule deduced from his work over the past four years the following result: *«Thus, the heat which can increase the temperature of 1 gram of water by 1 degree centigrade is equal to a mechanical force which can raise a weight of 432.1 grams to 1 meter in height»* . The following year, Julius Robert Mayer [100] claimed authorship for the principle of the energy conservation discovery, with reference to an article published in 1842, as well as *«the calculation of the number of heat equivalents »* in a memoir which he published in 1846.

Among other scientists whose names have remained associated with major achievements related to this issue are prominent figures such as Carnot, Helmholtz, Clausius, Rankine, Thomson (or Lord Kelvin). However we also meet less famous personalities.

In terms of anteriority, we will mention Benjamin Rumford [142] who had been measuring since 1798 the amount of heat released when drilling a bronze cylinder, for a known mechanical power.

Finally, for its direct relevance to the fluid motions which we are dealing with here, we will quote Gustave Adolphe Hirn [73] and [74]. In these papers, both published in 1855, this French manufacturer and physicist studied the relative efficiency of various engine lubricants as a function of the pressure, torque, etc. This led him to focus on the friction between two solid surfaces which are separated by a film of viscous fluid, which he calls the «*mediate friction*» , as opposed to friction by direct solid-solid contact («*immediate*» friction). He then notes that, unexpectedly, his measurements lead him to conclude that: «*The absolute amount of the caloric generated by the mediate friction is directly proportional soleley to the mechanical work absorbed by this friction. In addition if we express the work in kilograms of a weight elevation over a height of one meter and the caloric in calories, we find that the ratio of these two numbers is very close to 0.0027* [which corresponds to 0.370 kg.m/cal], *regardless of the velocity, temperature and the lubricant*» .

3.3.2.2 The Energy Balance According to the First Law of Thermodynamics

By applying the Eq. (3.21) to the moving fluid domain (see Fig. 3.6), we have:

$$\frac{d}{dt} \iiint_{\mathcal{D}(t)} (E_i + E_c)\, dv = \iiint_{\mathcal{D}(t)} \left[\mathcal{P}_{ext}^{\mathcal{W}} + \mathcal{P}_{ext}^{\mathcal{Q}} \right] dv. \qquad (3.22)$$

Let us explain the different terms of this equation as follows:

• The *total energy*: denoting by e the internal energy per unit mass of the fluid (m^2/s^2), the total energy of the fluid in the domain at time t is given by :

$$E_i + E_c = \iiint_{\mathcal{D}(t)} \rho \left(e + \frac{U_i U_i}{2} \right) dv. \qquad (3.23a)$$

• The *mechanical power*: this power results from the work of the forces acting both in volume (body forces) and throughout the surface of the domain,

$$\mathcal{P}_{ext}^{\mathcal{W}} = \mathcal{P}_{ext}^{v} + \mathcal{P}_{ext}^{s}.$$

Given the results previously obtained, Eqs. (3.16a) and (3.16b), we immediately obtain:

$$\mathcal{P}_{ext}^{\mathcal{W}} = \rho U_i F_i + \frac{\partial \left(\sigma_{ij} U_i \right)}{\partial x_j} .$$

The total amount for the domain is therefore:

$$\iiint_{\mathcal{D}(t)} \mathcal{P}_{ext}^{\mathcal{W}} dv = \iiint_{\mathcal{D}(t)} \left[\rho U_i F_i + \frac{\partial \left(\sigma_{ij} U_i \right)}{\partial x_j} \right] dv . \tag{3.23b}$$

• The *thermal power*: with regard to heat exchanges with the environment, we consider here the sole mode of transfer by *conduction* across the surface $\mathcal{S}(t)$ bounding the domain. Denoting by **q** the density vector of the heat conduction flux, whose components are q_i $(i = 1, 2, 3)$, this power will be expressed by:

$$- \iint_{\mathcal{S}(t)} q_i n_i d\sigma = - \iiint_{\mathcal{D}(t)} \frac{\partial q_i}{\partial x_i} dv, \tag{3.23c}$$

the negative sign resulting from the choice of the normal vector direction to the surface (*pointing outwards*), since by the thermodynamic convention, the flux is taken *positively* when heat *is supplied to* the system.

After substituting expressions (3.23a), (3.23b) and (3.23c) in equation (3.22), we obtain, by using Reynolds' formula:

$$\iiint_{\mathcal{D}(t)} \rho \left[\frac{d}{dt} \left(e + \frac{U_i U_i}{2} \right) \right] dv = \iiint_{\mathcal{D}(t)} \left[\rho U_i F_i + \frac{\partial \left(\sigma_{ij} U_i \right)}{\partial x_j} \right] dv - \iiint_{\mathcal{D}(t)} \frac{\partial q_i}{\partial x_i} dv . \tag{3.24}$$

Since the integration domain is arbitrary, we deduce, by simply taking the mesoscopic limit $\mathcal{D}(t) \to 0$, the *local* expression of the energy balance, according to the first law of thermodynamics for any fluid particle:

$$\boxed{\rho \frac{d}{dt} \left(e + \frac{U_i U_i}{2} \right) = \rho U_i F_i + \frac{\partial \left(\sigma_{ij} U_i \right)}{\partial x_j} - \frac{\partial q_i}{\partial x_i}} \tag{3.25}$$

3.3.3 The Internal Energy Transport Equation

The material derivative of the kinetic energy being given by Eq. (3.13), we deduce, by simple subtraction from Eq. (3.25) :

$$\rho \frac{de}{dt} = \sigma_{ij} \frac{\partial U_i}{\partial x_j} - \frac{\partial q_i}{\partial x_i} , \tag{3.26}$$

or again by introducing the pressure according to Eq. (3.10):

$$\rho \frac{de}{dt} = -P \frac{\partial U_i}{\partial x_i} + \tau_{ij} \frac{\partial U_i}{\partial x_j} - \frac{\partial q_i}{\partial x_i} \tag{3.27}$$

This last equation highlights the following result:

> In an *isovolume* evolution, the internal energy of a moving fluid can only be changed by the mechanical action of the stresses *other than pressure*, or by heat transfer.

3.3.4 The Entropy Transport Equation

By virtue of Gibbs' relation $T ds = de + P dv$, it is possible to rewrite the first law of thermodynamics using a new state function of the system, s the *specific entropy*. In terms of the transport equation and per unit mass of fluid, the result is:

$$T \frac{ds}{dt} = \frac{de}{dt} + P \frac{dv}{dt} \equiv \frac{de}{dt} - \frac{P}{\rho^2} \frac{d\rho}{dt}.$$

We can now transform the right hand side by taking into account, on the one hand, the continuity Eq. (3-2a) and, on the other hand, the result which has just been given (3.27). This results in:

$$\rho T \frac{ds}{dt} \equiv \rho \frac{de}{dt} - \frac{P}{\rho} \frac{d\rho}{dt} = \tau_{ij} \frac{\partial U_i}{\partial x_j} - \frac{\partial q_i}{\partial x_i}. \tag{3.28}$$

We will now see how to interpret this result in the light of the second law of thermodynamics.

3.3.5 The Second Law of Thermodynamics in a Moving Fluid

3.3.5.1 Irreversibility Sources and the Entropy Concept in a Moving Fluid

In any thermodynamic process where a system exchanges heat and work with the environment, Carnot's principle [26] leads to introducing a distinction with respect to the *nature* of the possible evolutions between the same two states of a given system. In a very broad sense, this principle postulates the existence of a physical criterion making it possible to distinguish the future from the past, a later state from

an earlier state in any *real* evolution. At the very root of this distinction is the notion
of *irreversibility* which, for a moving fluid, can have two distinct origins:

- A **macroscopic source**: The irreversibility is then associated with any *heat transfer* occuring in a fluid due to a non infinitely small thermal disequilibrium. A gas
undergoing strong pressure variations, a highly exothermic chemical reaction
taking place in an existing '*cold*' fluid, the mixture of fluids with high temperature contrasts, etc are such irreversibility types, referred to as a *thermodynamic
cycle* or, more generally, as *extrinsic* irreversibilities. Such irreversibilities necessarily involve one or more *external sources* whose temperature markedly differs
from that of the fluid;
- A **microscopic source**: The origin of such irreversibilities in a fluid is due to the
molecular agitation and therefore associated with the diffusive characteristics of
the fluid at a mesoscopic level. In all *real fluid flows*, the viscosity and thermal
conductivity are at the root of this type of irreversibility, which occurs in a
situation of a quasi-thermal equilibrium motion. This type is referred to as the
intrinsic irreversibility in a fluid motion.

Fluid mechanics, for which the motion study is a central concern, is therefore primarily involved in the second type of irreversibility. It is for this reason that, and whatever
the field of applications (aerodynamics or hydrodynamics), we will express the second law of thermodynamics by explaining that there exists a state function (S), called
the *entropy* such that :

$$\begin{cases} dS = \dfrac{\delta Q_{ext}}{T} + \delta f \\[2mm] \text{with } \delta f \geqslant 0 \end{cases} \tag{3.29}$$

In these two relations, the heat amount which the fluid exchanges with the environment is δQ_{ext}, T is the absolute temperature of the fluid under the conditions of this
exchange and δf stands for any **intrinsic** irreversibility of the fluid motion.

The evolution is called reversible when $\delta f = 0$. It becomes isentropic if it is also
adiabatic ($\delta Q_{rev} = \delta f = 0$, where δQ_{rev} refers to the heat exchange amount in this
case).

HISTORICAL NOTE—The Second Law of Thermodynamics—At the origin of
the second law of thermodynamics is Sadi Carnot's major intuition, according
to which, in order to convert heat into mechanical work, the system should be at
a different temperature from the environment. A thermomachine can therefore
only be operated with a temperature drop, just as a drop in level is required to
operate a hydraulic machine. This idea was put forward in his memoir [26] that
he published in 1824 **previously** to the equivalence principle (first law).
From that date, several alternative formulations have been published (see for
example Bruhat's work [21]), among which we will quote the contributions of

1. Lord Kelvin *"with a system which evolves in a cyclical manner and which
is in contact with only one heat source, no work can be produced"*.

2. Clausius [38] "*heat transfer from a cold body to a hot body never takes place spontaneously or without compensation*".

The entropy concept for a macroscopic system in thermal equilibrium was also identified by Clausius to differentiate a *reversible* cyclic transformation, for which $\oint \dfrac{dQ}{T} = 0$, from an *irreversible* transformation where $\oint \dfrac{\delta Q}{T} < 0$. In the first case, dQ/T is a total (or exact) differential, which allows us to introduce the differential of a *state* function of the system, *i.e.* independent of the transformation and such that

$$dS = \left(\frac{dQ}{T}\right)_{rev}.$$

In the second (irreversible) case, the conclusion is different, which means that the value of $\delta Q/T$ depends on the path followed by the system during the transformation. Nevertheless according to the second law, it can then be stated that:

$$dS = \left(\frac{dQ}{T}\right)_{rev} > \left(\frac{\delta Q}{T}\right)_{irr}.$$

It is essential to remember that, in the previous expressions, T is the absolute temperature of the *source* with which the system exchanges heat, not the temperature of the system.

It is also to Clausius [38] that we owe the very denomination of the term '*entropy*', as he writes, in 1865 : «*...I propose the name entropy, taken from the Greek the transformation. It is with intention that I formed the word entropy as similar as possible to the word energy, since the quantities thus designated have precisely this relationship, according to their physical meaning, that a certain analogy seemed to me conceivable in their denomination.*»

A mere three decades later in 1896 and 1898, Boltzmann [13] gave a new *microscopic* interpretation of the entropy concept.

3.3.5.2 The Local Formulation of the Second Law of Thermodynamics

Assuming the local thermodynamic equilibrium of any fluid particle, we will rewrite relation (3.29) in terms of the material derivative, viz:

$$\frac{ds}{dt} = \frac{\mathcal{P}_{ext}^{Q}}{T} + \frac{\mathcal{P}^{*}}{T}, \qquad (3.30)$$

where s denotes the specific (per unit mass) particle entropy and $\mathcal{P}^{*} \geqslant 0$ the intrinsic irreversibility amount in the motion, whose dimension is a power.

When applied to a finite material fluid domain such as $\mathcal{D}(t)$ (see Fig. 3.6), for which heat exchanges with the environment take place solely by conduction, relation (3.30) results in:

$$\frac{\mathrm{d}}{\mathrm{d}t}\left[\iiint\limits_{\mathcal{D}(t)} \rho s\, \mathrm{d}v\right] \equiv \iiint\limits_{\mathcal{D}(t)} \rho \frac{\mathrm{d}s}{\mathrm{d}t}\, \mathrm{d}v = \iint\limits_{\mathcal{S}(t)} -\frac{q_i n_i}{T}\, \mathrm{d}\sigma + \iiint\limits_{\mathcal{D}(t)} \frac{\mathcal{P}^*}{T}\, \mathrm{d}v.$$

By applying Ostrogradski's formula, we still have

$$\iiint\limits_{\mathcal{D}(t)} \rho \frac{\mathrm{d}s}{\mathrm{d}t}\, \mathrm{d}v = \iiint\limits_{\mathcal{D}(t)} \left[-\frac{\partial}{\partial x_i}\left(\frac{q_i}{T}\right) + \frac{\mathcal{P}^*}{T}\right]\mathrm{d}v,$$

which leads locally to:

$$\boxed{\rho \frac{\mathrm{d}s}{\mathrm{d}t} = -\frac{\partial}{\partial x_i}\left(\frac{q_i}{T}\right) + \frac{\mathcal{P}^*}{T}} \qquad (3.31)$$

3.3.5.3 The General Expression of the Intrinsic Irreversibilities in a Fluid Motion

By using relations (3.28) and (3.31), we are now able to specify the expression of the intrinsic irreversibility amount of a fluid motion. Indeed, by developing Eq. (3.31) it follows that:

$$\rho \frac{\mathrm{d}s}{\mathrm{d}t} = -\frac{1}{T}\frac{\partial q_i}{\partial x_i} + \frac{q_i}{T^2}\frac{\partial T}{\partial x_i} + \frac{\mathcal{P}^*}{T},$$

or again, after multiplying throughout by the absolute temperature:

$$\rho T \frac{\mathrm{d}s}{\mathrm{d}t} = -\frac{\partial q_i}{\partial x_i} + \frac{q_i}{T}\frac{\partial T}{\partial x_i} + \mathcal{P}^*. \qquad (3.32)$$

A direct comparison of this expression with Eq. (3.28) then results in:

$$\boxed{\mathcal{P}^* = \underbrace{\tau_{ij}\frac{\partial U_i}{\partial x_j}}_{mechanical} - \underbrace{\frac{q_i}{T}\frac{\partial T}{\partial x_i}}_{thermal}}_{\text{intrinsic irreversibilities}} \qquad (3.33)$$

Thus, the power of the intrinsic irreversibilities generated within a moving fluid consists of two additive contributions, which we will qualify respectively as *mechanical* dissipation and *thermal* dissipation, with reference to their corresponding expressions. The effect of these irreversibilities on the motion is a heat generation which is irrevocably dissipated in the fluid, *i.e.* in a *non-recoverable* way, both in the form of internal and kinetic energy. We will write:

$$\Phi_M = \tau_{ij}\frac{\partial U_i}{\partial x_j} \equiv \frac{1}{2}\tau_{ij}\left(\frac{\partial U_i}{\partial x_j} + \frac{\partial U_j}{\partial x_i}\right),\tag{3.34}$$

which introduces the mechanical dissipation function Φ_M. By changing the sign, this function is nothing other than the internal viscosity force power, according to Eq. (3.17).

Similarly, we will define:

$$\Phi_T = -\frac{q_i}{T}\frac{\partial T}{\partial x_i},\tag{3.35}$$

where Φ_T now refers to the heat dissipation function.

3.3.5.4 Recapitulation of the Energy Balance in a Moving Fluid

The following table groups the expressions of the various power types which are exchanged between a fluid particle as it moves and the environment. The horizontal entry identifies the different mechanical and thermal powers with, for the former, the contributions related to the body, pressure and viscous forces. The vertical entry allows us to differentiate the intrinsic irreversibility contributions in the entire energy transfer total.

Finally, it is important to stress that the energy balance is to be taken *as a whole*, in the sense that even if the fluid is *initially* set in motion by the sole action of *a single* type of forces— that of gravity for example—the motion is not, however, reversible, as the consideration of the *single* first line in Table 3.1 might suggest. Indeed, as soon as the motion is created, it should satisfy the momentum balance, Eq. (3.11), involving all the forces and in particular the viscous forces through the stress tensor σ_{ij}. The mechanically supplied energy by the work of the gravity forces (which are assumed to be the unique forces at the origin of the motion) is thus necessarily

Table 3.1 The various contributions to the energy balance of a fluid motion

				Reversible part	Irreversible part	Total value
Power Sources	Mechanical Power	Volume forces		$\rho U_i F_i$	0	$\rho U_i F_i$
		Surface forces	Viscosity	$\frac{\partial}{\partial x_j}(\tau_{ij}U_i)$	$\tau_{ij}\frac{\partial U_i}{\partial x_j}$	$U_i\frac{\partial \tau_{ij}}{\partial x_j}$
			Pressure	$-\frac{\partial (PU_i)}{\partial x_i}$	0	$-\frac{\partial (PU_i)}{\partial x_i}$
	Thermal Power			$-T\frac{\partial}{\partial x_i}(\frac{q_i}{T})$	$-\frac{q_i}{T}\frac{\partial T}{\partial x_i}$	$-\frac{\partial q_i}{\partial x_i}$

involved in the overall balance of *all acting forces in the motion*, and will therefore be subject, in part, to viscosity dissipation, as we will explain in the next Chapter.

3.4   The Eulerian Formulation of the Global Balance Equations

The balance equations have been established so far by adopting a Lagrangian approach, *i.e.*, by referring to a material domain which is followed when it moves. In applications, it is often more suitable to adopt an Eulerian point of view, considering a fixed control volume, arbitrarily chosen in the flow field, under the sole restriction of having a boundary surface entirely composed of fluid particles. In this section, we will therefore give the new forms of the balance equations appropriate to this type of rationalization.

3.4.1   Mass Conservation Through a Stream Tube

We assume that the regime is *stationary* and according to the *Eulerian point of view*, we consider a control volume Δ consisting of a stream tube portion (Σ) bounded by two cross-sections (A_e) and (A_s), as shown in Fig. 3.7. By integrating relation (3-2b) on this domain Δ it follows that:

$$\iiint_{\Delta} div\,(\rho \mathbf{V})\,dv \equiv \oint_{\Sigma + A_e + A_s} \rho \mathbf{V}.\mathbf{n}\,d\sigma = 0\,,$$

or, again, the scalar product $\mathbf{V}.\mathbf{n}$ being identically zero on (Σ):

$$\iint_{A_e} \rho \mathbf{V}.\mathbf{n}d\sigma + \iint_{A_s} \rho \mathbf{V}.\mathbf{n}d\sigma = 0\,.$$

This relation embodies the *mass flow rate* conservation through any cross-section of a stream tube in a steady flow. If the density is constant, we obviously retrieve the *volume flow rate* conservation.

Fig. 3.7 Schematic for the
mass conservation along a
stream tube

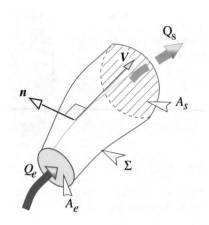

3.4.2 The Global Momentum Balance: Euler's Theorem

3.4.2.1 The General Formulation

A particularly important result for practical applications, known as *Euler's theorem*, states that:

> The momentum flux through a fixed control surface in a *stationary flow* is equal to the external resultant force applied to the fluid in the domain bounded by this surface.

This theorem is expressed in the following vector formulation:

$$\iint_{\Sigma} \rho \mathbf{V} \left(\mathbf{V.n} \right) d\sigma = \iiint_{\Delta} \rho \mathbf{F} dv + \iint_{\Sigma} \mathbf{T} d\sigma \qquad (3.36)$$

To prove this relation, we integrate Eq. (3.12) over the control domain Δ. We thus obtain, for a steady flow:

$$\iiint_{\Delta} \frac{\partial \left(\rho U_i U_j \right)}{\partial x_j} dv = \iiint_{\Delta} \rho F_i dv + \iiint_{\Delta} \frac{\partial \sigma_{ij}}{\partial x_j} dv.$$

By applying the divergence theorem, we can still write:

$$\iint_{\Sigma} \rho U_i \left(U_j n_j \right) d\sigma = \iiint_{\Delta} \rho F_i dv + \iint_{\Sigma} \sigma_{ij} n_j d\sigma.$$

Fig. 3.8 Control volume for
the calculation of the force
generated by a fluid on a
pipe corner

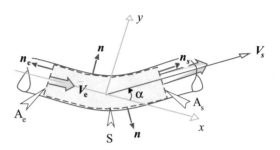

It is then sufficient to introduce relation (3.8) and to return to a vector notation to
obtain the expected result.

Remark It is possible, by similar reasoning, to derive a second balance theorem for
the resultant moment. It is written, always under the stationary motion assumption:

$$\iint_{\Sigma} \rho \mathbf{OM} \wedge \mathbf{V}\,(\mathbf{V.n})\,d\sigma = \iiint_{\Delta} \rho \mathbf{OM} \wedge \mathbf{F} dv + \iint_{\Sigma} \mathbf{OM} \wedge \mathbf{T} d\sigma. \quad (3.37)$$

3.4.2.2   Application Examples

In practice, Euler's theorem makes it possible to determine the external surface
resultant force whereas the *local force distribution* $\mathbf{T}\,(P, \mathbf{n})$ over the surface bounding the volume remains unknown. To do so, all that needs to be done is to choose a
control volume Δ which can allow determining the momentum flux throughout its
bounding surface. The joint use of the moment resultant theorem, Eq. (3.37) allows
us to find the application point of this resultant force. These theorems are widely
used for both hydraulic and aerodynamic applications. Some of them, among the
most classic, are suggested as complementary exercises in the following part of the
section.

EXERCISE **11** The force due to a flow in a bent-pipe.

Let us consider a pipe of constant cross-section with a corner bend α (Fig. 3.8). The
usual experience, carried out with flexible pipes, demonstrates that a fluid flowing in
such a geometric space generates forces on the pipe.

QUESTION Determine the resultant force acting on the bent pipe generated by the
fluid flowing in the corner, assuming a steady motion, with no accounts of viscous
and gravity effects and taking a constant density fluid.

SOLUTION: To apply Euler's theorem, the first requirement is to define the control
volume to be used for assessing the force balance. Here we choose the control volume
bounded by the surface $\mathbb{S} = A_e + S + A_s$ where A_e and A_s are two cross-sections
of the pipe, located respectively at the inlet and outlet of the corner, and S the lateral

surface formed by the peripheral *fluid particles* in contact with the inner surface of the tube. The unit normal vectors of these surfaces, oriented positively when pointing outwards, are denoted respectively as $\mathbf{n}_e$, $\mathbf{n}_s$ and $\mathbf{n}$.

By applying relation (3.36) to the domain which has just be defined, we have, under the problem assumptions (an inviscid and weightless fluid in particular):

$$\iint_{\mathbb{S}} \rho \mathbf{V}\,(\mathbf{V}.\mathbf{n})\,d\sigma = \iint_{\mathbb{S}} -P\mathbf{n}d\sigma\,,$$

or by specifying the surfaces and accounting for algebraically the velocities ($V_e = V_e \mathbf{n}_e$ and $V_s = V_s \mathbf{n}_s$):

$$\iint_{A_e} \rho \mathbf{V_e}\,(V_e)\,d\sigma + \iint_{A_s} \rho \mathbf{V_s}\,(V_s)\,d\sigma = -\left[\iint_{A_e} P_e \mathbf{n_e}d\sigma + \iint_{A_s} P_s \mathbf{n_s}d\sigma + \iint_{S} P\mathbf{n}d\sigma\right].$$

Denoting by $\mathbf{R}_S$ the pressure resultant force applied *by the pipe to the fluid* throughout the peripheral surface S, we obviously have:

$$\mathbf{R}_S = \iint_{S} -P\mathbf{n}d\sigma\,, \quad \text{so that the previous relation yields:}$$

$$\mathbf{R}_S = \iint_{A_s} \left(\rho V_s^2 + P_s\right)\mathbf{n}_s d\sigma + \iint_{A_e} \left(\rho V_e^2 + P_e\right)\mathbf{n}_e d\sigma\,.$$

This expression suggests introducing the function

$$J = P + \rho V^2\,, \tag{3.38}$$

called the *dynalpy* of the flow and shows that the required resultant force is none other than the flux of the dynalpy, namely:

$$\mathbf{R}_S = J_S\,A_s\,\mathbf{n}_s + J_e A_e \mathbf{n}_e\,,$$

when assuming, to simplify, uniform pressure and velocity distributions over the cross-sections A_e and A_S. The resultant force applied by *the inner fluid* to the S surface of the pipe is thus simply $-\mathbf{R}_S$.

This is not, however, the force which the pipe mounting flanges need to overcome. Indeed, the *surrounding external* atmospheric pressure applies a resultant force to S:

$$\mathbf{R}_a = -\iint_{S} P_a \mathbf{n}_s d\sigma \equiv -\iint_{\mathbb{S}} P_a \mathbf{n}\,d\sigma + \iint_{A_e + A_s} P_a \mathbf{n}d\sigma\,.$$

Now, since the surface $\mathbb{S}$ is closed and the pressure P_a is constant, the first integral of the right hand side is zero. It therefore follows that the resultant force to be applied

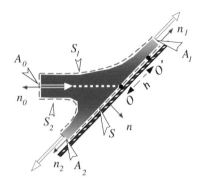

Fig. 3.9 The control volume for the impinging jet on an oblique plate

to maintain the pipe in a fixed position is:

$$\mathbf{R} = -\mathbf{R}_S + \mathbf{R}_a = (P_a - J_S)\, A_S \mathbf{n}_S + (P_a - J_e)\, A_e \mathbf{n}_e\,.$$

Remark It should be noted that the previous rationale deals successively with two different physical systems: (i) the fluid within the volume bounded by the surface $\mathbb{S}$, for applying Euler's theorem, and (ii) the solid, for the determination of the resultant force applied to the S portion of the pipe by both the inner and the outer fluid.

EXERCISE 12 The action of an impinging jet on a flat plate.

A plane jet, issuing from a nozzle with a very wide span-dimension and evolving in a constant pressure environment at rest (P_a) impinges on a wide flat plate inclined at an angle (α) with the initial jet direction, see Fig. 3.9. The motion is considered as steady and two-dimensional. The cross-sections of the flow are taken per unit length in the span-wise direction. Gravity effects are taken as negligible. After impinging on the plate, the fluid separates and discharges in two parallel streams whose cross-sections are respectively denoted as A_1 and A_2.

The inlet jet section is A_0. Putting aside viscosity effects, the velocity and pressure distributions across these sections are taken as uniform[6] (velocity module V_0, pressure P_a respectively).

QUESTION 1. Determine the pressure resultant force on the plate resulting from the sole impact of the jet.

QUESTION 2. Locate the application point of this resultant force on the plate.

---

[6] This twofold property can be demonstrated by applying Bernoulli's theorems. Considering firstly the streamline corresponding to the intersection of the surface S_1 with the plane of the figure for example, we can write *along this line*, in application of Bernoulli's *weak* theorem, Eq. (3.19): $P_0 + \frac{1}{2}\rho V_0^2 = P_1 + \frac{1}{2}\rho V_1^2$. Since, on this surface, the fluid is in contact with a constant pressure environment, we have $P_0 = P_1 = Pa$, which leads to the module uniformity of the velocity vector. As for the pressure uniformity over a cross-section, this results from Bernoulli's *strong* theorem,

SOLUTION: **1.** As before, we start by defining the control volume Δ. It is bounded by the three cross-sections A_0, A_1, A_2 which are connected by the free boundary surfaces S_1 and S_2 and supplemented with the fluid surface S in contact with the wall. The flow rate conservation on Δ results in:

$$A_0 = A_1 + A_2,$$

the free surfaces S_1 and S_2 being stream-surfaces and the mass flow being obviously zero through the solid wall S.

Euler's theorem applied to the same domain then implies that:

$$\iint_{A_0} \rho V_0^2 \mathbf{n}_0 d\sigma + \iint_{A_1} \rho V_1^2 \mathbf{n}_1 d\sigma + \iint_{A_2} \rho V_2^2 \mathbf{n}_2 d\sigma =$$

$$\iint_{A_0} -P_0 \mathbf{n}_0 d\sigma + \iint_{A_1} -P_1 \mathbf{n}_1 d\sigma + \iint_{A_2} -P_2 \mathbf{n}_2 d\sigma + \iint_{S_1+S_2} -P_a \mathbf{n}' d\sigma + \iint_{S} -P \mathbf{n} d\sigma.$$

Now, the pressure resultant force due to the *sole impact* of the jet, *i.e.*, regardless of the contribution of the constant environment pressure, is none other than

$$\iint_{S} (P_a - P) \mathbf{n} d\sigma .$$

Knowing that, for a constant pressure $(P_a = C')$ we have $\iint_{\Sigma} -P_a \mathbf{n} d\sigma \equiv 0$ on a closed, simply connected surface Σ, we obtain, by specifying $\Sigma = A_0 + A_1 + A_2 + S_1 + S_2 + S$:

$$\iint_{S_1+S_2} -P_a \mathbf{n}' d\sigma = \iint_{A_0} P_a \mathbf{n}_0 d\sigma + \iint_{A_1} P_a \mathbf{n}_1 d\sigma + \iint_{A_2} P_a \mathbf{n}_2 d\sigma + \iint_{S} P_a \mathbf{n} d\sigma.$$

After substituting in the previous momentum balance equation, the result is:

$$\rho V_0^2 (A_0 \mathbf{n}_0 + A_1 \mathbf{n}_1 + A_2 \mathbf{n}_2) = \left[\iint_{S} (P_a - P) \, d\sigma \right] \mathbf{n}.$$

(Chap. 5 Sect. 5.1), based on the main assumption of irrotational flow. This theorem then allows us to conclude that, in a flow field with a *uniform velocity profile*, $P + \frac{1}{2}\rho V^2 = C'$. For $V = V_0 = C'$ this leads to $P = C' = P_a$, which is the value on the free surface.

These features are only relevant if the viscosity effects are not taken into account, which is the case here.

The right hand side of this equation is the *force* applied *by* the solid *to* the fluid, through the wetted area (S) of the wall. Its fixed direction is normal to the plate. Denoting by F its module, we obtain, by projecting the previous equation on the normal and tangent directions to the plate:

$$F = \rho V_0^2 A_0 \sin \alpha \,,$$
$$A_1 - A_2 = A_0 \cos \alpha \,.$$

Joined to the continuity equation, the last relation makes it possible to determine the value of the cross-sections, namely $A_1 = A_0 \cos^2 \frac{\alpha}{2}$ and $A_2 = A_0 \sin^2 \frac{\alpha}{2}$.

2. To locate the application point of the resultant force F, we simply apply Eq. (3.37) and write the balance of moments with respect to a horizontal axis of the plate plane passing through point O' and normal to the figure, such that $OO' = h$. We thus have:

$$\iint_{\Sigma} \mathbf{O'M} \wedge (-P)\, \mathbf{n} d\sigma = \iint_{\Sigma} \mathbf{O'M} \wedge \rho \mathbf{V}\, (\mathbf{V}.\mathbf{n})\, d\sigma \,.$$

Now

$$\iint_{\Sigma} \mathbf{O'M} \wedge (-P)\, \mathbf{n} d\sigma = -P_a \iint_{\Sigma - S} \mathbf{O'M} \wedge \mathbf{n} d\sigma - \iint_{\Sigma} \mathbf{O'M} \wedge P \mathbf{n} d\sigma$$

$$= -P_a \iint_{\Sigma} \mathbf{O'M} \wedge \mathbf{n} d\sigma + \iint_{\Sigma} \mathbf{O'M} \wedge (P_a - P)\, \mathbf{n} d\sigma \,.$$

However, according to Green's formula $\iint_{\Sigma} \mathbf{O'M} \wedge \mathbf{n} d\sigma = \iiint_{\Delta} curl\, (\mathbf{O'M})\, dv \equiv 0$, so that the moment resulting in O' from the pressure forces on the plate solely remains. As for the moment resulting from the momentum flux, we have:

$$\iint_{\Sigma} \mathbf{O'M} \wedge \rho \mathbf{V}\, (\mathbf{V}.\mathbf{n})\, d\sigma = \int\!\!\!\int_{A_0 + A_1 + A_2} \mathbf{O'M} \wedge \left(\rho V_0^2 \mathbf{n}\right) d\sigma$$

$$= \rho V_0^2 A_0 h \sin \alpha - \rho V_0^2 \frac{A_1^2}{2} + \rho V_0^2 \frac{A_2^2}{2} \,.$$

If O' is taken at the application point of the resultant force we consequently have

$$0 = \rho V_0^2 A_0 h \sin \alpha - \rho V_0^2 \frac{A_1^2}{2} + \rho V_0^2 \frac{A_2^2}{2} \,, \text{ hence finally } h = \frac{A_0}{2} \cot \alpha.$$

EXERCISE 13 The shear-stress profile in a pipe flow.

We consider the steady motion of a non-weighing viscous fluid in a cylindrical duct of circular cross-section ($\phi = 2R$). The analysis is carried out at a sufficient distance

Fig. 3.10 Control volume
for the application of Euler's
theorem in a fully developed
pipe flow

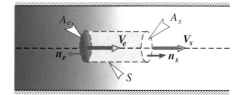

from the inlet section to consider the flow regime as fully developed, *i.e.* so that any
kinematic quantity is invariant in the motion direction.

   QUESTION. Show that the dimensionless shear profile $\tau\,(r)\,/\tau\,(R)$ is *linear*. (The
Euler theorem will be applied to a cylindrical volume around the duct axis, of radius
r and length L, assuming an axial symmetric motion and $P = P(x)$.)

SOLUTION: As mentioned above, we apply Euler's theorem to the control volume
bounded by the surface $\Sigma = A_e + A_s + S$, see Fig. 3.10. A_e and A_s are two cross-
sections of common area πr^2 and S is the peripheral section of area $2\pi r L$, where L
is the distance, along the duct axis, between the two cross-sections.
Since the flow is fully developed and axially symmetrical, the velocity vector reduces
to its sole axial component $V(r)$, which is independent of x, *i.e.* $\mathbf{V} = V(r).\mathbf{e}_x$, by
denoting $\mathbf{e}_x$ the motion unit vector pointing positively in the flow direction.
The application of Euler's theorem by projection on $\mathbf{e}_x$ implies that:

$$-2\pi\int_0^R \rho V_e^2(r)rdr - 2\pi\int_0^R \sigma_e(r)rdr + 2\pi\int_0^R \rho V_s^2(r)rdr + 2\pi\int_0^R \sigma_s(r)rdr =$$

$$-P_e A_e + P_s A_s + \tau\,(r)\,S.$$

In the above expression, σ_e and σ_s are possible *normal* viscous stresses which may
be applied to the lateral cross-sections of the control cylinder. They are algebraically
positive if they are pointing in the direction normal to the respective faces. The
(tangential) viscous *shear* stress which is present on the cylindrical boundary of
radius r is noted as $\tau\,(r)$. Finally, P_e and P_s denote the pressure values $P(x)$ at the
location of the A_e and A_s sections respectively.

   The flow regime being fully developed and assuming the viscous stresses to solely
depend on the velocity, the previous equality leads to :

$$0 = -P_e A_e + P_s A_s + \tau\,(r)\,S\,, \text{ that is } \tau\,(r) = \frac{P_e - P_s}{2} \times \frac{r}{L}\,, \text{ after introducing}$$

the respective aeras of the surfaces. It is easy to deduce that:

$$\frac{\tau\,(r)}{\tau\,(R)} = \frac{r}{R}\,,$$

which shows that the *dimensionless shear profile*, normalized by the friction value
at the wall is *linear*.

Fig. 3.11 Control volume
for Euler's theorem
application to the flow in a
diffuser

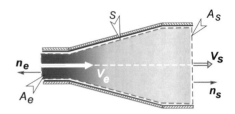

It should be noticed that this relation has been demonstrated with no requirement
for specifying the rheological behaviour of the fluid. In this respect, it appears as a
general result, *i.e.*, which applies to both Newtonian and non-Newtonian fluids.

EXERCISE **14** The flow action in a diffuser.

A weightless fluid is flowing through a diffuser *i.e.*, a diverging duct whose cross-
section gradually increases in the flow direction, from A_e to $A_s > A_e$. We put
$\alpha = A_e/A_s$ $(0 \leqslant \alpha \leqslant 1)$. This variation in velocity is accompanied by a variation
in pressure and therefore in the forces exerted by the diffuser on the through-flowing
fluid.

QUESTION. Determine the resultant force applied by the diffuser to the moving fluid,
tacking no account of the viscosity effects[7] and assuming, in addition, the velocity
and pressure profiles to be uniform in both sections A_e and A_s.

SOLUTION: The application of Euler's theorem to the domain bounded by the surface
outlined in Fig. 3.11 implies that:

$$\rho V_e^2 A_e \mathbf{n}_e + \rho V_s^2 A_s \mathbf{n}_s = -P_e A_e \mathbf{n}_e - P_s A_s \mathbf{n}_s - \iint_S P \mathbf{n} d\sigma.$$

The previous equality shows that the duct exerts a force on the fluid in the flow
direction whose module F is such that:

$$F = \left| \left(P_s + \rho V_s^2\right) A_s - \left(P_e + \rho V_e^2\right) A_e \right|.$$

We recognize the *dynalpy* difference between the outlet and inlet diffuser sections.
Assuming that the velocities and pressures satisfy Bernoulli's relation between the
upstream and downstream sections of the diffuser, the previous result can still be

---

[7] These viscosity effects include, on the one hand, a friction force on the side wall of the diffuser and,
on the other hand, the possibility of a profound change in the flow geometry within the divergent.
We will see in Chap. 11 Sect. 3.6 that in a *viscous fluid*, the pressure variation of the flow along the
diffuser can cause the separation of the fluid lines from the side wall. It is the phenomenon of *flow
separation* which invalidates the result which should be derived here. In practice, this situation is
prevented by limiting the aperture angle of the divergent (a half angle of about 7 to 8 degrees if the
flow regime is turbulent).

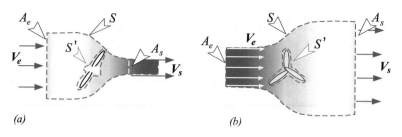

Fig. 3.12 The control volume for the flow past a propeller **a** and a turbine **b**

expressed by introducing the values taken, either in the inlet or outlet section. We then find:

$$F = \frac{1-\alpha}{\alpha}\left[P_e + \frac{1-\alpha}{2}\rho V_e^2\right]A_e,$$

$$F = \frac{\beta-1}{\beta}\left[P_s + \frac{\beta-1}{2\beta}\rho V_s^2\right]A_s.$$

where we put $\beta = 1/\alpha \equiv A_S/A_e$.

The above expressions are positive, so that the force exerted by the divergent on the fluid has the same direction as the flow, and by reaction, the fluid tends to make the divergent *move backwards*. We can, by similar reasoning, extend these conclusions to the case of a *convergent*.

EXERCISE 15 The motive / receiving action of a propeller / wind turbine.

An airscrew is a rotary device by which a given mechanical power can either be transferred to or recovered from a moving fluid. In the first case, it is used as a means of propulsion, the setting in motion of the fluid, being associated, by reaction, with a pulling force on the blades of the propeller. In the second case, it is used to convert part of the kinetic energy of the wind or water into work on its shaft. These two operation modes are shown in Fig. 3.12 and correspond to motive applications, for both aerodynamic and hydrodynamic propulsion on the one hand, and to energy supply by wind or marine turbine on the other hand.

By setting aside the fluid compressibility, viscosity effects and the swirling motion component, the passage through a motive propeller results, for the stirred fluid, in an *acceleration* and, for a receiving turbine, a *deceleration*. This results in the respective effects of contraction and expansion of the fluid vein which, occurring in a constant pressure environment, generate forces applied to or by the propeller.

QUESTION. Determine the resultant force applied by a propeller to the stirred fluid.

SOLUTION: Denoting by $\mathbf{F}$ the resultant force applied to the fluid by the propeller, the application of Euler's theorem to the domain—see Fig. 3.12—bounded by the surface $\Sigma = A_e + S + A_s + S'$ can be expressed by:

$$\iint_{A_e} \rho V_e^2 \mathbf{n}_e d\sigma + \iint_{A_s} \rho V_s^2 \mathbf{n}_s d\sigma = \mathbf{F} + \iint_{\Sigma} -P_a \mathbf{n} d\sigma,$$

or finally, assuming uniform velocity profiles $\mathbf{F} = \left(\rho V_s^2 A_s - \rho V_e^2 A_e \right) \mathbf{n}_s$.
In addition, the mass flow rate conservation implies that $\rho V_e A_e = \rho V_s A_s \equiv Q_m$, so that the resultant force is finally written as:

$$\mathbf{F} = Q_m \left(V_s - V_e \right) \mathbf{n}_s.$$

For a motive propeller, $V_s - V_e$ is positive and the force exerted by the fluid on the propeller $(-\mathbf{F})$ is applied in the opposite direction to the flow, the result reversing for a receiving propeller operating as a turbine.

EXERCISE 16 The thrust of a turbojet engine.

The operation of a turbojet engine can be described very schematically as follows: a mass flow rate of fluid Q_m is taken from the machine inlet; It is then compressed and mixed with a mass flow rate q_f of fuel; The mixture is then burned in a combustion chamber, which causes the temperature and therefore the energy (the total enthalpy in fact) to increase. One part of this energy is used to operate a turbine driving the compressor, the other part is discharged, after expansion through a nozzle, in kinetic and thermal modes into the outlet jet of the machine.

QUESTION. Determine the thrust generated by a turbojet engine ejecting the mass flow Q_m at velocity V_e, knowing that the inlet velocity of the fresh gas to the machine is V_a.

SOLUTION: Let us consider the control domain Δ outlined in Fig. 3.13, whose peripheral boundary consists of the surface S composed of the fluid particles in contact with the machine and the upstream and downstream stream tubes, as a whole referred to as S_0. By setting aside the fuel flow rate $(q_f \ll Q_m)$, everything happens, as far as the momentum balance over Δ is concerned, as if the mass flow rate Q_m undergoes an axial acceleration, the velocity increasing from the intake value V_a to that of the ejection $V_e \gg V_a$.
Euler's theorem applied to the control volume bounded by the surface $\Sigma = A_a + A_e + S_0 + S$ then gives:

$$\rho_a V_a^2 A_a \mathbf{n}_a + \rho_e V_e^2 A_e \mathbf{n}_e = -P_a A_a \mathbf{n}_a - P_e A_e \mathbf{n}_e + \iint_{S_0} -P \mathbf{n} d\sigma + \iint_{S} -P \mathbf{n} d\sigma,$$

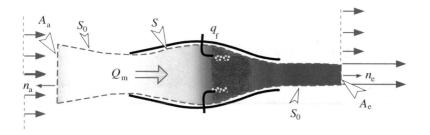

Fig. 3.13 The control volume for Euler's theorem application in a turbomachine

by taking the velocity and pressure profiles as uniform, at both the intake section (A_a) and the ejection section (A_e). The resultant force applied to the *internal* fluid by the machine is therefore:

$$\mathbf{F}_{int} \equiv -\iint_S P\mathbf{n}d\sigma = +\left(P_a + \rho_a V_a^2\right) A_a \mathbf{n}_a + \left(P_e + \rho_e V_e^2\right) A_e \mathbf{n}_e + \iint_{S_0} P\mathbf{n}d\sigma.$$

We will assume that, as a first approximation, the static pressure is the same on all the free boundaries of the volume Δ and equal to P_∞. The previous relation then becomes:

$$\mathbf{F}_{int} = \left(P_\infty + \rho_a V_a^2\right) A_a \mathbf{n}_a + \left(P_\infty + \rho_e V_e^2\right) A_e \mathbf{n}_e + P_\infty \iint_{S_0} \mathbf{n}d\sigma.$$

In addition, the fluid *from the outside* exerts, on the external face ($\mathbf{n}_{ext} = -\mathbf{n}$) of the surface S, pressure forces whose resultant force is equal to:

$$\mathbf{F}_{ext} \equiv \iint_S P_\infty \mathbf{n}_{ext}d\sigma = \iint_S -P_\infty \mathbf{n}d\sigma,$$

so that the global resultant force applied to the machine by both the internal and external fluid is:

$$\mathbf{R} = -\mathbf{F}_{int} + \mathbf{F}_{ext} = -\left(P_\infty + \rho_a V_a^2\right) A_a \mathbf{n}_a - \left(P_\infty + \rho_e V_e^2\right) A_e \mathbf{n}_e - P_\infty \iint_{S_0} \mathbf{n}d\sigma - P_\infty \iint_S \mathbf{n}d\sigma.$$

It thus appears that the global contribution involving the pressure P_∞ is zero since it is applied to the closed surface Σ.

We deduce that the engine thrust is finally:

$$\mathbf{R} = \left(\rho_e V_e^2 A_e - \rho_a V_a^2 A_a\right)\mathbf{n}_a \equiv Q_m \left(V_e - V_a\right)\mathbf{n}_a.$$

3.4.3   The Kinetic Energy Balance and Head Loss

3.4.3.1   The General Head Loss Concept

We limit ourselves to a constant density fluid. We start from the conservative form of Eq. (3.17), which we rewrite as follows:

$$\frac{\partial}{\partial x_i}\left(E_c U_i\right) = \rho U_i F_i - \frac{\partial}{\partial x_i}\left(P U_i\right) + \frac{\partial}{\partial x_j}\left(\tau_{ij} U_i\right) - \Phi_M,$$

where $E_c = \frac{1}{2} U_j U_j$ is the kinetic energy per unit mass and Φ_M the mechanical dissipation function (3.34). Then integrating over a control volume Δ which is fixed in the flow field, the result is:

$$\iiint_\Delta \frac{\partial}{\partial x_i}\left(E_c U_i\right) dv = -\iiint_\Delta \frac{\partial}{\partial x_i}\left(P U_i\right) dv + \iiint_\Delta \frac{\partial}{\partial x_i}\left(\tau_{ij} U_i\right) dv$$

$$+ \iiint_\Delta \rho U_i F_i dv - \iiint_\Delta \Phi_M dv.$$

By applying the divergence formula to each integral in the first line of the previous relation, we obtain:

$$\iint_\Sigma E_c U_i n_i d\sigma = -\iint_\Sigma P U_i n_i d\sigma + \iint_\Sigma \tau_{ij} U_i n_j d\sigma + \iiint_\Delta \rho U_i F_i dv - \iiint_\Delta \Phi_M dv,$$

or, again, in vector notations and making use of Eq. (3.8):

$$\underbrace{\iint_\Sigma E_c \mathbf{V}.\mathbf{n} d\sigma}_{\text{Kinetic energy flux}} = \underbrace{-\iint_\Sigma P\mathbf{V}.\mathbf{n} d\sigma}_{Pressure} + \underbrace{\iint_\Sigma \mathbf{T}_v.\mathbf{V} d\sigma}_{Viscosity} + \underbrace{\iiint_\Delta \rho \mathbf{F}.\mathbf{V} dv}_{Volume} - \underbrace{\iiint_\Delta \Phi_M dv}_{\text{Dissipation}}$$

$$\text{Power of the external forces}$$

$$(3.39)$$

where $\mathbf{T}_v$ stands for the viscous force associated with the viscous stress tensor $\overline{\overline{\tau}}$.

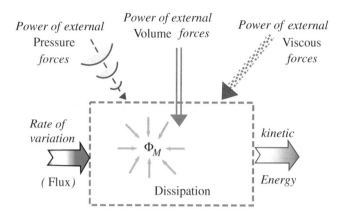

Fig. 3.14 Eulerian schematic for the energy balance in a steady motion of an incompressible fluid

We can therefore state that:

> In any steady motion of an incompressible fluid, the kinetic energy flux through a control domain is equal to the power supplied to the fluid by all the *external* forces minus the *mechanical dissipation* within the domain.

The above conclusions are outlined in Fig. 3.14. It should be noted that the role of viscosity is not, as is sometimes thought, limited solely to dissipation. This fluid property can also have a driving effect, by supplying energy to the motion, as a result of the *external* (viscous) force work, which is the case in Couette's flow, for example (see Chap. 9 Sect. 9.6.4).

According to the terminology commonly used, the integral of the dissipation rate on the domain under consideration leads directly to the notion of pressure loss ($\Delta\xi$) or head loss ($\Delta\xi/\rho g$), namely :

$$\iiint_\Delta \Phi_M dv = \Delta\xi \times Q_v , \qquad (3.40)$$

where Q_v is the volume flow rate through the domain. Hence, in the above relation (3.40), $\Delta\xi$ is homogeneous to an energy per unit volume or a pressure. The head loss ($\Delta\xi/\rho g$) is expressed in units of height.

3.4.3.2   The Pipe Head Loss

In the case of a fully developed steady flow of a viscous fluid in a constant section pipe, the previous theorem provides an expression for the pressure drop between two cross-sections *without the need* to define the local distribution of the dissipation rate Φ_M.

Referring to the situation outlined in Fig. 3.10, we consider a cylindrical control volume whose circular cross-section is equal to that of the pipe (radius R) and whose length is L.

Under the present study assumptions and assuming uniform velocity and pressure profiles in each cross-section, the application of the integral relation (3.39) to this domain leads to:

$$0 = P_e V_e A_e - P_s V_s A_s - \iiint_\Delta \Phi_M dv.$$

Indeed, the adherence (no slip) condition at the wall imposes a zero velocity on the inner surface of the pipe and consequently the power of the external viscosity forces is zero. The previous relation shows that:

$$\iiint_\Delta \Phi_M dv = (P_e - P_s)\, Q_v .$$

We immediately deduce, by comparison with relation (3.40) that:

$$\Delta\xi = P_e - P_s .$$

In the end, and for this flow, the counter-balancing of the intrinsic irreversibilities in the motion, the so-called dissipation, can only be achieved by a pressure drop all along the pipe length ($\Delta P/L$). In a dimensionless form, this pressure drop per unit length can be written as:

$$\Lambda = \frac{\Delta P}{\frac{1}{2}\rho U^2} \times \frac{D}{L}, \tag{3.41}$$

where U is a velocity scale (usually that of the average flow rate) and D a length scale (the actual diameter or hydraulic diameter as appropriate). Relation (3.41) introduces, according to the terminology commonly used in hydraulics, the pressure drop or head loss coefficient Λ. This is a pressure loss qualified as a "*major loss*" associated with energy loss per unit pipe length, as opposed to "*minor losses*[8]" resulting from singularities in the geometry of the duct due to various components or fittings (bends, tees, valves …)

For pipe flows under pressure, the above considerations lead to rewriting Eq. (3.19) in a more suitable form, better adapted to practical use and known as Bernoulli's generalized theorem:

$$\boxed{P_1 + \frac{1}{2}\rho V_1^2 + \rho g z_1 = P_2 + \frac{1}{2}\rho V_2^2 + \rho g z_2 + \xi_{12}} \tag{3.42}$$

---

[8] The qualifiers of "*major*" and "*minor*" do not in any way concern the respective *actual amounts* of the pressure losses: in the case of many duct flow singularities, minor losses can easily exceed major losses.

The above relation applies only in a steady flow. Indices 1 and 2 refer to two upstream and downstream sections of the same pipe and ξ_{12} is the pressure loss between these two sections.

Remark The validity of relation (3.42) is actually ascertained whenever the power of the external viscous forces is either zero or negligible. As such, it can still be applied to free surface gravity flows.

As previously mentioned, in hydraulics, pressures are usually expressed in water column height. As shown in relation (3.42) the same is necessarily true for the head loss. In this case too, the generalized Bernoulli equation is rewritten as follows:

$$\frac{P_1}{\rho g} + \frac{V_1^2}{2g} + z_1 = \frac{P_2}{\rho g} + \frac{V_2^2}{2g} + z_2 + \xi_{12}'.$$

HISTORICAL NOTE—**Pipe head loss**—The question of the pressure drop in a pipe flow was not originally asked in terms of energy. It was approached from practical considerations owing to the liquid "*resistance to flowing*" in a pipe. In 1801, Coulomb, [42], published an impressive paper, a major part of which was intended to clarify the resistance law's dependency on the flow velocity, whether linear and/or quadratic. In 1842, Poiseuille, [117], communicated to the Paris Academy of Sciences the results of a significant experimental study. He gave in this study a formula, which we know to be exact today ($V_{mean} \propto \frac{\Delta P}{L} D^2$), which however was in contradiction with the "theoretical result" of that time ($V_{mean} \propto \frac{\Delta P}{L} D$), and nevertheless without immediately calling into question the previous opinion according to which the hydraulicians in line with Coulomb's statement consider that the friction force is proportional to $AV + BV^2$, V being the flow velocity, A and B two constant coefficients. Thus, concepts had not greatly changed since 1801, and the question could only be incompletely resolved as long as the transition in the laminar-turbulent flow regime had not been identified as the physical origin of the phenomena being quantified. It was finally Reynolds, [135], who would take this decisive step forward in 1883.

3.4.3.3   The Minor Head Loss

In addition to the *major* head loss which we have just highlighted, there are also pressure drops due to changes in the pipe geometry. The latter are referred to as "*minor*" and are linked to singularities such as sudden enlargment, vein contraction, elbow, passage through grids, flow gates, valves, etc.

The theoretical calculation of the pressure drop through all these devices is generally not possible. Hence, semi-empirical data, such as those grouped by Idel'cik [80], are used in practice. We confine ourselves here, for example, to expressing the pressure drop caused by a sudden enlargement of a pipe cross-section.

Fig. 3.15 Internal flow
separation in a sudden
enlargement

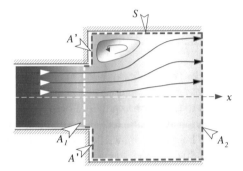

We thus consider (Fig. 3.15) the flow in a duct whose cross section aera suddenly
changes from a value of A_1 to a value of $A_2 > A_1$.

Immediately downstream of the enlargement, a detachment occurs of the fluid
threads which, by separating from the wall, enclose a part of fluid referred to as the
"*dead water*" zone. By this name, it is meant that there is no or little renewal of the
fluid particles in this area but not that the fluid is at rest. In fact, captive roll ups are
located in this area whose motion is sustained by an energy intake from the overall
main flow.

This is the physical origin of the minor head loss which we will now calculate.
For this purpose, let us define the control domain bounded by the surface $\Sigma =
A_1 + A' + A' + A_2 + S$ where A_1 and A_2 are the inlet and outlet cross-sections
respectively, S the lateral surface and $A' = A_2 - A_1$ the enlargement area.

Denoting by V_1 and V_2 the values of the velocity assumed to be uniformly distributed over the inlet/outlet sections and setting $\alpha = A_1/A_2$, the continuity equation
is $V_2 = \alpha V_1$.

Assuming a steady motion and setting aside gravity and viscosity[9] forces, the
application of Euler's theorem to the domain bounded by the surface Σ gives the
following result, when projected along the x axis of the duct:

$$\rho V_2^2 A_2 - \rho V_1^2 A_1 = P_1 A_1 - P_2 A_2 - \iint\limits_{A'} P d\sigma .$$

If the A_1 section is taken close enough to the enlargement in order to ignore the
major head loss, the experiment shows that the pressure P is fairly uniform over the
A' section and only slightly different from that, P_1, of A_1 section. In these conditions
and taking the continuity equation into account, the previous relation gives the result:

$$P_2 - P_1 = \rho \left(\alpha V_1^2 - V_2^2\right) = \rho \alpha (1 - \alpha) V_1^2.$$

---

[9] The role of the viscosity is decisive in the detachment phenomenon and within the flow itself in the
separated zone. It is therefore only its contribution in terms of parietal friction which is disregarded
here.

Now, according to the generalized Bernoulli theorem between the A_1 and A_2 sections, we have:

$$\xi_{12} = P_1 + \frac{1}{2}\rho V_1^2 - \left(P_2 + \frac{1}{2}\rho V_2^2 \right).$$

After substituting the pressure difference, we finally obtain the Borda formula:

$$\xi_{12} = \frac{1}{2}\rho (1 - \alpha)^2 V_1^2. \tag{3.43}$$

HISTORICAL NOTE—Head loss in a sudden enlargement: Borda's formula—
Historically, it was the notion of "minor" head loss which first emerged. The evidence of the non-conservation of mechanical energy was, in fact, explicitly addressed in the memorandum published by Borda [14], in 1766. Section 11 of this text is indeed entitled «On Hydrodynamic issues, in which we should admit a loss of vis viva.» The origin of this loss can be understood, according to Borda, by an analogy between the action of two liquid masses mixing together and that of two solids agglomerating after a mechanical shock. In this case, the kinetic energy[10] before impact is $\frac{1}{2}(m_1 V_1^2 + m_2 V_2^2)$ and becomes, for the consolidated body mass $m_1 + m_2$ after impact $\frac{1}{2}(m_1 + m_2)\left(\frac{m_1 V_1 + m_2 V_2}{m_1 + m_2} \right)^2$ hence the kinetic energy loss $\frac{m_1 m_2}{m_1 + m_2} \frac{(V_1 - V_2)^2}{2}$.

It is this idea which guides him in the calculation of «the motion that the fluid will have by entering a cylindrical vessel [through an orifice drilled at the bottom of the vessel held vertically] which is assumed to be pressed into an undefined fluid». This configuration corresponds very precisely to that of the sudden enlargement in a gravity driven free surface flow. Basing his calculation on a concept equivalent to that of the modern streamlines, which is radically different from Bernoulli's horizontal slice approach [10], he gave the exact theoretical head loss expression due to this singularity.

EXERCISE 17 The pressure loss optimization in a sudden pipe enlargement.

A pipe is considered to be undergoing a sudden enlargement from a cross-section area of A_1 to $A_2 > A_1$. To reduce the minor pressure drop, it is suggested to introduce an intermediate section expansion A_3 such that $A_1 < A_3 < A_2$.

QUESTION 1. Does such a design effectively reduce the pressure drop?
QUESTION 2. How to choose the aera A_3 to optimize the pressure drop?
Discuss the case of an intermediate section reducing the enlargement by half.

SOLUTION: 1. Let us set $\alpha \equiv A_1/A_2 = \beta\gamma$, with $\beta = A_1/A_3$ and $\gamma = A_3/A_1$. If there is no intermediate section, the head loss is $\xi_{12} = \frac{1}{2}\rho V_1^2 (1 - \alpha)^2$.

---

[10] The kinetic energy is half of the vis viva.

Fig. 3.16 Comparison of the optimization line of the head loss reduction **a** with that of the doubling of section **b**

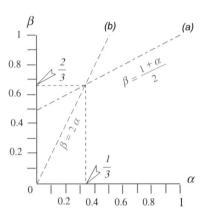

By inserting an intermediate section A_3, it is possible to set, for the new value, $\xi'_{12} = \xi_{13} + \xi_{32}$ with $\xi_{13} = \frac{1}{2}\rho V_1^2 (1 - \beta)^2$ and $\xi_{32} = \frac{1}{2}\rho V_3^2 (1 - \gamma)^2$. Since $V_3 = \beta V_1$ and $\alpha = \beta\gamma$, we obtain $\xi'_{12} = \frac{1}{2}\rho V_1^2 \left[2\beta^2 - 2 (1 + \alpha) \beta + \alpha^2 + 1 \right]$.

This trinomial in β is extremum for $\beta = \frac{1}{2} (1 + \alpha)$ and the corresponding value of ξ'_{12} is:

$$\left[\xi'_{12} \right]_{min} = \frac{1}{2}\rho V_1^2 \frac{(1 - \alpha)^2}{2}$$

Thus, introducing an intermediate section reduces the head loss of a single sudden expansion between the same two extreme sections by a factor which can be as much as double.[11]

    2. The optimal reduction of the head loss corresponds to the mid-section law $\beta = \frac{1}{2} (1 + \alpha)$. It is therefore not with a section equal to half of the final section ($\beta = 2\alpha$) that the reduction in head loss is optimized.

As plotted in Fig. 3.16, this result is only valid for the particular case where the final section is triple the initial section ($\alpha = 1/3$). Therefore we then have $\beta = 2/3 = 2\alpha$.

---

[11] This result is only valid insofar as the pressure drops for each enlargement can be considered as independent, so that they can be added for the sections connected in series.

Chapter 4
Fluid Motion Models

Abstract The closed system of equations governing the thermomechanical evolution of a fluid is established at the beginning of the chapter. Restricted models are then derived which apply under different groups of assumptions. To conclude, the dimensionless form of the equations is derived and the characteristic numbers of any fluid motion are discussed.

4.1 Chapter Objectives

The objectives of this chapter are threefold.

- Firstly, with reference to the equations developed in the previous chapter, it is necessary *to close* the system, i.e., to build a set of relations in a number equal to that of the functions to be determined. As we will see, this result cannot be obtained without including information inferred from experimental evidence. It is therefore dependent on a certain degree of empiricism. Given suitable initial and boundary conditions, this set of relations is, by construction, and at least *in theory*, able to provide the solution to various fluid flow problems which all sharing the same type of *physical conceptualization*. It is for this reason that we are speaking about *mathematical models* of fluid motion. Under the assumptions of this course, the "*general model*" that we will describe here is the Navier-Stokes model.
- In practice, some applications do not require the full use of the previous general model. Various simplifications can be made to this model, depending on considerations related to *(i)* the 'fluid material' itself, or *(ii)* to its mode of motion. This results in "*restricted models*" whose presentation is the second objective of the chapter.
- Finally, we will introduce the necessary elements for the analysis of the *adimensional* equations. In this way, we will identify the expression and the interpretation of the characteristic numbers of various types of flows. The relevance of such numbers will be briefly illustrated through their role in the π theorem and the simplifications of the *incompressible* flow model.

© The Author(s), under exclusive license to Springer Nature Switzerland AG 2022   147
P. Chassaing, *Fundamentals of Fluid Mechanics*,
https://doi.org/10.1007/978-3-031-10086-4_4

4.2   The Current State in the Equation Formulation

In Chap. 3, we derived *five* independent scalar equations by applying the laws and general principles of mechanics and thermodynamics, and using a *purely deductive* method, namely:

$$\text{Continuity} \quad \frac{d\rho}{dt} + \rho\frac{\partial U_i}{\partial x_i} = 0 \tag{4.1}$$

$$\text{Momentum} \quad \rho\frac{dU_i}{dt} = \rho F_i + \frac{\partial \sigma_{ij}}{\partial x_j} \quad (i = 1, 2, 3) \tag{4.2}$$

$$\text{Energy} \quad \rho\frac{d}{dt}\left(e + \frac{U_i U_i}{2}\right) = \rho U_i F_i + \frac{\partial\left(\sigma_{ij} U_i\right)}{\partial x_j} - \frac{\partial q_i}{\partial x_i} \tag{4.3}$$

To the previous relations, it is necessary to add the *two* equations characterizing the thermodynamic state of the fluid, which can be formally expressed as:

$$e = f_e\left(\rho, T\right), \qquad\qquad P = f_p\left(\rho, T\right). \tag{4.4}$$

We therefore have a system of *seven* equations for a total of *sixteen* unknown numerical functions, namely:

- *seven* *main* unknown functions:
  - the three velocity vector components U_i $(i = 1, 2, 3)$;
  - the four scalar functions: pressure (P), density (ρ), temperature (T) and internal energy (e);

- *nine* *intermediate* unknown functions:
  - the six components of the *symmetrical* stress tensor: σ_{ij} $(i = 1, 2, 3$ *and* $j = 1, 2, 3)$;
  - the three components of the heat flux density vector: q_i $(i = 1, 2, 3)$.

In the present state of theoretical formulation, the system is therefore *open*, with more unknown elements than equations. To 'close' it, additional relations are needed, linking the *intermediate* unknown functions σ_{ij} and q_i to the *main* unknown functions. They are not obtainable by virtue of general principles whose information has already been *fully* implemented. The closing procedure is therefore of a lower general level, which is reflected in the term *schematization* which is reserved for this process. We will now specify the *behaviour schemes* or "*constitutive laws*" of the fluid model which will be used in the rest of this course. This will result in a *mathematical model* which will be the *Newtonian* fluid model.

ADDITIONAL INFORMATION—The Constitutive law—By *constitutive law*, we mean a set of equations and inequalities which reflect or '*model*', in terms of the continuous media approach, the material's response to thermo-mechanical stresses. This modelling process is part of the thermodynamics of irreversible processes and statistical mechanics. For complex materials (see for example Chap. 1, Sect. 1.3.2.3), the elaboration of these laws may require developments largely outside the scope of this work. We will restrict ourselves here to a simple *description* of *linear* behaviour schemes which are valid for *typical* fluids such as air and water.

4.3 The Newtonian Behaviour Schemes

4.3.1 The Dynamic Behaviour Scheme

4.3.1.1 Newton's Scheme

As an extension of the concepts introduced in the first chapter Sect. 1.3.2, a fluid will be referred to as *Newtonian* if the relation between the stress tensor $(\overline{\overline{\sigma}})$ and the strain rate tensor $(\overline{\overline{S}})$ satisfies the following requirements:

1. The stress field in the fluid *at rest* is spherical, the pressure being equal to the thermodynamic pressure;
2. The viscous part of the stress tensor only depends on the strain rate tensor and is zero for any '*solidifying*' motion;
3. The relation between $(\overline{\overline{\tau}})$ and $(\overline{\overline{S}})$ is *linear*;
4. There is no preferential direction within the fluid, so that the relation between viscous stresses and strain rates is *isotropic*.

By distinguishing—Cf. Chap. 3 Eq. (3.10)—between the pressure and viscous stresses (τ_{ij}) we will set:

$$\sigma_{ij} = -P\delta_{ij} + \tau_{ij} . \tag{4.5}$$

It can then be shown that the previous Newtonian fluid hypotheses leads to the expression of the viscous stress tensor in the form:

$$\tau_{ij} = \mu \left(\frac{\partial U_i}{\partial x_j} + \frac{\partial U_j}{\partial x_i} \right) + \mu' \frac{\partial U_k}{\partial x_k} \delta_{ij} \tag{4.6a}$$

Equally, in tensor notations, this relation can still be written as:

$$\overline{\overline{\tau}} = 2\mu \overline{\overline{S}} + \mu' \vartheta \overline{\overline{I}} \tag{4.6b}$$

where $\vartheta = \partial U_k / \partial x_k$ is the trace of $\overline{\overline{S}}$.

As previously mentioned in the first chapter Sect. 1.3.2.2, μ and μ' refer to two scalars similar to *Lamé's coefficients* of the linear elasticity. The first is still called the dynamic viscosity. They are both positive, which we will only demonstrate for the dynamic viscosity.

The previous relations (4.6a) and (4.6b) can be expressed differently by introducing the decomposition into spherical and deviatoric parts of the tensor $\overline{\overline{\tau}}$. We have thus:

$$\overline{\overline{\tau}} = \underbrace{2\mu \left(\overline{\overline{S}} - \frac{\vartheta}{3} \overline{\overline{I}} \right)}_{Deviatoric\ part} + \underbrace{\left(\mu' + \frac{2}{3}\mu \right) \vartheta \overline{\overline{I}}}_{Spherical\ part} . \tag{4.7}$$

This form shows that the group

$$\mu'' = \mu' + \frac{2}{3}\mu , \tag{4.8}$$

plays the role of a *bulk viscosity* in the sense that it is associated with variations in the fluid particle volume. As a result, it disappears in any isovolume evolution.

Proof Mathematically, the linear dependence of the viscous stress tensor ($\overline{\overline{\tau}}$) and the strain rate tensor ($\overline{\overline{S}}$) results in a relation between the components of the form:

$$\tau_{ij} = A_{ijkl} S_{kl} .$$

A_{ijkl} is a fourth order tensor which can be demonstrated as being necessarily written under the *isotropy* assumption as follows:

$$A_{ijkl} = a\delta_{ik}\delta_{jl} + b\delta_{il}\delta_{jk} + c\,\delta_{ij}\delta_{kl} ,$$

where a, b, c are three real constants. The symmetry of the stress tensor ($\tau_{ij} = \tau_{ji}$) requires that $A_{ijkl} = A_{jikl}$, which leads to the condition $a = b$. We therefore obtain, under the isotropy assumption

$$\tau_{ij} = \left[a \left(\delta_{ik}\delta_{jl} + \delta_{il}\delta_{jk} \right) + c\delta_{ij}\delta_{kl} \right] S_{kl} = a \left(S_{ij} + S_{ji} \right) + c\vartheta\delta_{ij} ,$$

where we recall that $\vartheta = S_{ll}$. The last relation is directly identified with Eq. (4.6a), by simply setting $a = \mu$ and $c = \mu'$.

4.3.1.2 Stokes' Assumption

As already noted in Chap. 3 Sect. 3.2.2.3, the mechanical pressure, P', based on the stress tensor trace is linked to the thermodynamic pressure, P, by

$$P' = P - \frac{\tau_{ii}}{3}, \tag{4.9}$$

which, for a Newtonian fluid, means $P' = P - \frac{3\mu'+2\mu}{3}\vartheta$.

The two pressure values are therefore equal if and only if either of the following two conditions are satisfied:

- an isovolume evolution of any fluid;
- a fluid whose bulk viscosity is zero, in other words:

$$\boxed{2\mu + 3\mu' = 0} \tag{4.10}$$

Relation (4.10) is known as Stokes' condition and will subsequently be adopted as a hypothesis. The fluid behaviour scheme, now referred to as Newton-Stokes, is expressed in the relation :

$$\boxed{\overline{\overline{\tau}} = 2\mu\overline{\overline{S}} - \frac{2}{3}\mu\vartheta\overline{\overline{I}}} \tag{4.11}$$

ADDITIONAL INFORMATION—The bulk viscosity and Stokes' condition—For a monoatomic gas where the agitation energy of the molecules can only be attributed to the translation mode motion, the kinetic theory of gases can justify Stokes' condition. For gases of a more complex molecular structure, the molecule agitation energy combines translation, rotation and vibration modes. The velocity gradients, which occur in the macroscopic fluid motion, directly affect the translation mode. Vibration and rotation modes are only indirectly affected, as a consequence of the effects of the translation mode. This process, called relaxation, requires a finite time scale. At this time scale, molecular energies are not in equilibrium with each other and the results of the thermodynamics of equilibrium states do not directly apply. This explains the difference between the thermodynamic pressure (related only to the translation mode) and the actual pressure during relaxation, involving the other modes. The latter are also responsible for the absorption of sound when propagating in a compressible environment. Therefore, the *second viscosity coefficient* can be evaluated by measuring the attenuation of the acoustic energy in the fluid where it propagates, as shown by Marcy [95].

HISTORICAL NOTE—On the viscosity issue—The viscosity concept, as we understand it today through its microscopic conceptualization and mesoscopic schematization, has had a somewhat paradoxical history.

Indeed, as Prandtl [125] points out, the modern expression of the friction in a simple shear flow $\tau_{xy} = \mu\partial U/\partial y$ was already stated by Newton in his "*Principia*".[1] In fact, as far back as 1687 at the beginning of section 9 of Book II [106], Newton took as his study hypothesis that «*the resistance produced by friction*

---

[1] The first edition of this work dates back to 1687 [106]. It was followed by several others, notably in 1713 [107] and 1726 [108].

between parts of the fluid is proportional to the velocity difference between these parts».[2]

Newton's work therefore predated Euler's 1757 article [56], in which the fluid dynamics equations were established by including the viscosity effects in those of the pressure. The *«viscosity»* thus disappeared from the equations for nearly a hundred years, and its role in explaining the paradox reported by D'Alembert [44], in 1768, was also only be suspected much later.

Nevertheless, the manifestations of the *«internal resistance»* of fluid motions could not have eluded the hydrodynamicians of the time. Thus, at the end of the 18th century and the beginning of the 19th century, many experimental studies on pressure pipe flows and free surface channel flows were carried out, in France in particular by scientists and engineers, such as Du Buat, Girard, Prony, Coulomb and Poiseuille. Unfortunately, identifying and quantifying the role of the sole *molecular viscosity* in these flows would pose an unsuspected challenge to all experimenters whose test conditions combined laminar and turbulent regimes. Thus, the relevance of the measurements was not ensured and a considerable part of the analysis focused, as in Coulomb's work, on the form of the linear and/or quadratic expression to be given to the fluid resistance, as a function of the flow velocity.

The theoretical approach proved to be more conclusive and we owe it to Navier [105], for having taken the first step in this new direction in 1822. The *constant* ϵ, which Navier introduces into his mathematical development of the equations by interactions between the *«fluid molecules»*, is interpreted, *a posteriori*, as follows: *«Let us imagine a portion of fluid resting on a plane, and whose total number of molecules move along parallel lines between them and with this plane. Let us assume that the velocities of the fluid in the same layer parallel to the plane are equal to each other; and that the velocities of each layer, as they become more distant from the plane, increase gradually and uniformly, so that two layers whose distance is equal to the linear unit have velocities whose difference is also equal to the linear unit. Under this assumption, the constant ϵ represents in weight units the resistance resulting from the sliding of any two layers one over the other, for a surface area equal to the unit surface.»* This is explicitly the expression of the linear law of the shear stress, whose dynamic viscosity is ϵ. In Navier's study, the evolution being assumed isovolume, only one *constant of viscosity* appears. It was Poisson [121], who, in 1831, introduced the second viscosity coefficient, based on an approach related to a *«fluid molecule interaction»* conceptually identical to that of Navier.

Even if formally there is no difference between the expressions of the equations, Barré de Saint-Venant's approach was nevertheless a remarkable improvement in terms of modernity. In his note [1] of 1843, to be attached to his article in 1834, he reversed the problem and raised, a priori, the modern question of the *constitutive scheme* of fluid's behaviour. Referred to by him as the *«formulae of*

---

[2] The Latin quotation is *«Resistantiam, quæoritur ex defectu lubricitatis partium Fluidi, cæteris paribus, proportionalem esse velocitati, qua partes Fluidi separantur ab invicem.»*

the pressures in the interior of moving fluids», he addressed this question in a conceptual break with his predecessors, i.e., «*without making assumptions about the size of the attractions and repulsions of the molecules according to either their distances or their relative velocities.*» He finally gave the stress-strain relationships by specifying the second viscosity coefficient from the deviation at mechanical pressure of the normal stresses, in a *non-isovolume* situation.

Stokes' article, [149] in 1845, referred to Navier and Poisson's publications, but curiously made no reference to those of Saint-Venant, which had been published at an earlier date, even though it was exactly on the same theoretical line. The only difference is the reduction of the two viscosity coefficients to one, due to the relation known since then as *Stokes' condition*.

Once the *mechanical* aspect of the viscous forces had been clarified from the *theoretical* viewpoint, the use to experimentation became essential again to determine, by direct measurement, the value of the corresponding coefficient for different fluids. This question was the subject of prolific work, including that of Hirn [73] in 1855 and Reynolds [136] in 1886 for liquids used as lubricants. The viscosity of air and various gases was measured by Maxwell, [98] in 1866. In this article, Maxwell also returned to the problematic issue of the gas/solid boundary condition whether in adhesion or slip. His conclusion is that: «*. . . if there were any slipping, it is of exceedingly small amount; and that the evidence in favour of the indicated amount being real is very precarious. The results of the hypothesis, that there is no slipping, agree decidedly better with the experiments.*»

In parallel, but in an entirely new light, the molecular approach was to be revived, giving the concept of viscosity a microscopic interpretation, in the sense of what would become the kinetic theory of gases. Following Clausius, Maxwell [97, 99], in the 1860s, was to contribute to the development of the new reflection. More specifically, with regard to the viscosity, he put forward a mechanism by which he was able to derive an expression of the viscosity (μ_M), as a function of the mean free path ($< \ell >$) and the mean velocity ($< v >$) of the molecular agitation, which is:

$$\mu_M = \frac{1}{3}\rho < v >< \ell > .$$

The path thus opened would continue through Boltzmann and lead almost simultaneously to Chapmann, [33] in 1916 and Enskog, [54] in 1917, to improve the previous proposal by suggesting for the viscosity's new expression (μ_B), referred to as *Boltzmann's viscosity*:

$$\mu_B = 1,0162\frac{5}{16\sigma^2}\sqrt{\frac{mk_BT}{\pi}},$$

where m is the mass of the molecules, k_B the Boltzmann constant, T the absolute temperature and σ a length representative of the effective cross-section of the collisions between the molecules.

Nevertheless, the continuous medium approach was not discarded, and in 1901, Duhem [52] revisiting an article he wrote in a work of 1896 [51], showed that, to comply with the thermodynamic criterion of the dissipation positiveness, the two viscosity coefficients necessarily had to satisfy the condition $2\mu + 3\mu' \geq 0$. This result would prove all the more important as the work on the second viscosity coefficient, with the questioning of Stokes' condition coming to maturity, following Tisza's paper [160], in 1942.

We cannot conclude this history without mentioning the additional insights on the issue "viscosity & irreversibility", and particularly recommend to this extent reading Viscardy's article [166], published in 2006.

4.3.2 The Thermal Behaviour Scheme

The only heat transfer mode considered in this course is of the conductive type. The thermal conduction scheme is Fourier's which proportionally links the heat flux density vector to the temperature gradient:

$$\boxed{\mathbf{q} = -\lambda\, \boldsymbol{grad}\, T}\tag{4.12}$$

The coefficient λ whose dimension is $[M] \times [L] \times [T]^{-3} \times [\Theta]^{-1}$, is called the *thermal conductivity* of the fluid which we will prove to be positive.

4.3.3 The Thermodynamic Equation of State

In the rest of the course we will consider that, whenever the fluid is a gas, it *thermodynamically* behaves as an *ideal gas*. We will therefore adopt as the equation of state the relation

$$\frac{P}{\rho} = RT \ .\tag{4.13}$$

According to this same hypothesis, we will express the internal energy and the specific enthalpy[3] of the fluid as functions of the temperature, in compliance with both *Joule's laws*:

$$de = C_v\, dT \quad \text{and} \quad dh = C_p\, dT \ ,\tag{4.14}$$

where C_v and C_p denote the specific heat at constant volume and pressure respectively.

---

[3] They are functions expressed in J/Kg.

4.4 The Ideal Fluid Concept

4.4.1 The Intrinsic Irreversibilities in a Newtonian Fluid Motion

Using the previous dynamic and thermal behaviour schemes, the dissipation functions defined by the general terms of Chap. 3, Eqs. (3.34) and (3.35) respectively, can be made explicit. Thus, taking as the viscous stress tensor the scheme given by Newton-Stokes' formula (4.6a), we obtain for the *mechanical dissipation function* Φ_M:

$$\Phi_M = 2\mu S_{ij} S_{ij} + \mu' \vartheta^2. \tag{4.15}$$

In the rest of the course, we will note this dissipation function Φ_ν to remind ourselves that this is the restriction of the general relation to the case of a Newtonian viscous fluid. Similarly, Fourier's scheme (4.12) leads to expressing the thermal dissipation function Φ_T in the form:

$$\Phi_T = \frac{\lambda}{T} \frac{\partial T}{\partial x_i} \frac{\partial T}{\partial x_i}. \tag{4.16}$$

In the absence of any macroscopic motion, the heat transfer is purely conductive (diffusion by molecular agitation) so that the above expression can only be in agreement with the second law if and only if the thermal conductivity λ is positive or zero.

ADDITIONAL INFORMATION—The mechanical dissipation function—By specifying the derivatives of the velocity vector in an orthonormal reference frame, the dissipation function Φ_ν given by relation (4.15) is put in the form:

$$\begin{aligned}
\Phi_\nu = 2\mu & \left[\left(\frac{\partial U_1}{\partial x_1} \right)^2 + \left(\frac{\partial U_2}{\partial x_2} \right)^2 + \left(\frac{\partial U_3}{\partial x_3} \right)^2 \right] \\
+ \mu & \left[\left(\frac{\partial U_1}{\partial x_2} + \frac{\partial U_2}{\partial x_1} \right)^2 + \left(\frac{\partial U_2}{\partial x_3} + \frac{\partial U_3}{\partial x_2} \right)^2 + \left(\frac{\partial U_1}{\partial x_3} + \frac{\partial U_3}{\partial x_1} \right)^2 \right] \\
+ \mu' & \left(\frac{\partial U_1}{\partial x_1} + \frac{\partial U_2}{\partial x_2} + \frac{\partial U_3}{\partial x_3} \right)^2.
\end{aligned} \tag{4.17}$$

It is a quadratic form in velocity space derivatives, necessarily positive or zero by virtue of the second law, as we saw in Chap. 3, Sect. 3.3.5.3, also called *Rayleigh's dissipation function*. It can then be shown—see Duhem, [52], or more recently Bouttes, [18] —, that a necessary and sufficient condition to fulfill the relation $\Phi_\nu \geqslant 0$ is

$$2\mu + 3\mu' \geqslant 0 \qquad \text{and} \qquad \mu \geqslant 0.$$

We will not give here the details of the general mathematical demonstration, in order to limit ourselves to two specific physical observations concerning:

- an *isovolume* evolution: since the velocity field is divergence free, the dissipation function Φ_ν is proportional to the dynamic viscosity alone μ and to a grouping of clearly positive or null terms. The condition $\mu \geqslant 0$ results directly from this;
- an *isotropic bulk deformation*: in this case the velocity gradient tensor is reduced to the following components only $\partial U_1/\partial x_1 = \partial U_2/\partial x_2 = \partial U_3/\partial x_3 = \alpha$. We deduce from this:

$$\Phi_\nu = 2\mu \left(3\alpha^2\right) + \mu' \left(3\alpha\right)^2 = 3 \left(2\mu + 3\mu'\right) \alpha^2 ,$$

from which the condition $2\mu + 3\mu' \geqslant 0$ immediately follows. This last result provides a better understanding of the physical meaning of Stokes' condition (4.10), namely *when this relation is satisfied, any isotropic bulk deformation of the fluid particles occurs with no intrinsic irreversibility*.

4.4.2 The Ideal Fluid Concept

We have already mentioned the concept of an *'ideal'* or *inviscid* fluid in the first chapter Sect. 1.3.2.3 and Fig. 1.7. It was then a purely *mechanical* approach to this concept. We will extend it here to a *thermodynamic* meaning by stating that:

> The concept of an *ideal* fluid is that of a fluid which develops no intrinsic irreversibility during any *discontinuity-free* motion. According to the Newton-Stokes and Fourier schemes, this concept corresponds to an environment which is *inviscid* ($\mu = 0$) and *non-heat conductive* ($\lambda = 0$).

Remark 1. To avoid any confusion, care should be taken to distinguish the concept of an ideal *fluid* from that of an ideal *gas*.

2. Since the ideal fluid has no thermal conductivity, any temperature variation in such an environment can only result from the variation in pressure and/or volume, in accordance with the law $P/\rho = RT$ if the fluid behaves like an ideal gas.

3. Any ideal fluid motion refers, by nature, to an infinite Reynolds number. The question of whether such a situation can be representative of a real fluid flow at very high Reynolds numbers will be considered at a later stage.

4.5 The General Navier-Stokes Model

We will explain in detail here the equations and boundary conditions which govern any motion of a Newtonian viscous fluid which together constitute the Navier-Stokes *model*.

4.5.1   The Governing Equations

The Navier-Stokes equations are understood here as those resulting from the introduction of the Newton-Stokes and Fourier schemes in the mass, momentum and
energy general balance equations, (4.1) to (4.3). Accordingly these are written

$$Continuity \quad \frac{d\rho}{dt} + \rho\vartheta = 0 \tag{4.18}$$

$$Dynamic \quad \rho\frac{dU_i}{dt} = \rho F_i - \frac{\partial P}{\partial x_i} + 2\frac{\partial\left(\mu S_{ij}\right)}{\partial x_j} - \frac{2}{3}\frac{\partial\left(\mu\vartheta\right)}{\partial x_i} \tag{4.19}$$

$$Energy \quad \rho\frac{d}{dt}\left(C_v T + \frac{U_i U_i}{2}\right) = \rho U_i F_i - \frac{\partial\left(PU_i\right)}{\partial x_i} + 2\frac{\partial\left(\mu S_{ij}U_i\right)}{\partial x_j}$$

$$- \frac{2}{3}\frac{\partial\left(\mu\,\vartheta\,U_i\right)}{\partial x_i} + \frac{\partial}{\partial x_i}\left(\lambda\frac{\partial T}{\partial x_i}\right) \tag{4.20}$$

In these equations, the velocity, pressure, density and temperature are taken as the
main unknown functions. The dynamic viscosity, using Stokes' condition, is noted as
μ, λ refers to the thermal conductivity and C_v to the specific heat at constant volume
of the fluid. Let us recall that the strain rate tensor, noted S_{ij} is by definition:

$$S_{ij} = \frac{1}{2}\left(\frac{\partial U_i}{\partial x_j} + \frac{\partial U_j}{\partial x_i}\right) \quad with \quad \vartheta = \frac{\partial U_k}{\partial x_k}.$$

The previous relations form a set of *five* equations for *five* unknown numerical
functions, the physical properties of the fluid (μ, λ, C_v) being considered as specific
problem parameters. It is a coupled system of partial differential equations of *first*
order in time, *second* order in space, *non-linear* and with *non-constant* coefficients.
To adapt the model to a particular flow, it is therefore necessary to add a set of initial
conditions, with reference to the time variable, and boundary conditions for the space
variables.

HISTORICAL NOTE—**The Navier-Stokes equations**—The dual authorship of
these equations refers to Navier's memoir [105], published in 1823, and Stokes'
article [149], published in 1845.

In the founding paper of 1823, the equations are not in any way derived from
a continuous media approach. Following Laplace's ideas, Navier applied the
d'Alembert principle to discrete material points, described as *«fluid molecules»*,
subject to external forces and mutual interactions. Its basic assumption is that *«by
the effect of the fluid motion, the repulsive actions of the molecules are increased
or decreased by an amount proportional to the speed at which the molecules
approach or move away from each other.»*

A few years later, in an article read at the French Academy in 1829 and
published in the journal of the École Polytechnique in 1831, Poisson, [121],

remaining fundamentally on the same conceptual line, restated these equations within the more general framework of a non-isovolume evolution.

The break with the approach from the *"molecular phenomenology"* seems to be the work of Barré de Saint-Venant. In a paper [1] written in 1843, in addition to his article on fluid dynamics in 1834, Saint-Venant was perfectly clear this breaking of new ground, when writing: «*The purpose of this paper ... is to search for formulae about pressures inside moving fluids, without making assumptions about the amount of the attractions and repulsions of molecules as a function of either their distances or their relative velocities.*»

He then asserted the linearity hypothesis, assuming that «*... the friction, generated by the slip condition, and which should depend on it since a slip resistance acts in the same direction as this slip.*» On this basis, he specified the stress and strain rate dependency for the three normal and three tangential components, through relations which are in all respects in line with the modern formulation.

History has not retained the name of Barré de Saint-Venant for his contribution to the establishment of the fluid dynamic equations, but rather that of Stokes. Stokes' approach is also free of any *"molecular phenomenology"*. In his article published in 1845, Stokes [149] first established the result on the velocity space variation in small displacements at the continuous medium scale: «*Hence the most general instantaneous motion of an elementary portion of a fluid is compounded of a motion of translation, a motion of rotation, a motion of uniform dilatation, and two motions of shifting of the kind just mentioned.*»

From this analysis, he derived the modern equivalent of the strain rates to which he linked the stresses, referred to as *pressure* in his founding principle statement: «*The difference between the pressure on a plane in a given direction passing through any point P of a fluid in motion and the pressure which would exist in all directions about P if the fluid in its neighbourhood were in a state of relative equilibrium depends only on the relative motion of the fluid immediately about P ; and that the relative motion due to any motion of rotation may be eliminated without affecting the differences of the pressures above mentioned.*»

The application of D'Alembert's principle led Stokes to give the modern expression of the general equations (non-constant density) where only one viscosity coefficient is present.

4.5.2 Initial and Boundary Conditions

4.5.2.1 The Initial Conditions

Apart from the steady motion situation, the flow field determination at any time $t \geq t_0$ requires knowing the same field at a given time t_0. The values of the functions $U_i\left(x_j, t_0\right)$, $P\left(x_j, t_0\right)$, $\rho\left(x_j, t_0\right)$, $T\left(x_j, t_0\right)$ are the initial conditions for calculating the flow over the time interval $[t_0, +\infty[$.

4.5.2.2 The Boundary Conditions

In this course, the flow takes place in a domain Ω of $\mathfrak{R}^3$ whose boundary δ_Ω can only be composed, in whole or in part, of a solid wall and a free surface. The conditions to be prescribed on this boundary, depending on whether the velocity or temperature field is considered, are as follows:

- *The equality of the velocities at coincident fluid/solid points*: the macroscopic adhesion condition of the viscous fluid is indeed reflected for any fluid particle in contact with a solid surface, by the equality of the local velocity of this particle and that of the solid at the contact point;
- *A given value of the temperature or heat flux at the wall*: the heat conduction within a fluid in contact with a solid under thermal imbalance can be defined, contrary to the momentum transfer, by prescribing either (*i*) a parietal temperature, or (*ii*) a heat flux at the fluid-solid contact.
- *The continuity of the pressure across a free boundary* in the incompressible flow regime.

HISTORICAL NOTE—**The boundary condition according to Navier and Stokes**
—Regarding the boundary condition on a solid wall, Navier [105] assumes a flow situation with a slip condition. The velocity tangent to the wall, which is therefore not zero, is taken in proportionality to the local value of the parietal shear, namely $U_p \propto \mu(\partial U/\partial n)_p$.

As for Stokes, the first condition he considers is indeed that of zero velocity at the wall, to reject it after evaluation: «*But having calculated, according to the conditions which I have mentioned, the discharge of long straight circular pipes and rectangular canals, and compared the resulting formulae with some of the experiments of Bossut and Dubuat, I found that the formulae did not at all agree with the experiment.*»

He therefore reverts to a slip condition to finally establish, by integrating his local equations, the velocity profile in a circular duct assuming a finite *non-zero* velocity at the wall.

4.6 The Simplified Models

4.6.1 The Concept of Restricted Models

The use of the general Navier-Stokes model presents a number of problems, both from a theoretical and application point of view.

– From the fundamental viewpoint first of all, the fundamental issues are about the *existence*, *unicity* and *stability* of the solutions of such a model. In particular, the questions of existence and unicity have still not been currently answered in the

Table 4.1 The restricted reference models

Evolution ⇒ Fluid ⇓	Compressible	Incompressible
Real	**R.C.F.** $--\rightarrow$	**R.I.F.**
Ideal	**I.C.F.** $--\rightarrow$	**I.I.F.**

general case. Nevertheless, partial results exist under some simplifications of the general case, such as the two-dimensional flow for example.
- From a more applicative perspective, not all the phenomena potentially included in the general model, necessarily have comparable levels of influence in different flow situations. For example, the compressibility of the air is not a determining factor in the trajectories of balls or balloons for recreational activities!
- Finally, the numerical resolution efficiency of a partial differential set of equations still depends on the performance of algorithms which fit the mathematical properties of the operators to be processed. In this respect, Navier-Stokes' general equations conceal the various types of difficulties associated with many "physical models", in the sense of Dautray & Lions [47].

At least for these three reasons, it seems appropriate to reduce the degree of generality of the Navier-Stokes model by applying *simplifying assumptions* which limit its use to specific *fluid flow situations*. This protocol results in the definition of *restricted models* of fluid motions. This concept is therefore closely linked to different flow specifying sets, each of them embracing a common problematic regarding the physical motion identity, the type of approach to the numerical resolution and, where applicable, some partial answers to the existence, uniqueness and stability issue. The various **restricted models** of fluid motions which we will define are based on three criteria relating to:

- the *fluid*: viscous or ideal;
- the *evolution*: incompressible or compressible;
- the *inertial to the viscosity force ratio*: asymptotically zero or infinite.

As shown in Table 4.1, the first two criteria can be combined in pairs to produce four distinct models, which will be use as a reference.

These models are organized in order of decreasing complexity referring to the:

- Real Compressible Fluid (**R.C.F.**)
- Real Incompressible Fluid (**R.I.F.**)
- Ideal Compressible Fluid (**I.C.F.**)
- Ideal Incompressible Fluid (**I.I.F.**)

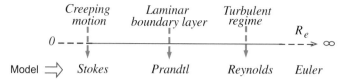

Fig. 4.1 Flow models with respect to the Reynolds number

We will then provide the detailed balance equations for each of these four restricted reference models in the continuation of this section.

As for the third classification criterion, we will see at a later stage that it will be closely dependent on the value of a dimensionless parameter called the Reynolds number (R_e). This will lead to the identification of different models depending on the relative influence of the viscous effects, as shown in the diagram of Fig. 4.1.

The presentation of the models derived from the Reynolds number classification will be discussed in Sect. 4.8.

4.6.2 The Real Compressible Fluid Flow Model

This **RCF** model is that of Navier-Stokes for a Newtonian viscous fluid ($\mu \neq 0$), heat conducting in the Fourier sense ($\lambda \neq 0$) and compressible in the sense of the *ideal gas*. We will add in this course the condition that the coefficients μ, λ, C_p and C_v are constant. The transport equations of the model are then:

$$
\begin{array}{lll}
Continuity & \dfrac{d\rho}{dt} + \rho \dfrac{\partial U_j}{\partial x_j} = 0 & (4.21) \\[3mm]
Momentum & \rho \dfrac{dU_i}{dt} = \rho F_i - \dfrac{\partial P}{\partial x_i} + \mu \left[\dfrac{\partial^2 U_i}{\partial x_j \partial x_j} + \dfrac{1}{3} \dfrac{\partial}{\partial x_i} \left(\dfrac{\partial U_l}{\partial x_l} \right) \right] & (4.22) \\[3mm]
Energy & \rho C_v \dfrac{dT}{dt} = -P \dfrac{\partial U_j}{\partial x_j} + \Phi_\nu + \lambda \dfrac{\partial^2 T}{\partial x_j \partial x_j} & (4.23)
\end{array}
$$

In the last equation, Φ_ν is the viscous dissipation rate according to relation (4.17). If we use the enthalpy h, Eq. (4.14), then we can note that:

$$
\frac{dh}{dt} = \frac{de}{dt} + \frac{1}{\rho} \frac{dP}{dt} - \frac{P}{\rho^2} \frac{d\rho}{dt},
$$

which, given the continuity equation, is also written as:

$$
\rho \frac{dh}{dt} = \rho \frac{de}{dt} + \frac{dP}{dt} + P \frac{\partial U_j}{\partial x_j}.
$$

In the enthalpic form, the energy equation is therefore still equivalent to:

$$\rho C_p \frac{\mathrm{d}T}{\mathrm{d}t} = \frac{\mathrm{d}P}{\mathrm{d}t} + \Phi_v + \lambda \frac{\partial^2 T}{\partial x_j \partial x_j} \qquad (4.24)$$

Finally, it can be verified that the dynamic equation is also written, in vector notation, as follows:

$$\frac{\partial \mathbf{V}}{\partial t} + \mathbf{grad}\left(\frac{V^2}{2}\right) + curl\mathbf{V} \wedge \mathbf{V} = \mathbf{F} - \frac{1}{\rho}\,\mathbf{grad}\,P + \nu\left(\Delta\mathbf{V} + \frac{1}{3}\mathbf{grad}\,(div\mathbf{V})\right)$$
$$(4.25)$$

The *RCF* model is intended to handle some of the most complex physical situations which can be considered in this course.

In particular, it allows us to include both viscosity and compressibility effects. An illustration of the manifestation of these phenomena is shown in Fig. 4.2.

Figure 4.2 is a false colour rendering of the iso-Mach curves of the flow field at Mach 3 around an obstacle which simulates a space shuttle forebody. This data has been obtained by F. Pavie [113] through the numerical resolution of the local Navier-Stokes equations in a two-dimensional plane geometry.

The first phenomenon which noticeably appears is the presence of shock waves. As we will see in Chap. 7, it is basically the consequence of the compressibility of the fluid moving at a high speed in the presence of a fixed solid obstacle. It is firstly illustrated here by the formation of a detached shock ahead of the shuttle's nose. A second wave system originates further downstream, as a result of the flow recompression which occurs here, due to the angular change in the cabin fuselage. This results in the presence of an oblique shock, downstream of the previous shock.

Thanks to the model used for this simulation, it is possible to observe a second phenomenon revealing the influence of the fluid viscosity. As can be observed in the figure, the foot of the oblique shock wave affects the so-called characteristic "λ-shape", which in particular causes the location of the recompression region to be located upstream of the variation in the fuselage angle (the white arrows in the figure). Such an observation can only be explained by the contribution of the viscosity effects in the vicinity of the wall, in the shock wave/boundary layer interaction which we will briefly describe in Chap. 7 Sect. 7.5.2.

4.6.3 The Ideal Compressible Fluid Flow Model

The *ICF* model derives from the previous model by substituting the ideal fluid assumption for those of the real, viscous Newtonian and heat conductive fluid, therefore inviscid and non-heat conductive. It is worth noticing that the *ideal gas* thermodynamic behaviour remains unchanged.

Accordingly, the transport equations of the *ICF* model are simply obtained from those of the previous case by setting $\mu = \lambda = 0$ and are written as follows:

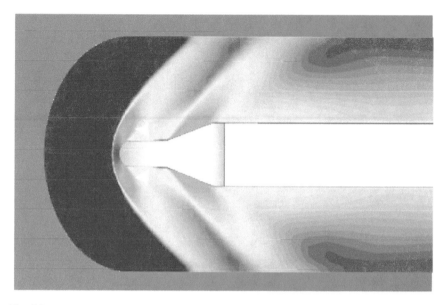

Fig. 4.2 Mach 3 flow on the forebody of a space shuttle as a solution of the **RCF** model. F. Pavie [113], courtesy of ENSICA

$$Continuity \quad \frac{\mathrm{d}\rho}{\mathrm{d}t} + \rho\frac{\partial U_j}{\partial x_j} = 0 \tag{4.26}$$

$$Momentum \quad \rho\frac{\mathrm{d}U_i}{\mathrm{d}t} = \rho F_i - \frac{\partial P}{\partial x_i} \tag{4.27}$$

$$Energy \quad \rho C_v\frac{\mathrm{d}T}{\mathrm{d}t} = -P\frac{\partial U_j}{\partial x_j} \tag{4.28}$$

The previous equations constitute the compressible extension of Euler's model (see Sect. 4.6.5). When no consideration is given to viscosity effects, this compressible Euler model is fully appropriate. In particular, it may be useful to determine the pressure field, in addition to the location of possible shock waves in the supersonic regime. An example of a flow simulation by the numerical computation solving these equations is shown in the following figure (Fig. 4.3).

It is the same flow configuration as previously described in Sect. 4.6.2 which is now investigated by numerically solving the equations of the **ICF** model. Compared to the flow field in Fig. 4.2, it can be firstly observed that the influence of the viscosity is negligible as far as the location and geometry of the detached shock upstream of the obstacle are concerned. On the other hand, the viscosity is decisive on the oblique shock in the vicinity of the wall, which, for the ideal fluid, originates directly at the angular variation point along the fuselage, i.e., in conjunction with the compression

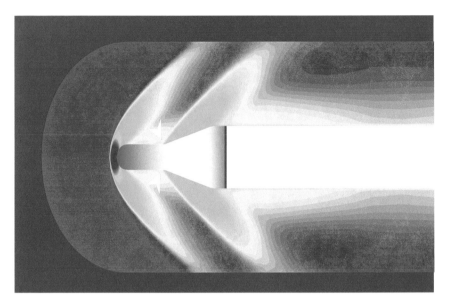

Fig. 4.3 Mach 3 flow on the forebody of a space shuttle as a solution of the compressible Euler model (*ICF*). F. Pavie [113], courtesy of ENSICA

ramp. Hence the shock exhibits no "λ" shape here, which confirms the viscous origin of this phenomenon.

4.6.4 *The Real Incompressible Fluid Flow Model*

The incompressibility character should be understood here as the absence of any fluid density change during the motion. As we have mentioned in the first chapter, Sect. 1.3.4.3, this type of situation is that of hydraulics and aerodynamics at a low Mach number, in practice $M < 0.2$ for the air, under the standard temperature and pressure conditions. This hypothesis should not be mistaken for that of the isovolume evolution, which defines a class of motions which includes but is not restricted to the incompressible case.

Considering the case of a homogeneous mass environment, the transport equations of the *RIF* model are deduced from those of the *RCF* model by setting $\rho = C^t$ and $div\mathbf{V} = 0$. Hence they are written:

$$Continuity \quad \frac{\partial U_j}{\partial x_j} = 0 \qquad (4.29)$$

$$Momentum \quad \rho\frac{dU_i}{dt} = \rho F_i - \frac{\partial P}{\partial x_i} + \mu\frac{\partial^2 U_i}{\partial x_j \partial x_j} \qquad (4.30)$$

$$Energy \quad \rho C_v \frac{dT}{dt} = +\Phi_\nu^i + \lambda\frac{\partial^2 T}{\partial x_j \partial x_j}, \qquad (4.31)$$

$$with \quad \Phi_\nu^i = \frac{\mu}{2}\left(\frac{\partial U_l}{\partial x_j} + \frac{\partial U_j}{\partial x_l}\right)^2 \qquad (4.32)$$

Two very important remarks should be made with regard to the previous equations:

- The *density* has a fixed prescribed value for this type of motion. Hence the first two continuity and momentum equations provide a closed set of four equations for the four scalar functions $U_i\,(M,t)$, $i = 1, 2, 3$ and $P\,(M,t)$. As a result, the determination of the temperature field can be decoupled from that of the velocity and pressure fields.
- The *pressure* now only plays a purely mechanical role, exclusively associated with the trace of the stress tensor. As an immediate physical consequence, the speed of sound propagation takes an infinite value.

In summary, it should be noted that:

In any incompressible flow, the pressure has no thermodynamic interpretation; The dynamic problem (velocity & pressure) is decoupled from the thermal problem, which can therefore be solved independently.

Finally we will note the vectorial formulation of the dynamic equation:

$$\frac{\partial \mathbf{V}}{\partial t} + \mathbf{grad}\left(\frac{V^2}{2}\right) + curl\,\mathbf{V} \wedge \mathbf{V} = \mathbf{F} - \frac{1}{\rho}\,\mathbf{grad}\,P - \nu\,curl\,(curl\,\mathbf{V}) \qquad (4.33)$$

To illustrate the balance between inertial, viscosity and pressure forces which governs the dynamics of this model in the absence of gravity, we consider, for example, the sudden enlargement of a two-dimensional plane flow. The velocity field in the laminar regime, as obtained by numerically solving the Navier-Stokes equations of the incompressible fluid, is plotted in Fig. 4.4. The Reynolds number based on the step height and the upstream velocity is $Re = 600$. The false colour rendering is obtained from the iso-values of the stream function according to a non-uniform law chosen so as to concentrate the colour on the extremum values of this function in the recirculating zones.

As we mentioned briefly in Chap. 3 Sect. 3.4.3.3, a first recirculation zone (here in yellow and red) appears due to the flow detachment immediately downstream of the downward facing step. This phenomenon is a direct consequence of the fluid viscosity and would not occur in an ideal fluid. A second recirculating zone (in pink

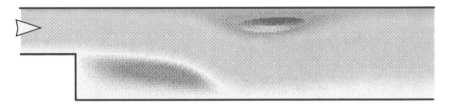

Fig. 4.4 Incompressible flow in a sudden plane expansion, as the numerical resolution of the isovolume Navier-Stokes equations. J.-B. Cazalbou & T. Lili [30], courtesy of ENSICA

and blue) can be seen on the flat upper wall of the duct, downstream of the previous region. Indeed, as a result of the slowing down of the fluid in the main stream (green-coloured in the figure), an adverse pressure gradient is created in the longitudinal direction. Located near the upper wall (bluish region) where the viscous forces are high, it induces the separation of the fluid threads in the pink marked region.

4.6.5   The Ideal Incompressible Fluid Flow Model

This model is obtained by setting $\mu = \lambda = 0$ in the previous model. The thermal problem then becomes trivial. Indeed, the energy equation is reduced to $dT/dt = 0$, which means that the temperature field remains *"frozen"* in its initial value along the paths of any fluid particle. In particular, if the initial temperature distribution is homogeneous throughout the flow field, then the evolution is isothermal. Physically, this is easily conceivable, since the heat transfer modes by diffusion and/or work–heat thermodynamic conversion are both incompatible with the **IIF** model assumptions. For these flows, only the dynamic field has therefore to be determined from the following transport equations:

$$
\begin{array}{lll}
Continuity & \dfrac{\partial U_j}{\partial x_j} = 0 & (4.34) \\[3mm]
Momentum & \rho \dfrac{dU_i}{dt} = \rho F_i - \dfrac{\partial P}{\partial x_i} & (4.35)
\end{array}
$$

It is worth remembering that the ideal fluid models we have presented are usually refered to as Euler models, respectively compressible and incompressible.

As we will justify in Chap. 11, the viscosity effects can be decoupled from those of the pressure in high Reynolds number flows without separation. This property then justifies determining the pressure as a solution of the Euler model. An example of such a pressure field for the two-dimensional steady flow past an airfoil is plotted in Fig. 4.5. It is an *Aerospace profile* at 15 degrees of incidence in a uniform velocity flow at infinity.

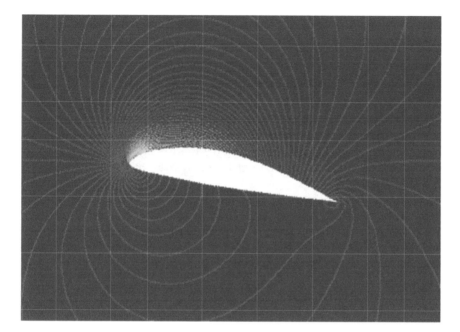

Fig. 4.5 Pressure field around an airfoil as the solution of the incompressible 2D Euler equations, J.-B. Cazalbou [31], courtesy of ENSICA

The iso-value pressure curves of the numerical simulation, plotted here in regular colour increments, clearly highlight the "*non-transportable*" character of this function. We will indeed see in Chap. 9 Sect. 9.2.5 that the pressure is a solution of a *Poisson equation*, irrespective of the viscous or ideal nature of the fluid in any isovolume evolution.

In this visualization, we can also discern the high pressure area (in pink) near the flow stagnation point at the leading edge of the wing. On the upper surface of the wing, a depression zone settles almost all along the trailing edge (orange-yellow curves). It is evidence of the flow acceleration on this upper surface with a resulting lifting force. In addition, the intrados is slightly over-pressurized, as demonstrated by the more pronounced spacing of the purple curves in this case. All these comments will be explained in detail in Chap. 6.

Remark Since the previous presentation was restricted to equations only, it will be necessary, in order to be able to rigorously speak about "models", to associate the initial and boundary conditions. In this respect, it should be noted that Euler models are characterized by the diminution of one unit in the order of the space derivatives, as compared to that of the Navier-Stokes equations. This degeneration in the model order will lead, as we will see subsequently, to the impossibility to fulfilling all the boundary conditions reflecting the viscous adhesion at a solid wall. A choice will therefore have to be made, in coherence with the Euler model order, which will result, on a solid, in a slip condition identical to that of an ideal (inviscid) fluid motion.

EXERCISE 18 Euler's equations in the Lagrangian formulation.

The Euler variables are referred to as (x, y, z, t) and the Lagrange variables as (a, b, c, t), with reference to the same orthonormal Cartesian reference frame.

QUESTION 1. Recall the expressions, in projections, of the Euler equations related to an incompressible fluid motion, by taking the velocity and pressure as functions of the Euler variables.

QUESTION 2. How are these equations written when the Lagrangian variables are used?

SOLUTION: **1.** Noting F_x, F_y and F_z as the components of the external body forces, the Euler equations are written as follows:

$$\begin{cases} \dfrac{\partial U}{\partial t} + U\dfrac{\partial U}{\partial x} + V\dfrac{\partial U}{\partial y} + W\dfrac{\partial U}{\partial z} = F_x - \dfrac{1}{\rho_0}\dfrac{\partial P}{\partial x}, \\[2mm] \dfrac{\partial V}{\partial t} + U\dfrac{\partial V}{\partial x} + V\dfrac{\partial V}{\partial y} + W\dfrac{\partial V}{\partial z} = F_y - \dfrac{1}{\rho_0}\dfrac{\partial P}{\partial y}, \\[2mm] \dfrac{\partial W}{\partial t} + U\dfrac{\partial W}{\partial x} + V\dfrac{\partial W}{\partial y} + W\dfrac{\partial W}{\partial z} = F_z - \dfrac{1}{\rho_0}\dfrac{\partial P}{\partial z}. \end{cases}$$

2. By using the Lagrangian variables, it is easy to directly transform the acceleration term in the previous relations and to write immediately:

$$\begin{cases} \dfrac{\partial^2 x}{\partial t^2} = F_x - \dfrac{1}{\rho_0}\dfrac{\partial P}{\partial x}, \\[2mm] \dfrac{\partial^2 y}{\partial t^2} = F_y - \dfrac{1}{\rho_0}\dfrac{\partial P}{\partial y}, \\[2mm] \dfrac{\partial^2 z}{\partial t^2} = F_z - \dfrac{1}{\rho_0}\dfrac{\partial P}{\partial z}, \end{cases}$$

where the pressure gradient components are still expressed as a function of the Euler variables. To turn to Lagrange's expressions, it is simply necessary to note that:

$$\frac{\partial P}{\partial a} = \frac{\partial P}{\partial x}\frac{\partial x}{\partial a} + \frac{\partial P}{\partial y}\frac{\partial y}{\partial a} + \frac{\partial P}{\partial z}\frac{\partial z}{\partial a},$$

and similarly for $\partial P/\partial b$ and $\partial P/\partial c$.

By multiplying the first equation in Euler variables by $\partial x/\partial a$, the second by $\partial y/\partial a$, the third by $\partial z/\partial a$ and adding throughout, we can express the first component of the pressure gradient $(\partial P/\partial a)$ with respect to the Lagrangian variables. A similar process leads to the other two components. Finally, the equations are written as follows:

$$
\begin{cases}
\left(\dfrac{\partial^2 x}{\partial t^2} - F_x\right)\dfrac{\partial x}{\partial a} + \left(\dfrac{\partial^2 y}{\partial t^2} - F_y\right)\dfrac{\partial y}{\partial a} + \left(\dfrac{\partial^2 z}{\partial t^2} - F_z\right)\dfrac{\partial z}{\partial a} + \dfrac{1}{\rho_0}\dfrac{\partial P}{\partial a} = 0, \\[2mm]
\left(\dfrac{\partial^2 x}{\partial t^2} - F_x\right)\dfrac{\partial x}{\partial b} + \left(\dfrac{\partial^2 y}{\partial t^2} - F_y\right)\dfrac{\partial y}{\partial b} + \left(\dfrac{\partial^2 z}{\partial t^2} - F_z\right)\dfrac{\partial z}{\partial b} + \dfrac{1}{\rho_0}\dfrac{\partial P}{\partial b} = 0, \\[2mm]
\left(\dfrac{\partial^2 x}{\partial t^2} - F_x\right)\dfrac{\partial x}{\partial c} + \left(\dfrac{\partial^2 y}{\partial t^2} - F_y\right)\dfrac{\partial y}{\partial c} + \left(\dfrac{\partial^2 z}{\partial t^2} - F_z\right)\dfrac{\partial z}{\partial c} + \dfrac{1}{\rho_0}\dfrac{\partial P}{\partial c} = 0.
\end{cases}
$$

4.7 The Characteristic Numbers of a Fluid Motion

4.7.1 The Dimensionless Formulation of the Local Equations

We address here quite a widespread approach in fluid mechanics, aiming to simplify the models using order of magnitude considerations, by retaining, in the equations, only the most dominant terms. The procedure is based on a dimensionless formulation which identifies dimensionless groupings or *characteristic numbers* to allow the relative importance of the different terms to be compared with each other. The dimensionless quantities are defined using *reference scales*, as specified in the following table.

With regard to the time variable, it should be observed in Table 4.2 that no specific scale is introduced for this variable. The corresponding dimensionless quantity will therefore be defined from the velocity and length scales,[4] according to $t^* = t\,\mathcal{U}/\mathcal{L}$.

Considering the scales as *fixed* parameters with respect to the space/time coordinates, we will now establish the dimensionless form of the **RCF** model's local equations, assuming that the physical properties μ, λ, C_p, C_v etc, are constant.

Table 4.2 Scaling of the main physical variables

Physical variable	Length (x_i)	Velociy (U_i)	Pressure (P)	Temperature (T)
Scale	$\mathcal{L}$	$\mathcal{U}$	$\mathcal{P}$	Θ
Dimensionless variable	$x_i^* = x_i/\mathcal{L}$	$U_i^* = U_i/\mathcal{U}$	$P^* = P/\mathcal{P}$	$T^* = T/\Theta$

---

[4] This choice is not mandatory and may even be inappropriate if the flow instability results from factors which are independent of the length and velocity scales. In other words, the analysis adopted here can be considered as representative of the sole steady motion. We will consider the unsteady flow regime in Chap. 9 Sect. 9.3.

4.7.2   The Dimensionless Continuity Equation

By introducing the dimensionless quantities, the continuity equation is written as:

$$\frac{\mathcal{U}}{\mathcal{L}}\frac{\partial \rho}{\partial t^*} + \frac{\mathcal{U}}{\mathcal{L}}\frac{\partial \rho U_j^*}{\partial x_j^*} = 0\,.$$

Thus, provided a density reference is specified, the dimensionless continuity equation is identical to its dimensional formulation.

4.7.3   The Dimensionless Momentum Equation

By applying the same operation to the momentum equation we obtain:

$$\rho\frac{\mathcal{U}^2}{\mathcal{L}}\left(\frac{\partial U_i^*}{\partial t^*} + U_j^*\frac{\partial U_i^*}{\partial x_j^*}\right) = \rho g_i - \frac{\mathcal{P}}{\mathcal{L}}\frac{\partial P^*}{\partial x_i^*} + \mu\frac{\mathcal{U}}{\mathcal{L}^2}\left(\frac{\partial^2 U_i^*}{\partial x_j^*\partial x_j^*} + \frac{1}{3}\frac{\partial}{\partial x_i^*}\left(\frac{\partial U_l^*}{\partial x_l^*}\right)\right),$$

or again, after dividing throughout by $\rho\mathcal{U}^2/L$ and introducing the module g of the gravity acceleration:

$$\left[\frac{\partial U_i^*}{\partial t^*} + U_j^*\frac{\partial U_i^*}{\partial x_j^*}\right] = \frac{\mathcal{L}g}{\mathcal{U}^2}\left[\frac{g_i}{g}\right] - \frac{\mathcal{P}}{\rho\mathcal{U}^2}\left[\frac{\partial P^*}{\partial x_i^*}\right] + \frac{\mu/\rho}{\mathcal{U}\mathcal{L}}\left[\frac{\partial^2 U_i^*}{\partial x_j^*\partial x_j^*} + \frac{1}{3}\frac{\partial}{\partial x_i^*}\left(\frac{\partial U_l^*}{\partial x_l^*}\right)\right].$$
$$(4.36)$$

Since the bracketed terms are dimensionless, the same applies to their respective multiplicative coefficients, which leads to introducing of the following numbers:

- the Froude number $\qquad\qquad F_r = \mathcal{U}^2/g\mathcal{L}\,,$ (4.37)
- the Euler number $\qquad\qquad\quad E_u = \mathcal{P}/\rho\mathcal{U}^2\,,$ (4.38)
- the Reynolds number $\qquad\quad R_e = \mathcal{U}\mathcal{L}/\nu\,.$ (4.39)

Equation (4.36) is then written as follows:

$$\left[\frac{\partial U_i^*}{\partial t^*} + U_j^*\frac{\partial U_i^*}{\partial x_j^*}\right] = \frac{1}{F_r}\left[\frac{g_i}{g}\right] - E_u\left[\frac{\partial P^*}{\partial x_i^*}\right] + \frac{1}{R_e}\left[\frac{\partial^2 U_i^*}{\partial x_j^*\partial x_j^*} + \frac{1}{3}\frac{\partial}{\partial x_i^*}\left(\frac{\partial U_l^*}{\partial x_l^*}\right)\right].$$
$$(4.40)$$

4.7.4   The Dimensionless Energy Equation

The analogous process for the energy equation, in the form of the enthalpy balance, leads to:

$$\rho C_p \Theta \frac{\mathcal{U}}{\mathcal{L}}\left(\frac{\partial T^*}{\partial t^*}+U_j^*\frac{\partial T^*}{\partial x_j^*}\right)=\frac{\mathcal{U}P}{\mathcal{L}}\left(\frac{\mathrm{d}P^*}{\mathrm{d}t^*}\right)+\mu\frac{\mathcal{U}^2}{\mathcal{L}^2}\Phi_D^*+\frac{\lambda\Theta}{\mathcal{L}^2}\left(\frac{\partial^2 T^*}{\partial x_j^*\partial x_j^*}\right),$$

or again, after division throughout by $\rho C_p \Theta \mathcal{U}/\mathcal{L}$:

$$\left[\frac{\partial T^*}{\partial t^*}+U_j^*\frac{\partial T^*}{\partial x_j^*}\right]=\frac{P}{\rho C_p \Theta}\left[\frac{\mathrm{d}P^*}{\mathrm{d}t^*}\right]+\frac{\mu}{\rho C_p \Theta}\frac{\mathcal{U}}{\mathcal{L}}\Phi_D^*+\frac{\lambda}{\rho C_p}\frac{1}{\mathcal{U}\mathcal{L}}\left[\frac{\partial^2 T^*}{\partial x_j^*\partial x_j^*}\right].$$
(4.41)

As before, the multiplying coefficients of the bracketed terms have no dimension and can be expressed on the basis of numbers which have been adopted by usage, namely:

- the Prandtl number $\qquad Pr = \mu C_p/\lambda,$ (4.42)
- the Eckert number $\qquad E_C = \mathcal{U}^2/C_p\Theta.$ (4.43)

We will then easily ascertain that Eq. (4.41) actually takes the following dimensionless form:

$$\left[\frac{\partial T^*}{\partial t^*}+U_j^*\frac{\partial T^*}{\partial x_j^*}\right]=E_u E_C\left[\frac{\mathrm{d}P^*}{\mathrm{d}t^*}\right]+\frac{E_C}{Re}\Phi_D^*+\frac{1}{Pr\,Re}\left[\frac{\partial^2 T^*}{\partial x_j^*\partial x_j^*}\right].$$
(4.44)

4.7.5   The Physical Interpretation of the Characteristic Numbers

We summarize in Table 4.3 the expressions of the main characteristic numbers with their classical physical meanings.

   In addition, we will note that the dimensionless groupings associated with the formulation of Eq. (4.33) and (4.40), do not allow us to define the "characteristic numbers" in a *unique* way, which explains some duplicating with, among other elements:

- the Brinkman number $\quad B_n = Pr \times E_c,$ (4.45)
- the Peclet number $\qquad P_e = P_r \times R_e.$ (4.46)

Table 4.3 The main characteristic numbers with their corresponding physical interpretations

Number	Expression	Interpretation
R_e	$\dfrac{\rho \mathcal{U}^2/\mathcal{L}}{\mu \mathcal{U}/\mathcal{L}^2}$	$Inertial\ Force$ / $Viscous\ Force$
$2E_u$	$\dfrac{\mathcal{P}}{\frac{1}{2}\rho \mathcal{U}^2}$	$Static\ Pressure$ / $Dynamic\ Pressure$
F_r	$\dfrac{\rho \mathcal{U}^2/\mathcal{L}}{\rho g}$	$Inertial\ Force$ / $Gravity\ Force$
Gr/R_e^2	$\dfrac{\rho g \alpha \Theta}{\rho \mathcal{U}^2/\mathcal{L}}$	$Buoyancy\ Force$ / $Inertial\ Force$
P_e	$\dfrac{\rho C_p \Theta \mathcal{U}/\mathcal{L}}{\lambda \Theta/\mathcal{L}^2}$	$Convective\ Heat\ Transport$ / $Diffusive\ Heat\ Transfer$
$\frac{1}{2}E_c$	$\dfrac{\frac{1}{2}\rho \mathcal{U}^2}{\rho C_p \Theta}$	$Kinetic\ Energy$ / $Enthalpy$
B_r	$\dfrac{\mu\,(\mathcal{U}/\mathcal{L})^2}{\lambda \Theta/\mathcal{L}^2}$	$Mechanical\ Dissipation$ / $Heat\ Diffusion$

Finally, in the presence of buoyancy forces due to the thermal fluid expansion, let us remember that we have already introduced in the first chapter, (Eq. 1.34)

- the Grashof number $Gr = g\alpha\,\Theta\mathcal{L}^3/\nu^2$,   (4.47)

where g is the gravity module and α the thermal dilatation coefficient of the fluid.

The selection of the *relevant* dimensionless parameters for a given flow is a major issue to which we will return in Chap. 9 Sect. 9.3.2. Apart from choices of use, we will see that it can be based, in general, on a time scale analysis of the transport equations.

4.7.6 Applications of the Dimensionless Analysis

The dimensionless analysis of the fluid motion's local transport equations can be used in several extensions. We will mention two main applications.

- **The flow similarity**: If two *different fluids* are set in motion under *different scale* conditions, but in such a way that the *characteristic numbers* of the two situations are *identical*, then these flows are governed by the *same dimensionless equations*. Therefore, if the initial and boundary conditions are similar, these equation solutions will also be similar. It is on this principle that all small scale model studies are based, both in hydraulics and aerodynamics. It should be noted, however, that the practical implementation of this *simple* principle, raises some difficult problems, insofar as the equality of all characteristic numbers between the model and the actual situation may not be possible. It is then necessary to adjust the *partial* representativeness of the model to the equality of the dimensionless parameters which are relevant for the preponderant effects, an assessment which requires the ability to make judicious compromises.

- **The general model simplification**: In many applications, it often happens that not all the effects present in the equations are of equal importance. Therefore, the dimensionless processing makes it possible to derive a simplification adapted to a given flow configuration or to a given region of a flow. This can be simply done by considering in the model, the predominant terms of the equations, i.e. which account for the major effects of each particular situation. It is on the basis of such analysis that new flow models can be defined, corresponding to the same balance type between forces and/or energies. In the following section (Sect. 4.8), we will specify in detail the classification which results from the relative importance of inertial and viscous forces.

4.7.7 The Vaschy-Buckingham or 'π' Theorem

The final consequence of using a *dimensionless* formulation mentioned in this section is the use of a specific theoretical tool. This is a very general theorem whose exploitation in fluid mechanics makes it possible to derive some key results in many applications.

4.7.7.1 The Theorem Statement

The *Vaschy–Buckingham* theorem is so called in reference to Vaschy's article, [163] published in 1892 and Buckingham's article [22] published in 1914. It is still called the « π » theorem. It consists of the following proposals:

Assumptions:	Let there be a physical phenomenon governed by a **dimensional** relation $f(G_1, G_2, ...G_n) = 0$ between n **dimensional** quantities $G_1, G_2, ... G_n$.
Conclusion 1 :	The dimensional relation between the n dimensioned physical quantities can be reduced to a **dimensionless** relation $\phi(\pi_1, \pi_2...\pi_p) = 0$ between the p **dimensionless** products $\pi_1, \pi_2...\pi_p$ formed from the n dimensional variables $G_1, G_2, ... G_n$.
Conclusion 2 :	The number of the dimensionless products (p) is equal to the total number of the dimensional quantities (n) minus the rank (r) of their dimension matrix, $p = n - r$.

4.7.7.2   Implementation of the π Theorem in Practice

We simply limit ourselves here to explaining how a set of dimensionless products can be derived from the significant dimensional physical quantities of a given problem. As far as the proof of the theorem is concerned, we can refer to the founding articles [22, 163]. Huerre's work [76] can also be used for a more recent demonstration.

The dimension $[G]$ of any physical quantity G can be expressed by a monome-type *dimensional function*, as the power of the fundamental units associated with the phenomenon. In (thermo)fluid mechanics, these are usually the length $[L]$, mass $[M]$, time $[T]$ and temperature $[\Theta]$. We can therefore set for any quantity G_j, $j = 1...n$:

$$[G_j] = [L]^{a_{1j}} \times [M]^{a_{2j}} \times [T]^{a_{3j}} \times [\Theta]^{a_{4j}}, \quad j = 1 \text{ to } n,$$

or still in matrix form:

	$[G1]$	$[G2]$ ...	$[Gj]$ ...	$[Gn]$
$[L]$	a_{11}	a_{12} ...	a_{1j} ...	a_{1n}
$[M]$	a_{21}	a_{22} ...	a_{2j} ...	a_{2n}
$[T]$	a_{31}	a_{32} ...	a_{3j} ...	a_{3n}
$[\Theta]$	a_{41}	a_{42} ...	a_{4j} ...	a_{4n}

The entire set of the $r \times n$ coefficients (a_{ij}) can be grouped in a matrix $(\mathbf{A})$, called the *dimensional matrix* of the problem, with:
- $i = 1, ...r$, where r is the number of the fundamental units involved (here $r = 4$),
- $j = 1, ...n$, where n is the number of the dimensional quantities of the problem.

The expressions of the dimensionless products π_α, $\alpha = 1$ to p, are determined from a subset $\{G_1, G_2, ... G_r\}$ of the n initial quantities, by relationships of the form:

$$\pi_1 = \frac{G_{r+1}}{G_1^{b_{11}} \times G_2^{b_{12}}, ...\times G_r^{b_{1r}}}, \quad ... \quad \pi_\alpha = \frac{G_{r+\alpha+1}}{G_1^{b_{\alpha1}} \times G_2^{b_{\alpha2}}, ...\times G_r^{b_{\alpha r}}}, \quad ...$$

$$\pi_p = \frac{G_n}{G_1^{b_{p1}} \times G_2^{b_{p2}}, ...\times G_r^{b_{pr}}}.$$

A practical method for determining the coefficients b_{kl}, $k = 1$ to p and $l = 1$ to n is to solve p linear systems in matrix form:

$$(\mathbf{A})(\mathbf{B}_k) = 0, \quad k = 1 \text{ to } p,$$

where $(\mathbf{A})$ is the $(r \times n)$ dimensional matrix of the problem and $(\mathbf{B}_k)$ the p column vectors whose components are defined by :

$$(\mathbf{B}_1) = \begin{pmatrix} b_{11} \\ b_{12} \\ . \\ . \\ b_{1r} \\ 1(l \equiv r + 1) \\ 0 \\ 0 \\ . \\ . \\ 0 \end{pmatrix} \quad , \quad (\mathbf{B}_2) = \begin{pmatrix} b_{21} \\ b_{22} \\ . \\ . \\ b_{2r} \\ 0 \\ 1(l \equiv r + 2) \\ 0 \\ . \\ 0 \end{pmatrix} \quad \dots \quad (\mathbf{B}_k) = \begin{pmatrix} b_{k1} \\ b_{k2} \\ . \\ . \\ b_{kr} \\ 0 \\ 0 \\ . \\ 1(l \equiv k) \\ . \\ 0 \end{pmatrix}$$

4.8   The Incompressible Flow Models According to the Reynolds Number

In any incompressible flow, we have seen that the dynamic and thermal problems are decoupled. This leads us to mainly focus on the momentum equation in order to consider possible simplifications according to the relative importance of inertia and viscous forces. As we have explained in the previous section, this amounts to making a flow configuration classification according to the values of the corresponding Reynolds number.

Under the incompressibility assumption, the dimensionless equation (4.39) is simply:

$$\left[\frac{\partial U_i^*}{\partial t^*} + U_j^* \frac{\partial U_i^*}{\partial x_j^*} \right] = \frac{1}{Fr} \left[\frac{g_i}{g} \right] - Eu \left[\frac{\partial P^*}{\partial x_i^*} \right] + \frac{1}{Re} \frac{\partial^2 U_i^*}{\partial x_j^* \partial x_j^*} . \qquad (4.48)$$

To make the analysis, we assume that the characteristic scales have been chosen such that the dimensionless groups in square brackets are of the same order of magnitude.

Thus, all terms in Eq. (4.48) are of comparable magnitude if $Fr \sim Eu \sim Re \sim 1$, with reference to the unit coefficient of the inertia term.

4.8.1   The Creeping Motion: Stokes' Model

This first model applies to high-viscosity fluid flows, whose characteristic Reynolds number, within the limit $\mu \to \infty$, is asymptotically zero. In practice, this situation concerns fluid motions where the velocity is very low and the fluid viscosity very high. Under these conditions, the Euler number is very high and the Froude number very small. Taking these three conditions into account, ($Re \ll 1$, $Fr \ll 1$, $Eu \gg 1$), the inertia terms can be legitimately disregarded, so that the dimensional momentum equation is reduced to:

$$\rho g_i - \frac{\partial P}{\partial x_i} + \mu \frac{\partial^2 U_i}{\partial x_j \partial x_j} = 0 \tag{4.49}$$

The model we have just set out is known as the Stokes model and applies to creeping motions. Some of its properties will be discussed in Chap. 10.

4.8.2 The Infinite Reynolds Number Flow: Euler's Model

The incompressible ideal fluid model (*IIF*—Sect. 4.6.5) can be considered as the second limit case for $R_e = \infty$. Assuming that the characteristic scale of the pressure forces can be estimated from the dynamic pressure based on the velocity reference scale, it can be deduced that $E_u \sim 1$. We also have $F_r^{-1} \ll 1$ and obviously $R_e^{-1} \ll 1$, so that the momentum equation, as reduced to the main order terms, is written as:

$$\rho \left(\frac{\partial U_i}{\partial t} + U_j \frac{\partial U_i}{\partial x_j} \right) = -\frac{\partial P}{\partial x_i} \tag{4.50}$$

which is indeed the Euler model of the incompressible inviscid fluid motion.

4.8.3 The Thin Shear Layer Laminar Flow: Prandtl's Model

When the Reynolds number is high but not infinite, the previous model leads to some paradoxes for viscous fluid flows past solids at significant velocity values. It is no longer justified, in such cases, to neglect the viscosity effects, and in particular the diffusion by the molecular agitation, *throughout the entire flow field*. If the characteristic time scale of the diffusive transfer is equivalent to that of the advective transport, we will see that it is then possible to restrict the viscosity-influencing region to a particular area of the flow field, called the *boundary layer*. The motion is then governed by the Prandtl model which we will describe in detail in Chap. 9.

One of the major characteristics of the boundary layer type motions is the existence of a preferred advection direction. This physical characteristic is mathematically reflected in a model consisting of partial differential equations of a *parabolic type* with respect to the space variables. Numerical resolution methods can take advantage of this feature to reduce the memory size required for computing such flows with appropriate algorithms, which confirms, as noted in the introduction to the chapter, the close interconnection between the physical flow properties, the mathematical characteristics of the equations and the specificities of the solving methods which are associated within the very '*flow model concept*'.

Chapter 5
Flow Classes

Abstract After introducing distinctive classes of fluid flows, the main properties of each of them are discussed in detail. The fundamental theorems of fluid mechanics are then established. Application examples are dealt with at the end of the chapter.

5.1 Introduction

A well-known specific feature in teaching fluid mechanics is the segmentation of the discipline into various parts whose relevance is not always obviously apparent at first glance and often only becomes clear after the fact, in relation to its applications.

The result is a certain pedagogical difficulty for the teacher and often for the student, a learning difficulty due to the impression of a lack of coherence which makes the overall view difficult to grasp at first reading.

In addition, the introduction of '*various different assumptions*' sometimes gives the impression of an '*approximative construction*', where the deductive approach seems to be deliberately abandoned in favour of an '*ad hoc*' progression.

In an attempt to avoid this type of criticism it seems worthwhile to briefly provide some methodological information and to firstly specify the criteria of the classification which is the basis of this clarification.

In order to avoid any confusion among the criteria discussed here, it is important to distinguish two concepts: on the one hand, the *fluid motion* **model** and, on the other hand, the *flow* **class**. As we have discussed in the previous chapter, the model basically refers to a set of equations. Flows which are part of the same model are therefore solutions of the same equations, for a variety of initial and boundary conditions.

A class, on the other hand, includes flows which may belong to different models, but which share the same common characteristic, constituting the *distinctive class feature*. The *model* and *class* are therefore to be considered as two '**crossed**' concepts.

In this chapter, we will focus more specifically on those classes whose distinctive feature is one of the following:

- an *irrotational* flow;
- an *isovolume* flow;
- a *stationary* flow;
- an *irrotational* **and** *isovolume* flow.

© The Author(s), under exclusive license to Springer Nature Switzerland AG 2022
P. Chassaing, *Fundamentals of Fluid Mechanics*,
https://doi.org/10.1007/978-3-031-10086-4_5

5.2   The Irrotational Flow Class

5.2.1   Definition

The distinctive property of this class is obviously expressed by:

$$\forall M, \quad \forall t, \quad curl\, \mathbf{V}\,(M,t) = \mathbf{0}\,. \qquad (5.1)$$

It is therefore a *local* condition which, let us recall, expresses the fact that the motion
of any fluid particle is carried out *without rotation* of the latter on itself. In other
words, the fluid particles can move with deformation but along axes simply moving
by translation between the different particle positions along its trajectory, as outlined
in the diagram below (Fig. 5.1).

Important properties are associated to *irrotational* velocity flow fields. We will
have the opportunity to study them in more detail for the two-dimensional plane
configuration in Chap. 6. We will limit ourselves here to introducing some general
results which do not require any particular assumption on the viscous or inviscid
nature of the fluid, nor on its compressibility or incompressibility. This confirms
the specificity of the *class concept*, in the sense which has just been defined, since
this class embraces several of the restricted models which have been introduced in
Chap. 4. The irrotational motion study is important in itself, but also in reference to
the opposite situation, from which the specificities can be more clearly distinguished.

5.2.2   The Velocity Potential

The velocity vector irrotationality is a necessary and sufficient condition for the
existence of a scalar field function $\phi\,(M,t)$ such that, at each time t and any M point
in the flow field, we have :

$$\boxed{\mathbf{V}\,(M,t) = \mathbf{grad}\,\phi\,(M,t)} \qquad (5.2)$$

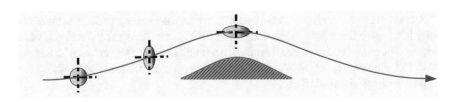

Fig. 5.1 Fluid particle deformation in an irrotational flow

Fig. 5.2 Schematic for the orthogonality of the equipotential surfaces and streamlines

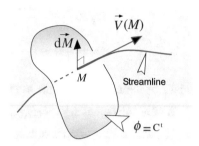

The function $\phi(M, t)$, whose dimension is $[L]^2 \times [T]^{-1}$, is called the *velocity potential* function.

Noting that $d\phi = \mathbf{grad}\phi . \mathbf{dM}$, we immediately deduce from Eq. (5.2) that at any time:

$$d\phi = \mathbf{V}.\mathbf{dM} .$$

Thus, when the displacement vector $\mathbf{dM}$ is taken in the plane orthogonal to the streamline at the M point, we have $d\phi = 0$. In other words, the streamlines normally cross the iso-ϕ surfaces called the *equipotential surfaces* (Fig. 5.2).

We will see at a later stage (Sect. 5.5.1) how to determine, in general, the potential function for the isovolume flow class which we will examine in the following section.

EXERCISE 19 The velocity potential and irrotational flow.

Let us consider a steady flow, whose velocity components, in a rectangular coordinate system, are:

$$U (x, y, z) = -x + y + z ,$$
$$V (x, y, z) = x - y + z ,$$
$$W (x, y, z) = x + y - z .$$

QUESTION 1. Show that this field is that of an irrotational flow.
QUESTION 2. Find the velocity potential function, whose value at the origin is zero.

SOLUTION: **1.** It is straightforward to verify that:

$$\Omega_x = \frac{\partial W}{\partial y} - \frac{\partial V}{\partial z} = 1 - 1 = 0 ,$$

$$\Omega_y = \frac{\partial U}{\partial z} - \frac{\partial W}{\partial x} = 1 - 1 = 0 ,$$

$$\Omega_z = \frac{\partial V}{\partial x} - \frac{\partial U}{\partial y} = 1 - 1 = 0 ,$$

which by definition proves the irrationality of the flow.

2. The corresponding velocity potential is such that: $\dfrac{\partial \phi\,(x, y, z)}{\partial x} = U = -x +$ $y + z$, whose integration results in $\phi\,(x, y, z) = -\frac{1}{2}x^2 + xy + xz + f\,(y, z)$.

Now, by taking the derivative with respect to y, we obtain:
$\dfrac{\partial \phi\,(x, y, z)}{\partial y} = x + \dfrac{\partial f\,(y, z)}{\partial y} \equiv V = x - y + z$ hence, after integrating in y:
$f\,(y, z) = -\frac{1}{2}y^2 + yz + g\,(z)$.

It is again deduced that $\dfrac{\partial \phi\,(x, y, z)}{\partial z} = x + y + \dfrac{dg}{dz} \equiv W = x + y - z$, whose integration finally leads to $\phi\,(x, y, z) = -\frac{1}{2}\left(x^2 + y^2 + z^2\right) + xy + yz + zx$, by taking $\phi\,(0, 0, 0) = 0$.

5.2.3   The Acceleration Potential

> The acceleration at any point of a flow with a twice continuously differentiable velocity potential, derives from a potential function.

This property is the direct consequence of the sole irrotational nature of the flow (see demonstration below). If we refer to $\mathcal{A}(M, t)$ as the potential acceleration function, we have indeed:

$$\frac{d\mathbf{V}}{dt} = \mathbf{grad}\ \mathcal{A} \quad \text{with} \quad \mathcal{A} = \frac{\partial \phi}{\partial t} + \frac{V^2}{2} .$$

Proof As we have stated in the first chapter, the general expression of the acceleration in terms of the Eulerian variables is written in vectorial form:

$$\frac{d\mathbf{V}}{dt} = \frac{\partial \mathbf{V}}{\partial t} + \mathbf{grad}\left(\frac{V^2}{2}\right) + curl\ \mathbf{V} \wedge \mathbf{V} .$$

In an irrotational flow, the expected result is straightforward since:

$$\frac{d\mathbf{V}}{dt} = \frac{\partial}{\partial t}\,(\mathbf{grad}\ \phi) + \mathbf{grad}\left(\frac{V^2}{2}\right) \equiv \mathbf{grad}\left(\frac{\partial \phi}{\partial t} + \frac{V^2}{2}\right) .$$

5.2.4   The Velocity Circulation Expression

As we have introduced in Chap. 2 Sect. 2.4.3, the elementary velocity circulation along an elementary oriented arc $\mathbf{t}\,d\ell$ is, by definition:

$$d\Gamma = \mathbf{V}.\mathbf{t}\,d\ell .$$

In consideration of relation (5.2), we still have $d\Gamma = \mathbf{grad}\,\phi.\mathbf{t}\,d\ell \equiv d\phi$, where the variation of the function is taken along the curve element. In an integral form, it should be noted that:

$$\boxed{\Gamma_{A'B} = \int_A^B d\Gamma = \phi_B - \phi_A} \qquad (5.3)$$

According to Stokes' theorem (see Chap. 2 Sect. 2.4.3), we have, along any closed contour of an irrotational flow field:

$$\Gamma_O \equiv \oint d\Gamma = 0 .$$

EXERCISE 20 The velocity potential in a two-dimensional plane flow.

Let there be the potential function of a two-dimensional stationary plane flow defined by $\phi(x, y) = \frac{1}{2}k\left(x^2 - y^2\right)$, where k is a non-zero real number.

QUESTION 1. Express the corresponding velocity field.

QUESTION 2. Determine the streamline equations and interpret the result.

QUESTION 3. Verify that the flow is irrotational. Deduce that the velocity field is written $\mathbf{V} = \mathbf{grad}\,[\psi(x, y)] \wedge \mathbf{k}$, where $\mathbf{k}$ is the unit vector orthogonal to the flow plane and ψ a function whose expression is to be given, prescribing that $\psi(0, 0) = 0$.

SOLUTION: **1.** By definition of the potential function we have:

$$U \equiv \frac{\partial \phi}{\partial x} = kx \;\; and \;\; V \equiv \frac{\partial \phi}{\partial y} = -ky .$$

2. The general differential equations of the streamlines $dx/U = dy/V$ can be written here, in any point different from the origin:

$$\frac{dx}{kx} = -\frac{dy}{ky} \;\; or \;\; k\,(ydx + xdy) = 0 ,$$

whose integration immediately gives $xy = C'$.
The streamlines of this flow are therefore equilateral hyperbolas.

3. The requested relation will be subsequently demonstrated (see Sect. 5.3.3.1) and we will only calculate here the expression of the function $\psi\,(x, y)$, knowing then that:

$$U = \frac{\partial \psi}{\partial y} \quad \text{and} \quad V = -\frac{\partial \psi}{\partial x}\,.$$

The integration of the first relation provides $\psi\,(x, y) = kxy + f\,(y)$ which, by substitution in the second relation, leads to $f\,(y) = C^t$. Taking into account the condition $\psi\,(0, 0) = 0$, the solution is finally written $\psi\,(x, y) = kxy$. Hence we retrieve, for the iso-ψ curves, the streamline equations. We will demonstrate this result in a more general way in Sect. 5.3.3 and will also revisit this flow somewhat later (see Exercise 21).

5.2.5 The Barotropy Relation in an Ideal Fluid

Assuming that there is an external body force potential, the following result can be stated:

> Any *irrotational* flow of an ideal fluid is necessarily barotropic.

Proof Noting $\mathcal{F}$ as the potential function of the external body forces, the momentum equation of an *ideal fluid* in an irrotational flow results in

$$\frac{\partial \mathbf{V}}{\partial t} + \mathbf{grad}\left(\frac{V^2}{2}\right) = \mathbf{grad}\,\mathcal{F} - \frac{1}{\rho}\mathbf{grad}\,P\,.$$

Remembering that there is also a velocity potential, we can still write:

$$\frac{1}{\rho}\mathbf{grad}\,P = \mathbf{grad}\left(\mathcal{F} - \frac{\partial \phi}{\partial t} - \frac{V^2}{2}\right).$$

Then taking the curl throughout, we deduce:

$$curl\left(\frac{1}{\rho}\mathbf{grad}\,P\right) \equiv \frac{1}{\rho}curl\,(\mathbf{grad}\,P) - \mathbf{grad}\,P \wedge \mathbf{grad}\left(\frac{1}{\rho}\right) = \mathbf{0}\,.$$

The first term on the right hand side being identically zero, it can be deduced that there is a barotropic relation $f\,(P, \rho) = 0$ in the form:

$$\mathbf{grad}\,P \wedge \mathbf{grad}\,\rho = 0.$$

The previous equation physically reflects the fact that the isobaric and the iso-density surfaces are identical. We will see later in Chap. 8 that the isentropic flow of an ideal compressible fluid is an example of a barotropic situation, the corresponding relations being nothing more than Laplace's law $P/\rho^\gamma = C^t$.

5.3 The Isovolume Flow Class

5.3.1 Definition

Another important class of flows is where the following condition is fulfilled *at any time*:

$$\forall M, \ \forall t, \quad div\mathbf{V}(M,t) = 0. \tag{5.4}$$

Such a velocity vector field, whose divergence is zero, is called a *solenoidal* field. It is worth recalling that, by definition, the flow class concept encompasses various fluid motion models. This is particularly the case here, where the results obtained in the present section will equally apply to both viscous and ideal fluids.

As we have explained in the first chapter Sect. 1.3.4.2, condition (5.4) refers to the conservation of the elementary fluid particle volume during its motion. In other words, any fluid particle translation, rotation or deformation takes place without reducing or increasing its elementary volume.

5.3.2 The Velocity Vector Potential

The divergence-free condition $(div\mathbf{V} = 0)$ of the velocity vector field in an isovolume flow ensures the existence of a vector function $\mathbf{\Psi}(M,t)$ such that:

$$\mathbf{V}(M,t) = curl\,[\,\mathbf{\Psi}(M,t)\,]. \tag{5.5}$$

The vector function $\mathbf{\Psi}(M,t)$, whose dimension is $[L]^2 \times [T]^{-1}$, is called the *velocity potential vector*.[1] It generally depends on the Eulerian variables, reduced to the space coordinates in a *stationary motion*.[2]

As shown by the definition relation (5.5), the potential vector function exists even in a three-dimensional situation.

---

[1] This function is not to be confused with that of the *scalar* velocity potential introduced in the previous section.

[2] In the unsteady case, the considered properties will apply *at any time* to streamlines which can *vary over time*.

5.3.3   The Two-Dimensional Case Restriction

5.3.3.1   The Vector Potential Expression

We are concerned here with *two-component* velocity fields in a *two-dimensional geometry*, as previously defined[3] (see Additional Information in Chap. 2 Sect. 2.1.1.3). In such cases, it is easy to verify, that *at any time*,[4] the vector Ψ (x_1, x_2) can be taken as:

$$\Psi (x_1, x_2) = \psi (x_1, x_2)\, \mathbf{k}\,,$$

where $\mathbf{k}$ denotes the unit normal vector to the flow plane (x_1, x_2).
The vector potential is therefore reduced to a single scalar component $\psi(x_1, x_2)$, with, from relation (5.5):

$$\mathbf{V} = curl\,(\psi \mathbf{k}) = \mathbf{grad}\;\psi \wedge \mathbf{k}\,. \tag{5.6}$$

Table 5.1 specifies in detail relation (5.6) for various two-dimensional geometry conditions and two-component velocity fields, namely:

– the 2D *Cartesian plane* or invariance by translation on a space coordinate, z for instance, with in this case $\frac{\partial}{\partial z} \equiv 0$;
– the 2D *polar plane* or invariance by translation on the axial coordinate z, similar to the previous situation in polar coordinates (r, θ);
– the 2D *axisymmetric* or invariance by translation on an angular coordinate, which corresponds geometrically to an invariance by rotation about an axis, hence, in cylindrical (x, r, θ) or spherical (r, θ, φ) coordinates: $\frac{\partial}{\partial \theta} \equiv 0$, for instance.

5.3.3.2   The Stream Function: Definition and Properties

In any two-dimensional flow, since the vectors $\mathbf{grad}\;\psi$ and $\mathbf{grad}\;\psi \wedge \mathbf{k}$ are necessarily orthogonal, it immediately follows from relation (5.6) that:

$$\mathbf{V} \cdot \mathbf{grad}\psi = 0\,.$$

The space variation of the function ψ is thus orthogonal to the velocity vector. In other words, this function remains constant along any streamline whose generic equation is none other than $\psi = C^t$, at any time. This is the reason for calling the ψ function the *stream function*.

---

[3] The *two-dimensional flow* specification of the section title is a simple '*standard shortcut*', consecrated by usage, but which can be confusing.

[4] In the rest of this paragraph, the time dependency will no longer be made explicit in the arguments of the functions.

Table 5.1 Components of the velocity vector potential in various 2D-geometries

| | 2D-plane flow | | 2D—axisymmetric flow | |
	Cartesian	Polar	Cylindrical	Spherical
Velocity components	$U(x,y)$ $V(x,y)$	$V_r(r,\theta)$ $V_\theta(r,\theta)$	$V_r(r,z)$ $V_z(r,z)$	$V_r(r,\varphi)$ $V_\varphi(r,\varphi)$
Stream function	$\psi(x,y)$	$\psi(r,\theta)$	$\psi(r,z)$	$\psi(r,\varphi)$
Relations	$U = \dfrac{\partial\psi}{\partial y}$ $V = -\dfrac{\partial\psi}{\partial x}$	$V_r = \dfrac{1}{r}\dfrac{\partial\psi}{\partial\theta}$ $V_\theta = -\dfrac{\partial\psi}{\partial r}$	$V_r = \dfrac{1}{r}\dfrac{\partial\psi}{\partial z}$ $V_z = -\dfrac{1}{r}\dfrac{\partial\psi}{\partial r}$	$V_r = \dfrac{1}{r^2\sin\varphi}\dfrac{\partial\psi}{\partial\varphi}$ $V_\varphi = -\dfrac{1}{r\sin\varphi}\dfrac{\partial\psi}{\partial r}$

We will now present two important properties related to:

• The *Stream function and Flow rate*:
From the result established in Chap. 2 Sect. 2.2.3, it can be stated that:

> The flow rate between two streamlines is independent of the path connecting the points on these lines and is simply obtained by the difference of the (constant) stream function values referring to each of these lines.

In other words, if A and B are two points taken respectively on two separate streamlines $\psi = \psi_A$ and $\psi = \psi_B$, then:

$$Q^v_{AB} = \int_A^B \mathbf{V}.\mathbf{n}\,dl = \psi_B - \psi_A. \tag{5.7}$$

• The *Stream function and Rotational*[5]:
Given relation (5.6), the curl of the velocity vector field is expressed by:

$$\mathbf{\Omega} \equiv curl\,\mathbf{V} = curl\,(curl\,(\psi\mathbf{k})) = \mathbf{grad}\,(div\,(\psi\mathbf{k})) - \Delta\psi\,\mathbf{k}.$$

Now, according to the vector identities mentioned in the appendix, we have:

$$div\,(\psi\mathbf{k}) = \psi\,div\mathbf{k} + \mathbf{k}.\mathbf{grad}\psi.$$

The first term on the right hand side is zero since $\mathbf{k}$ is a fixed vector; the second term is also zero, due to the vector orthogonality in the scalar product.
The final result is now written:

---

[5] The curl of the velocity can be equally referred to as the velocity rotational, or simply the rotational. A vector field whose curl is zero is called irrotational.

$$\boxed{\Omega = -\Delta\psi} \tag{5.8}$$

where Ω refers to the only non-zero component of the vorticity field.

Hence, it can be stated that:

> In any two-dimensional flow, the only non-zero component of the vorticity is, at each time, opposite to the Laplacian of the stream function.

Equation (5.8) also provides a significant way to determine the stream function in any irrotational flow, where this function can be obtained by solving the Laplace equation $\Delta\psi = 0$ in the domain. We will have the opportunity to return to this property at a later stage.

5.4   The Stationary Flow Class

5.4.1   Definition

By definition (see first chapter Sect. 1.6.10), the *stationary flow* class is such that any field function $f(M, t)$ fulfills at any point the condition:

$$\forall M, \quad \frac{\partial f(M, t)}{\partial t} = 0. \tag{5.9}$$

Under this assumption, the continuity equation is reduced to:

$$div(\rho\mathbf{V}) = 0, \tag{5.10}$$

which applies to any fluid of *any density*.

5.4.2   The Mass-Weighted Velocity Vector Potential

Relation (5.10) provides a necessary and sufficient condition for the existence of a vector potential, $\mathbf{\Psi}_\rho(M)$, for the *mass-weighted* velocity.

Denoting by ρ_0 a fixed density reference, we can therefore write:

$$\rho(M)\mathbf{V}(M) = curl\left[\rho_0\mathbf{\Psi}_\rho(M)\right] = \rho_0 curl\left[\mathbf{\Psi}_\rho(M)\right],$$

where the dimension $[L]^2 \times [T]^{-1}$ of the function $\mathbf{\Psi}_\rho(M)$ remains the same as that of the *isovolume* flow vector potential. The velocity vector is now given by

$$\mathbf{V}(M) = \frac{\rho_0}{\rho(M)} \; curl \; [\mathbf{\Psi}_\rho(M)]. \tag{5.11}$$

This relation presumes nothing about the fluid's actual density value and therefore applies, in particular, to *compressible* flows.

To summarize, by grouping all the elements introduced in Sects. 5.3.2 and 5.4.2, we can conclude that the velocity vector potential exists in either of the following two cases:

> - an *isovolume* motion with $\mathbf{\Psi}(M, t)$, so that:
> $$\mathbf{V}(M, t) = curl \, [\mathbf{\Psi}(M, t)];$$
> - a *stationary* motion with $\mathbf{\Psi}_\rho(M)$, so that:
> $$\mathbf{V}(M) = \frac{\rho_0}{\rho(M)} \; curl \, \big[\mathbf{\Psi}_\rho(M)\big].$$

5.4.3 The Two-Dimensional Restriction

- The **stream function**

In a two-dimensional situation, the vector potential of the mass-weighted velocity field $\mathbf{\Psi}_\rho(M)$ in any steady flow is reduced, as in the isovolume case, to a scalar function, i.e.

$$\mathbf{\Psi}_\rho(M) = \psi_\rho(M) \, . \, \mathbf{k}, \tag{5.12}$$

where $\mathbf{k}$ stands for the unit vector in the geometric invariance direction. This function provides access here to the *mass flow rate* (Q^m).

Indeed, as an immediate extension of relation (5.7), if A and B are two points on two separate streamlines, with respectively $\psi_\rho = \psi_{\rho A}$ and $\psi = \psi_{\rho B}$, then:

$$Q_{AB}^m = \int_A^B \rho \mathbf{V}.\mathbf{n} \, dl = \psi_{\rho B} - \psi_{\rho A}. \tag{5.13}$$

The function ψ_ρ is therefore again equivalent to the flow *stream function*.
- The **velocity rotational**

By designating Ω as the only non-zero component of the vorticity, the relation equivalent to Eq. (5.8) in the isovolume case, is now written:

$$\boxed{\Omega = -div\left(\frac{\rho_0}{\rho}\mathbf{grad} \, \psi_\rho\right) \mathbf{k}} \tag{5.14}$$

Proof By definition of the stream function ψ_ρ we have $curl \, \mathbf{V} = curl \left[\frac{\rho_0}{\rho} curl \left(\psi_\rho \mathbf{k}\right)\right]$. Using the vector identity expressing $curl(p\mathbf{A})$ (see Appendix C), this expression can be developed as:

$$curl\ \mathbf{V} = \frac{\rho_0}{\rho}\ curl\ \left[curl\ \left(\psi_\rho \mathbf{k}\right)\right] + \mathbf{grad}\left(\frac{\rho_0}{\rho}\right) \wedge curl\ \left(\psi_\rho \mathbf{k}\right)$$

As previously, see Sect. 5.3.3.2, $curl\ \left[curl\ \left(\psi_\rho \mathbf{k}\right)\right] = -\Delta\psi_\rho\ \mathbf{k}$, hence:

$$curl\ \mathbf{V} = -\frac{\rho_0}{\rho}\Delta\psi_\rho\ \mathbf{k} + \mathbf{grad}\left(\frac{\rho_0}{\rho}\right) \wedge curl\ \left(\psi_\rho \mathbf{k}\right)\ .$$

Now developing, $curl\ \left(\psi_\rho \mathbf{k}\right)$ it can be deduced (see the identity in the Appendix):

$$curl\ \left(\psi_\rho \mathbf{k}\right) \equiv \psi_\rho curl\ \mathbf{k} + \mathbf{grad}\ \psi_\rho \wedge \mathbf{k} = \mathbf{grad}\ \psi_\rho \wedge \mathbf{k}\ ,$$

the first term of the right hand side being identically zero for a constant vector. Using this last equality, we can therefore express:

$$curl\ \mathbf{V} = -\frac{\rho_0}{\rho}\Delta\psi_\rho\ \mathbf{k} + \mathbf{grad}\left(\frac{\rho_0}{\rho}\right) \wedge \left(\mathbf{grad}\ \psi_\rho \wedge \mathbf{k}\right)\ .$$

Knowing that $\mathbf{U} \wedge (\mathbf{V} \wedge \mathbf{W}) = (\mathbf{U}.\mathbf{W})\mathbf{V} - (\mathbf{U}.\mathbf{V})\mathbf{W}$ (see for example Coirier [40]), we obtain:

$$curl\ \mathbf{V} = -\frac{\rho_0}{\rho}\Delta\psi_\rho\ \mathbf{k} + \left[\mathbf{grad}\left(\frac{\rho_0}{\rho}\right).\mathbf{k}\right]\mathbf{grad}\ \psi_\rho - \left[\mathbf{grad}\left(\frac{\rho_0}{\rho}\right).\mathbf{grad}\ \psi_\rho\right]\mathbf{k}\ .$$

Now, $\mathbf{grad}\left(\frac{\rho_0}{\rho}\right).\mathbf{k} = -\frac{\rho_0}{\rho^2}\mathbf{grad}\ \rho.\mathbf{k} = 0$, because of the two-dimensional condition. Hence finally

$$curl\ \mathbf{V} = -\left[\frac{\rho_0}{\rho}\Delta\psi_\rho + \mathbf{grad}\left(\frac{\rho_0}{\rho}\right).\mathbf{grad}\ \psi_\rho\right]\mathbf{k} \equiv div\left(\frac{\rho_0}{\rho}\mathbf{grad}\ \psi_\rho\right)\mathbf{k}\ .$$

ADDITIONAL INFORMATION—Mathematical comments on the velocity and vector potential—The velocity potential function $\phi(M, t)$, Eq. (5.2), can only be defined to within an *additive constant* ϕ_0, since:

$$\mathbf{grad}(\phi + \phi_0) \equiv \mathbf{grad}\ \phi\ .$$

Similarly, relations (5.5) or (5.11) do not allow us to define a *vector potential* in a unique way. Thus for example, if $\boldsymbol{\psi}$ is the vector potential defined by Eq. (5.5), i.e., $\mathbf{V} = curl\ \boldsymbol{\psi}$, then we also have $\mathbf{V} = curl\ \boldsymbol{\psi}^*$, with $\boldsymbol{\psi}^* = \boldsymbol{\psi} + \mathbf{grad}\ f$, where f is any scalar function. The vector potential is thus defined to within an additive gradient.

Fig. 5.3 Simple shear flow: translation, rotation and deformation of a fluid particle

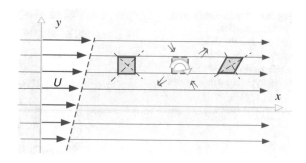

5.4.4 Flow Examples Defined by a Stream Function

5.4.4.1 The Simple Shear Flow $\psi(x, y) = cy^2$

The velocity vector components,[6] as deduced from the stream function $\psi(x, y) = cy^2$, where c denotes a non-zero real, are $U = 2cy$ and $V = 0$. The streamlines are therefore straight lines parallel to the x axis, as shown in Fig. 5.3.

In this flow, the fluid particles have a general straight motion, but undergo a local rotation and deformation resulting in an elongation and compression along slanted directions with an inclination of $\pi/4 \pm \pi/2$ from that of the general motion.

5.4.4.2 The Pure Strain Plane Flow $\psi(x, y) = \dfrac{a}{2}(y^2 - x^2)$

The velocity vector deduced from the stream function $\psi(x, y) = \frac{a}{2}(y^2 - x^2)$, where a stands for a non-zero real, has the following components $U = ay$ and $V = ax$. Its *module* is therefore constant on any circle centred at the origin.

The streamlines are equilateral hyperbolas whose asymptotes are the bisectors of the reference frame.

The velocity field also satisfies the condition $curl\mathbf{V} = \mathbf{0}$, so that the flow is irrotational. As for the deformation rate tensor, it is reduced to the sole diagonal component $S_{12} = 2a$. The fluid particles are thus simply undergoing a pure deformation when moving in the flow direction, see Fig. 5.4.

---

[6] Troughout this section,we restrict ourselves to stationary isovolume 2D-flows.

Fig. 5.4 A pure strain plane flow in Cartesian coordinates: the fluid particles move without rotation about their axis along hyperbolic trajectories

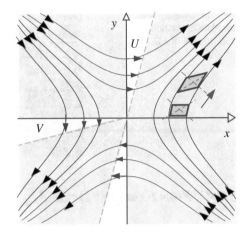

5.4.4.3   The Pure Strain Axisymetric Flow $\psi(r, \theta) = \dfrac{\Gamma}{2\pi} \ln r$

The velocity vector derived from the stream function $\psi = \frac{\Gamma}{2\pi} \ln r$, where Γ is a non-zero real, has the following components $V_r = 0$ and $V_\theta = -\frac{\Gamma}{2\pi} \frac{1}{r}$. It is therefore a decreasing function in modulus with the distance from the origin according to a hyperbolic law. The streamlines are circles centered at the origin of the reference frame on which the velocity module remains constant.

We can verify that the condition *curl* $\mathbf{V} = \mathbf{0}$ is also satisfied for this flow, which is another *irrotational* motion. The deformation rate tensor has a non-zero component $S_{r\theta} = \dfrac{\Gamma}{\pi r^2}$. The fluid particles are again stretched in the motion direction, in line with the elongation deformation outlined in Fig. 5.5.

Fig. 5.5 The pure deformation circular plane flow: the fluid particles move, without rotation on their axis, along circular trajectories

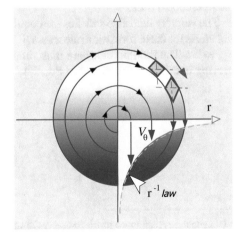

Remark From the previous results, it is possible to directly calculate the velocity vector circulation along a circle with a radius R_0 centred at the origin:

$$\oint \mathbf{Vdl} = \int_0^{2\pi} V_\theta\,(R_0)\,R_0 d\theta = C' = -\Gamma .$$

Thus the circulation is not zero. Now according to Stokes' theorem, this circulation is equal to the surface integral, over the disc πR_0^2, of the normal component of the curl of $\mathbf{V}$, which, in other words, means that the flow is not irrotational, with a vorticity equal to $-\Gamma/\pi R_0^2$. This apparent contradiction stems from the fact that the function $\psi\,(r, \theta) = \frac{\Gamma}{2\pi}\log r$ is not defined in $r = 0$, which makes it impossible to calculate the circulation along a path including the origin. Under this restriction, the potential function of this flow can then be found.—Answer: $\phi\,(r, \theta) = \frac{\Gamma}{2\pi}\theta.$

EXERCISE 21 The flow past a stagnation point.

In a half plane restricted by the first bisector (see Fig. 5.4), the pure deformation plane flow in Sect. 5.4.4.2 can be physically interpreted as the motion in the vicinity of a stagnation point. As an exercise, it can be demonstrated that the function $\psi\,(x, y) = kxy$, where k is a non-zero real, defines the same flow after an overall $\pi/4$ angle rotation. The motion being purely deformational is therefore irrotational. Hence there is a velocity potential, whose expression is to be determined. *Solution:* $\phi\,(x, y) = \frac{1}{2}k\left(x^2 - y^2\right).$

The method suggested here will be compared with that of the solution in Exercise 19.

5.4.4.4 The Axisymmetric Flow Past a Sphere

The last two examples which we will present are intended to make it clear that the stream function concept is independent of the viscous or inviscid nature of the fluid, as well as the rotational or irrotational nature of the motion.

To this end, we consider, in the spherical coordinates, the following two functions:

$$\psi\,(r, \varphi) = \frac{U_\infty}{2}\left(r^2 - \frac{a^3}{r}\right)\sin^2\varphi , \tag{5.15a}$$

$$\psi\,(r, \varphi) = U_0\left(\frac{3ar}{4} - \frac{a^3}{4r}\right)\sin^2\varphi , \tag{5.15b}$$

where U_∞, U_0 and a denote three positive real numbers.

The velocity field

• Firstly, let us consider Eq. (5.15a). We note that the streamline $\psi = 0$ is obtained for $r = a$ and $\varphi = 0 \pm \pi$. It is therefore formed by both the sphere of radius a, centered at the origin and the x-axis. The velocity vector components are $V_r (r, \varphi) = U_\infty \left(1 - \dfrac{a^3}{r^3}\right) \cos \varphi$, $V_\varphi (r, \varphi) = -U_\infty \left(1 + \dfrac{a^3}{2r^3}\right) \sin \varphi$ and of course $V_\theta = 0$. It can thus be verified that the V_r velocity component is zero on the sphere, but not the V_φ component which is $V_\varphi = -\frac{3}{2} U_\infty \sin \varphi$. The velocity field to the right of the main cross section is given by:

$$V_\varphi = -U_\infty \left(1 + \frac{a^3}{2r^3}\right) , \quad V_r = 0 .$$

The velocity is therefore parallel to the x-axis and decreases according to a power law in r^{-3} with the distance to the sphere center, with the maximum value on the sphere surface equal to one and a half times the value to infinity. Finally, for r tending to infinity, it can be immediately obtained that $V_r \sim U_\infty \cos \varphi$ and $V_\varphi = -U_\infty \sin \varphi$.

Thanks to these elements, and as outlined in Fig. 5.6a, the velocity field derived from relation (5.15a) can be interpreted as that of an **ideal fluid** flow, with a uniform velocity U_∞ at infinity, around a sphere of radius a.

• Let us turn now to the flow configuration defined by the potential function (5.15b). The only non-zero components of the velocity vector are:
$V_r (r, \varphi) = U_0 \left(\dfrac{3a}{2r} - \dfrac{a^3}{2r^3}\right) \cos \varphi$ and $V_\varphi (r, \varphi) = -U_0 \left(\dfrac{3a}{4r} + \dfrac{a^3}{4r^3}\right) \sin \varphi$. We deduce in particular that on the sphere of radius a, we have:

$$V_r (a, \varphi) = U_0 \cos \varphi ,$$
$$V_\varphi (a, \varphi) = -U_0 \sin \varphi ,$$

which corresponds to the vector field $\mathbf{V} (a) = U_0 \mathbf{i}$, where $\mathbf{i}$ denotes the unit vector of the x-axis. As for the infinite behaviour now we have $V_r (\infty, \varphi) \sim V_\varphi (\infty, \varphi) \sim 0$, with a power decay law of r^{-1} near the sphere and r^{-3} at a distance from it.

With these elements, and as outlined in Fig. 5.6b, the field derived from relation (5.15b) can be interpreted as being that generated by the *translation, at a constant speed U_0, of a sphere with radius a in a **viscous fluid**, at rest at infinity.*

We will see in Chap. 10 Sect. 10.4.3 that such a field is a solution of the Navier-Stokes equations *when discarding the inertial forces*. It therefore only applies to very low Reynolds number flows, still referred to as creeping motions.

To obtain a clear idea with regard to the difference between the ideal and viscous fluid situations, the flow field of the latter was rendered in a reference frame linked to the sphere, which is equivalent to vectorially adding $-U_0 \mathbf{i}$ to the velocity field previously found. The result, after symmetry ($\mathbf{i}$ versus $-\mathbf{i}$) is plotted in Fig. 5.6b'. Due to the adherence of the *viscous* fluid to the wall, the velocity is now zero on the sphere. The result in the main cross-section is an increasing velocity profile with the

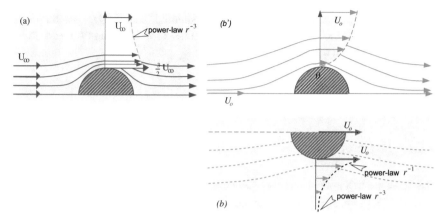

Fig. 5.6 Velocity fields and streamlines of potential flows around a sphere: **a** the ideal fluid, **b** and b' the creeping flow of a viscous fluid

distance to the sphere, from 0 to U_0, unlike the case of the *ideal* fluid where it is decreasing from $\frac{3}{2}U_\infty$ to U_∞.

Finally, due to the flow rate conservation condition between two adjacent streamlines, the comparison of their patterns in both flow configurations clearly exhibit a major geometric difference: the spacing, at infinity, between two streamlines increases in the main cross-section for the viscous fluid motion, while it decreases in an ideal fluid flow.

The vorticity field

From the vector operator expressions in spherical coordinates (see Appendix), it is easy to calculate the curl of the velocity fields defined by relations (5.15a) and (5.15b) respectively. As a result of the two-dimensional nature of the flows, it is clear that $\Omega_r = \Omega_\varphi = 0$. As for the component Ω_θ it is expressed as:

$$\Omega_\theta = \frac{1}{r}\left[\frac{\partial\left(rV_\varphi\right)}{\partial r} - \frac{\partial V_r}{\partial\varphi}\right].$$

This results in an ideal fluid flow:

$$\Omega_\theta = -U_\infty\left(1 - \frac{a^3}{r^3}\right)\sin\varphi + U_\infty\left(1 - \frac{a^3}{r^3}\right)\sin\varphi \equiv 0,$$

and for the viscous fluid flow:

$$\Omega_\theta = U_0 a^3\frac{\sin\varphi}{2r^3} + \frac{3}{2}U_0 a\frac{\sin\varphi}{r^2} - U_0 a^3\frac{\sin\varphi}{2r^3} = \frac{3}{2}U_0 a\frac{\sin\varphi}{r^2} \neq 0.$$

Thus, for the first flow, there will be a velocity potential function $\phi\,(r,\varphi)$, not for the second, where the motion is rotational. We will have the opportunity to revisit this flow in Chap. 10 Sect. 10.4.3.

5.5 The Irrotational and Isovolume Flow Class

5.5.1 Definition

The flows we will now discuss are those whose velocity field satisfies both relations (5.1) and (5.4), i.e. $div\mathbf{V} = 0$ (solenoidal) and $curl\,\mathbf{V} = \mathbf{0}$ (irrotational).

5.5.2 The Kinematic Properties

5.5.2.1 The Velocity Potential Equation

By applying the isovolume (solenoidal) condition to the velocity field $\mathbf{V} = \mathbf{grad}\,\phi$, we immediately deduce:

$$0 = div\mathbf{V} = div\,(\mathbf{grad}\phi)\,,$$

or again

$$\boxed{\Delta\phi = 0} \tag{5.16}$$

Thus:

> In any isovolume and irrotational flow, the velocity potential function is the solution of a Laplace equation in the flow domain.

We will subsequently explain how the pressure can then be deduced from the velocity field by using Bernoulli's second theorem.

5.5.2.2 Unicity of the Velocity Potential

As a solution of Laplace's equation, the ϕ function fulfills the following property:

> The velocity potential function of a flow in a simply connected bounded domain is unique (to within an additive constant).

Proof Since the flow is irrotational and the domain simply connected, Stokes' theorem makes it possible to write that, on any closed contour in the domain:

Fig. 5.7 Paths on a closed
contour in a simply
connected domain

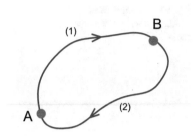

$$\oint \mathbf{V}.td\ell \equiv \iint_S curl\,\mathbf{V}d\sigma = 0.$$

Applied to a contour such as outlined in Fig. 5.7 connecting the $\widehat{AB}_{(1)}$ and $\widehat{BA}_{(2)}$ arcs, this relation immediately provides:

$$\int_{AB_{(1)}} \mathbf{V}.td\ell + \int_{BA_{(2)}} \mathbf{V}.td\ell = 0, \quad \text{or} \quad \int_{AB_{(1)}} \mathbf{V}.td\ell = \int_{AB_{(2)}} \mathbf{V}.td\ell.$$

In the case of a simply connected domain, all paths between A and B can be grouped in pairs into reducible closed curves, so that the previous integrals do not depend on the path in question. They are therefore *point functions*, which are simply expressed as:

$$\int_{AB} \mathbf{V}.td\ell = \phi(\mathbf{x}_B) - \phi(\mathbf{x}_A),$$

or again $d\phi = Udx + Vdy$ along a path element.

If there were another potential function ϕ' we would have, on this same element, $d\phi' = Udx + Vdy = d\phi$ hence finally $\phi' = \phi + C'$.

5.5.2.3   The Unicity of the Velocity Field

In a confined configuration i.e., when the motion occurs within a closed domain bounded by a solid wall, the velocity field of any irrotational and solenoidal flow is unique.

Proof To derive this result, let us first observe that, in particular, due to the solenoidal nature of the motion, we have identically:

$$div\,(\phi\,\mathbf{V}) = \phi\,div\mathbf{V} + \mathbf{V}.\,\mathbf{grad}\phi \equiv \mathbf{V}.\mathbf{V}. \tag{5.17}$$

Let us suppose, in a proof by contradiction, that there are two distinct velocity fields $\mathbf{V}_1 = \mathbf{grad}\,\phi_1$ and $\mathbf{V}_2 = \mathbf{grad}\,\phi_2$ satisfying Laplace's equation in the same

Fig. 5.8 Example of an
arbitrary doubly-connected
domain

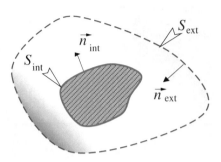

domain $\mathcal{D}$ bounded by a solid wall $\mathcal{S}$. This domain is not necessarily simply connected and can be taken as a doubly-connected domain, as illustrated in Fig. 5.8.

By linearity, the velocity difference field $\mathbf{V} = \mathbf{V}_2 - \mathbf{V}_1$ fulfills Laplace's equation with, as a potential function, $\phi = \phi_1 - \phi_2$. It also satisfies the condition $div\mathbf{V} = 0$. We can therefore apply the identity introduced at the beginning of this demonstration, Eq. (5.17) and hence deduce by integration on the domain $\mathcal{D}$:

$$\iiint_{\mathcal{D}} \mathbf{V}.\mathbf{V}\,dv = \iiint_{\mathcal{D}} div\,(\phi\mathbf{V})\,dv \equiv \iint_{\mathcal{S}} \phi\mathbf{V}.\mathbf{n}d\sigma,$$

the last equation resulting simply from Ostrogradski's relation.

However, on any finite domain *bounded by solid boundaries*, such as shown in Fig. 5.8, we have, on both external S_{ext} and internal S_{int} boundaries $\mathbf{V}_1.\mathbf{n}_{ext} = \mathbf{V}_2.\mathbf{n}_{ext} = \mathbf{V}_1.\mathbf{n}_{int} = \mathbf{V}_1.\mathbf{n}_{int} = 0$. The integral $\iiint_{\mathcal{D}} (\mathbf{V}_2 - \mathbf{V}_1)^2\,dv$ is therefore zero, which means that the velocity fields are identical, since the domain can be taken arbitrarily.

ADDITIONAL INFORMATION—**The velocity potential in external flows**—The previous unicity of the potential does not directly apply to free flows around obstacles where the external boundary is rejected to infinity. In this case, the conclusion cannot be obtained without additional information on the condition $\lim_{r \to \infty} \phi$. If the limit to infinity on the domain obeys a quadratic law with respect to the distance (r^2), namely proportional to the spherical surface element, it can be demonstrated, Cf. Guyon et al. [62] that the infinite behaviour of $\varphi\,(r)$ is as r^{-3}, so that the result is still valid.

HISTORICAL NOTE—**The velocity potential flow in Lagrange's work**—The theoretical study of flows from a velocity potential function ϕ was founded by Lagrange in 1781. In his pioneering article [85], Lagrange focused on the flows for which «*the quantity* $pdx + qdy + rdz$ *is an exact differential of any function* ϕ *of* x, y, z *and* t. » (Here, p, q and r refer to the velocity components). He demonstrated that, as a result of the «*incompressibility equation*» , the function ϕ is governed by the equation $\frac{\partial^2\phi}{\partial x^2} + \frac{\partial^2\phi}{\partial y^2} + \frac{\partial^2\phi}{\partial z^2} = 0$. He also provided the result referred to in the following as Bernoulli's third theorem, Cf. Sect. 5.6.2.2, Eq. (5.23).

5.5.3   The Dynamical Properties

The flow properties we are going to demonstrate will assume that the external body forces can be derived from a potential.

If we refer to this potential function[7] as $\mathcal{F}$, by definition the external body forces are expressed as :

$$\mathbf{F} = \mathbf{grad}\,\mathcal{F}\,.$$

5.5.3.1   The Viscosity Independence in a Constant Density Fluid

By introducing the condition $curl\,\mathbf{V} = \mathbf{0}$ in the dynamic equation, under the assumption of an *incompressible motion*,[8] we immediately deduce:

$$\frac{\partial \mathbf{V}}{\partial t} + \mathbf{grad}\left(\frac{V^2}{2}\right) = \mathbf{F} - \frac{1}{\rho_0}\,\mathbf{grad}\,P\,.$$

This reveals the following significant result:

> In any *isovolume* evolution of a *constant density* fluid, the momentum equation of an irrotational flow is the same, whether the fluid is viscous or not.

In particular, it follows that the isovolume irrotational flows of *ideal* and *viscous* fluids can only be differentiated by the *boundary conditions* which can be adopted in each case.

5.5.3.2   The Barotropy Relation

As a direct consequence of the "*viscosity transparency*" of the isovolume and irrotational flow equations, the *barotropic* character of the evolution, as introduced in

---

[7] The potential force function defines a *scalar field* homogeneous to an *energy*, thus qualified as *potential energy*. We will not use this term here to reserve the term *potential* for functions relating to a *flow*.

The forces deriving from a potential are still called *conservative*, to express the fact that their work does not depend on the path followed by the location point where the force is applied, but only on the starting (initial) and ending (final) positions. The electrostatic and gravitational forces are conservative. The friction forces in a viscous fluid are not.

[8] This term is to be understood here in the sense of the isovolume evolution of a homogeneous fluid (see Chap. 3 Sect. 3.1.3). In this case, the density is a constant *throughout the entire flow field*.

Sect. 5.2.5, now applies without distinction to the fluid, whether it is viscous or not. We can therefore state:

In the presence of conservative external body forces, in any isovolume and irrotational flow, the fluid behaviour is *barotropic*.

Proof Denoting by $\mathcal{F}$ the external body force potential, the general momentum equation, as derived in Chap. 4 Sect. 4.5.1, Eq. (4.19), can be written in vector notations and assuming a constant viscosity as follows:

$$\frac{\partial \mathbf{V}}{\partial t} + \mathbf{grad}\left(\frac{V^2}{2}\right) + curl\mathbf{V} \wedge \mathbf{V} =$$

$$\mathbf{grad}\,\mathcal{F} - \frac{1}{\rho}\mathbf{grad}\,P + \nu\,(\mathbf{grad}(div\mathbf{V}) - curl(curl\mathbf{V})\,) - \frac{2\nu}{3}\mathbf{grad}(div\mathbf{V}),$$

which, under the double assumption $div\mathbf{V} = 0$ and $curl\mathbf{V}$ is reduced to:

$$\frac{1}{\rho}\mathbf{grad}\,P = \mathbf{grad}\left(\mathcal{F} - \frac{\partial \phi}{\partial t} - \frac{V^2}{2}\right).$$

We precisely retrieve, but under different assumptions, the same equation as that which has led to the barotropic relation in paragraph 5.2.5.

5.5.4   The Energetic Properties

In any isovolume evolution, the study of the energetic flow properties can be focused on the kinetic energy alone, whose value per unit volume is $\frac{1}{2}\rho V^2$. We will now demonstrate important results with regard to some specificities of this function in irrotational and isovolume flows.

5.5.4.1   The Energy Minimization

The first property relating to the kinetic energy of this flow class can be stated as follows:

The irrotational flow of any incompressible fluid which takes place in a closed domain bounded by solid walls is that whose overall kinetic energy total is minimum in the entire domain.

Proof We will demonstrate this result using proof by contradiction. Let there be a closed domain $\mathcal{D}$ bounded by a set of solid surfaces Σ (see Fig. 5.8 for instance). We assume that in this domain there is a solenoidal flow (index 1), whose velocity field $\mathbf{V}_1$ derives from a potential ϕ_1. Let us imagine that there could be in *this same domain*, a different flow (index 2), of a velocity field $\mathbf{V}_2$. The latter is not necessarily a potential flow but is supposed:

– to be *isovolume*;
– to fulfill the *same boundary conditions* as the field (1), i.e., $\mathbf{V}_2.\mathbf{n} = 0$ on Σ.

Finally, in line with the type of reasoning stated, we will assume that the total kinetic energy of the flow (2) over the domain is *lower* than that of the potential flow (1), namely:

$$\iiint_{\mathcal{D}} V_2^2 \, \mathrm{d}v < \iiint_{\mathcal{D}} V_1^2 \, \mathrm{d}v \quad \text{or} \quad A \equiv \iiint_{\mathcal{D}} \left(V_1^2 - V_2^2 \right) \mathrm{d}v > 0 .$$

As we have already observed—see relation (5.17)—, we can note that under the study assumptions:

$$div\,(\phi_1 \mathbf{V}_2) \equiv \phi_1\,div\mathbf{V}_2 + \mathbf{V}_2.\,\mathbf{grad}\phi_1 = \mathbf{V}_1.\mathbf{V}_2 ,$$
$$div\,(\phi_1\,(\mathbf{V}_1 - \mathbf{V}_2)) \equiv \phi_1\,div\,(\mathbf{V}_1 - \mathbf{V}_2) + (\mathbf{V}_1 - \mathbf{V}_2).\,\mathbf{grad}\phi_1 = \mathbf{V}_1.\,(\mathbf{V}_1 - \mathbf{V}_2) .$$

Now, $V_1^2 - V_2^2 \equiv \mathbf{V}_1\,(\mathbf{V}_1 - \mathbf{V}_2) + \mathbf{V}_1.\mathbf{V}_2 - V_2^2$, so that:

$$A = \iiint_{\mathcal{D}} div\,[\phi_1\,(\mathbf{V}_1 - \mathbf{V}_2)]\,\mathrm{d}v + \iiint_{\mathcal{D}} div\,(\phi_1 \mathbf{V}_2)\,\mathrm{d}v - \iiint_{\mathcal{D}} V_2^2 \mathrm{d}v ,$$

or again

$$A = \underbrace{\iint_{\Sigma} \phi_1\,(\mathbf{V}_1 - \mathbf{V}_2)\,.\mathbf{n}\mathrm{d}\sigma}_{= 0} + \underbrace{\iint_{\Sigma} \phi_1 \mathbf{V}_2.\mathbf{n}\mathrm{d}\sigma}_{= 0} - \iiint_{\mathcal{D}} V_2^2 \mathrm{d}v ,$$

where the cancellation of the first two integrals on the right hand side results from the boundary conditions $\mathbf{V}_1.\mathbf{n} = 0$ and $\mathbf{V}_2.\mathbf{n} = 0$. This results in $A < 0$, a conclusion which is in contradiction with the initial assumption, *Q.E.D.*

5.5.4.2   The Mechanical Energy Conservation

As we will demonstrate in the next paragraph (Sect. 5.6.1), the isovolume and irrotational flow class, with conservative external body forces, obeys the mechanical energy conservation, in terms of Bernoulli's "*strong*" theorem.

This feature can be stated as follows:

> Any irrotational flow of a constant density fluid, in the presence of conservative body forces is mechanically homoenergetic. This satisfies at any time and at any field point the relation:
> $$\frac{\partial \phi}{\partial t} + \frac{V^2}{2} + \frac{P}{\rho} + \mathcal{F} = C',$$
> where ϕ is the flow velocity potential and $\mathcal{F}$ the potential function of the external body force.

It should be noted that the ideal fluid hypothesis is not required, so that the relation expressing this first Bernoulli theorem[9] applies whether or not the fluid is viscous.

ADDITIONAL INFORMATION—The physical interpretation of the energy conservation in a steady flow—In a stationary motion, the previous result expresses the *conservation* of the total mechanical energy per unit volume, the sum of the kinetic, pressure and potential energies. If the physical interpretation of this result is straightforward in an ideal fluid, it is not the case for a viscous fluid. Indeed, apart from the uniform flow case, any velocity gradient corresponds, even with no vorticity, to a non-zero deformation rate and therefore to an equally non-zero dissipation rate. *The energy conservation cannot therefore be considered as the removal of the dissipative character of the viscous fluid flow.*

In fact, in any irrotational flow of an isovolume viscous fluid, the reversibility of the conversion between the different energetic contributions in Bernoulli's theorem is due to the fact that the viscous effects can be decoupled from the entire energy balance. Indeed, the latter satisfies a specific balance, consisting in the exact local equilibrium of the dissipation rate by the external viscous force power.

To demonstrate this last proposal, it is only necessary to recall that in any isovolume evolution, the viscous forces are locally expressed by:

$$\nu \Delta \mathbf{V} \equiv \nu \, curl \, (curl \, (\mathbf{V})) \ .$$

Returning to index notations, we therefore have, by expanding this relation, for any irrotational and isovolume flow of a viscous fluid:

$$0 = \nu \Delta \mathbf{V} \equiv \nu \frac{\partial^2 U_i}{\partial x_j \partial x_j} \equiv \underbrace{\frac{\partial}{\partial x_j}\left[\nu U_i\left(\frac{\partial U_i}{\partial x_j} + \frac{\partial U_j}{\partial x_i}\right)\right]}_{\text{Power of ext. vis. forces}} - \underbrace{\frac{\nu}{2}\left(\frac{\partial U_i}{\partial x_j} + \frac{\partial U_j}{\partial x_i}\right)^2}_{\text{Dissipation rate}}.$$

This last relation clearly shows that the balance in question is of the "*production = dissipation*" type and that it is very specific to the mechanical power of the viscous forces alone.

---

[9] Hence the alternative name of Bernoulli's strong theorem.

5.6 The Fundamental Local Theorems

As we have seen in the previous chapters, the relations between the flow field functions are, in most cases, expressed in *partial differential equations*. Under specific conditions, however, it is possible to derive *explicit algebraic* expressions applying to the local values of these functions. Such relations, referred to here as "*local theorems*" are obviously only valid by virtue of a number of assumptions which "*interconnect*" the concepts of motion models and flow classes. We will review a number of some of the most noteworthy of these theorems.

Remark In the presentation of these theorems, the assumptions are at least as important as the "formulae" themselves. This observation deserves the reader's attention since it often goes against the most common memorization tendency encountered among a majority of students!

5.6.1 Bernoulli's First Theorem for an Irrotational Flow

The assumptions for this theorem, *which do not distinguish the viscous and ideal fluid cases*, are as follows[10]:

- an incompressible motion,
- an irrotational flow,
- a conservative body force.

As we have seen, the momentum equation for the irrotational and isovolume flow class is in the form of

$$\frac{\partial}{\partial t}(\mathbf{grad}\ \phi) + \mathbf{grad}\frac{V^2}{2} + \mathbf{grad}\mathcal{F} = -\frac{1}{\rho}\mathbf{grad}\ P\ ,$$

since by definition, the potential function ϕ is such that $\mathbf{V} = \mathbf{grad}\ \phi$. It is worth observing the absence of the viscosity term by the cancellation of $\Delta\mathbf{V}$ under the study assumptions. The previous relation therefore leads to the conclusion that:

$$\boxed{\frac{\partial\phi}{\partial t} + \frac{V^2}{2} + \mathcal{F} + \frac{P}{\rho} = C'}$$
(5.18)

In this relation, the value of the constant is to be taken at any *given time* and over the *entire flow field*.

---

[10] Let us remember that the incompressible motion (the first following assumption) is a short cut by which we mean the isovolume evolution of a constant density fluid throughout the flow field and at any time.

To differentiate this *first theorem* from Bernoulli's later theorems, we have qualified it as *strong*, in reference to the *constant value uniformity in the domain*. When adding to the above assumptions the condition of:

| | • a stationary motion,

the expression of Bernoulli's first theorem takes the simplified form:

$$\frac{V^2}{2} + \mathcal{F} + \frac{P}{\rho} = C^t \ in \ the \ entire \ flow \ field \qquad (5.19)$$

5.6.2 The Ideal Fluid Theorems

5.6.2.1 Bernoulli's Second Theorem in an Incompressible Flow

This theorem, previously mentioned in Chap. 3 Sect. 3.3.1.3, applies in any situation where the following assumptions are satisfied:

| • an incompressible flow,
| • an ideal fluid,
| • a stationary motion,
| • a conservative body force.

This theorem then states that:

$$\frac{V^2}{2} + \mathcal{F} + \frac{P}{\rho} = C^t \ along \ any \ streamline \qquad (5.20)$$

Proof Under the assumptions above, the momentum equation is written

$$\mathbf{grad} \left(\frac{V^2}{2} \right) + curl \ \mathbf{V} \wedge \mathbf{V} + \mathbf{grad} \mathcal{F} = -\frac{1}{\rho} \mathbf{grad} \ P \ .$$

By projecting this equation on a streamline, i.e., by scalar multiplication throughout by $\mathbf{V}$, we obtain:

$$\mathbf{V}. \left[\mathbf{grad} \left(\frac{V^2}{2} \right) + \mathbf{grad} \mathcal{F} + \frac{1}{\rho} \mathbf{grad} P \right] = -\mathbf{V}. (curl \ \mathbf{V} \wedge \mathbf{V}) \equiv 0 \ .$$

This last equality results directly from the vector orthogonality in the scalar product and demonstrates the announced proposal.

5.6.2.2 The Barotropic Fluid Theorems

Definition and property—We have previously introduced the barotropic evolution concept. It will be specified here by demonstrating how this concept makes it possible to consider some results of the isovolume evolution of a constant density fluid ("*incompressible flow*") as *restrictions* of the barotropic case. It should be noted, however, that the extension to the barotropic case is sometimes associated with changes in the assumptions which are required for validating the isovolume evolving formulations. It is for this reason that it is necessary to differentiate between the theorems for these two flow classes.

By definition, an evolution is qualified as barotropic if there is a direct link between the pressure and density. Consequently, the expression of the external pressure force term in the momentum equation can be transformed as follows:

$$\frac{1}{\rho}\mathbf{grad}\,P \equiv \mathbf{grad}\left(\int \frac{\mathrm{d}P}{\rho}\right). \tag{5.21}$$

Proof Formally expressing the pressure-density relation in the form of $\rho = h\,(P)$, it is clear that, provided that the functions can be integrated:

$$\int \frac{\mathrm{d}P}{\rho} = \int \frac{\mathrm{d}P}{h\,(P)} \equiv F\,(P)\,.$$

We actually deduce the result (5.21) since $\mathbf{grad}\,[F\,(P)] \equiv F'(P)\,\mathbf{grad}\,P = \dfrac{1}{\rho}\mathbf{grad}\,P$.

The acceleration potential—The following statement can be immediately deduced from the barotropy property:

> In the presence of conservative external body forces, there is an acceleration potential for any ideal barotropic fluid motion, whether or not the flow is irrotational, since:
> $$\frac{\mathrm{d}\mathbf{V}}{\mathrm{d}t} = -\mathbf{grad}\left(\mathcal{F} + \int \frac{\mathrm{d}P}{\rho}\right). \tag{5.22}$$

Bernoulli's third theorem—This third Bernoulli theorem can be seen as an "*extension*" to the barotropic case in the first theorem. However, it differs from the latter in that the *ideal fluid* assumption is now required in addition to that of the *irrotational motion*, which was not necessary for the isovolume evolution of a constant density fluid. Thus, the assumptions of this third theorem are:

- an ideal fluid,
- a barotropic evolution,
- an irrotational flow,
- a stationary motion,
- a conservative body force.

Under these conditions:

$$\boxed{\frac{V^2}{2} + \mathcal{F} + \int \frac{\mathrm{d}P}{\rho} = C^t \ in \ the \ entire \ field.}$$ (5.23)

Proof The derivation of relation (5.23) is straightforward from the momentum equation which, given the assumptions, can be reduced to:

$$\frac{\partial}{\partial t} (\mathbf{grad} \, \phi) + \mathbf{grad} \left(\frac{V^2}{2} \right) = -\mathbf{grad} \mathcal{F} - \mathbf{grad} \left(\int \frac{\mathrm{d}P}{\rho} \right).$$

Saint–Venant's theorem—The relation expressing Saint-Venant's theorem is that of Bernoulli's third theorem for an isentropic evolution of an ideal gas. In this case, the barotropy condition is expressed according to Laplace's law, as $P/\rho^\gamma = C^t$ where $\gamma = C_p/C_v$ is the isentropic coefficient of the gas.
With the following assumptions

- an ideal fluid,
- an isentropic evolution,
- an irrotational flow,
- a stationary motion,
- a conservative body force.

we have:

$$\boxed{\frac{V^2}{2} + \mathcal{F} + \frac{\gamma}{\gamma - 1} \frac{P}{\rho} = C^t \ in \ the \ entire \ field.}$$ (5.24)

Proof The derivation of Saint-Venant's relation from Bernoulli's third theorem, Eq. (5.23), is simply reduced to the calculation of the following integral, with $P/\rho^\gamma = K$, namely:

$$\int \frac{\mathrm{d}P}{\rho} \equiv K \int \gamma \rho^{\gamma - 2} \mathrm{d}\rho = \frac{\gamma}{\gamma - 1} \frac{P}{\rho}.$$

5.6.2.3   The Compressible Fluid Motion Theorems

The dynalpy conservation—The continuity, momentum and energy equations of an ideal compressible fluid motion (see Chap. 4) can be integrated thanks to the following assumptions:

- an ideal fluid,
- a one-directional flow,
- a stationary motion,
- no external body forces

Under these assumptions, these equations are written as follows

$$\frac{d\,(\rho U)}{dx} = 0\,,$$

$$\rho C_p U \frac{dT}{dx} = U \frac{dP}{dx}\,,$$

$$\rho C_p U \frac{dT}{dx} = U \frac{dP}{dx}\,,$$

where x stands for the unique space coordinate. The integration of the second equation, under the condition $\rho U = C'$ because of the continuity equation, leads to:

$$\boxed{P + \rho U^2 = C'} \tag{5.25}$$

The *dynalpy* is therefore indeed constant.

The stagnation enthalpy conservation—Under the same assumptions, it is possible to obtain, by eliminating the pressure gradient between the differential momentum and energy equations

$$C_p \frac{dT}{dx} = -U \frac{dU}{dx}\,,$$

whose integration results in:

$$\boxed{C_p T + \frac{U^2}{2} = C' = C_p T_i} \tag{5.26}$$

The quantity $C_p T_i$ refers to the *isentropic stagnation enthalpy* which corresponds to the value which the total enthalpy $h = C_p T + U^2/2 \equiv e + P/\rho + U^2/2$ takes at a point where the motion is brought to a stop from the velocity U in an adiabatic and reversible way. We will return to these concepts in more detail in Chap. 8. It can be simply noted here that:

> In any one-dimensional stationary flow of a compressible ideal gas, the isentropic stagnation enthalpy is retained during the motion.

This theorem does not mean that the stagnation enthalpy is *constant throughout the field*, but simply that it does not change as a given fluid particle is followed along its trajectory.

Proof First of all, it should be noted that $\rho \dfrac{d}{dt}\left(\dfrac{P}{\rho}\right) \equiv \dfrac{dP}{dt} - \dfrac{P}{\rho}\dfrac{d\rho}{dt}$. However according to the continuity equation $d\rho/dt \equiv -\rho\,div\mathbf{V}$, so that the previous relations become, in a *stationary flow*:

$$\rho \frac{d}{dt}\left(\frac{P}{\rho}\right) \equiv \mathbf{V}.\mathbf{grad}\,P + P\,div\,\mathbf{V}.$$

Nevertheless, under the above-mentioned assumptions, the energy equation is written (see Chap. 4):

$$\rho \frac{\mathrm{d}}{\mathrm{d}t}\left(e + \frac{V^2}{2}\right) = -div\,(P\mathbf{V}) \equiv -\left[Pdiv\,\mathbf{V} + \mathbf{V}.\mathbf{grad}\,P\right]. \qquad (5.27)$$

It is then only necessary to add throughout the two previous equations to obtain the expected result.

Crocco's theorem—The theorem published by Crocco [43] in 1937 stated that:

In any compressible ideal fluid flow with conservative external body forces, the entropy and stagnation enthalpy gradients are linked by:

$$T\mathbf{grad}\,s = \mathbf{grad}\,h_i + curl\,\mathbf{V} \wedge \mathbf{V} + \frac{\partial \mathbf{V}}{\partial t} + \mathbf{grad}\mathcal{F}. \qquad (5.28)$$

Proof We start from the following classical thermodynamic relations:

$$Tds = de - \frac{P}{\rho^2}\mathrm{d}\rho = \mathrm{d}h - \frac{\mathrm{d}P}{\rho}.$$

By applying the last equality to the space variations, we can write

$$T\mathbf{grad}\,s = \mathbf{grad}\,h - \frac{1}{\rho}\mathbf{grad}\,P.$$

Introducing the isentropic stagnation enthalpy $h_i = h + V^2/2$, the previous relation becomes

$$T\mathbf{grad}\,s = \mathbf{grad}\,h_i - \mathbf{grad}\left(\frac{V^2}{2}\right) - \frac{1}{\rho}\mathbf{grad}\,P.$$

However from the momentum equation, we know that

$$\frac{\partial \mathbf{V}}{\partial t} + \mathbf{grad}\left(\frac{V^2}{2}\right) + curl\,\mathbf{V} \wedge \mathbf{V} = -\mathbf{grad}\mathcal{F} - \frac{1}{\rho}\mathbf{grad}\,P.$$

By eliminating the square velocity gradient between the two previous expressions, Crocco's equality is actually obtained.

5.6.3 Summary

The main results and theorems drawn up so far are summarized in the following table, with reference to their respective validity conditions.

Table 5.2 Fundamental theorems associated with different flow classes

	Irrotational		Rotational	
	Ideal fluid	Viscous fluid	Ideal fluid	Viscous fluid
		Potential vector ψ or		
		Stream function ψ (2D)		
Isovolume	Potential function ϕ $\Delta\phi = 0$ Energy minimization Bernoulli (i)			
	Bernoulli's th. (ii)		Bernoulli's th. (ii)	
	Kelvin's th. Lagrange's th.		Kelvin's th.	
Barotropic	Potential function ϕ			
	Bernoulli's th. (iii) Saint-Venant's th.			
	Kelvin's th. Lagrange's th.		Kelvin's th.	
Compressible	Potential function ϕ			
	Stagnation enthalpy conservation		Crocco's Theorem	

As pointed out at the beginning of this section, the assumptions required for the various general theorems interconnect the concepts of fluid motion models and flow classes, as illustrated by the double entry in Table 5.2.

ADDITIONAL INFORMATION—The isentropic evolution of a compressible fluid—
The last two theorems provide a better understanding of the physical origin of the isentropic nature of perfect compressible fluid flows.
• By firstly projecting Crocco's relation on a stream line we obtain:
$$T\,\mathbf{V}.\mathbf{grad}\,s = \mathbf{V}.\mathbf{grad}\,h_i + \mathbf{V}.\mathbf{grad}\,\mathcal{F} + \frac{1}{2}\frac{\partial V^2}{\partial t}\,.$$
The right hand side of this equality is generally not zero, therefore the flow is not isentropic along the streamlines. It becomes so if the motion is stationary and the external body forces are negligible. In this case, there only remains the term $\mathbf{V}.\mathbf{grad}\,h_i \equiv \mathrm{d}h_i/\mathrm{d}t$ in a steady flow. The stagnation enthalpy conservation then implies the isentropic character of the evolution.
• Regardless of the previous result, Crocco's relation clearly proves that the entropy can vary from one streamline to another when the flow is not irrotational. We can therefore conclude that the vorticity is directly associated with the *entropy generation* in an ideal compressible fluid flow.

Fig. 5.9 Schematic for the gravity discharge of a free surface reservoir

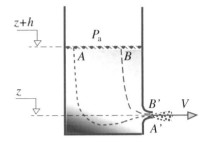

5.7 Examples of Local Theorem Application

5.7.1 Toricelli's Formula

The calculation of the gravity discharge velocity of a free-surface tank through a small cross-section hole (Fig. 5.9) is one of the simplest applications of Bernoulli's theorems. By keeping the tank filling level constant, the fluid motion can be considered as stationary.

Assuming that the fluid is ideal and incompressible, Bernoulli's second theorem applied along streamlines such as AA', BB' results in:

$$P_A + \frac{1}{2}\rho V_A^2 + gz_A = P_{A'} + \frac{1}{2}\rho V_{A'}^2 + gz_{A'} .$$

If the section ratio between A and A' is high enough, it can be assumed that $V_A \ll V_{A'} \equiv V$. In addition, the experimental evidence is that across the jet at the orifice, the static pressure is very close to the atmospheric pressure P_a.

Then writing $h = z_A - z_{A'}$, the previous equality finally becomes:

$$V = \sqrt{2gh} , \tag{5.29}$$

an expression known as Torricelli's formula.

5.7.2 Pitot and Prandtl Probes

Two devices, known as Pitot and Prandtl probes or "*tubes*", can be used to deduce the flow velocity from pressure measurements (see Fig. 5.10).

In their standard geometries, these are cylindrical bodies of circular cross-section with a locally hemispherical shaped front end. Schematically, the Pitot tube has only one pressure measurement point at the hemispherical face center, while the Prandtl

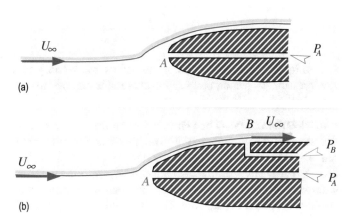

Fig. 5.10 Schematics for the Pitot tube (**a**) and Prandtl probe (**b**) operating

tube has a second point located at the periphery of the cylindrical body, at a suitable distance from the upstream end.[11]

Positioning the previous device in a flow parallel to the streamlines, two pressure values are provided at points A and B, noted respectively P_A and P_B.

Assuming an incompressible flow, the application of Bernoulli's second theorem along the streamline between points A and B produces, for the steady motion of an ideal fluid with negligible external body forces:

$$P_A + \frac{1}{2}\rho V_A^2 = P_B + \frac{1}{2}\rho V_B^2 .$$

Denoting by U the velocity at point B and noting that point A is a flow stagnation point ($V_A = 0$), the following result is finally obtained:

$$U = \sqrt{2\frac{P_A - P_B}{\rho}} . \tag{5.30}$$

Thus the value of the velocity is derived from the measurement of the *dynamic* pressure, i.e., the difference between the *total* pressure (or *stagnation* pressure) P_A and the *static* pressure P_B.

If we now assume that the flow is *compressible*, the previous reasoning can be repeated using this time Saint-Venant's formula, provided that the evolution remains isentropic between points A and B. Then we obtain:

---

[11] Originally, the Pitot probe consisted of two separate tubes fitted to measure the total and static pressures respectively. The lateral pressure measurement of the Prandtl probe is sometimes replaced by a serie of orifices arranged in a ring on the cylindrical body.

$$U = \sqrt{2\frac{\gamma}{\gamma - 1}s\left(\frac{P_B}{\rho_B} - \frac{P_A}{\rho_A}\right)}. \tag{5.31}$$

It is therefore no longer possible in this case to deduce the velocity value by only measuring the pressure. A second piece of information is required to determine the density value, which is usually based on a temperature measurement.

ADDITIONAL INFORMATION—The velocity measurement in a real fluid flow—
(a) At this stage of the presentation, it is not possible to justify that the Prandtl probe can be used as a means of measuring the velocity of a *real* fluid flow, i.e., with a non-zero viscosity. This justification will be given after the boundary layer concept has been presented, namely when the viscosity effects and their impact on the previous results is explained in high Reynolds number flows (Chap. 11).
(b) In addition to the above limitation, relation (5.31) is only valid in *subsonic* flows. In the supersonic regime, the stopping of the fluid on a fixed obstacle generates a shock wave, as we will see in Chaps. 7 and 8.

HISTORICAL NOTE—The Pitot tube—As previously mentioned, the static and total pressures are measured by two separate tubes, vertically immersed in the fluid vein, in the machine designed by Pitot [115] in 1732. The tube sensing the total pressure is bent at 90 degrees to be oriented in the course run of the river. The other tube is a simple vertical tube opening downwards. With regard to the use of his device, Pitot reported that «*It is easy to direct the tube opening towards the course run of the river, because by turning the machine slowly, we will see the point where the water rises the most in the first tube. Then if we turn the opening to the side opposite the current, as soon as we pass the perpendicular to the direction, the water will remain at the same height in both tubes.*»
To conclude on Pitot's discovery, we cannot resist the desire to share with the reader the inventor's communicative enthusiasm: «*The idea of this machine is so simple & so natural, that as soon as it came to me, I ran at once to the river to have a first try with a simple glass tube, & the effect met my expectations perfectly.*»
    As Brown notes [20], the "*machine*" designed by Pitot nevertheless had a number of defects. Several improvements have been made to the original device by various scientists. We can mention, in particular, Darcy, in a series of papers during the 1850s, the last of which, [46] was published posthumously in 1858, and Prandtl & Tietjens, [159], to whom we owe the modern, almost definitive geometry of the probe, from 1929 onwards.
    Consequently, subsequent research was only devoted to improving the conditions for using of the probe. Thus, in 1939, Camichel, [24] experimentally calibrated the position influence of the downstream (static) pressure sensing point on the measurement accuracy and was also interested in the question of the probe operating in low Reynolds number flows. Nevertheless, the creeping flow case at a *very low* Reynolds number ($R_e < 1$) will only be considered in 1990, by Chebbi and Tavoularis [37]. Finally, the device's capacities have been extended

to the measurement of various properties, such as, for instance: the compressible gas flow velocity, in the sub and supersonic regime (see Chap. 8 Sect. 8.4.3), the parietal friction in a turbulent boundary layer, according to the arrangement designed by Preston, [126] in 1954 and calibrated by Patel [112], in 1965.

EXERCISE 22 The compressibility effect on the velocity measurement.

We intend to measure an air flow velocity using a Prandtl probe. It is not known whether the motion conditions are such that the fluid compressibility should be or not be taken into consideration. The velocity values, derived from the pressure measurements provided by the probe, are referred to as $U^{(i)}$ and $U^{(c)}$, depending on whether the fluid is incompressible or compressible respectively.

QUESTION 1. Using Saint-Venant's formula, express the velocity value $U^{(c)}$ as a function of the pressure difference ΔP, the static pressure P_S, the sound celerity $a = \sqrt{\gamma P/\rho}$ and the isentropic coefficient γ.

QUESTION 2. By a second order Taylor expansion in $\Delta P/P_S$, derive the relation between the velocity $U^{(c)}$ and the value obtained by assuming that the fluid is incompressible $U^{(i)}$.

QUESTION 3. Estimate the error in the velocity calculation when considering the air as an incompressible fluid flowing at a velocity of 66 m/s and 93 m/s under the standard temperature and pressure conditions.

SOLUTION: **1.** By introducing the speed of sound at the static pressure point ($a^2 = \gamma P_S/\rho_S$) Saint-Venant's relation is still written as:

$$U^{(c)2} = \frac{2}{\gamma - 1} a^2 \left[\left(1 + \frac{\Delta P}{P_S} \right)^{\frac{\gamma-1}{\gamma}} - 1 \right],$$

where the notation $U^{(c)}$ has been introduced to recall that this velocity value corresponds to a calculation taking into account the fluid compressibility.

2. By a Taylor expansion of $(1 + \Delta P/P_S)^n$ for small $\Delta P/P_S$, we obtain:

$$\left(1 + \frac{\Delta P}{P_S} \right)^{\frac{\gamma-1}{\gamma}} \simeq 1 + \frac{\gamma - 1}{\gamma} \frac{\Delta P}{P_S} - \frac{\gamma - 1}{2\gamma^2} \left(\frac{\Delta P}{P_S} \right)^2 + \mathcal{O}\left(\frac{\Delta P}{P_S} \right)^3,$$

which leads to:

$$U^{c2} \simeq 2 \frac{\Delta P}{\rho_S} \left[1 - \frac{1}{2\gamma} \frac{\Delta P}{P_S} + \mathcal{O}\left(\frac{\Delta P}{P_S} \right)^2 \right].$$

By introducing the value $U^{(i)}$ of the velocity derived from an incompressible fluid calculation (5.30), the previous relation is still in the form of:

$$U^{(c)} \simeq U^{(i)} \left(1 - \frac{1}{4\gamma} \frac{\Delta P}{P_S} + \mathcal{O}\left(\frac{\Delta P}{P_S} \right)^2 \right) .$$

The estimation of the relative error between the values derived from compressible and incompressible calculations is deduced as follows:

$$\frac{U^{(i)} - U^{(c)}}{U^{(i)}} \approx \frac{1}{4\gamma} \frac{\Delta P}{P_S} .$$

It is easy to infer that this error is equal to $M'^2/8$, where M' denotes the Mach number $U^{(i)}/a$.

3. For the air under standard temperature and pressure conditions, the relative gas compressibility effect on the velocity measurement is 0.5% at 66 m/s and 1% at 93 m/s. This is the reason why, *in practice*, the air compressibility is neglected in most aerodynamic applications, as long as the velocity does not exceed a value in the order of 70 m/s, under standard temperature and pressure conditions.

5.7.3 Venturi's Flow Meter

The Venturi device (or tube) plays, for the velocity measurement in duct flows, a role equivalent to that of Prandtl's probe in free flows. It simply consists in fitting a calibrated reduction in the useful pipe cross-section, as outlined in Fig. 5.11.

Assuming a stationary incompressible flow and an irrotational motion with uniform velocity profiles in both cross sections A and B, the static pressure is constant in each of these sections, according to Bernoulli's first theorem. Applying this same theorem between the two sections A and B, we obtain, by discarding gravity forces

$$P_A + \frac{1}{2}\rho V_A^2 = P_B + \frac{1}{2}\rho V_B^2 .$$

In addition, the continuity equation ($S_A V_A = S_B V_B$) results in $V_B = \sigma V_A$, denoting as $\sigma = S_A/S_B$ the section ratio. By eliminating the velocity V_B between these two relations, the velocity value V_A is derived from the section ratio and the measurement of the static pressure difference $\Delta P = P_A - P_B$, according to the relation:

Fig. 5.11 Schematic for the Venturi device

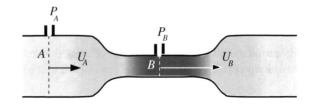

$$V_A = \sqrt{\frac{2\Delta P}{\rho\left(\sigma^2 - 1\right)}} \; . \tag{5.32}$$

ADDITIONAL INFORMATION—The **Venturi flow meter in the compressible regime**—(a) In the *incompressible* regime, the continuity equation results in that $V_A < V_B$. Therefore, according to Bernoulli's relation, the pressure is lower in the narrowing section. Such a property is used in various practical applications, such as the vacuum suction by a water jet pump, the air-fuel mixture suction into a motor vehicle carburetor, etc.

(b) In the *compressible* regime, we will explain that, if the direction of the velocity variation remains unchanged in the converging part of the tube, it can be opposed to that of the incompressible case in the diverging part, according to the Mach number value.

HISTORICAL NOTE—The **Venturi-Herschel flow meter**—The property of a convergent-divergent tube to produce a sucking action through holes bored into its narrowest section was demonstrated by Venturi. His experiments on the subject [164], were published in Paris in 1797. However, the design of the tube as a water flow meter was due to Herschel [70] in 1888. Hence, in a more historically suitable way, the device, usually termed as the Venturi tube, should be named after both Giovanni Baptista Venturi and Clemens Herschel.

5.7.4 The Isentropic Ejection Velocity of a Compressed Gas

The last example we will deal with is the equivalent of the first for a compressible environment. It concerns the ejection velocity determination of a gas exhausting from a reservoir under the sole pressure force action (assuming an inviscid fluid with no external body forces). We will also only consider the steady flow regime. The gas is contained, at rest, in a tank under the pressure P_i and temperature T_i corresponding to the stagnation values denoted by the index (i). The motion is generated when connecting this tank to an external environment using a conduit designed to ensure an isentropic flow at any velocity. The receiving environment—index (o)—is assumed to be infinitely large at a constant pressure P_0.

The application of Saint-Venant's relation, Eq. (5.24), between the states (i) and (o) leads to:

$$\frac{\gamma}{\gamma - 1}\frac{P_i}{\rho_i} = \frac{V_0^2}{2} + \frac{\gamma}{\gamma - 1}\frac{P_0}{\rho_0},$$

hence the expression of the ejection velocity V_0:

$$V_0 = \sqrt{\frac{2\gamma}{\gamma - 1}\left(\frac{P_i}{\rho_i} - \frac{P_0}{\rho_0}\right)}. \qquad (5.33)$$

Remark The previous expression proves that the ejection velocity is maximum when the expansion occurs in the absolute vacuum ($P_0 = 0$). We will see at a later stage that $\gamma P_i/\rho_i = \mathbf{a}_i^2$ where $\mathbf{a}_i$ is nothing other than the speed of sound value under the stagnation conditions. The maximum ejection velocity for an isentropic expansion is therefore:

$$V_{0Max} = \sqrt{\frac{2}{\gamma - 1}}\,\mathbf{a}_i .$$

To give an order of magnitude, in the case of the air ($\gamma = 1.4$), this velocity is slightly greater than twice the speed of sound in stagnation conditions (2.24).

Part II
Ideal Fluid Motions

Chapter 6
Irrotational 2D-Plane Motions of an Incompressible Fluid

Abstract This chapter is devoted to the potential motion of an inviscid fluid under stationary and isovolume conditions. After the statement of the assumptions, the mathematical tools are explained in detail and applied to various examples. The flow theory around sharp trailing-edge airfoils concludes the chapter.

6.1 The General Assumptions and Their Consequences

We have previously introduced in Chap. 5, Sect. 5.5.1 the irrotational isovolume flow class. Here we will focus on those class motions which also fulfill the following assumptions:

– a two-dimensional plane geometry,
– a stationary motion,
– an ideal (inviscid) fluid.

By virtue of the flow irrationality, we know that there is a *potential function* $\phi(M, t)$, such that at any point in the field and at any time the velocity vector is given as:

$$\forall M, \ \forall t, \quad \mathbf{V}(M, t) = \mathbf{grad}\,\phi(M, t) \ . \tag{6.1}$$

Since the evolution is also isovolume, we are aware that there is a vector potential function $\mathbf{\Psi}(M, t)$, such that at any point in the field and at any time, the velocity vector verifies the condition:

$$\forall M, \ \forall t, \quad \mathbf{V}(M, t) = curl\,\mathbf{\Psi}(M, t) \ .$$

In a two-dimensional plane geometry, we have also demonstrated that the vector function $\mathbf{\Psi}(M, t)$ is in fact reduced to a single component $\psi(M, t)$, called the *stream function*.

From these results, the velocity field expression can be deduced as:

$$\mathbf{V} = \mathbf{grad}\,\psi \wedge \mathbf{k} \ , \tag{6.2}$$

where **k** denotes the unit normal vector to the flow plane. In components, relations (6.1) and (6.2) are again written as:

$$U = \frac{\partial \phi}{\partial x} = \frac{\partial \psi}{\partial y} \, , \tag{6.3a}$$

$$V = \frac{\partial \phi}{\partial y} = -\frac{\partial \psi}{\partial x} \, . \tag{6.3b}$$

6.1.1 The Complex Potential

6.1.1.1 The Existence of the Complex Potential Function

By simply taking the derivatives of the previous relations, it is a straightforward conclusion that, at any given time,

$$\boxed{\Delta \phi = \Delta \psi = 0} \tag{6.4}$$

It should be recalled that these relations have already been demonstrated, in another way, in Chap. 5. Thus, the potential function ϕ and the stream function ψ are at any time two conjugate harmonic functions. They can therefore be considered respectively as the real and imaginary part of a single holomorphic function of the complex variable $z = x + iy$. It is therefore appropriate to write:

$$\boxed{f(z, t) = \phi(x, y, t) + i \psi(x, y, t)} \tag{6.5}$$

where the function $f(z, t)$ is called, by definition, the *complex potential* of the flow.

6.1.1.2 The Geometrical Interpretation

The equipotential lines ($\phi = C^t$) and the streamlines ($\psi = C^t$) form a network of locally orthogonal curves, as outlined in Fig. 6.1.

As we have demonstrated in Chap. 2, Eq. (2.8), the flow rate between two streamlines remains constant in space. Thus the spacing between these lines provides direct information on the flow velocity amplitude: the regions where the distance between the lines increases correspond to slowing-down areas of the fluid motion and vice versa.

HISTORICAL NOTE—Rankine's geometrical interpretation of the iso– ϕ and iso– ψ lines—We have already mentioned Lagrange's historical contribution to the development of the potential theory of irrotational flows, as early as 1781.

Fig. 6.1 Orthogonal
network of the iso-ϕ and
iso-ϕ lines

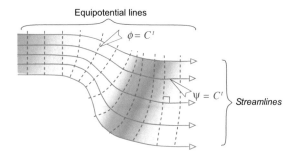

But rather to Rankine [127] that we owe the geometric interpretation of the
streamlines and equipotential lines, in 1864. Although quoting Stokes' publication of 1845 as a basic reference for his work in fluid mechanics, Rankine did
not explicitly associate the velocity potential function ϕ to the irrotational nature
of the motion, but rather to the «*condition of perfect fluidity*», which is that of
a *quasi*-parallel and uniform flow present at a sufficient distance from an obstacle. He then identified the streamline geometric properties in a plane geometry
(the «*plane water-lines in two dimensions*») and their orthogonal trajectories. He
applied these results to the hull shape optimization, the first theoretical approach
to an issue which until this time had only been addressed in a purely empirical
way.

6.1.1.3   The Complex Velocity

The complex potential being a holomorphic function has a complex-valued derivative
function which can be expressed, at any time, as

$$\frac{\partial}{\partial z} f(z,t) = \frac{\partial \phi}{\partial x} + i \frac{\partial \psi}{\partial x} = \frac{\partial \psi}{\partial y} - i \frac{\partial \phi}{\partial y}.$$

We will refer to this derivative function as $w(z,t)$ and write, returning to the velocity
vector components

$$\boxed{w(z,t) \equiv \frac{\partial f(z,t)}{\partial z} = U(x,y,t) - iV(x,y,t)} \tag{6.6}$$

The function of the complex variable $w(z,t)$ is called the *complex velocity* of the flow.
At any time this function is also holomorphic. Indeed, under the study assumptions:

$$div\,\mathbf{V} = 0 \iff \frac{\partial U}{\partial x} = \frac{\partial(-V)}{\partial y}$$

$$curl\mathbf{V} = \mathbf{0} \iff \frac{\partial U}{\partial y} = -\frac{\partial(-V)}{\partial x},$$

which are nothing more than the Cauchy conditions for the real part (U) and the imaginary part $(-V)$ of the complex velocity. In particular, we deduce that:

$$\Delta U = \Delta V = 0 , \qquad (6.7)$$

and that the complex potential function is at least twice differentiable[1].

HISTORICAL NOTE— **Euler and Cauchy's contributions to the complex variable theory**—It is not intended here, even briefly, to trace the genesis of the mathematical concepts of the complex variable theory, but simply to highlight a few points regarding Euler and Cauchy's contributions. In his work published in 1751 on the imaginary roots of the equations, Euler [55] explicitly defined the *imaginary* number concept: *«We call the imaginary quantity, that which is neither greater than zero, nor smaller than zero, nor equal to zero; it will therefore be something impossible, such as $\sqrt{-1}$, or in general $a + b\sqrt{-1}$; since such a quantity is neither positive, negative, nor zero.»*

Cauchy holds a central place in the theoretical analysis of the complex-valued function. In a report to the Paris Academy of Sciences [29], published in 1851, a hundred years after Euler's paper, he began by very clearly defining the *function of a complex variable*: *« If, as is naturally required, the definitions adopted in the case of real variables are extended to imaginary variables, u should be assumed to be a **function** of z, when the z value determines the u value. Now, for this to happen, it is sufficient that v and w are specified functions of x, y. Then in addition, considering the real variables x and y enclosed in z, or the real variables v and w enclosed in u, as suitable to represent the rectilinear and rectangular coordinates of a moving point Z or U, the position of the moving point Z will still determine the position of the moving point U.»* He then turned to the *«differential ratio of u to z »* (the rate of variation of the $u(z)$ function in the present vocabulary) and established the independence of this ratio to the path in the complex plane, provided that the conditions between the partial derivatives of the real and imaginary parts of the function, now bearing his name, are verified, see relations (6.3a) and (6.3b).

ADDITIONAL INFORMATION—**Basic reminders on holomorphic functions**—A *holomorphic* function is a complex-valued function of one or more complex variables, defined on a given complex domain and which is complex-differentiable in a neighbourhood of any point of the domain. The derivative function can then be defined by an extension of the real function case with real value, i.e.,

$$f'(z) = \lim_{z \to z_0} \frac{f(z + z_0) - f(z)}{z - z_0} ,$$

---

[1] See the next mathematical Additional Information.

whose value does not depend on the path taken in the complex plane between points z and z_0. As Cauchy demonstrated, this definition is closely related to the property of any holomorphic function $f(z)$ to be expandable in a convergent power series on any disk included in its definition domain. Any holomorphic function is therefore *analytic* and in particular indefinitely differentiable. Let us recall that an analytical function is a function which can be expressed *locally* as a *convergent power series*.

For an overview of *analytical functions* and *complex-valued functions*, the reader may refer to Volume II of Bass' work [4].

6.1.2 The Initial Condition

In the continuation of the chapter we will only deal with the case of stationary flows for which, and by definition (see Chap. 1, Sect. 1.6.10), we have

$$\frac{\partial}{\partial t} \equiv 0 .$$

The determination of the flow functions is then reduced to resolving an exclusively space field problem. Thus, it is no longer required to prescribe the initial condition since the potential and complex velocity functions now only depend on the single complex variable z.

6.1.3 The Boundary Conditions

Under the study assumptions, the dynamic problem (I) as it "naturally" presents itself, consists in solving the set of the continuity and momentum equations, reduced to:

$$\text{Problem } (I): \qquad \left[\begin{array}{c} div\,\mathbf{V} = 0 \\ \mathbf{grad}\left(\dfrac{V^2}{2}\right) = \mathcal{F} - \dfrac{1}{\rho}\,\mathbf{grad}\,P \end{array} \right] ,$$

under suitable boundary conditions still to be defined.
By introducing the complex potential and complex velocity functions, we can reformulate this problem (I) to that of resolving a Laplace equation (II):

$$\text{Problem } (II): \qquad [\ \Delta\phi = 0 \text{ or } \Delta\psi = 0\ ] ,$$

which can still be reduced to the calculation of a holomorphic function (f or w) which fulfils the boundary conditions specific to the flow geometry. Let us recall, in fact,

that once the velocity is determined, the pressure can be deduced using Bernoulli's
first theorem, as we have demonstrated in Chap. 5.

It should be noted that the ideal fluid assumption is not required for the problem
formulations (I) and (II), due to the removal of the viscous terms from the momentum equation in any irrotational motion of an incompressible fluid. This is not the
case, however, for boundary conditions. Indeed, for a viscous fluid, the *adherence* to
the surface of a solid immersed in the flow results, in a local reference frame, (see
Fig. 6.2) in

$$\mathbf{V}_n = \mathbf{0} \quad \text{and} \quad \mathbf{V}_t = \mathbf{0}, \tag{6.8}$$

at any point of the fluid/solid contact.

However, the prescription of both zero normal and tangential components $\mathbf{V}_t$ and
$\mathbf{V}_n$ of the viscous flow velocity on a solid wall, in a local reference frame attached to
the latter, is not compatible with the equation *order* in (I) or (II) problems, which
makes it possible to satisfy only *one* of the two previous requirements. Since it would
not be realistic to imagine the fluid passing through the solid, the chosen option is
therefore to impose the *impermeability* condition:

$$\mathbf{V}_n = \mathbf{0}. \tag{6.9}$$

Consequently, the velocity component in the tangent plane at any point of the solid
which as a solution to problem (I) or (II) is not *necessarily* zero. Physically, this
can be interpreted as a frictionless fluid slipping along the solid, a situation which
corresponds to the inviscid or ideal fluid concept, *in a purely kinematic sense*. The
conclusions of this discussion are outlined in Fig. 6.2 and can be summarized as
follows:

> In the presence of solid obstacles, any irrotational and isovolume plane flow
> admits the solid contours as streamlines.

Fig. 6.2 Viscous and ideal
fluid/solid boundary
conditions in a reference
frame attached to a fixed
solid

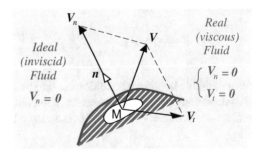

6.2 Some Mathematical Aspects of the Potential Flow Theory

6.2.1 The Direct and Inverse Problem

For a given set of boundary conditions defining a particular flow pattern, the *direct* problem consists in finding one of the functions ϕ, ψ, f or w which ultimately governs the velocity field of this particular flow pattern.

In the case where the motion takes place in a simply connected domain, we are dealing with a Dirichlet/Neumann problem[2] for which we know that the solution exists. It should be remembered, as we have demonstrated in Chap. 5, that the potential function is unique and that, under appropriate assumptions, the solution is that which minimizes the total of the kinetic energy throughout the flow domain. Despite these "theoretical guarantees", apart from domains bounded by circular contours, the analytical determination of the solution expression in a field of arbitrary geometry is not always easy in practice.

A far easier case to solve is the inverse problem, which consists in interpreting, in terms of flow field properties, a complex potential function of a given analytical expression. The question then simply amounts to *(i)* defining the domain(s) on the plane on which the function is defined and holomorphic and *(ii)* identifying all or part of the streamlines as solid obstacle contours.

It is this inverse problem we will address in the following two-step approach, with:

- the inventory of "*basic flow*" complex potential functions;
- the presentation of mathematical tools extending these basic functions to more general configurations.

The first issue will be addressed in the following section (Sect. 6.3). The second is discussed hereafter.

6.2.2 The Materialization Principle

The materialization principle is a direct consequence of the *slip condition* at the boundary of a solid wall for this type of flow (cf. Sect. 6.1.3). It can be stated as follows:

> Any stationary potential flow of an incompressible ideal fluid is not altered when materializing, i.e., substituting all or part of any streamline with a coinciding solid wall.

---

[2] The boundary condition for the velocity is of Dirichlet type (prescription of the function itself) for Problem *(I)* and of Neuman type (prescription of its derivative), for Problem *(II)*.

We will see, in the course of applications, that this property is very useful to give alternative interpretations to some potential flow fields, by identifying the presence of obstacles whose contours are those of some streamline portions.

6.2.3   The Superposition Method

The superposition method is in fact only an application-oriented consequence of the equation linearity governing the velocity field, as clearly appears in the formulation of problem (II). It can be stated as follows:

> If $f_1(z)$ and $f_2(z)$ are two complex potential functions of two flows in the same domain $\mathcal{D}$, any linear combination of these functions is also a complex potential of a new flow in the same domain.

Let us take as an example the simple sum of two functions $f(z) = f_1(z) + f_2(z)$. The resulting velocity field is directly obtained by the *vector additivity* of the component fields, namely
$$\mathbf{V} = \mathbf{V}_1 + \mathbf{V}_2 \quad \text{or} \quad w(z) = w_1(z) + w_2(z).$$

However, it is important to note that this velocity *vectorial* additivity does not imply, in general, the pressure *algebraic* additivity, hence $P \neq P_1 + P_2$. Indeed for each component flow, we have:
$$P_1 + \frac{1}{2}\rho V_1^2 + \mathcal{F}_1 = C_1 \quad \text{and} \quad P_2 + \frac{1}{2}\rho V_2^2 + \mathcal{F}_2 = C_2.$$

By simply adding the previous relations we deduce:
$$(P_1 + P_2) + \frac{1}{2}\rho\left(V_1^2 + V_2^2\right) + \mathcal{F} = C, \text{ bysetting } \mathcal{F} = \mathcal{F}_1 + \mathcal{F}_2 \text{ and } C = C_1 + C_2.$$

This relation does not express Bernoulli's theorem for the flow with the superposed velocity $\mathbf{V} = \mathbf{V}_1 + \mathbf{V}_2$ which, for its part, is written as:
$$(P_1 + P_2) + \frac{1}{2}\rho\left(\mathbf{V}_1 + \mathbf{V}_2\right)^2 + \mathcal{F} = C'$$

By comparing the last two relations, it can be concluded that the pressures are algebraically additive if and only if the scalar product $\mathbf{V}_1 . \mathbf{V}_2$ is zero at any point, which corresponds to the streamline orthogonality of the component flows, as outlined in Fig. 6.3.

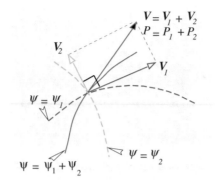

Fig. 6.3 Superposition of flow fields with orthogonal streamlines

6.2.4 The Hydrodynamic Image Method

The method of images consists in using, when available, some specific features of the domain boundary geometry $\mathcal{D}$ where the complex potential function $f(z)$ of the flow is to be determined.

More precisely, we define a *hydrodynamic image* of a singularity S (source, vortex, dipole, see below Sect. 6.3.1) located in a part of the complex plane $\mathcal{D} \subset \mathfrak{R}^2$, whose boundary is $\Sigma_{\mathcal{D}}$, as a set of singularities arranged in the complementary part of the complex plane $\complement_{\mathcal{D}}$ such that the complex potential resulting from the superposition of this set of singularities in the entire complex plane $(\mathcal{D} \cup \complement_{\mathcal{D}})$ has a streamline coinciding with the boundary $\Sigma_{\mathcal{D}}$.

The method of images is used in conjunction with the materialization principle, by considering the solid wall boundaries of the *unknown* flow in the domain $\mathcal{D} \subset \mathfrak{R}^2$ as the streamlines of a *known* flow in $\mathfrak{R}^2$. An example illustrating the method will be presented in Sect. 6.3.3.

6.2.5 The Conformal Mapping

The idea here is to generate, from the complex potential of a *known* flow in a given domain of the complex plane (z), a new potential function for another flow identified in a domain of a complex plane (Z), using a mapping between both domains, symbolically noted as the coordinate transformation $z \rightarrowtail Z$.

For the new transformed function to be considered as a *complex potential*, the mapping should fulfill some requirements. In this case, it is referred to as a *conformal transformation* or *conformal mapping*.

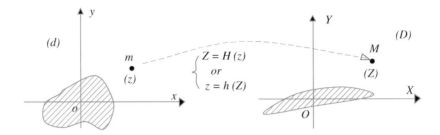

Fig. 6.4 The correspondence between domains (d) and (D) in z and Z planes

6.2.5.1 Definition

Let $Z = H(z)$ be a function defined so as to map a domain (d) of the (x, y) plane
to the domain (D) of the (X, Y) plane, Fig. 6.4.

The transformation of the domain (d) in the domain (D) will be said to be *conformal* if and only if the function $H(z)$ is

(a) *defined* and *continuous* in (d),
(b) *bijective* (one-to-one correspondence),
(c) *holomorphic*, with a *non-zero derivative* in the entire (d) domain.

This set of assumptions ensures the existence of a reverse correspondence of (D) on
(d), noted $z = h(Z)$, which is holomorphic throughout the domain (D).

6.2.5.2 The Characteristic Geometrical Property

This property can be simply stated as follows:

> A conformal map locally preserves orientation and angles: Two infinitesimal
> homologous triangles by conformal transformation are similar.

This result is a direct consequence of the non-nullity of the first derivative. Since the
local angle value is preserved, the streamline and equipotential line orthogonality is
also preserved by conformal mapping. We will discuss at a later stage how to include
some *singular points* in the field, i.e., where the first derivative is zero. The following
proof explains why the angular correspondence directly depends on the order of the
first non-zero derivative at such points.

Proof **The series expansion of the function** $Z = H(z)$ **about** $z = z_0$ **where** $Z = H(z_0) \equiv Z_0$ **is written:**

$$dZ \equiv Z - Z_0 = (z - z_0)\frac{dH}{dz}\Big|_{z=z_0} + \frac{(z-z_0)^2}{2!}\frac{d^2 H}{dz^2}\Big|_{z=z_0} \cdots + \frac{(z-z_0)^n}{n!}\frac{d^n H}{dz^n}\Big|_{z=z_0} + \ldots$$

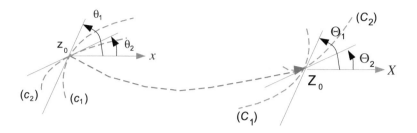

Fig. 6.5 Conformal mapping between homologous curve elements

If the $H(z)$ function derivatives for $z = z_0$ are zero until the $(n-1)$ order included, the previous relation is reduced to:

$$dZ \simeq \frac{(z-z_0)^n}{n!} \frac{d^n H}{dz^n}|_{z=z_0} + \mathcal{O}(z-z_0)^{n+1} .$$

Now, let us consider—see Fig. 6.5—two curves (c_1) and (c_2) intersecting at the affix point $z = z_0$.

We will focus on the variations near this point, according to the infinitesimal displacements along each of these curves, by denoting $z_1 - z_0 = r_1 e^{i\theta_1}$ and $z_2 - z_0 = r_2 e^{i\theta_2}$, the affix points (z_1) and (z_2) on the (c_1) and (c_2) curves respectively.

By applying the previous relation to the displacements taken along each curve, we obtain:

$$dZ_1 \simeq \frac{(z_1 - z_0)^n}{n!} \frac{d^n H}{dz^n}\bigg|_{z=z_0} \qquad \text{and} \qquad dZ_2 \simeq \frac{(z_2 - z_0)^n}{n!} \frac{d^n H}{dz^n}\bigg|_{z=z_0} .$$

The ratio of these two expressions directly leads to the following results between the modules and arguments:

$$R_1/R_2 = (r_1/r_2)^n ,$$
$$\Theta_1 - \Theta_2 = n(\theta_1 - \theta_2) .$$

These relations express all the expected results. The case of the *regular* conformal transformation corresponds, *by definition*, to $n = 1$.

6.2.5.3 The Application to the Potential Flow Study

We are concerned here with the conformal mapping of domains where flows are present. Thus, let us assume that in the domain (d) there is a flow whose complex potential function is $f(z)$. The question is to know if, by a conformal mapping of this domain in the domain (D), this flow is transformed into a new flow in (D) and how this occurs. To answer the question, we will determine the correspondences which exist between the complex potentials, complex velocities, flow rates and velocity circulations of these two motions.

- **The complex potential correspondence**

If $F\,(Z)$ denotes the complex potential function of a mapped flow obtained by applying the conformal representation $z = h\,(Z)$ to the complex potential function $f\,(z)$, then we have:

$$\boxed{F\,(Z) = f\,[h\,(Z)]}\qquad(6.10)$$

Proof   We limit ourselves to demonstrating the result for the potential function ϕ, the case of the function ψ being identical in all respects. Therefore, let us consider, by hypothesis, a function $\phi\,(x,\,y)$ defined, continuous, twice differentiable and solution on the domain (d) of the Laplace equation:

$$\frac{\partial^2 \phi}{\partial x^2} + \frac{\partial^2 \phi}{\partial y^2} = 0\,.$$

We introduce the change of variables $(x,\,y) \longmapsto (X,\,Y)$ which transforms the function $\phi\,(x,\,y)$ in the function $\Phi\,(X,\,Y)$. By assumption, since this change is that of a conformal mapping, the correspondence between the coordinates is holomorphic and the partial derivatives verify Cauchy's relations. Thus we have

$$\frac{\partial X}{\partial x} = \frac{\partial Y}{\partial y} \quad \text{and} \quad \frac{\partial X}{\partial y} = -\frac{\partial Y}{\partial x}\,.\qquad(6.11)$$

By applying this change of variables to the partial derivatives of the function $\phi\,(x,\,y)$, we obtain:

$$\frac{\partial^2 \phi}{\partial x^2} = \frac{\partial^2 \Phi}{\partial X^2}\left(\frac{\partial X}{\partial x}\right)^2 + 2\frac{\partial^2 \Phi}{\partial X \partial Y}\left(\frac{\partial X}{\partial x}\right)\left(\frac{\partial Y}{\partial x}\right) + \frac{\partial^2 \Phi}{\partial Y^2}\left(\frac{\partial Y}{\partial x}\right)^2,$$

$$\frac{\partial^2 \phi}{\partial y^2} = \frac{\partial^2 \Phi}{\partial X^2}\left(\frac{\partial X}{\partial y}\right)^2 + 2\frac{\partial^2 \Phi}{\partial Y \partial X}\left(\frac{\partial X}{\partial y}\right)\left(\frac{\partial Y}{\partial y}\right) + \frac{\partial^2 \Phi}{\partial Y^2}\left(\frac{\partial Y}{\partial y}\right)^2.$$

Now introducing these relations into Laplace's equation and taking into account Cauchy's conditions, we deduce

$$0 = \frac{\partial^2 \Phi}{\partial X^2}\left[\left(\frac{\partial X}{\partial x}\right)^2 + \left(\frac{\partial X}{\partial y}\right)^2\right] + \frac{\partial^2 \Phi}{\partial Y^2}\left[\left(\frac{\partial X}{\partial x}\right)^2 + \left(\frac{\partial X}{\partial y}\right)^2\right] +$$

$$2\frac{\partial^2 \Phi}{\partial X \partial Y}\left[\frac{\partial X}{\partial x}\frac{\partial Y}{\partial x} - \frac{\partial Y}{\partial x}\frac{\partial X}{\partial x}\right],$$

or finally, $0 = \left(\dfrac{\partial^2 \Phi}{\partial X^2} + \dfrac{\partial^2 \Phi}{\partial Y^2} \right) \left[\left(\dfrac{\partial X}{\partial x} \right)^2 + \left(\dfrac{\partial X}{\partial y} \right)^2 \right]$.

As a result, the transformed function $\Phi\,(X, Y) \equiv \phi\,[X\,(x, y)\,,\,Y\,(x, y)]$ is a solution of the equation $\Delta \Phi = 0$ on the domain (D), the bracketed groups not being equal to zero on the entire domain without specifying the function $h\,(Z)$.

- **The complex velocity correspondence**

The relation between the complex velocities $w\,(z)$ and $W\,(Z)$ of two conformal mapped flows is written:

$$\boxed{w\,(z)\,\mathrm{d}z = W\,(Z)\,\mathrm{d}Z} \tag{6.12}$$

Proof This property directly results from the complex potential correspondence. Indeed, by taking the derivative of Eq. (6.10) with respect to z, we obtain:

$$\frac{\mathrm{d}f}{\mathrm{d}z} = \frac{\mathrm{d}F}{\mathrm{d}Z}\frac{\mathrm{d}Z}{\mathrm{d}z},$$

which directly leads to the expected result by simply introducing the respective complex velocities of the two flows:

$$w\,(z) = \frac{\mathrm{d}f}{\mathrm{d}z} \quad \text{and} \quad W\,(Z) = \frac{\mathrm{d}F}{\mathrm{d}Z}.$$

Since any potential flow is characterized by either its complex potential or complex velocity, relation (6.12) shows that any conformal mapping can be equally defined by simply expressing the derivative $\mathrm{d}Z/\mathrm{d}z$. This expression may not be analytically integrable, as long as it complies with the definition requirements (see Sect. 6.2.5.1).

Finally, and as can also be seen from relation (6.12), both flow velocities at homologous points are not, in general, equal. When this is the case for the points *at infinity*, supposed to belong to each homologous domain, the transformation is termed as *canonical*. Hence, in this case and by definition, we have:

$$w\,(z)\,|_{z \to \infty} = W\,(Z)\,|_{Z \to \infty}.$$

- **The flow rate and velocity circulation correspondence**

The flow rates through, and the velocity circulations along *infinitesimal* homologous curve elements are preserved in conformal mapped flows.

Proof Both results are derived from relation (6.12), as a direct interpretation of the complex potential differential. Indeed, denoting as $\mathbf{t}\,\mathrm{d}l$ any oriented curve element in the domain (d) and $\mathbf{n}$ its unit normal vector, the flow rate per unit length in the spanwise direction through this element is:

$$\mathrm{d}q = \mathbf{V}.\mathbf{n}\,\mathrm{d}l = \left[U\frac{\mathrm{d}y}{\mathrm{d}l} + V\left(-\frac{\mathrm{d}x}{\mathrm{d}l} \right) \right]\mathrm{d}l \equiv \mathrm{d}\psi.$$

Similarly, the circulation of the velocity along this curve element is expressed as:

$$d\gamma = \mathbf{V}.\mathbf{t}\,dl = \left[U\frac{dx}{dl} + V\frac{dy}{dl}\right]dl \equiv d\phi\,.$$

Thus we have $df \equiv w\,(z)\,dz = d\phi + id\psi = d\gamma + idq$, so that relation (6.12) still takes the form $d\gamma + idq = d\Gamma + idQ$. Hence the expected results can be simply deduced by identifying the real and imaginary parts.

6.2.5.4 Examples of Conformal Transformations

The Joukowsky transform—Denoting by k a non-zero real, the function

$$\boxed{Z = z + \frac{k^2}{z}} \tag{6.13}$$

defines a conformal map, called the Joukowsky transformation, on *any* of the following *four domains*:

– the interior or exterior of the circle centered at the origin and of radius k, provided that the points $(0, 0)$ and $(\pm k, 0)$ are excluded;
– the upper or lower half plane, again excluding the same points $(0, 0)$ and $(\pm k, 0)$.

Proof   The function $H\,(z) = z + k^2/z$ is defined, continuous and differentiable in the entire complex plane with the exception of the origin. The derivative is equal to zero at the affix points $z = \pm k$ which should therefore also be excluded from the conformal domain. The last requirement to be fulfilled is that of bijectivity. By specifying relation (6.13) as a polynomial in z, we obtain:

$$z^2 - Zz + k^2 = 0\,.$$

This second degree equation proves that any point in the plane (Z) has *two* pre-images in the plane (z). Therefore, the correspondence is not generally bijective. The bijectivity or one-to-one correspondence can be satisfied if and only if the domains in the complex plane (z) are defined such that, when one of the two pre-images belongs to the domain, the other is excluded. This can be obtained in several ways.

The affixes z_1 and z_2 of the two pre-images of the same point in the plane (Z) are, in fact, such that $z_1 z_2 = k^2$. Their modules and arguments are therefore linked by

$$r_1 r_2 = k^2 \quad and \quad \theta_1 + \theta_2 = 0\,,$$

so that:

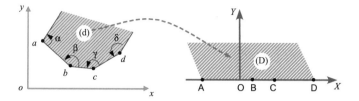

Fig. 6.6 Correspondence diagram of the Schwarz–Christoffel transform

$$\text{if } r_1 \in (0, k), \text{ then } r_2 \in (k, +\infty),$$
$$\text{if } \theta_1 \in (0, \pi), \text{ then } \theta_2 \in (-\pi, 0).$$

The first condition leads to separating the domains by the circle of radius k, centred at the origin, since the area inside (respectively outside) the circle can only be described by one pre-image at a time.

Similarly, the second condition leads to the distinction between the upper and lower half planes, on each of which the Joukowsky transformation operates as a conformal mapping.

The Kármán–Trefftz transform—We will see at a later stage that at the *singular point* $z = k$, the mapping of the radius k circle by the Joukowsky transform is a curve with a cusp at the corresponding point. Such a mathematical singularity cannot be physically representative of an actual airfoil trailing edge, whose contour is more like a cornered curve. The Kármán-Trefftz transform, which allows a non-zero angle at the trailing edge, is more appropriate in this respect. It is defined as:

$$\boxed{\frac{Z-1}{Z+1} = \left(\frac{kz-1}{kz+1}\right)^k} \tag{6.14}$$

This relation extends the Joukowsky transform (6.13) to which it is reduced for $k = 2$, with then $Z = z + 1/4z$.

The Schwarz–Christoffel transform—This conformal transformation maps the interior of a simple polygonal contour of the plane (x, y) onto the half plane $Y \geqslant 0$ or $Y \leqslant 0$. Denoting by K a real constant and with the notations of Fig. 6.6, the Schwarz–Christoffel transform is expressed by the following differential relation:

$$\boxed{\frac{dz}{dZ} = K\,(Z - Z_A)^{\frac{\alpha}{\pi} - 1} \times (Z - Z_B)^{\frac{\beta}{\pi} - 1} \times (Z - Z_C)^{\frac{\gamma}{\pi} - 1} \times \ldots} \tag{6.15}$$

6.3   Examples of Complex Potentials

6.3.1   Elementary Flows

We begin with an inventory of the complex potential functions of some *basic flows*. The results, given without a demonstration, can be usefully retrieved as an exercise. We will then derive some more complex potential flows by *superposition* of some of the previous basic functions. Finally, we will give an application example of the hydrodynamic image method.

6.3.1.1   The Uniform Flow

The complex potential

$$f(z) = V_0 z e^{-i\alpha}$$ (6.16)

where V_0 and α are two real constants, is that of a uniform rectilinear flow whose velocity, equal in modulus to $|V_0|$, is inclined at a trigonometric angle α on the Ox axis (Fig. 6.7).

6.3.1.2   The Source/Sink Line

With the restriction $arg(z) \in \,]0.2\pi[$, the complex potential

$$f(z) = \frac{D}{2\pi} Log \, (z - z_0)$$ (6.17)

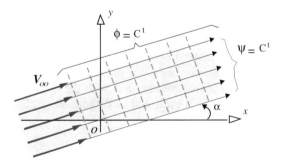

Fig. 6.7 Uniform flow at an incidence angle α

Fig. 6.8 Line source/sink at the origin

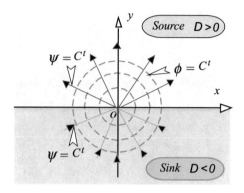

where D is a real constant and z_0 a complex parameter, is that of a source ($D > 0$) or a sink ($D < 0$) line normal to the z-plane at the affix point $z = z_0$. D is the total algebraic flow rate (Fig. 6.8).

It should be noted that a branch cut along the half axis Ox (the positive real axis) for instance, has to be introduced due to the multi-valued complex logarithm.[3]

6.3.1.3 The Doublet Flow

The doublet flow[4] is obtained by superimposing a source and a sink of the same absolute flow rate $|D|$, and taking the limit under the condition $\lim\limits_{|D| \to \infty,\, \epsilon \to 0} (D \times \epsilon) = \mathcal{M}$, where ϵ refers to the distance between the source and sink. The doublet complex potential whose strength is $\mathcal{M}$ which is located at point $z = z_0$ is written (Fig. 6.9):

$$f(z) = \frac{\mathcal{M}}{z - z_0} \tag{6.18}$$

6.3.1.4 The Straight Line Vortex

With the restriction $arg(z) \in \,]0.2\pi[$, the complex potential

$$f(z) = -i\frac{\Gamma}{2\pi} Log(z - z_0) \tag{6.19}$$

---

[3] The Log symbol denotes the logarithm of a complex number, namely, with $z = re^{i\theta}$, $Log(z) = \ln r + i\theta$, where $\ln$ refers to the natural (Napierian) logarithm of a real.

[4] In fluid mechanics, the doublet flow corresponds to the dipole field in electrostatics. The same name is sometimes used in both cases.

Fig. 6.9 Doublet flow at the
origin

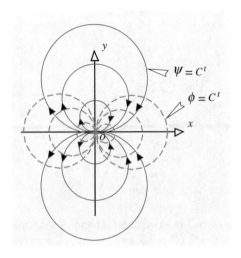

Fig. 6.10 The line vortex at
the origin

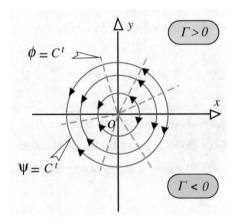

where Γ is a real constant and z_0 a complex parameter, represents the line vortex flow
rotating in the trigonometric direction ($\Gamma > 0$) or in the opposite direction ($\Gamma < 0$)
positioned at $z = z_0$. It can be shown that due to the branch cut of the half Ox-axis,
the velocity circulation around any contour *surrounding* the logarithmic singularity
is equal to Γ (Fig. 6.10).

Nevertheless, the motion remains irrotational, a result which does not contradict
Stokes formula in Chap. 2, Sect. 2.4.3, which cannot be applied to a closed disk
centered on the vortex, due to the branch cut imposed to define a single-valued
logarithm function $Log(z)$.

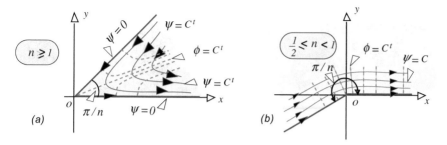

Fig. 6.11 Flows in a corner (**a**) and past a wedge (**b**)

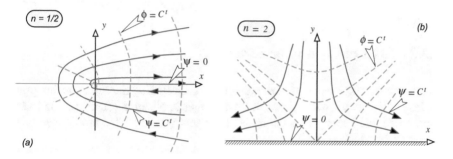

Fig. 6.12 Flow around a half-plane tip (**a**), impinging flow on an infinite flat plate (**b**)

6.3.1.5 Corner Flows

The complex potential

$$f(z) = K (z - z_0)^n \tag{6.20}$$

where K and n are two real numbers, generate a set of corner flows of angle π/n whose vertex is positioned at the affix point $z = z_0$. As outlined in the above figures, various configurations are possible. Depending on the exponent value, we can distinguish between (Fig. 6.11):

1. flows in a corner, for $n \geqslant 1$;
2. flows past a wedge, for $\frac{1}{2} \leqslant n \leqslant 1$.

The limit value $n = 1/2$ corresponds to the flow around a half plane tip, Fig. 6.12(a). Another important specific case, see Fig. 6.12(b) is obtained for $n = 2$. It is that of a flow normal to a plane, which can be locally identified as the motion in the proximity of a stagnation point, as we will see at a later stage.

Fig. 6.13 Streamlines of a
source in a uniform flow

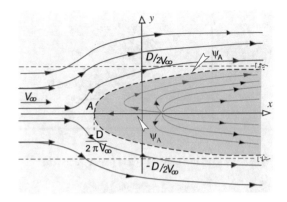

6.3.2   Superposition Flow Examples

In this section three examples are presented for illustrating the superposition principle. The first two are to be considered as simple application exercises. The last is more important because it is the basis of the sharp trailing edge profile theory, as introduced by Joukowsky.

6.3.2.1   A Source in a Uniform Flow

This first flow example is obtained by superposing a line source $D > 0$ at the reference frame origin and a uniform flow whose velocity V_∞ at infinity is parallel to the x axis. As we have explained in Sect. 6.2.3, the complex potential of the resulting flow is written as:

$$f(z) = V_\infty z + \frac{D}{2\pi} \, \mathrm{Log} z \,. \tag{6.21}$$

The streamline equations are $V_\infty r \sin\theta + \dfrac{D}{2\pi}\theta = C^t$, and the complex velocity is

$$w(z) = V_\infty + \frac{D}{2\pi z} \,.$$

This function is zero at the single point A, whose abscissa is $x_A = -D/2\pi V_\infty$, on the real axis. This point is by definition referred to as a stagnation point.

The equation of the streamline passing through this stagnation point is obtained by requiring that the general streamline equation is satisfied for $z = x_A$, viz $V_\infty r \sin\theta + \frac{D}{2}\left(\frac{\theta}{\pi} \pm 1\right) = 0$, the negative sign being associated with positive y.

Thus, we can deduce that the real axis belongs to the streamline passing through the stagnation point, which also consists of two symmetrical branches with respect to Ox, which exhibit two horizontal asymptotes of ordinates $\pm D/2V_\infty$, as outlined in the general flow pattern in Fig. 6.13.

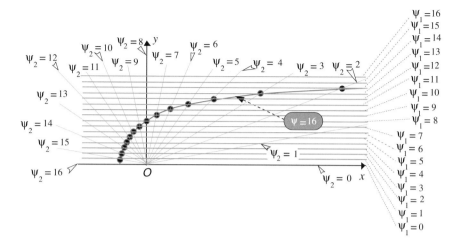

Fig. 6.14 Schematic for the geometric construction of the streamline passing through the stagnation point

This example also illustrates the streamline construction using the superposition principle of the two fields $\psi_1 = C_1^t$ and $\psi_2 = C_2^t$, by adding the constants $\psi = C_1^t + C_2^t$.

For this purpose, and with reference to Fig. 6.14, let us assume that the streamline constant values of both the uniform flow and source flow are labelled from 0 to 16. The unit corresponds to an arbitrary increment of the vertical metric scale, for the first flow, and the radial angular scale, for the second. Now let us consider drawing the streamline passing through the stagnation point. Since this point is located on the x axis, we have, due to the previous constant convention for this streamline, $\psi_A = 16$. The geometrical drawing of the streamline $\psi = \psi_A$ is simply done by joining the intersections of the horizontal-valued lines $\psi = \psi_1$ with the radial-valued lines $\psi = \psi_2$, so that $\psi_1 + \psi_2 = 16$. This plotting is outlined in Fig. 6.14.

HISTORICAL NOTE— Rankine's half body—The materialization of the ruled surface (Σ) spanning the asymptotic branches of the streamline passing through the stagnation point in Fig. 6.13, generates a blunt-nosed solid called the Rankine half body. As an exercise, it can be verified that the Rankine solid of revolution, axisymmetric by rotation around the flow direction at infinity is obtained from the stream function whose expression, in spherical coordinates is:

$$\psi (r, \varphi) = V_\infty \frac{r^2}{2} \sin^2 \varphi - \frac{D}{4\pi} \cos \varphi .$$

In this case, the stagnation point coordinates are $(-\sqrt{\dfrac{D}{4\pi V_\infty}}, 0, 0)$. The equation of the streamline passing through this stagnation point is

$$r^2 = \frac{D}{2\pi V_\infty} \left(1 + \cos\varphi\right) / \sin^2\varphi .$$

The mention of Rankine's name for the solid in question is a reference to the paper [127], this scientist published in 1864, where a geometric construction diagram which is perfectly in line with that of Fig. 6.14 is described. Nevertheless, the general principle of this construction is not due to Rankine who quoted in this regard Maxwell's contribution «*The general process of constructing a series of curves whose equations is* $\phi(x, y) + \psi(x, y) = constant$, *by drawing lines diagonally through a network consisting of two sets of curves whose equations are respectively* $\phi(x, y) = constant$ *and* $\psi(x, y) = constant$, *is due to Professor* CLERK MAXWELL.»

Remark The external flow action on the streamline passing through the stagnation point. Materializing the surface (Σ) spanning the contour which coincides with the streamline passing through the stagnation point (Cf. line ψ_A in Fig. 6.13), it is easy to understand that the uniform flow to infinity applies pressure forces, whose resultant force is, for symmetric reasons, parallel to the x -axis.

This result can be rigorously demonstrated by applying Euler's theorem to the fluid domain between, on the one hand, the outside of the surface $\Sigma + S$, (where S is the cross-section of height D/V_∞, closing the contour Σ) for $x \to \infty$) and, on the other hand, the interior of an infinitely wide circle centered at the origin. The resultant action on this domain, as defined at a later stage (cf. Sect. 6.4.1) is equal to $\rho V_\infty D$, generating a pressure drag on the sole surface (Σ) equal to $P_\infty D/V_\infty$.

The reader may take the opportunity to directly demonstrate these results as an exercise. They can also be re-demonstrated from the general expression of the pressure forces exerted by a fluid in a potential motion on any obstacle, which we will see later (cf. Sect. 6.4.4).

6.3.2.2 Rankine's Oval Cylinder

When, in the previous situation, the source's flow rate does not escape to infinity, but is drawn by a sink of the same strength, symmetrically arranged with respect to the y axis, the streamline passing through the upstream stagnation point closes on itself to form an oval, called Rankine's oval (see the following figure).

The corresponding complex potential is written as:

$$f(z) = V_\infty z + \frac{D}{2\pi} Log \left(\frac{z+a}{z-a}\right) , \tag{6.22}$$

where $\pm a$ denotes the abscissa of the sink and source respectively. It is easy to see that the flow pattern, as outlined in Fig. 6.15, has a double plane symmetry, with respect to $x = 0$ and $y = 0$.

Due to the symmetry with respect to the x-axis, the pressure resultant force of the external flow on the oval cylinder in the y direction (*lift*), is zero. Similarly, with

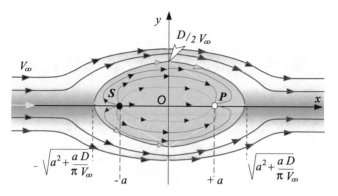

Fig. 6.15 Flow streamlines around Rankine's oval

respect to the y-axis symmetry, there is also no pressure resultant force in the flow direction (*drag*).

This double conclusion, which we will come back to at a later stage, is a first illustration of a potential theory paradox, known as *d'Alembert's paradox*.

6.3.2.3 The Flow Around a Circular Cylinder

We consider the potential flow composed by the superposition of a uniform velocity motion ($V_0 > 0$, parallel to the x-axis) and a positive strength doublet ($\mathcal{M} > 0$) located at the reference frame origin. The corresponding complex potential is written as follows:

$$f(z) = V_0 z + \frac{\mathcal{M}}{z}, \quad \text{with} \quad z \neq 0. \tag{6.23}$$

- **The flow pattern**

In this case, the equipotential line and streamline equations are respectively:

$$\phi : \quad V_0 x + \frac{\mathcal{M}}{r} \cos \theta = C', \qquad \psi : \quad V_0 y - \frac{\mathcal{M}}{r} \sin \theta = C',$$

and the complex velocity is expressed as $w(z) = V_0 - \mathcal{M}/z^2$.

We can easily deduce that there are two stagnation points A and A', whose affixes (z_A and $z_{A'}$) are equal to $\pm\sqrt{\mathcal{M}/V_0}$. These points are located on the real axis, symmetrically with respect to the origin. The streamline equation passing through these stagnation points is written as $\psi_A = \psi_{A'} = 0$, namely:

$$\left(V_0 r - \frac{\mathcal{M}}{r} \right) \sin \theta = 0.$$

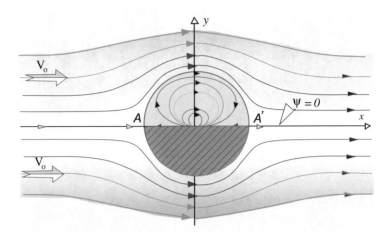

Fig. 6.16 Superposition of a doublet and a uniform flow. Upper part: streamlines of the whole flow pattern. Lower part: streamlines outside the circle after materialization

The corresponding curve consists of two elements: (i) *the real axis*, for $\sin \theta = 0$ and (ii) the *circle centered at the origin* with *the radius* $a = \sqrt{\mathcal{M}/V_0}$, for $r^2 = \mathcal{M}/V_0$. The general flow pattern, as outlined in the $y > 0$ part of Fig. 6.15, reveals two distinct motion regions, located respectively inside and outside the circle of radius a (Fig. 6.16).

These two regions are characterized by the separation of the flow rates between the motions *inside* and *outside* the circle. By giving explicit details about the doublet strength as a function of the circle radius, namely $\mathcal{M} = V_0 a^2$, the complex potential is still written:

$$f(z) = V_0 \left(z + \frac{a^2}{z} \right) \tag{6.24}$$

In line with the materialization principle, this *same* function, under the *sole restriction* $\|z\| \geqslant a$, is none other than that of a uniform velocity stream V_0 at infinity around a circular cylinder centered at the origin whose radius is a. This configuration is outlined in the $y < 0$ part of Fig. 6.15.

- **The velocity and pressure on the cylinder**

It is easy to deduce from the previous results that the complex velocity on the circle is $2V_0 \sin \theta \, e^{-i\theta}$, which corresponds to a vector locally tangent to the circle, whose module varies with the polar angle according to:

$$V_{r=a}(\theta) = 2V_0 \sin \theta \, .$$

The application of Bernoulli's first theorem then leads to the following expression of the pressure distribution along the circle:

Fig. 6.17 Velocity modulus and pressure distribution on the cylinder in a potential flow without circulation

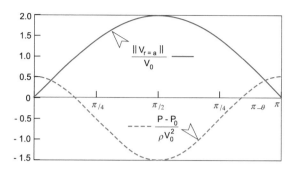

$$P_{r=a}(\theta) = P_0 + \frac{1}{2}\rho V_0^2 (1 - 4\sin^2\theta),$$

where P_0 is the pressure at infinity.

It is easy, by integration from this last result, to determine the x and y components of the pressure resultant force applied to the cylinder. We then ascertain, by this direct analytical calculation, that no effort is applied to the cylinder by the flow, which is an explicit demonstration of the previously mentioned *d'Alembert's paradox* (Fig. 6.17).

6.3.2.4 The Lifting Flow Over a Circular Cylinder

We will now examine what happens when superimposing on the infinite uniform flow around a cylinder, a vortex of strength Γ centered at the origin. With the previous notations, the complex potential is written:

$$f(z) = V_0\left(z + \frac{a^2}{z}\right) - i\frac{\Gamma}{2\pi}Log\,z \tag{6.25}$$

Taking into account the singularity of the multivalued logarithmic function, a cut infinitesimally below the positive real axis, for instance, including the origin, is made in the complex plane. The complex velocity is then expressed as:

$$w(z) = V_0\left(1 - \frac{a^2}{z^2}\right) - i\frac{\Gamma}{2\pi z}.$$

It is zero at the affix points z_A such that $z_A^2 - i\frac{\Gamma}{2\pi V_0}z_A - a^2 = 0$, which provides two stagnation points:

$$z_A = z_{A'} = i\frac{\Gamma}{4\pi V_0} \pm \frac{1}{2}\sqrt{4a^2 - \frac{\Gamma^2}{4\pi^2 V_0^2}}.$$

Several cases are to be discussed, depending on the sign of the discriminant.

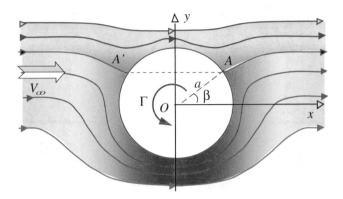

Fig. 6.18 Flow around a cylinder with a low (sub-critical) circulation

- **First case** : $\boxed{0 \leqslant \Gamma < 4\pi a V_0}$

The discriminant is then positive and the modules of the two stagnation point affixes are identical and equal to:

$$\sqrt{x_A^2 + y_A^2} = \sqrt{\frac{16\pi^2 a^2 V_0^2 - \Gamma^2}{16\pi^2 V_0^2} + \frac{\Gamma^2}{16\pi^2 V_0^2}} = a \,.$$

Both stagnation points are therefore located on the circle of radius a, in symmetrical positions with respect to the y-axis. They are identified by the polar angles β and $\pi - \beta$ respectively, with

$$\boxed{\sin \beta = \frac{\Gamma}{4\pi a V_0}} \qquad\qquad (6.26)$$

The general flow pattern is outlined in the following figure, where only the external flow field $r \geqslant a$ is shown (Fig. 6.18).

   Without circulation $\beta = 0$, we retrieve the location of the two stagnation points at the intersection of the circle and the real axis. It can therefore be noted that the circulation effect consists in shifting the two stagnation points along the circle, *symmetrically about* Oy, by an ordinate proportional to the circulation intensity.

ADDITIONAL INFORMATION—**The flow with an attack angle**—In the case where the uniform flow at infinity has an incident angle α with respect to the x-axis, the previous symmetric position shifting of both stagnation points with respect to the y-axis is transposed directly to the direction orthogonal to that of the flow at infinity, as outlined in the following figure.

   Equation (6.26) then takes the general form (Fig. 6.19):

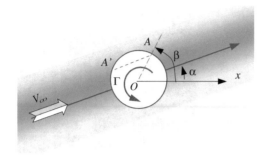

Fig. 6.19 Position of the stagnation points in a non-zero incidence flow

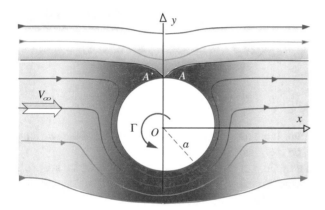

Fig. 6.20 Flow around a cylinder at the critical circulation

$$\sin (\beta - \alpha) = \frac{\Gamma}{4\pi a V_0}$$

- **_Second case_**: $\boxed{\Gamma = 4\pi a V_0}$

For this so-called "*critical*" value of the circulation, the two stagnation points merge with the intersection of the circle and the y-axis ($\beta = \beta' = \pi/2$). This results in the pattern in Fig. 6.20.

- **_Third case_**: $\boxed{\Gamma > 4\pi a V_0}$

The discriminant in the affix equation of the stagnation points is then negative, so that the roots can be put in the form:

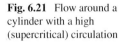

Fig. 6.21 Flow around a
cylinder with a high
(supercritical) circulation

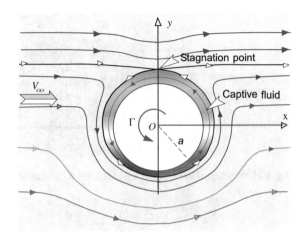

$$z_A = z_{A'} = i\,\frac{\Gamma \pm \sqrt{\Gamma^2 - 4\pi^2 a^2 V_0^2}}{4\pi a V_0}.$$

These are pure imaginary numbers, which means that the two stagnation points
are located on the y-axis.

Since the module of the root product is a^2, one of the stagnation points lies
inside the circle of radius a and the other lies outside. This results in the pattern
outlined in Fig. 6.21.

Remark It can be interpreted that in this case, the vortex strength, as compared
with the velocity value of the flow at infinity, is high enough to drive a captive part
of the external fluid to rotate around the cylinder. Therefore, the streamline passing
through the stagnation point outside the circle separates the flow field into two areas,
depending on whether or not they include a flow rate contribution from the motion
at infinity.

6.3.3   The Method of Images

For illustrating the implementation of the method of images defined in Sect. 6.2.4,
we consider the flow generated, in the half-plane $x \leq 0$, by a source (flow rate D),
located at the coordinate point $(-a, 0)$, a denoting a positive real. The line $x = 0$ is
supposed to materialize a *solid* wall on which the source stream flows, as outlined
in Fig. 6.22a.

To determine the complex potential function of the flow in the half plane $x \leq 0$,
we extend it, by symmetry with respect to $x = 0$, to a new flow in the entire plane,
in which the $x = 0$ line is a streamline. For this purpose, it is just necessary to

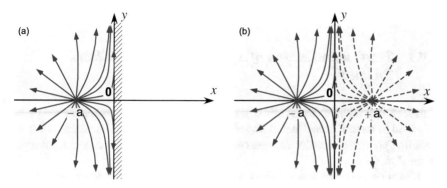

Fig. 6.22 Source flow on a wall (**a**), equivalent reconstruction using the method of images (**b**)

add in the half-plane $x \geq 0$ the "*image of the source*", i.e. a source with the same flow rate D located at the coordinate point $(+a, 0)$. The complex potential of these two combined source flows in the entire plane is immediately obtained using the superposition principle, namely:

$$f(z) = \frac{D}{2\pi} Log(z + a) + \frac{D}{2\pi} Log(z - a) \equiv \frac{D}{2\pi} Log[(z + a)(z - a)] \,.$$

Putting $z + a = r_1 e^{i\theta_1}$ and $z - a = r_2 e^{i\theta_2}$, this complex potential is still written:

$$f(z) = \ln(r_1 r_2) + i(\theta_1 + \theta_2) \,.$$

Thus, the streamline equations are $\psi = \theta_1 + \theta_2 = C^t$, from which it is easy to deduce that, for $\theta_1 + \theta_2 = \pi$, the ordinate axis is indeed a streamline. Then we have more than applying the materialization principle to the flow defined by:

$$f(z) = \frac{D}{2\pi} Log[(z + a)(z - a)] \equiv \frac{D}{2\pi} Log(z^2 - a^2) \quad \text{with} \quad z \in \mathbb{C} \,,$$

to obtain, by simply restricting the variation range of z, the expected potential function of the source flow in the bounded half-plane, i.e., $f(z) = \frac{D}{2\pi} Log(z^2 - a^2)$, with $\mathcal{R}(z) \leq 0$.

6.4 Efforts of a Potential Flow on an Immersed Solid

6.4.1 The General Action of a Moving Fluid on a Closed Contour

This section is intended to determine the pressure efforts which an ideal incompressible fluid in an irrotational steady 2D-plane motion exerts on an immersed solid. As a starting point, we consider a closed curve (C) bounding a simply connected domain of the flow field.

This curve is not, *in general terms*, a streamline, hence we will denote $\mathbf{R}_M$ (C) the momentum resultant flux crossing through (C). The pressure resultant force, applied by the fluid *outside the domain* bounded by (C), is noted $\mathbf{R}_P$ (C).

We define the *external fluid action* on the contour (C) as the quantity:

$$\boxed{\mathbf{A}\,(\mathrm{C}) = \mathbf{R}_M\,(\mathrm{C}) - \mathbf{R}_P\,(\mathrm{C})}$$

To calculate this action, we introduce—see Fig. 6.23—a second surface (Σ), external to (C) which includes no other singularities than those possibly present in the domain bounded by the first contour.

Then let us apply Euler's theorem to the fluid domain (Δ) between (Σ) and (C). Along the contour (C), the outward-pointing normal of the domain (Δ) is $\mathbf{n}\,' = -\mathbf{n}$, so that the momentum resultant flux of the fluid crossing out of the domain through (C) is $-\mathbf{R}_M$ (C) and similarly for the resultant force due to the external pressure applied to (Δ) through (C), *viz* $-\mathbf{R}_P$ (C). By designating $\mathbf{R}_M$ (Σ) the sum of the momentum resultant flux crossing out (Σ) and by $\mathbf{R}_P$ (Σ) the external pressure resultant force applied to the fluid in (Δ) along the same surface, we have:

$$\mathbf{R}_M\,(\Sigma) - \mathbf{R}_M\,(\mathrm{C}) = \mathbf{R}_P\,(\Sigma) - \mathbf{R}_P\,(\mathrm{C})\ ,$$

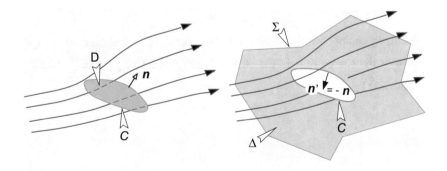

Fig. 6.23 Application domain of Euler's theorem for the action calculation on an arbitrary contour (C)

or again $\mathbf{A}\,(\text{C}) = \mathbf{R}_M\,(\Sigma) - \mathbf{R}_P\,(\Sigma)$. By noting (X, Y), (X_M, Y_M) and (X_P, Y_P) the respective components of the vectors $\mathbf{A}$, $\mathbf{R}_M\,(\Sigma)$ and $\mathbf{R}_P\,(\Sigma)$ in the complex plane, we will rewrite the previous result in the form :

$$\overline{A} \equiv X - iY = (X_M - X_P) - i\,(Y_M - Y_P)\,. \tag{6.27}$$

We will now provide details relating to the vector $\mathbf{R}_M\,(\Sigma)$ and $\mathbf{R}_P\,(\Sigma)$ components.

- The components of the momentum resultant flux through (Σ) are:

$$X_M = \int_\Sigma \rho U\,(\mathbf{V}.\mathbf{n})\,d\sigma = \int_\Sigma \rho U d\psi\,,$$

$$Y_M = \int_\Sigma \rho V\,(\mathbf{V}.\mathbf{n})\,d\sigma = \int_\Sigma \rho V d\psi\,.$$

Thus $X_M - iY_M = \int_\Sigma \rho\,(U - iV)\,d\psi = \int_\Sigma \rho w d\psi$, by introducing the complex velocity $w\,(z)$ of the flow. However, it can be noted that $wdz = d\varphi + id\psi$ and $\overline{w}\overline{dz} = d\varphi - id\psi$, so that $d\psi = \frac{i}{2}\left(\overline{w}\overline{dz} - wdz\right)$. Hence, the momentum resultant flux can be finally written:

$$X_M - iY_M = i\frac{\rho}{2}\int_\Sigma w\overline{w}\overline{dz} - w^2 dz\,. \tag{6.28}$$

- To express the external pressure resultant force applied through (Σ) to the fluid domain (Δ), let us first note (see Fig. 6.24) that the outward-pointing unit normal vector is derived from the unit tangent vector to the contour oriented in the trigonometric direction, by a rotation of the angle $-\pi/2$. We then have:

$$X_P = -\int_\Sigma P\mathbf{n}.\mathbf{i}\,d\sigma = -\int_\Sigma P dy \quad \text{and} \quad Y_P = -\int_\Sigma P\mathbf{n}.\mathbf{j}\,d\sigma = \int_\Sigma P dx\,.$$

Now, according to Bernoulli's first theorem and discarding the external body forces, we know that $P + \frac{1}{2}\rho w\overline{w} = C'$, so that, since (Σ) is a closed contour:

$$X_P = \frac{\rho}{2}\int_\Sigma w\overline{w}dy \quad \text{and}$$

$$Y_P = -\frac{\rho}{2}\int_\Sigma w\overline{w}dx$$

Thus, the final result is:

Fig. 6.24 Schematic for the
unit vector directions along
the contour (Σ)

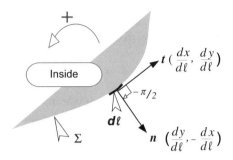

$$X_P - iY_P = \frac{\rho}{2} \int_\Sigma w\overline{w}\,(dy + idx) = i\frac{\rho}{2} \int_\Sigma w\overline{w}\overline{dz}. \tag{6.29}$$

By substituting expressions (6.28) and (6.29) in equation (6.27), we obtain the
expected result:

$$\boxed{X - iY = -i\frac{\rho}{2} \int_\Sigma w^2 dz} \tag{6.30}$$

It is worth recalling that this formula expresses the external pressure resultant force
and momentum resultant flux through any closed contour (C) located within the flow
domain bounded by an arbitrary contour (Σ).

6.4.2 Application to a Closed Streamline: Blasius' Formulas

In the particular case where the contour (C) coincides with a closed streamline,
the action on this contour reduces to the sole external pressure resultant force so
that $\mathbf{A}\,(C) = -\mathbf{R}_P\,(C)$. Denoting as $(\widetilde{X}, \widetilde{Y})$ the components of the external pressure
resultant force, we obtain by directly applying the previous general relation:

$$\boxed{\widetilde{X} - i\widetilde{Y} = i\frac{\rho}{2} \int_\Sigma w^2 dz} \tag{6.31}$$

Relation (6.31) is Blasius' first formula. The second results from the application of
Euler's theorem to the resultant moment. For a closed streamline surrounding the
origin and considering the resultant moment $\widetilde{M}$ with respect to this point, we have:

$$\boxed{\widetilde{M} = -\frac{\rho}{2}\mathcal{R}_e \left[\int_\Sigma z w^2 dz \right]} \tag{6.32}$$

Fig. 6.25 Velocity
circulation induced lift on a
circular cylinder

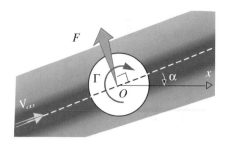

6.4.3   Kutta-Joukowsky's Theorem for the Circular Cylinder

For a uniform flow at infinity around a cylinder with circulation, Kutta-Joukowsky's
theorem can be stated as follows:

> The pressure resultant force of a uniform potential flow (V_0) at infinity on a
> circular cylinder with a velocity circulation Γ:
> – is normal to the flow direction at infinity,
> – has a direction derived from that of the uniform flow by a $\pi/2$ rotation opposite
> to the circulation,
> – has as module
>
> $$F = \rho V_0 |\Gamma| \qquad (6.33)$$

If, with reference to Fig. 6.25, the abscissa axis x is assumed to be "horizontal", the
force in question is directed towards the positive y side, when the velocity circulation
is *opposite to the trigonometric direction* with, for the complex potential of the vortex
$-i\frac{\Gamma}{2\pi} Log\, z$. It is then referred to as a *lift force*—see Additional Information—to
specify that this direction is opposed to that of gravity (in the negative y direction).

Thus, in terms of the potential flow theory of an ideal fluid, any steady irrotational
motion imparts no force in the direction of the fluid velocity at infinity (*drag*). Since
in the absence of circulation there is also no lift, as a direct consequence of Kutta-
Joukowsky's theorem, such a flow ultimately exerts no force on an immersed solid.
This result, which disagrees with common experiments of *viscous* fluid motions, is
only a new manifestation of the previously mentioned d'Alembert paradox. We will
return to this issue at a later stage (Sect. 6.4.5.1).

Proof   The Kutta-Joukowsky formula directly results from Blasius' formula (6.31)
for the flow whose complex potential is $f(z) = V_0\left(z + \frac{a^2}{z}\right) - i\frac{\Gamma}{2\pi} Log\, z$. It is then
easy to verify that the square of the complex velocity is

$$w^2(z) = V_0^2 - i\frac{V_0\Gamma}{\pi}\frac{1}{z} - \left(2V_0 a^2 + \frac{\Gamma^2}{4\pi^2}\right)\frac{1}{z^2} + i\frac{V_0\Gamma}{\pi}\frac{a^3}{z^3} + V_0^2\frac{a^4}{z^4}. \qquad (6.34)$$

The value of the complex function integral $w^2(z)$ is obtained using the residue theorem or Cauchy's integral formula in the form of

$$\int_\Sigma w^2(z)\,dz = 2i\pi\,[Residue]_{z=0}\,.$$

Now the value of the residue is obtained directly from relation (6.34) by noticing that this is the Laurent expansion of the function to be integrated around the pole $z_0 = 0$. Thus, the residue at point $z = 0$ is none other than the coefficient of the term in $1/z$ so that:

$$\int_\Sigma w^2(z)\,dz = 2i\pi \times \left(-i\frac{V_0\Gamma}{\pi}\right) = 2V_0\Gamma\,.$$

The components of the pressure resultant force are therefore $\widetilde{X} = 0$, $\widetilde{Y} = -\rho V_0\Gamma$. It is indeed a lift force. Finally, in application of relation (6.32) and by the same type of calculation, it can be demonstrated that the resultant moment is

$$\int_\Sigma zw^2(z)\,dz = -2i\pi\left(2V_0a^2 + \frac{\Gamma}{4\pi a^2}\right),$$

which is a pure imaginary number. The efforts are therefore solely reduced to a lift force.

ADDITIONAL INFORMATION—The Lift and Drag—In a uniform velocity air or water flow at infinity, any solid obstacle experiences aero-hydro-dynamic actions, characterized by a resultant force and moment.

– The *drag* is the resultant force component in the flow direction at infinity (or the relative wind direction in the case of a moving machine).

– The *lift* is the resultant force component normal to the flow direction at infinity (or the relative wind direction in the case of a moving machine), and pointing "upward". It was originally introduced in the context of flight mechanics of objects heavier than air, to balance the gravity force, which explains its orientation. Therefore, we still speak today of "*downforce*" or *negative lift*, when this force acts in the opposite way to press down the vehicle on the ground.

6.4.4 The General Theorem for a Logarithmic Singularity

The Kutta-Joukowsky result can be generalized to the case of a flow with any logarithmic singularity. The corresponding theorem is then expressed as follows:

The flow whose complex potential is:

$$f(z) = V_0 z + \frac{D}{2\pi} Log\, z - i \frac{\Gamma}{2\pi} Log\, z + g(z),$$

where D and Γ are two reals and $g(z)$ any complex function with no linear contribution nor logarithmic singularities, applies to any closed contour surrounding the origin:

1. a *drag* of algebraic value $\rho V_0 D$, where D stands for the flow rate crossing through the contour;
2. a *lift* of algebraic value $-\rho V_0 \Gamma$, where Γ stands for the velocity circulation along the contour.

Proof For the considered flow, the complex velocity is written:

$$w(z) = V_0 + \frac{D - i\Gamma}{2\pi} \frac{1}{z} + \frac{dg}{dz}.$$

Given the assumptions about the complex function $g(z)$, the term in $1/z$ in the $w^2(z)$ expression can only come from the first two contributions of the above expression. Hence we have $w^2(z) = V_0 \dfrac{D - i\Gamma}{\pi} \dfrac{1}{z} + \ldots$

In application of the residue theorem, Blasius' first formula (6.31) then leads to

$$\widetilde{X} - i\widetilde{Y} = -i\frac{\rho}{2} \times 2i\pi \times V_0 \frac{D - i\Gamma}{\pi} \equiv \rho V_0 (D - i\Gamma).$$

It would be easy to demonstrate that the resultant moment has no real component, which completes the demonstration of the proposal.

6.4.5 *Physical Interpretations*

6.4.5.1 D'Alembert's Paradox

We have previously reported, on several occasions, paradoxical deductions relating to the absence of *drag* (Sect. 6.3.2.2) and lift (Sects. 6.3.2.2, 6.4.3) on solid obstacles immersed in a potential ideal fluid flow. As we have then pointed out, these are particular manifestations of *d'Alembert's paradox* which can now be stated in the following general form:

The steady, 2D-plane, irrotational flow of an ideal incompressible fluid exerts no forces on any closed contour in the flow field if and only if the flow rate through this contour and the velocity circulation along this same contour are zero.

This is the direct consequence of the general theorem on the flow action, in terms of the Potential Flow theory (Sect. 6.4.4). Such a result could lead to a definitive rejection of such a theory along with the ideal fluid concept to cope with practical applications of necessarily *viscous* fluid motions. However, if this statement applies to the drag[5], we have demonstrated, on the other hand, that it is possible to remove the paradox for the lift force, thanks to Kutta-Joukowsky's theorem. It should be recalled that this major conclusion is reached in a purely theoretical way by mathematically introducing a logarithmic singularity associated with a vortex within the domain bounded by the external contour of the obstacle.

In this regard, it is important to remember that such a singularity generates a motion:

• which is irrotational,
• which has concentric circular streamlines,
• whose velocity decreases, in modulus, according to a hyperbolic law with the vortex center distance.

However, as previously observed in Chap. 5, Sect. 5.4.4.3 and as we will demonstrate in the continuation of this chapter, such a velocity field does not require the *inviscid* nature of the environment and may very well be encountered in viscous fluid flows.

As a result, such an observation leads to the double question:

• Can we induce, in a *viscous* fluid flow, a lift effect on an immersed body, by locally generating around this body, a motion with a similar kinematics to that of the vortex in an *ideal* fluid?
• If this is the case, can a *quantitative* correspondence be derived between the resultant force in real and ideal fluid flows?

In the following paragraph Sect. 6.4.5.2, we will discuss the first part of the question, the second part being addressed in section 5, for specific shape obstacles.

HISTORICAL NOTE—D'Alembert's paradox—In one of the memoirs of his 1768 opuscules, d'Alembert [44], analysed the resistance that a rectilinear flow exerts on a symmetric body. He was led to conclude from his study: «*Hence I don't see, I confess, how the resistance of fluids can be satisfactorily explained by the theory. On the contrary, it seems to me that this theory, developed and deepened with all possible rigour, gives, at least in several cases, absolutely zero resistance; a singular paradox which I leave the geometers to clarify.*»

It was only about 80 years later, in 1847, that Barré de Saint-Venant [2] removed the paradox, by clearly identifying the origin of the contradiction: «*But we find another result if, instead of the ideal fluid, the calculation object of the last century's geometers, we restore a real fluid composed of a finite number of molecules, and exerting, in the state of motion, unequal pressures or which have tangential components to the surfaces through which they act; components which we will refer to as the fluid friction, a name which has been given to them from*

---

[5] In the absence of separation and vortex shedding in the wake past a bluff body.

Descartes and Newton to Venturi. By introducing this friction in the calculation ...we find that the fluid impetus, in the motion on the body immersed in it, is equal to the total friction work which its presence causes, on the part of the fluid, both on this solid body and on itself, per unit of the space which the ambient fluid flows through.»

6.4.5.2 The Robins-Magnus Effect

The common practice of many sports such as soccer, golf, tennis, table tennis etc, clearly demonstrates the existence of aerodynamic forces when projectiles moving in the air also rotate on themselves.

Some changes in the "expected" trajectories of a ball, due to its spinning motion, are outlined in the following figure, where it is assumed that the ball rotation axis is normal to the plane of its trajectory. Highly diversified path deviations can result from different spin orientations with, in particular, curved paths in both vertical and horizontal projections.

A shot such that the ball rotates forwards as it is moving, called a topspin, imparts a downward force that causes the ball to drop, see Fig. 6.26a. In the opposite case, a ball propelled through the air with a slice or backspin effect rotates backwards and experiences an upward force that lifts the ball (b).

HISTORICAL NOTE—**The Robins-Magnus effect**—This phenomenon was described as early as 1742 by Robins [139], although history has more readily associated it with Magnus' study [93] published in 1853.

   When analyzing the path deviation of military projectiles, Robins, on pages 196-198, remarkably clearly identified the presence of an aerodynamic force due to the spinning motion of the projectiles. Following Magnus' work, various applications of this force have been considered to aerodynamically propel ships, operate windmills or simply modify the ball or balloon trajectories in various ludic applications. For instance, in 1927, Anton Flettner, an engineer and captain

Fig. 6.26 Ludic illustration of the Robins-Magnus effect

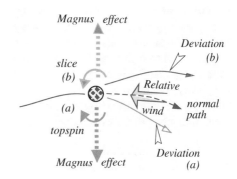

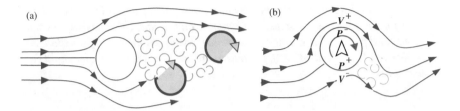

Fig. 6.27 Schematic for the viscous fluid flow around a fixed (**a**) and rotating (**b**) cylinder

of the German navy crossed the North Atlantic ocean at an average speed of 7.5 knots, on a 90-metre-long boat powered by rotating cylinders.

6.4.5.3   The Viscous Effects in Real Fluid Flows Past Bluff Bodies

Let us consider, for simplicity, the ideal case of an infinitely long cylindrical body, with a circular cross-section of radius R_0. Immersed in a viscous fluid motion of uniform velocity V_∞ at infinity, this solid only undergoes a drag force, i.e., whose direction is the same as that of the flow at infinity.

Indeed, at a sufficiently high Reynolds number $R_e = V_\infty R_0/\nu$, where ν is the fluid kinematic viscosity, the flow remains *on average* symmetric in the direction orthogonal to the motion at infinity. As a result, the *mean lift* is zero[6].

On the other hand, and contrary to the ideal fluid situation, the viscous fluid flow separates downstream of the cylinder beyond the creeping motion regime (see Chap. 10). As we have discussed in Chap. 2, this is a complex region where, depending on the Reynolds number value, $V_\infty R_0/\nu$ a vortex shedding with a periodic tendency can take place for moderate Reynolds numbers, with the presence, at higher values, of a disordered agitation, characteristic of the turbulent regime.

Thus, the lack of symmetry in the pressure field between the upstream and downstream half faces of the cylinder generates a force in the flow direction (*drag*). We will explain, at a later stage, that this drag is also due, but in a smaller proportion, to the skin friction of the attached upstream part of the flow along the obstacle.

To generate, in a viscous fluid flow, a *non-zero mean lift*, it is necessary to *steadily* induce, a pressure asymmetry in the direction normal to that of the flow at infinity. This is carried out, for instance, by simply introducing a circulation around the cylinder, by rotating it around its axis. Indeed, it can be proved (see Chap. 9) that, by steadily rotating a circular cylinder around its axis in a viscous fluid at rest to infinity, the motion driven by viscosity for $r \geq R_0$, is *identical* to that resulting of a vortex line along this axis in an ideal fluid.

---

[6] As already mentioned, a vortex shedding appears in the wake of the cylinder for $R_e > 45$ to 50. These vortex structures, as outlined in Fig. 6.27a induce the *instantaneous* lift to oscillate, but its mean value is zero.

If the fluid is moving at a uniform velocity to infinity, the resulting flow pattern is similar to that outlined in Fig. 6.27b.

It is then clear that this double asymmetry, parallel and orthogonal to the flow direction results in both lift and drag forces in the viscous fluid situation. For the rotation direction shown in the figure, the upper part undergoes a depression due to the increase in the local velocity driven by the rotation, the opposite phenomenon taking place on the lower part of the cylinder; Hence the lift generation.

By comparing the real and ideal fluid situations, it can be concluded from the previous discussion that:

(a) the existence of a velocity circulation around the obstacle is at the source of the lift force, both in real and ideal fluids;

(b) the mechanisms responsible for this circulation are specific to each type of fluid.[7]

In summary, we will note that:

> The superposition of a vortex type singularity in the irrotational flow around a circular cylinder can be understood as a theoretical means for *simulating*, in an ideal fluid, the lift effect driven by rotating the obstacle on itself in a viscous fluid.

Thus, to make the potential theory relevant to the lift calculation, all that remains is to know whether it is possible to account quantitatively for the velocity circulation effects, with no regard to the physical origin of this circulation.

It is this issue which we will now address for a particular geometry of obstacles.

6.5 The Theory of Sharp Trailing Edge Airfoils

6.5.1 The Airfoil Shape Design

By using the Joukowsky transformation (cf. Sect. 6.2.5.4), it is possible to generate, from circles correctly positioned in the plane (x, y), a family of airfoil shapes in a transformed plane (X, Y), known as Joukowsky foils.[8] They are geometrically designed, as outlined in the following Figure, as a domain bounded by a closed curve, whose contour mirrors the cross-section of a turbine blade or an airplane wing, i.e., with a *rounded* leading edge and a *sharp* trailing edge.

As an exercise, one may prove that a configuration of the type outlined in Fig. 6.28 is obtained by applying the Joukowsky mapping—Eq. (6.13)—to the exterior of a so-called generating circle whose radius is greater than k, passing through the downstream critical point b $(0, k)$ of the transformation and whose centre coordinates are taken such that $x_c < 0$ and $y_c > 0$.

---

[7] Indeed, while the viscosity can drive the fluid rotation in layers near the cylinder in a real fluid, this method is totally inoperable in an ideal fluid.

[8] This shortcut actually refers to two-dimensional bodies of an infinite extension in a span-wise or transverse direction, with "streamlined" cross-sections.

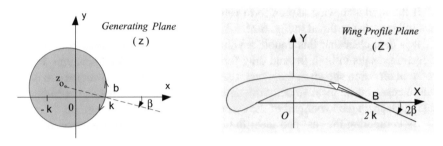

Fig. 6.28 Generating circle for Joukowsky's airfoil with a sharp trailing edge

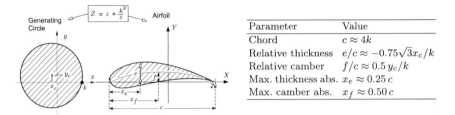

Parameter	Value
Chord	$c \approx 4k$
Relative thickness	$e/c \approx -0.75\sqrt{3}x_c/k$
Relative camber	$f/c \approx 0.5\,y_c/k$
Max. thickness abs.	$x_e \approx 0.25\,c$
Max. camber abs.	$x_f \approx 0.50\,c$

Fig. 6.29 Characteristic parameters of a Joukowsky airfoil: definition (left) and values (right) within the limit of minor thickness ($x_c \to 0$) and minor camber ($y_c \to 0$)

ADDITIONAL INFORMATION—Joukowsky airfoils—As outlined in Fig. 6.28, the thickness and camber of a Joukowsky airfoil are associated respectively with the abscissa x_c and ordinate y_c of the generating circle center.

Thus, a circle centered on the x-axis generates a thick, symmetrical airfoil with no camber. A circle centered on the y-axis leads to a thin airfoil, reduced to a circular arc passing through points $-2k$ and $+2k$ of the plane (X, Y).

We give in Fig. 6.29 the values of some characteristics parameters for *thin airfoils*, i.e., those obtained for a low offset from the origin in the coordinates (x_c, y_c) of the generating circle center.

The common feature of the Joukowsky airfoil family is a cusp at the trailing edge, unlike real wings which take the form of a finite angle dihedral.

As an explanation, it is simply necessary to expand the transformation expression around the singular point $z_b = k$, as we have done in Sect. 6.2.5.2. The first non-zero derivative at this point is of order two, which immediately provides the correspondence between the arguments of the tangent directions on either side of the singular point, namely:

$$\arg\,(Z - Z_B) \approx 2\arg\,(z - z_b) \ .$$

Now, in the vicinity of point b, the polar angles of the two half-tangents to the circle are $\pm\pi/2$. By denoting these angles as θ^+ and θ^- respectively, we have $\theta^+ - \theta^- = \pi$. Accordingly, the polar angles Θ^+ and Θ^- of the two oriented half-tangent directions associated with the paths B^- and B^+ along the airfoil are such that $\Theta^+ - \Theta^- = 2\pi$, which corresponds indeed to a cusp at the trailing edge (a zero dihedral angle at point B).

To avoid a physical inconsistency caused by this mathematical singularity, specific conditions should be prescribed to the flow at this point.

6.5.2 The Kutta Condition

As we have just pointed out, the sharp trailing edge of the Joukowsky airfoil is associated with a *geometrical* singularity of the transformation, since

$$\left(\frac{dZ}{dz}\right)_{z=z_b} = 0 \quad \text{or} \quad \left(\frac{dz}{dZ}\right)_{Z=Z_B} = \infty.$$

This geometrical singular point can be at the origin of a *dynamical* singularity for the flow around the airfoil. Indeed, by referring to $w(z)$ and $W(Z)$ as the respective complex velocities of the flows around the circle and airfoil, we know that from Eq. (6.12):

$$[W(Z)]_{Z=Z_B} = \left(\frac{dz}{dZ}\right)_{Z=Z_B} \times [w(z)]_{z=z_b}.$$

Thus, the velocity at the airfoil trailing edge is infinite in any case of non-zero velocity $w(z_b)$. Such a dynamical singularity has no physical relevance and it is therefore necessary to investigate whether it could be avoided by means of the mathematical condition:

$$\boxed{[w(z)]_{z=z_b} = 0} \tag{6.35}$$

The complex velocity at the airfoil trailing edge is then of the indeterminate form $W(Z_B) = \infty \times 0$ and it can be demonstrated (see Additional Information in Sect. 6.5.4 hereafter) that it is possible to remove this indetermination and obtain a finite non-zero value at this point.

Relation (6.35) is the mathematical expression of Kutta's condition namely and in explicit terms:

The pre-image of the trailing edge tip of a Joukowsky airfoil is the downstream stagnation point of the flow around the generating circle.

HISTORICAL NOTE— Joukowsky's theorem and Kutta's condition—According to Lamb [86], the *theorem* in question would be attributed to the German mathematician Kutta, for an article published in 1910. The historical authorship issue of this theorem would, however, be somewhat more complex, according to the study published by Zeytounian [169] in 2001.

In fact, in a paper dated 1910, Joukowsky [81], restated the result $\ll P = \rho J V \gg$ (with author's symbols), while also mentioning that he had previously published it in 1906. He also points out that this theorem appears in Kutta's dissertation for a tenure-track faculty position ("*Habilitationschrift*"), an unprinted document dating back to 1902.

Henri Villat [165] evoked, with great subtlety and delicacy, how, having himself drawn up this theorem in the early 1900s and having submitted his contribution to a French Sciences Academy member for review, he had received the answer that the work was uncontroversibly correct, but of little interest, and in any event not of sufficient relevance for publication by the Academy. He added that five years later, his discovery object had been about to be famous under the name of *Jukowsky's theorem*.

As far as the *condition* is concerned, it was Chaplygin, who put it forward in 1909, according to Loitsyanskii [91]. Discussing a report by Zhukovskii,— Joukowsky in romanized form—, he had suggested the following hypothesis as an experiment fact: «*out of the infinite number of theoretically possible flows past a profile with a sharp trailing edge, the flow which actually occurs is the one with a finite velocity at the trailing edge.*»

It is also in an article by Chaplygin [32], published in 1910, that we find the general expressions of the resultant force and moment on an airfoil, according to a formulation very close to that discussed in Sect. 6.4.2.

EXERCISE 23 Some specific Joukowsky airfoils

The Joukowsky transformation $Z = z + k^2/z$, where k is a non-zero real, is applied to the following three contours of the complex plane (z):

(a) a circle centered at the origin whose radius is $a > k$;
(b) a circle centered at the origin whose radius is $a = k$;
(c) a circle centered on Oy and passing through the coordinate point $(k, 0)$.

QUESTION 1.
What are the profiles obtained by mapping these contours?
QUESTION 2.
Can Kutta's condition be applied to each of them?

ANSWER: **1.** (a) *An ellipse centered at the origin and half axes $a \pm k^2/a$.*

(b) *A X axis segment, whose length is $4k$, centered at the origin.*

(c) *A circular arc centered on the Y axis whose ends are the points $\pm 2k$ on the real axis.*

2. *Only the last two foils have a tip at the trailing edge, as a result of the downstream singular point mapping. Kutta's condition can therefore be applied to them. This will prevent a physically unacceptable flow behaviour at the trailing edge. These foils also have a singularity at the leading edge which Kutta's condition, on the other hand, is powerless to avoid. In fact, it is the very shape of the leading edge with a cusp which is unrealistic.*

6.5.3 The Adapted Circulation Around an Airfoil

The application of Kutta's condition to the flow around a Joukowsky airfoil, leads to prescribing a single constant value for the circulation around the airfoil, the so-called "adapted circulation". Its value is given by:

$$\boxed{\Gamma = -4\pi V_\infty a \sin(\alpha - \beta)} \tag{6.36}$$

Proof The flow around the airfoil is the mapping of that around a circle, of uniform velocity V_∞ and incidence α at infinity, with a circulation Γ. By designating a as the radius of the circle, centred at the affix point $z = z_0$, the complex potential of this flow is written as:

$$f(z) = V_\infty \left[(z - z_0) e^{-i\alpha} + \frac{a^2}{(z - z_0) e^{-i\alpha}} \right] - i\frac{\Gamma}{2\pi} Log(z - z_0) .$$

The corresponding complex velocity is expressed by

$$w(z) = V_\infty \left[e^{-i\alpha} + \frac{a^2}{(z - z_0)^2 e^{-i\alpha}} \right] - i\frac{\Gamma}{2\pi}\frac{1}{(z - z_0)}.$$

Recalling that the singular point b is defined by the argument β such as $ae^{i\beta} = k - z_0$, Kutta's condition (6.35) is then expressed as:

$$0 = w(z_b) = V_\infty \left[e^{-i\alpha} + e^{i(\alpha - 2\beta)} \right] - i\frac{\Gamma}{2\pi a} e^{-i\beta} ,$$

which easily leads to the specific constant value of the adapted circulation (6.36).

6.5.4 The Lift of an Infinite Span Airfoil

Using Joukowsky's theorem *along with* Kutta's condition, we can now calculate the lift of a two-dimensional infinite span airfoil. Indeed, by introducing relation (6.36) in the lift expression, the following proposition can finally be stated:

> The Joukowsky airfoil lift in a two-dimensional steady plane flow of an ideal incompressible fluid *with an adapted circulation* is unique. Its module, per unit span, is equal to
> $$|F_\perp| = 4\pi \rho V_\infty^2 a \, |\sin(\alpha - \beta)| \tag{6.37}$$

Thus, Kutta-Joukowsky theory of sharp trailing edge airfoils makes it possible to prescribe *one value and one only* of the lift *for a given angle of incidence*. This conclusion is outlined in the following figure (Fig. 6.30).

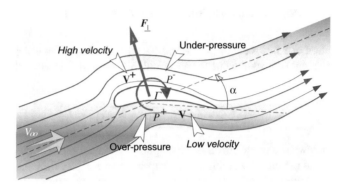

Fig. 6.30 Schematic for the flow past a Joukowsky airfoil with an adapted circulation

ADDITIONAL INFORMATION—**The finite value of the velocity at the trailing edge tip**—We are now able to justify the mathematical indeterminacy removal for the velocity value at the trailing edge tip by means of Kutta's condition. Indeed, taking for simplicity's sake a zero incidence at infinity, the complex potential of the flow around the cylinder with an *adapted circulation* is written as follows:

$$f(z) = V_\infty \left(z + \frac{a^2}{z - z_0} \right) - 2iaV_\infty \sin \beta \, Log \, (z - z_0).$$

The associated complex velocity is then:

$$w(z) = V_\infty \left(1 - \frac{a^2}{(z - z_0)^2} \right) - aV_\infty \frac{e^{i\beta} - e^{-i\beta}}{z - z_0}.$$

The expansion of this velocity around $z = k$ is written as

$$w(z) = w(k) + (z - k) \frac{dw}{dz}\bigg|_{z=k} + \frac{(z-k)^2}{2!} \frac{d^\Theta w}{dz^2}\bigg|_{z=k} + \cdots$$

Now, thanks to the adapted circulation property, $w(k) = 0$ and $\frac{dw}{dz}\big|_{z=k} = \frac{V_\infty}{a}$ $\left(e^{-i\beta} + e^{-3i\beta} \right)$, so that, within the limit $z \to k$:

$$w(z) \approx \frac{V_\infty}{a} \left(e^{-i\beta} + e^{-3i\beta} \right) (z - k) + \mathcal{O}(z - k)^2.$$

In addition, the differential element of the Joukowsky transformation can be put in the form $\dfrac{dz}{dZ} = \dfrac{z^2}{(z+k)(z-k)} \approx \dfrac{k}{2(z-k)}$ $for \ z \to k$.

We therefore deduce from all these results that:

$$\lim_{Z \to 2k} W(Z) \equiv \lim_{z \to k} \left[w(z) \frac{dz}{dZ} \right] = \lim_{z \to k} \left[\frac{V_\infty}{a} \left(e^{-i\beta} + e^{-3i\beta} \right) (z - k) \times \frac{k}{2(z-k)} \right],$$

or, all calculations done, $k \dfrac{V_\infty}{a} \cos \beta \, e^{-2i\beta}$ which is indeed a finite value, Q.E.D.

6.5.5 The Kutta-Joukowsky Theory Relevance

6.5.5.1 The Lift Mechanism Analysis of an Infinite Span Airfoil

The circulation which, according to Kutta-Joukowsky's theory, is at the origin of the lift of a sharp trailing edge airfoil, is obtained, in an ideal fluid, by superposing a logarithmic singularity to the flow around the wing. Such a circulation can be seen as the effect of a bound vortex, as we will explain now.

In fact, in a viscous fluid, we still need to understand which mechanisms could also justify the existence of a velocity circulation around the wing.

The rotation of the solid, discussed in Sect. 6.4.5.2, obviously cannot be invoked for the flow situation being considered here.

Let us consider, as a way to illustrate, that the motion results from the displacement at $t = t_0$ of a wing, initially motionless in an infinite fluid at rest. To simplify, we will take a uniform rectilinear translation displacement for any given time $t > t_0$. Such a body displacement generates, from time t_0, that of the fluid adhering to the wall which then extends by the viscous diffusion, to the surrounding fluid. As outlined in Fig. 6.31, three different flow patterns can be considered at the trailing edge.

For times which are very close to the wing displacement onset in the fluid, the experiment reveals the existence of a transient phase, where a circumvention from the intrados is observed, which then resorts over time by the emission of a '*starting vortex*' which separates from the trailing edge, see Fig. 6.32.

On the other hand, in the stationary regime for $t \gg t_0$, there is no longer any circumvention of the trailing edge by the fluid threads, neither from the intrados nor extrados, as outlined in Fig. 6.31a, b. The Kutta condition which corresponds to the

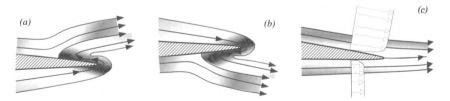

Fig. 6.31 Possible flow patterns at the trailing edge tip: **a** circumvention from intrados, **b** circumvention from extrados, **c** smooth flow-off

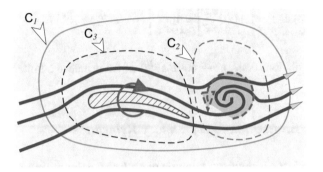

Fig. 6.32 The starting vortex and the induced circulation around the airfoil

flow pattern (c) in Fig. 6.31 is therefore perfectly in line with the physics of the real fluid steady flow past a wing.

It is the asymmetry in the flow conditions on the extrados and intrados which is at the origin of the velocity circulation around the airfoil in a viscous fluid. This can be understood as the counterpart of the transient trailing edge vortex resulting from any change in the flow conditions around the airfoil: starting from rest or any change in the incidence angle in a steady-state regime, for example. We will now examine the case of the starting vortex in more detail.

The starting vortex, which leaves the trailing edge when the airfoil is set in motion, takes away a given velocity circulation Γ. However, since the fluid is initially at rest, the velocity circulation along any contour surrounding the airfoil, such as C_1 in Fig. 6.32, is zero at any time $t < t_0$. We will then see in Chap. 11 that the fluid viscosity does not alter the irrotational nature of the flow away from the obstacle at a high Reynolds number. As a result, Lagrange's theorem of the previous chapter applies to any contour far enough away from the obstacle such as C_1, which implies that the circulation along the contour remains constantly zero. However, the presence of the starting vortex moving away within a contour such as C_2, inside the domain bounded by C_1, generates at any time $t > t_0$ a non-zero circulation $+\Gamma$ on the latter, which contradicts the theorem. To fulfill Lagrange's theorem, we should therefore consider that there is, on a complementary contour such as C_3, necessarily surrounding the profile, an opposite circulation $-\Gamma$. This is precisely what Kutta-Joukowsky's theory accounts for.

In summary, it can be concluded that:

> The Kutta-Joukowsky theory aims at *simulating*, by an *ideal* fluid approach, the lift capacity of an airfoil in a *viscous* flowing fluid. Kutta's condition, whereby the flow leaves smoothly at the trailing edge, provides a *constant circulation* value around the airfoil, which is supposed to balance that of the starting vortex carried away in the flow field, and whose strength is assumed to be that of the logarithmic singularity in the potential flow theory.

Remark One explanation sometimes put forward to justify, by "common sense" considerations, the lift phenomenon of an airfoil in an ideal fluid flow, is as follows:

Let us consider two fluid particles P_E and P_I, moving along paths infinitely close to each other and reaching at the same time the upstream stagnation point. It is assumed that the subsequent trajectory of the first, P_E, coincides with the extrados airfoil line and the second with the intrados. When merging together at the trailing edge, and due to the flow geometry around the airfoil, the distance travelled by P_E will be greater than that of P_I. The velocity of the first path (extrados) will therefore be higher than that of the second (intrados). By applying Bernoulli's theorem, this will result in a lower pressure on the extrados and a higher pressure on the intrados, hence the wing lift effect.»

The reader is left to detect the weak point in the reasoning method. He or she will be able to refine his or her judgment through a certain number of considerations gained from the following two exercises.

EXERCISE 24 The path times along a circle

We consider the potential flow around a circle of radius a, centered at the affix point z_0 of the complex plane, in the presence of a vortex centered at the origin with the circulation Γ and whose velocity at infinity is uniform (V_0 in module), and makes the angle α with the x axis. We set $z - z_0 = \zeta \equiv R e^{i\theta}$.

QUESTION 1. Give the expressions of the complex potential $f(\zeta)$ and complex velocity $w(\zeta)$ of the flow as a function of ζ.

QUESTION 2. In the case where stagnation points exist on the circle, we denote by β the polar angle of the downstream point. Express the value of $\sin(\beta - \alpha)$ and specify the condition for the existence of stagnation points on the circle.

QUESTION 3. Give the expression of the complex velocity on the circle as a function of the polar angle θ and the other parameters of the problem.

QUESTION 4. Deduce the path time dt_c along an elementary arc $d\theta$ of a fluid particle moving on the circle. Deduce the formal expression (without calculating the integral) of the path time along a finite arc, which ends at the polar angles θ_1 and θ_2, such that $0 < \theta_1 < \theta_2 < 2\pi$.

QUESTION 5. Considering the extrados and intrados circle arcs as defined in Fig. 6.33, formally express (without calculating the integral) the path times on each of these arcs assuming $\beta < 0$ and $|\beta| < \alpha < \pi/2$.

QUESTION 6. Calculate the difference in the extrados and intrados path times of the fluid particles moving along the respective circle arcs. Discuss the physical meaning of the result.

(We give $\displaystyle \int \frac{dx}{a + b \sin x} = \frac{1}{\sqrt{b^2 - a^2}} \ln \left\{ \frac{a \, tg(x/2) + b - \sqrt{b^2 - a^2}}{a \, tg(x/2) + b + \sqrt{b^2 - a^2}} \right\}$).

SOLUTION: **1.** By superposition, the complex potential is written:

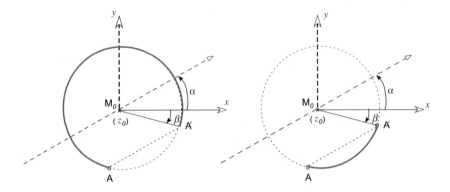

Fig. 6.33 Extrados (left) and intrados (right) arcs for the flow on the circle

$$f(z) = V_0\left((z - z_0)e^{-i\alpha} + \frac{a^2}{z - z_0}e^{i\alpha}\right) - i\frac{\Gamma}{2\pi}Log(z - z_0)\,,$$

either by introducing the ζ variable, $f(\zeta) = V_0(\zeta e^{-i\alpha} + \dfrac{a^2}{\zeta}e^{i\alpha}) - i\dfrac{\Gamma}{2\pi}Log(\zeta)$.

We immediately deduce the complex velocity expression as follows:

$$w(z) = \frac{df}{dz} = \frac{df}{d\zeta} \equiv w(\zeta) = V_0\left(e^{-i\alpha} - \frac{a^2}{\zeta^2}e^{i\alpha}\right) - i\frac{\Gamma}{2\pi} \times \frac{1}{\zeta}\,.$$

2. Denoting by ζ_A a stagnation point affix, we have, by definition:

$$0 = V_0(e^{-i\alpha} - \frac{a^2}{\zeta_A^2}e^{i\alpha}) - i\frac{\Gamma}{2\pi}\frac{1}{\zeta_A}\,.$$

Putting $\zeta_A = ae^{i\beta}$, where the angle β identifies the position of the possible stagnation point(s) on the circle, we can deduce:

$$0 = V_0(e^{-i\alpha} - \frac{a^2}{a^2 e^{2i\beta}}e^{i\alpha}) - i\frac{\Gamma}{2\pi}\frac{1}{a}e^{-i\beta}\,,$$

or, after multiplication throughout by $e^{i\beta}$

$$V_0(e^{i(\beta-\alpha)} - e^{-i(\beta-\alpha)}) = i\frac{\Gamma}{2\pi a}\,,$$

from which the well-known result is derived $\sin(\beta - \alpha) = \dfrac{\Gamma}{4\pi V_0 a} \leq 1\,,$ (6.38) yielding two values of β under the above-mentioned condition.

3. Noting that the affix of any point on the circle is $\zeta_c = ae^{i\theta}$ for $0 < \theta < 2\pi$, the complex velocity on the circle is directly deduced from its previous general expression, namely:

$$w_c(\theta) = V_0 \left(e^{-i\alpha} - \frac{a^2}{a^2 e^{2i\theta}} e^{i\alpha} \right) - i \frac{\Gamma}{2\pi} \frac{1}{a} e^{-i\theta}.$$

Now, factorizing by $e^{-i\theta}$ and simplifying, we easily obtain:

$$w_c(\theta) = 2i V_0 e^{-i\theta} [\sin(\theta - \alpha) - \sin(\beta - \alpha)], \tag{6.39}$$

after using relation (6.38) to express the circulation term.

4. Denoting by $\mathbf{e}_\theta$ the unit tangent vector to the circle, the elementary displacement time dt_c of a fluid particle along the circle is such that $ad\theta \mathbf{e}_\theta = \mathbf{V}_\theta dt_c$. As the circle is a streamline, the velocity is collinear with the circle, so that $\mathbf{V}_\theta = ||w_c(\theta)||\mathbf{e}_\theta$. We deduce from this $dt_c = \dfrac{ad\theta}{||w_c(\theta)||}$. Hence, taking into account the previous result, Eq. (6.39), the elementary path time is:

$$dt_c = \frac{ad\theta}{2V_0|\sin(\theta - \alpha) - \sin(\beta - \alpha)|},$$

by assuming the consideration of the reference velocity at infinity as positive.

On a finite arc extending from θ_1 to θ_2, we directly have:

$$t_{c\,\theta_1\theta_2} = \frac{a}{2V_0} \int_{\theta_1}^{\theta_2} \frac{d\theta}{|\sin(\theta - \alpha) - \sin(\beta - \alpha)|} = \frac{a}{2V_0} \int_{\theta_1 - \alpha}^{\theta_2 - \alpha} \frac{du}{|\sin u - \sin(\beta - \alpha)|}, \tag{6.40}$$

the last expression resulting from the change of variable $u = \theta - \alpha$ in the first integral.

5. The extrados arc extends, by definition, from the upstream A to the downstream A' stagnation point and includes the $Y > 0$ part of the circle (see Fig. 6.33, on the left). The polar angle of point A' is by definition $\beta < 0$. The polar angle of point A, by moving along the extrados line, is calculated as $(-\beta + \alpha) + \pi + (-\beta + \alpha)$ and is therefore equal to $\pi + 2\alpha - 2\beta$. It is always greater than β, which leads to adopting, for the integral angular limits:

$$\text{and} \quad \theta_2 = \pi + 2\alpha - 2\beta.$$

Moreover, with the conditions[9] imposed here for the angles α and β, we have for this path $\sin(\theta - \alpha) > \sin(\beta - \alpha)$. Therefore, the path time along the extrados circle

---

[9] These conditions are defined so as to anticipate the flow situation around the airfoil in the following exercise.

arc, t_{ce}, can be calculated with the formula:

$$t_{ce} = \frac{a}{2V_0} \int_{\beta-\alpha}^{\pi+\alpha-2\beta} \frac{du}{\sin u - \sin(\beta - \alpha)}. \tag{6.41}$$

As for the intrados arc—Fig. 6.33, on the right—, the angular location of point A is now given by the value $-\pi + \alpha + (-\beta + \alpha) - \beta$, namely $-\pi + 2\alpha - 2\beta$. Owing to the conditions on α and β, this value is lower than β, which leads to taking:

$$\theta_1 = -\pi + 2\alpha - 2\beta \quad \text{and} \quad \theta_2 = \beta.$$

In addition, on this intrados path, we have with the same conditions $\sin(\theta - \alpha) < \sin(\beta - \alpha)$, hence, for the corresponding path time:

$$t_{ci} = \frac{a}{2V_0} \int_{-\pi+\alpha-2\beta}^{\beta-\alpha} \frac{du}{\sin(\beta - \alpha) - \sin u}. \tag{6.42}$$

6. The difference in the path times between the upstream and downstream stagnation points, depending on whether the fluid particle moves along the extrados or intrados circle arc, is given by

$$\Delta t_c = t_{ce} - t_{ci} = \frac{a}{2V_0} \left\{ \int_{\beta-\alpha}^{\pi+\alpha-2\beta} \frac{du}{\sin u - \sin(\beta - \alpha)} - \int_{-\pi+\alpha-2\beta}^{\beta-\alpha} \frac{du}{\sin(\beta - \alpha) - \sin u} \right\}$$

$$= \frac{a}{2V_0} \left\{ \int_{\beta-\alpha}^{\pi+\alpha-2\beta} \frac{du}{\sin u - \sin(\beta - \alpha)} + \int_{-\pi+\alpha-2\beta}^{\beta-\alpha} \frac{du}{\sin u - \sin(\beta - \alpha)} \right\},$$

$$\text{or finally} \quad \Delta t_c = \frac{a}{2V_0} \int_{-\pi+\alpha-\beta}^{\pi+\alpha-2\beta} \frac{du}{\sin u - \sin(\beta - \alpha)}. \tag{6.43}$$

By putting $I = \int \frac{du}{\sin u - \sin(\beta - \alpha)}$, we have, by identification with the given formula and with $a = -\sin(\beta - \alpha)$ and $b = 1$:

$$I = \frac{1}{\cos(\beta - \alpha)} \ln \left\{ \frac{\sin(\alpha - \beta)tg(x/2) + 1 - \cos(\alpha - \beta)}{\sin(\alpha - \beta)tg(x/2) + 1 + \cos(\alpha - \beta)} \right\},$$

for $\cos(\beta - \alpha) > 0$. Switching to the half arc, it should be noted that:

$$\sin(\alpha - \beta)tg(x/2) + 1 - \cos(\alpha - \beta) = 2\sin\frac{\alpha - \beta}{2} \cos\frac{\alpha - \beta}{2} \left(tg(x/2) + tg(\frac{\alpha - \beta}{2}) \right),$$

and likewise

$$\sin(\alpha - \beta)tg(x/2) + 1 + \cos(\alpha - \beta) = 2\sin\frac{\alpha - \beta}{2}\cos\frac{\alpha - \beta}{2}\left(tg(x/2) + 1/tg(\frac{\alpha - \beta}{2})\right).$$

After substitution and simplification, we obtain:

$$I = \frac{\ln\left(tg(\frac{\alpha - \beta}{2})\right)}{\cos(\beta - \alpha)} + \frac{1}{\cos(\beta - \alpha)}\ln\left\{\frac{tg(x/2) + tg(\frac{\alpha - \beta}{2})}{1 + tg(x/2)\,tg(\frac{\alpha - \beta}{2})}\right\}.$$

It can be deduced that $\Delta t_c = \dfrac{a}{2V_0}[I]_{-\pi+\alpha-2\beta}^{\pi+\alpha-2\beta}$, which results in:

$$\frac{2V_0}{a}\Delta t_c = \frac{1}{\cos(\beta - \alpha)}\left[\ln\left\{\frac{tg(\pi + \alpha - 2\beta)/2) + tg(\frac{\alpha - \beta}{2})}{1 + tg(\pi + \alpha - 2\beta)/2)\,tg(\frac{\alpha - \beta}{2})}\right\}\right.$$

$$\left. - \ln\left\{\frac{tg(-\pi + \alpha - 2\beta)/2) + tg(\frac{\alpha - \beta}{2})}{1 + tg(-\pi + \alpha - 2\beta)/2)\,tg(\frac{\alpha - \beta}{2})}\right\}\right]$$

Now, $tg\left(\frac{\alpha - 2\beta}{2} + \frac{\pi}{2}\right) = tg\left(\frac{\alpha - 2\beta}{2} - \frac{\pi}{2}\right)$, so that both subtraction terms in the right hand side are equal. As a result, $\Delta t_c = 0$.

Physical comment: In the absence of circulation ($\beta = \alpha$), it is obvious that the flow symmetry around the axis passing through the stagnation points provides equal path times on the two semicircles. The result which we have just derived demonstrates that the same is true *with a non-zero circulation*, shifting the downstream stagnation point by an angle $\beta < 0$, lower than the incidence angle α, in absolute value. Thus, two fluid particles which simultaneously reach the upstream stagnation point and later move along the extrados (upper) and intrados (lower) arcs respectively, will meet at the same time at the downstream stagnation point. This conclusion, which leads to the over-speeding (and underpressure) on the extrados as well as the under-speeding (and overpressure) on the intrados due to the difference in the distances travelled during the same time, makes the 'common-sense' explanation become legitimate after calculations!.

EXERCISE 26 The path times along an airfoil

We consider a Joukowsky profile obtained by applying the conformal mapping:
z (circle) $\rightarrow$ Z (airfoil) with:

$$Z = z + \frac{k^2}{z}, \quad k \text{ non zero real}, \tag{6.44}$$

outside a *generating* circle centred at the affix point z_0 and passing through the critical point $(k, 0)$ of the transformation.

QUESTION 1. Determine the relation between the path times on homologous curve elements in the circle and airfoil planes respectively. (The result will be expressed in terms of the module r and the argument θ of a point in the complex plane of the circle).

QUESTION 2. Is it possible to conclude that the path times along the extrados and intrados from the upstream stagnation point on the airfoil are equal?

SOLUTION: **1.** The respective path times of fluid particles moving on elementary contours dz and dZ which correspond to each other by the Joukowsky transformation are given by:

- circle plane $d\bar{z} = w(z)dt$

- airfoil plane $d\bar{Z} = W(Z)dT$,

by noting $\bar{z}$ (*resp.* $\bar{Z}$) as the complex conjugate of z (*resp.* Z).

We deduce that $dzd\bar{z} = w(z)dzdt$ and $dZd\bar{Z} = W(Z)dZdT$. However, from the characteristic property of the conformal mapping, we know that, for homologous elements, $w(z)dz = W(Z)dZ$. Hence, the time correspondence, at any regular point in Joukowsky's transformation, can be derived as follows:

$$\frac{dT}{dt} = \frac{d\bar{Z}\,dZ}{d\bar{z}\,dz} = \left(1 - \frac{k^2}{\bar{z}^2}\right)\left(1 - \frac{k^2}{z^2}\right).$$

By putting $z = re^{i\theta}$, the previous relation leads to:

$$\frac{dT}{dt} = \left(1 - \frac{k^2}{r^2}e^{-2i\theta}\right)\left(1 - \frac{k^2}{r^2}e^{2i\theta}\right) = 1 - 2\frac{k^2}{r^2}\cos 2\theta + \frac{k^4}{r^4} \equiv \left(1 - \frac{k^2}{r^2}\right)^2 + 4\frac{k^2}{r^2}\sin^2\theta.$$

$$(6.45)$$

2. Let us refer to $[\theta]_e$ and $[\theta]_i$ as the value ranges of θ along the extrados and intrados circle arcs respectively. By applying the previous relation, this leads directly to:

$$\frac{dT_{pe}}{dt_{ce}} = \left(1 - \frac{k^2}{r_{ce}^2}\right)^2 + 4\frac{k^2}{r_{ce}^2}\sin^2\theta \quad \text{for } \theta \in [\theta]_e,$$

relating the elementary path times of the extrados motions along the circle (dt_{ce}) and airfoil (dT_{pe}).

Thus

$$T_{pe} = \int_{[\theta]_e} \left[\left(1 - \frac{k^2}{r_{ce}^2} \right)^2 + 4 \frac{k^2}{r_{ce}^2} \sin^2 \theta \right] dt_{ce} = \int_{[\theta]_e} \frac{\left[\left(1 - \frac{k^2}{r_{ce}^2} \right)^2 + 4 \frac{k^2}{r_{ce}^2} \sin^2 \theta \right]}{2 V_0 \, |\sin(\theta - \alpha) - \sin(\beta - \alpha)|} \, a d\theta$$

by introducing in the last equality the expression of the elementary path time along the circle as obtained in the previous exercise.

A similar expression can be derived for T_{pi} which shows that the difference in the path times on the extrados and intrados along the airfoil ($T_{pe} - T_{pi}$) is not proportional to that in the path times along the circle ($t_{pe} - t_{pi}$). It is therefore no longer possible to conclude that the path times along the airfoil are equal.

6.5.5.2 The Lift Coefficient

The lifting mechanism analysis of an infinite span wing (see Sect. 6.5.5.1) has demonstrated the qualitative value of the ideal fluid approach. However, it is legitimate to wonder whether this is also the case from a quantitative point of view.

To express the forces applied to an obstacle by a moving fluid, it is common practice in aero-hydrodynamics to introduce specific dimensionless coefficients. The lift coefficient, with respect to the incidence angle $C_L (\alpha)$, is defined as

$$C_L (\alpha) = \frac{F_L(\alpha)}{\frac{1}{2} \rho S V_\infty^2} \tag{6.46}$$

where $F_L(\alpha)$ is the lift force at the incidence α, S a relevant surface area, ρ and V_∞ the fluid density and velocity at infinity.

Taking as reference width $4a$, a value roughly equal to the *chord* within the thin airfoil limit—see Fig. 6.29—and as reference length the span unit, we have $S = 4a \times 1$. It then immediately results from relation (6.37) that, for a Joukowsky airfoil

$$C_z (\alpha) = 2\pi \sin (\alpha - \beta) \tag{6.47}$$

In Fig. 6.34 this result is compared to the experimental measurement for the NACA 0012 airfoil.

This is a symmetrical airfoil ($\beta = 0$) with a maximum relative thickness of 12% at about 30% of the leading edge (see Fig. 6.34). The experimental lift coefficient value is measured in a wind tunnel at a Reynolds number of 3.2×10^6.

For incidence angles less than 16 degrees, an almost perfect agreement can be observed between the experiment and the previous theoretical formula. Beyond this value, there is a clear difference between the measurement and theory. The reason for this discrepancy, as we will explain at a later stage in Chap. 11, is a new phenomenon: the boundary layer separation, fundamentally dependent on the viscous nature of the fluid and which cannot be taken into account by the present ideal fluid theory.

HISTORICAL NOTE—The NACA airfoil specification—The National Advisory
Committee for Aeronautics (NACA, 1915–1958) is the American predecessor of
the National Aeronautics and Space Administration (NASA) (Fig. 6.35).

One of its actions was to carry out systematic wind tunnel tests on an extensive
variety of airfoils. These are identified by the acronym NACA followed by a
numerical code which lists them into several major families qualified as four-digit,
five-digit, six-digit, seven-digit series. For the first family (the 4 digit series), the
coding convention is as follows.

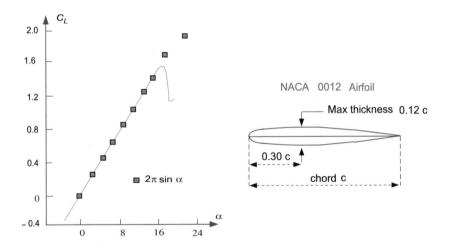

Fig. 6.34   NACA 0012 lift coefficient: Experiment and theory comparison

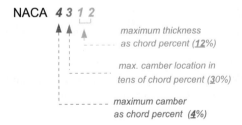

Fig. 6.35   Explanation of the four-digit series NACA airfoil classification

Chapter 7
Sound Wave Propagation and the Shock Phenomenon

Abstract This chapter is the first of two which, in this book, are devoted to flows where the fluid compressibility is responsible for major effects. It opens with the study of infinitesimal pressure wave propagation, allowing a linearized treatment of the problem in an ideal fluid. Although the phenomenon occurs in both liquids and gases, it is only for the latter, under the assumption of the ideal gas, that the results will be given in detail. We will then discuss the case of finite amplitude waves along with the shock phenomenon, whose relations will be derived for both normal and oblique waves.

7.1 The Propagation of Small Pressure Disturbances

7.1.1 The Propagation Speed in a Still Environment

7.1.1.1 Introducing the Issue

We consider a fluid at rest in a space domain which has at least one infinite direction (say x, by convention). Any limit in other directions are assumed to be without influence on the analysis along the preferential direction. The fluid is taken as homogeneous, with a constant pressure, temperature and density, respectively referred to as P_0, T_0, ρ_0, with of course $U_0 = 0$. At a given time, chosen as the time origin (t_0), the pressure in the vicinity of a given section x_0 is very slightly disturbed (see next Fig. 7.1), by increasing its value to $P = P_0 + P'$. Since the fluid is supposed to be *compressible*, we have explained, in the first chapter, that such a pressure disturbance is necessarily associated with a density disturbance, and we will therefore put $\rho = \rho_0 + \rho'$.

This same compressibility of the environment will also propagate the previous disturbances. The general equations set out in Chap. 4 prove that the phenomenon

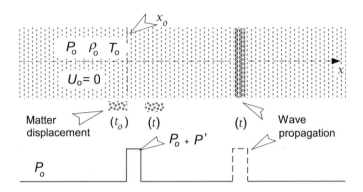

Fig. 7.1 Schematic of slight disturbances in a unidirectional propagation $(x > x_0)$

cannot occur without variations in the velocity and temperature, so that the values of these functions corresponding to the disturbed state become $U = 0 + u'$ and $T = T_0 + T'$.

Subsequently, the problem consists in determining the disturbed values $\rho'(x, t)$, $T'(x, t)$, $u'(x, t)$ as a function of the pressure disturbance $P'(x, t)$ under the following assumptions:

1. negligible external body forces;
2. an ideal fluid, i.e., inviscid and non-heat conductive;
3. an isentropic evolution, i.e., adiabatic and reversible;
4. small disturbances, allowing the equations to be linearized.

Remark The last hypothesis corresponds in practice to pressure fluctuations which occur in a sound propagation which can be tolerated by the human ear without damage. The acoustic pressure, which is the local deviation in the pressure disturbance from the ambient, is in fact in the range of 10^{-4} to 1 Pa.   For the atmospheric pressure of about $10^5\,Pa$, the relative pressure fluctuation lies, in absolute value, between 10^{-9} and 10^{-5}.

7.1.1.2   The Governing Equations

Under the study assumptions, the continuity and momentum equations are written respectively

$$\frac{\partial \rho}{\partial t} + \frac{\partial (\rho U)}{\partial x} = 0 \quad and \quad \rho \left(\frac{\partial U}{\partial t} + U \frac{\partial U}{\partial x} \right) = -\frac{\partial P}{\partial x}.$$

By introducing the fluctuation decompositions, these equations directly yield

$$\frac{\partial \rho'}{\partial t} + \rho_0 \frac{\partial u'}{\partial x} + \rho' \frac{\partial u'}{\partial x} + u' \frac{\partial \rho'}{\partial x} = 0 \,,$$

$$\rho_0 \left(\frac{\partial u'}{\partial t} + u' \frac{\partial u'}{\partial x} \right) + \rho' \left(\frac{\partial u'}{\partial t} + u' \frac{\partial u'}{\partial x} \right) = -\frac{\partial P'}{\partial x} \,.$$

Considering that the *first order terms* in the fluctuations are preponderant, we will reduce these equations to the following *linearized form*:

$$\frac{\partial \rho'}{\partial t} + \rho_0 \frac{\partial u'}{\partial x} = 0 \,, \tag{7.1a}$$

$$\rho_0 \frac{\partial u'}{\partial t} + \frac{\partial P'}{\partial x} = 0 \,. \tag{7.1b}$$

To close the model, thermodynamic relations should be added to the previous equations. Assuming that the propagation is isentropic in nature, pressure and density fluctuations can be interrelated using the isentropic compressibility coefficient χ_S as defined in the first Chapter, Eq. (1.25). We will therefore take:

$$\rho' = \chi_S \rho_0 P' \tag{7.2}$$

It is worth recalling that for an ideal gas, an assumption which we will adopt for the rest of the study, we have $\chi_S = (\gamma P)^{-1}$, γ as the isentropic coefficient which is equal to the heat capacity ratio C_p / C_v.

Then we can notice that the three previous equations form a closed system for the three functions P', ρ', u'. In other words, the calculation of the temperature fluctuation which is thus decoupled from that of these three functions, is simply derived by using the equation of state. It is for this reason that we will no longer be interested in this fluctuation in the following sections.

By substituting relation (7.2) in the Eq. (7.1a and b), the following system of two equations for the two unknown functions P' and u' is deduced:

$$\rho_0 \chi_S \frac{\partial P'}{\partial t} + \rho_0 \frac{\partial u'}{\partial x} = 0 \,, \tag{7.3a}$$

$$\rho_0 \frac{\partial u'}{\partial t} + \frac{\partial P'}{\partial x} = 0 \,. \tag{7.3b}$$

Remark For an ideal gas, the expression of the isentropic compressibility coefficient demonstrates that the relative value of the density fluctuations corresponding to the acoustic pressure of the usual sound waves is $\rho'/\rho_0 \simeq \gamma^{-1} (P'/P_0)$. It is therefore also in line with the small disturbance approximation, since the coefficient γ is of the order of unity for the air at the atmospheric pressure.

7.1.1.3   The Propagation in an Infinite Environment

By adding throughout the equations for the fluctuations as transformed after applying the following operators[1]:

$$\rho_0\chi_s \frac{\partial P'}{\partial t} + \rho_0 \frac{\partial u'}{\partial x} = 0 \qquad \begin{vmatrix} \dfrac{\partial}{\partial t} & -\dfrac{\partial}{\partial x} \\[2mm] \dfrac{\partial P'}{\partial x} + \rho_0 \frac{\partial u'}{\partial t} = 0 \qquad -\dfrac{\partial}{\partial x} & \rho_0\chi_s \dfrac{\partial}{\partial t} \end{vmatrix}$$

we obtain:

$$\rho_0\chi_s \frac{\partial^2 P'}{\partial t^2} - \frac{\partial^2 P'}{\partial x^2} = 0 \qquad (7.4a)$$

$$\rho_0\chi_s \frac{\partial^2 u'}{\partial t^2} - \frac{\partial^2 u'}{\partial x^2} = 0 \qquad (7.4b)$$

The above relations are of the one-dimensional wave equation type, also referred to as d'Alembert equation. It is known that the general solution then comes in the form (see Additional Information hereafter):

$$u'(x,t) = f_1(x - \mathbf{a}_0 t) + g_1(x + \mathbf{a}_0 t),$$
$$P'(x,t) = f_2(x - \mathbf{a}_0 t) + g_2(x + \mathbf{a}_0 t).$$

The functions f_1 and f_2 represent a progressive plane wave, namely which propagates in the positive x direction; the functions g_1 and g_2 are associated with a regressive wave propagating in the negative x direction. Both wave systems have the same propagation speed or celerity[2] which is equal to:

$$\mathbf{a}_0 = \frac{1}{\sqrt{\rho_0\chi_s}} \qquad (7.5)$$

Several expressions can be derived from this relation. By firstly using *Reech's* formula $\gamma = \chi_T/\chi_S$, we can express the celerity as a function of the isothermal compressibility coefficient χ_T

$$\mathbf{a}_0 = \sqrt{\frac{\gamma}{\rho_0\chi_T}}. \qquad (7.6)$$

The previous relation is known as Laplace's formula.

---

[1] The first Eq. (7.4a) results from the addition of the equations in the first column after applying the operators outlined in the second column; A similar process is carried out for Eq. (7.4b) with the operators in the last column.

[2] Note that this italic bold type font does not imply here and in the continuation, a vector character.

As far as an ideal gas is concerned, the celerity of sound wave propagation, in short the *speed of sound*, is expressed as

$$\mathbf{a}_0 = \sqrt{\gamma \frac{P_0}{\rho_0}} \equiv \sqrt{\gamma r T_0}, \tag{7.7}$$

which clearly demonstrates that the speed of sound value depends upon the local temperature and pressure conditions.[3]

ADDITIONAL INFORMATION—The one-dimensional wave equation:

D'Alembert's solution—(a) The resolution of d'Alembert's one-dimensional equation of the form $\dfrac{\partial^2 F(x,t)}{\partial x^2} - \dfrac{1}{\mathbf{a}_0^2} \dfrac{\partial^2 F(x,t)}{\partial t^2} = 0$, is commonly carried out by introducing the change of variables $p = x + \mathbf{a}_0 t$ and $q = x - \mathbf{a}_0 t$.

Hence $\dfrac{\partial^2 F}{\partial x^2} = \dfrac{\partial^2 F}{\partial p^2} + \dfrac{\partial^2 F}{\partial q^2} + 2\dfrac{\partial^2 F}{\partial p \partial q}$ and $\dfrac{\partial^2 F}{\partial t^2} = \mathbf{a}_0^2 \left(\dfrac{\partial^2 F}{\partial p^2} + \dfrac{\partial^2 F}{\partial q^2} \right) - 2\mathbf{a}_0^2 \dfrac{\partial^2 F}{\partial p \partial q}$, so that the equation to be solved simply becomes $\dfrac{\partial^2 F}{\partial p \partial q} = 0$.

From this, we deduce $\dfrac{\partial F}{\partial q} = f'(p)$ and as a result $F(p, q) = f(p) + g(q)$, Q.E.D.

(b) In the case of a free 3D-space propagation, one can easily obtain that the generic form of the equations is $\rho_0 \chi_S \dfrac{\partial^2 F}{\partial t^2} - \Delta F = 0$, where the Laplacian operator now substitutes the second derivative in space. This is called the three-dimensional d'Alembert equation.

7.1.1.4 The Physical Interpretation

To simplify, we limit ourselves to progressive waves only.

By setting $q = x - \mathbf{a}_0 t$, we can deduce that

$$\frac{\partial u'}{\partial t} = \frac{d f_1}{dq} \frac{\partial q}{\partial t} = -\mathbf{a}_0 \frac{d f_1}{dq} \quad \text{and} \quad \frac{\partial P'}{\partial x} = \frac{d f_2}{dq} \frac{\partial q}{\partial x} = \frac{d f_2}{dq}.$$

After substitution in Eq. (7.3b) we derive that

$$-\rho_0 \mathbf{a}_0 \frac{d f_1}{dq} + \frac{d f_2}{dq} = 0,$$

whose integration results in $f_2 = \rho_0 \mathbf{a}_0 f_1$. The integration constant is zero since, for $f_1 = 0$, we necessarily have $f_2 = 0$. For an ideal gas, this result can also be written

---

[3] The speed of sound in dry air at 20 °C under the standard sea-level atmospheric pressure is about 343 m/s.

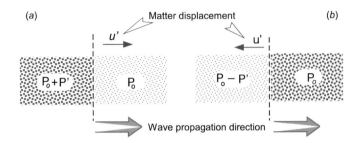

Fig. 7.2 Schematic for the wave propagation direction and matter displacement

in the form:

$$\frac{u'(x,t)}{\mathbf{a}_0} = \frac{P'(x,t)}{\gamma P_0}. \tag{7.8}$$

Thus, the velocity and pressure fluctuations have the same sign. Physically, this result demonstrates that, in the case of a progressive compression wave ($P' > 0$), the fluid particles tend to move in the direction of the acoustic wave propagation (here, that of the positive x). Conversely, for an expansion wave ($P' < 0$), the matter displacement is in the opposite direction to that of the wave propagation. These conclusions are schematically outlined in Fig. 7.2.

Remark For the acoustic pressure range previously mentioned ($P'/P_0 \simeq 10^{-9}$ to 10^{-4}), the relation between the pressure and velocity fluctuations proves that the acoustic Mach number u'/a_0 is also very small. In other words, and for a standard atmosphere, the velocity fluctuations induced by an acoustic disturbance within the previous range lie between 3 cm/s and 3×10^{-5} cm/s.

HISTORICAL NOTE—**Waves in fluids: d'Alembert and Cauchy**—D'Alembert played a prominent role in the formulation of physical problems in terms of partial differential equations or *PDE*, according to modern terminology. The analysis of his highly numerous publications, made public at different periods of his life in an '*encyclopedic*' scientific production, bordered on epistemology. With reference to Guilbaud's thesis [61] in 2007, dedicated to the Hydrodynamic contributions in this scientist's work, there were no less than fourteen articles on *PDE*, five of which dealt with string vibrations in relation to the sound propagation issue. It was in memory of this part of D'Alembert's work around 1761, that the wave equation was named after him.

  Another great French scientist's name deserves to be also mentioned here, even if his study does not directly fall within the scope of this book. This name is that of Augustin Louis Cauchy [27], whose "*Theory of the propagation of waves on the surface of a heavy fluid*", won him the Mathematical Analysis Prize of the Academy of Sciences in Paris, in 1815.

EXERCISE 26 The speed of sound in the upper air

In standard meteorological conditions, the atmospheric temperature varies linearly, up to about 10 km, as a function of the altitude according to the law $T = T_0 - \alpha z$, with $\alpha = 6,5 \ 10^{-3}$ K/m.

QUESTION Calculate, in the earth's surface reference frame, the speed of a bisonic aircraft (cruise Mach number $M = 2$), when (i) flying low to the ground ($z \simeq 0$) and (ii) at an altitude of 10 000 m.

(It will be taken $\gamma = 1.4$, $r = 287$ J.kg^{-1}.K^{-1}, $T_0 = 290$ K)

SOLUTION: The speed of sound is obtained by relation (7.7). At the ground level ($T_0 = 290$ K), its value is $a_0 = 341$ m/s. At an altitude of 10 000 m, where $T_1 = 225$ K, it is equal to $a_1 = 300$ m/s. A flying aircraft at Mach 2 therefore moves at a speed of 2 455 km/h at the ground level and 2 160 km/h at 10 km height.

7.1.2 Energetic Considerations

Acoustic fluctuations in pressure, density and velocity are obviously coupled with fluctuations in the internal and kinetic energy, which we will specify here. In direct application of the results in Chap. 3, the local variation of the total energy (internal energy + kinetic energy), given by equation (3.25), is expressed, under the study assumptions as

$$\rho \frac{\mathrm{d}}{\mathrm{d}t} \left(e' + \frac{u'^2}{2} \right) = -div \left(P\mathbf{u}' \right) .$$

The material variation of the internal energy alone—Eq. (3.27)—becomes:

$$\rho \frac{\mathrm{d}e'}{\mathrm{d}t} = -P div \, \mathbf{u}' .$$

According to the interpretation of $div \ V$ which we have explained in the first Chapter, the internal energy variation is a direct function of the volume variation of any fluid particle since we now have:

$$div \, \mathbf{u}' = -\frac{1}{\rho_0} \frac{\mathrm{d}\rho'}{\mathrm{d}t} .$$

Therefore, taking into consideration the environment compressibility, Eq. (7.2), the internal energy variation is still expressed as

$$\rho \frac{\mathrm{d}e'}{\mathrm{d}t} = \chi_S \left(P_0 + P' \right) \frac{\mathrm{d}P'}{\mathrm{d}t} .$$

As part of the linear acoustic approximations ($\frac{d}{dt} \equiv \frac{\partial}{\partial t}$), this equation is integrated in the form:

$$\rho_0 e' = \chi_S P_0 P' + \frac{1}{2}\chi_S P'^2 \qquad (7.9)$$

This results in a quadratic term $\frac{1}{2}\chi_S P'^2$ which is called the *acoustic potential energy* density.

7.1.3 The Propagation from a Moving Source

The previous simplified analysis is limited to the unidirectional propagation of infinitesimal disturbances originating from a *stationary* source. The generalization, all other things being equal, to the propagation in a homogeneous and isotropic three-dimensional environment leads to spherical waves spreading with a constant celerity equal to the speed of sound in the environment, that is, for an ideal gas $\mathbf{a} = \sqrt{\gamma P/\rho}$.

When the source moves in a three-dimensional space, three cases can be distinguished according to whether its velocity is in modulus less than, greater or equal to the celerity $\mathbf{a}$. To simplify, we will assume that the motion of the source is a linear and uniform translation at a speed of U.

1. In the first case $U < \mathbf{a}$, the source has changed its location at a given time t by the distance Ut, *lower* than that the wave front spreading ($\mathbf{a}t$). The result is a nested sphere pattern, outlined by the diagram (a) in Fig. 7.3, where the source is continuously surrounded by the waves it radiates at any time.
2. In the second case $U > \mathbf{a}$, the source moves a greater distance than the distance by which the wave front has spread during the same time interval. The pattern then becomes of the type outlined in the diagram (c) in the same Figure. In this case at any time, the spherical fronts of the spherical disturbances radiated by the

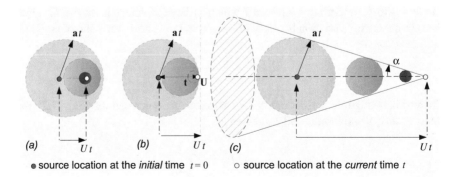

$\bullet$ source location at the *initial* time $t = 0$ $\circ$ source location at the *current* time t

Fig. 7.3 Propagation from a moving source at a subsonic (**a**), sonic (**b**) and supersonic velocity (**c**)

source during its motion exhibit a conical envelope, whose vertex is the location of the source at time t and the base radius is the distance spread by the wave front radiated at the initial time. This envelope is called the *Mach cone*, its trace in a plane of symmetry being termed as Mach's line or Mach's wave. The vertex half angle (α), called the Mach angle, is such that:

$$\sin \alpha = \frac{\mathbf{a}}{U} = \frac{1}{M}. \tag{7.10}$$

3. The intermediate situation for which $U = \mathbf{a}$ is outlined in Fig. 7.3b. It can easily be regarded as the common limit of the two previous situations, the nesting of the spheres being reduced to a point and the conical envelope reducing to the tangent plane at this point.

The three cases which we have just described correspond respectively and by definition to the source motion at a subsonic, supersonic and sonic velocity. The distinction is obvious since in the subsonic regime, the source never overtakes the disturbances it radiates, whereas it moves ahead of them in the supersonic regime.

Remark For a standing observer located on the source path, the previous regimes can be easily discriminated: in the subsonic regime, he or she will be informed of the coming source by the perception of the sound disturbances previously radiated by the source. On the other hand, in the supersonic regime, the mobile source will reach the observer without any acoustic signal having alerted him or her. However, and as is the case when the disturbances are focused in a shock wave, this observer will receive, some time later, quite an explicit audible confirmation of the source crossing!

7.2 Finite Amplitude Waves, Shock

Returning to the unidirectional analysis of the second section, we assume that the pressure disturbance consists of an accumulation of identical fluctuations generated, from an initial time $t = t_0$, at regular time intervals τ_0. We will explain that such a wave train propagation can give rise to different phenomena, depending on whether compression or expansion waves are concerned.

7.2.1 The Propagation of Successive Compression Waves

Let us firstly consider the case of a compression wave train which corresponds to the successive emission of N positive elementary pressure variations ($P' > 0$) at times $t_0 + n\tau_0$, $0 \leqslant n \leqslant N - 1$, the time interval τ_0 being common to all elementary waves for the sake of simplification.

In the situation as considered here, these elementary variations are assumed to constitute the *increments* of a finite amplitude pressure variation, namely from P_0 to P_1. Thus, it should be understood that the elementary waves do not follow one another in an *independent* way but rather 'stack up' until they reach the finite value of the pressure jump. In other words, in the first case, the environment recovers, after the elementary wave has crossed through, its own characteristics before any disturbance. In the second case, on the other hand, the incremental wave propagation of n rank takes place in an environment whose properties are those acquired after the wave of $n - 1$ rank has propagated.

Thus, by referring to P_0, ρ_0, T_0 as the pressure, density and temperature in the assumed undisturbed environment at rest, these values will be altered after each passage of an incremental pressure wave. Let us examine now how this change is being executed and study its consequences.

- At time t_0, a first pressure increment is issued in the x_0 section. Then, it propagates at the celerity $\mathbf{a}_0 = \sqrt{\gamma r T_0}$. After crossing through the environment, the properties have changed to respectively $P_1' = P_0 + P'$, $\rho_1' = \rho_0 + \rho'$, $T_1' = T_0 + T'$. Considering relation (7.2), we obviously deduce that, for a compression wave $\rho' > 0$, the density after this first perturbation crossing is $\rho_1' > \rho_0$.
- Thus, the next elementary wave following it at $t_0 + \tau_0$ propagates in an environment with different characteristics and therefore a different celerity. From the speed of sound expression, it is easy to deduce that $\mathbf{a}_1 > \mathbf{a}_0$.
- Hence, by considering a section $x = x_1$, downstream of the section x_0 in the propagation direction, we deduce that the amount of time taken by the first pressure increment to propagate over the distance $x_1 - x_0$ is *higher* than that taken by that of the increment which follows it and so on. In other words, the time interval between two successive waves tends to decrease as the propagation process continues.

The final consequence of this result is the straightening of the compression wave front, since, as outlined in Fig. 7.4, each incremental wave propagates at a faster rate than the preceding wave: $\mathbf{a}_N > \ldots \mathbf{a}_n > \mathbf{a}_{n-1} \ldots \mathbf{a}_1 > \mathbf{a}_0$.

Then referring to $\Delta_0 = (N - 1) \tau_0$ as the total compression time from P_0 to P_1 in the x_0 section, this duration will tend to decrease as the propagation progresses, with, in a $x_1 > x_0$ section, $\Delta_1 = (N - 1) \tau_1 < \Delta_0$ and so on. For a sufficient time lag, this straightening mechanism of successive compression wave fronts leads to the propagation of a *finite* amplitude pressure jump $P_1 - P_0$, as outlined in the diagram on the right in Fig. 7.4. This is the phenomenon of the plane shock wave, which we will further analyze in more detail at a later stage in this chapter.

7.2.2 The Successive Expansion Wave Propagation

By applying the same analysis to the succession of elementary expansion waves ($P' < 0$), it is clear that the propagation of the elementary pressure steps takes place at different celerities such as:

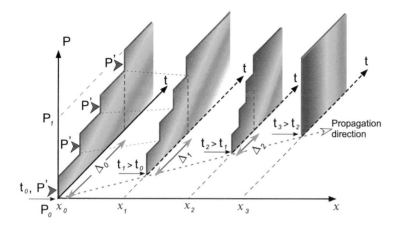

Fig. 7.4 Straightening process of the incremental compression wavefront during the propagation

$$\mathbf{a}_N < \dots \mathbf{a}_n < \mathbf{a}_{n-1} \dots \mathbf{a}_1 < \mathbf{a}_0 .$$

Thus, during this propagation, the time interval between two successive waves tends to increase. In other words, any time lag occurring in an expansion wave increases during the propagation leading to the eventual wave front spreading in an expansion fan. Therefore, the opposite phenomenon to the shock wave cannot occur for a pressure drop of finite amplitude.

Remark It should not be concluded from the previous discussion that the only mechanism for a plane shock wave formation is that which was schematically introduced in the previous comparative analysis. In fact, we will see that such a phenomenon occurs whenever a supersonic flow is abruptly decelerated in a non-isentropic manner.

7.3 The Plane Shock Wave

7.3.1 The Study Assumptions

The problem we address here is that of a plane shock wave, orthogonal to the flow direction, as outlined in the following Fig. 7.5. We will consider, to derive the equations, the plane shock wave as a *discontinuity* surface *on the macroscopic scale of motion*. In other words, the local flow functions, in the *mesoscopic sense* of the term,

Fig. 7.5 Notation
convention through a normal
shock crossing

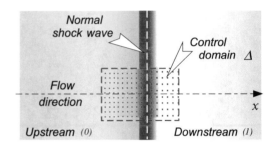

(see first chapter) remain *defined* on either side of the shock, but undergo, during its crossing, abrupt variations in their respective values[4] which have to be calculated.

For this purpose, we consider the fluid, *on both sides of the shock*, as a continuous medium which fulfills the conditions of local thermodynamic equilibrium, and whose dynamic and energetic state is modified as it crosses through the shock due to momentum and energy transfers under the following assumptions:

H1 : a steady motion;
H2 : a one-dimensional flow[5];
H3 : an ideal fluid i.e., inviscid and non-heat conductive;
H4 : an ideal gas behaviour;
H5 : negligible body forces.

The H2 assumption results from the simple fact that the flow characteristics relating to the upstream and downstream sides of the shock wave front are taken at sufficiently close points.

Thus the flow functions only depend on the single variable (x)—see Fig. 7.5—the values being identified by the index $_0$ or $_1$ depending on whether they refer to the upstream (before) or downstream (after) of the shock.

Finally, according to assumption H3, the viscosity and heat transfer effects by conduction *within the flow* can only occur inside the shock thickness itself. Such phenomena are excluded from the present analysis model.

---

[4] Whether such an abrupt variation or "jump" corresponds to a function *discontinuity* in the usual mathematical sense of the term, is a complex issue. Indeed, this questions the validity of the formulation of the Euler and Navier-Stokes equations to represent, at the *mesoscopic* scale, the flow crossing through a normal shock. A first insight into this subject, which any in-depth discussion is beyond the scope of this course, will be given as Additional Information in Sect. 7.3.7.1.

[5] Strictly speaking, this hypothesis means that the velocity vector has only one component and that all the functions of the flow field depend on only one space variable.

7.3.2 The Local Equations for a Normal Shock Crossing

By directly applying the mass, momentum and energy balances on a control domain such as Δ, outlined in Fig. 7.5, we obtain the algebraic equations governing the four unknown functions $U(x)$, $P(x)$, $\rho(x)$ and $T(x)$ on both sides of the shock. They are written:

Continuity
$$\rho_0 U_0 = \rho_1 U_1 \,, \tag{7.11}$$

Momentum
$$P_0 + \rho_0 U_0^2 = P_1 + \rho_1 U_1^2 \,, \tag{7.12}$$

Energy
$$C_p T_0 + \frac{U_0^2}{2} = C_p T_1 + \frac{U_1^2}{2} \,, \tag{7.13}$$

State
$$\frac{P_0}{\rho_0 T_0} = \frac{P_1}{\rho_1 T_1} \,. \tag{7.14}$$

- The first relation is none other than the *mass conservation* in the domain Δ, in the absence of any flux through the lateral surface (normal shock and assumption H2 in particular).

- The next, Eq. (7.12) states that the momentum flux is only balanced by the existing pressure forces, which leads under hypotheses H1 and H2 in particular to the *dynalpy conservation* across the shock.

- Eq. (7.13) expresses the energy conservation as the *enthalpy* balance. Resulting from the first law of thermodynamics, it does not in any way prejudge the reversibility or not of the energy exchanges by the fluid compression when crossing through the shock. It can be written in a different form, noting that:

$$C_p T = C_p \frac{\gamma r T}{\gamma r} = C_p \frac{a^2}{\frac{C_p}{C_v}(C_p - C_v)} = \frac{a^2}{\gamma - 1} \,.$$

Thus we obtain:
$$\frac{a_0^2}{\gamma - 1} + \frac{U_0^2}{2} = \frac{a_1^2}{\gamma - 1} + \frac{U_1^2}{2} \,.$$

Let us denote by a_c the speed of sound value of a flow whose velocity is $U_c = a_c$, and whose Mach number is therefore equal to one. The enthalpy conservation then takes the following form:

$$\frac{a_0^2}{\gamma - 1} + \frac{U_0^2}{2} = \frac{a_1^2}{\gamma - 1} + \frac{U_1^2}{2} = \frac{\gamma + 1}{\gamma - 1} \frac{a_c^2}{2} \,. \tag{7.15}$$

7.3.3   Prandtl's Formula for the Normal Shock

From the previous algebraic equations, it is possible to derive a relation between the velocities on either side of the shock. This is *Prandtl's relation*, which regarding the normal shock is written as:

$$\boxed{U_0 U_1 = \mathbf{a}_c^2} \tag{7.16}$$

where $\mathbf{a}_c$ stands for the speed of sound for $M = 1$.

Proof Rewriting the momentum equation in the form $\rho_1 U_1^2 - \rho_0 U_0^2 = P_0 - P_1$, we can deduce, by dividing throughout by $\rho_0 U_0$ ($= \rho_1 U_1$, according to the continuity equation):

$$U_1 - U_0 = \frac{P_0}{\rho_0 U_0} - \frac{P_1}{\rho_1 U_1} .$$

By introducing the local celerities $\mathbf{a}_0^2 = \gamma P_0 / \rho_0$ and $\mathbf{a}_1^2 = \gamma P_1 / \rho_1$, this equation becomes

$$U_1 - U_0 = \frac{1}{\gamma} \left(\frac{\mathbf{a}_0^2}{U_0} - \frac{\mathbf{a}_1^2}{U_1} \right) .$$

In addition, the energy Eq. (7.15) provides $U_1^2 - U_0^2 = \frac{2}{\gamma - 1} \left(\mathbf{a}_0^2 - \mathbf{a}_1^2 \right)$.

By eliminating the celerity $\mathbf{a}_1^2$ between these last two relations, we derive that:

$$U_1 - U_0 = \frac{1}{\gamma} \left[\frac{\mathbf{a}_0^2}{U_0} - \frac{1}{U_1} \left(\mathbf{a}_0^2 - \frac{\gamma - 1}{2} \left(U_1^2 - U_0^2 \right) \right) \right] ,$$

that is

$$U_1 - U_0 = \frac{1}{\gamma} \left[\mathbf{a}_0^2 \left(\frac{U_1 - U_0}{U_0 U_1} \right) + \frac{\gamma - 1}{2 U_1} (U_1 - U_0)(U_1 + U_0) \right] ,$$

or, after simplifying $\dfrac{\gamma + 1}{\gamma - 1} \dfrac{U_0 U_1}{2} = \dfrac{U_0^2}{2} + \dfrac{\mathbf{a}_0^2}{\gamma - 1}$, which, given the enthalpy conservation (7.15) leads to the expected result.

7.3.4   The Mach Number Relation on Either Side of a Shock

The Mach number downstream of the shock varies as a function of the upstream flow Mach number ahead of the shock according to

$$\boxed{M_1^2 = \frac{(\gamma - 1) M_0^2 + 2}{2\gamma M_0^2 + 1 - \gamma}} \tag{7.17}$$

This relation between the Mach numbers on either side of the normal shock provides several significant insights. This is plotted graphically in Fig. 7.6.

As displayed by the graph in question, the function $f(M_0)$ for $M_0 \geqslant 0$ has two asymptotes, respectively vertical and horizontal, both positioned at the same distance $\sqrt{(\gamma - 1)/2\gamma}$ from the origin. Moreover, we will justify at a later stage (Cf. Sect. 7.3.7.4) that the shock phenomenon cannot *physically* appear in the subsonic regime. Thus, the upstream Mach number should be taken such that $M_0 \geqslant 1$. Under these conditions, the possible variation range of M_1, as outlined in Fig. 7.6, is confined to the interval $\left[\sqrt{(\gamma - 1)/2\gamma}, 1\right]$.

We will therefore retain that:

> Downstream of a normal shock wave, the flow is subsonic, with a Mach number value which cannot be lower than $\sqrt{(\gamma - 1)/2\gamma}$.

Proof By dividing Eq. (7.15) throughout by the local velocity we obtain, whatever the position and after introducing the Mach number $M = U/\mathbf{a}$

$$\frac{1}{(\gamma - 1)M^2} + \frac{1}{2} = \frac{\gamma + 1}{2(\gamma - 1)} \frac{\mathbf{a}_c^2}{U^2}.$$

This relation is easily expressed in the form of

$$\left(\frac{U}{\mathbf{a}_c}\right)^2 = \frac{(\gamma + 1)M^2}{(\gamma - 1)M^2 + 2}. \tag{7.18}$$

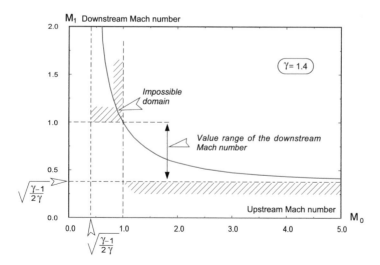

Fig. 7.6 Relation between Mach numbers on either side of a normal shock

By writing it for the conditions upstream and downstream of the shock, we directly obtain:

$$\left(\frac{U_0}{\mathbf{a}_c}\right)^2 = \frac{(\gamma + 1)\, M_0^2}{(\gamma - 1)\, M_0^2 + 2} \quad \text{and} \quad \left(\frac{U_1}{\mathbf{a}_c}\right)^2 = \frac{(\gamma + 1)\, M_1^2}{(\gamma - 1)\, M_1^2 + 2}.$$

By multiplying throughout these two expressions, we obtain, taking into account Prandtl's formula:

$$\left(\frac{U_0 U_1}{\mathbf{a}_c^2}\right)^2 \equiv 1 = \frac{(\gamma + 1)^2\, M_0^2 M_1^2}{\left[(\gamma - 1)\, M_0^2 + 2\right]\left[(\gamma - 1)\, M_1^2 + 2\right]},$$

from which the expected result is directly deduced.

7.3.5 Static Value Variations Through a Normal Shock

The relations which we intend now to demonstrate aim to express the various static quantity ratios, P_1/P_0, ρ_1/ρ_0, T_1/T_0, solely in term of the upstream Mach number.

7.3.5.1 The Static Pressure Ratio

The static pressure ratio on either side of a normal shock is:

$$\boxed{\frac{P_1}{P_0} = 1 + \frac{2\gamma}{\gamma + 1}\left(M_0^2 - 1\right)} \qquad (7.19)$$

Proof Starting from the dynalpy conservation, we can write, using the continuity equation:

$$P_0 - P_1 = \rho_1 U_1^2 - \rho_0 U_0^2 = \rho_0 U_0 U_1 - \rho_0 U_0^2 = \rho_0 U_0\,(U_1 - U_0),$$

or by dividing throughout by P_0 and factorizing the right hand side by U_0

$$1 - P_1/P_0 = \frac{\rho_0 U_0^2}{P_0}\left(\frac{U_1}{U_0} - 1\right) = \gamma M_0^2\left(\frac{U_1}{U_0} - 1\right),$$

after introducing the speed of sound of the upstream flow $\mathbf{a}_0^2 = \gamma P_0/\rho_0$. Now, from relations (7.15) and (7.16), it can be observed that:

$$\frac{U_1}{U_0} \equiv \frac{U_1 U_0}{U_0^2} = \frac{\mathbf{a}_c^2}{U_0^2} = \frac{(\gamma - 1)\, M_0^2 + 2}{(\gamma + 1)\, M_0^2},$$

the last equality being a direct consequence of result (7.18).

All that remains is to substitute this equality in the previous pressure ratio expression to reach the expected result (7.19).

7.3.5.2 The Density Ratio

For a normal shock wave, the fluid density ratio on either side of the shock is equal to:

$$\frac{\rho_1}{\rho_0} = \frac{(\gamma + 1) M_0^2}{(\gamma - 1) M_0^2 + 2} \tag{7.20}$$

Proof According to the continuity equation, we have $\rho_1/\rho_0 = U_0/U_1$. By involving relations (7.15) and (7.16), we obtain:

$$\frac{\rho_1}{\rho_0} = \frac{U_0^2}{U_1 U_0} \equiv \frac{U_0^2}{a_c^2} = \frac{(\gamma + 1) M_0^2}{(\gamma - 1) M_0^2 + 2} \qquad (Q.E.D.).$$

7.3.5.3 The Static Temperature Ratio

The static temperature ratio on either side of a normal shock is expressed by

$$\frac{T_1}{T_0} = \frac{\left(2\gamma M_0^2 + 1 - \gamma\right)\left[(\gamma - 1) M_0^2 + 2\right]}{(\gamma + 1)^2 M_0^2} \tag{7.21}$$

which is simply obtained from Eqs. (7.19) and (7.20), by using the ideal gas equation of state.

7.3.6 Hugoniot's Relation Through a Normal Shock

The relation between the pressure and density ratios of the *adiabatic* motion of an ideal gas crossing through a normal shock, is usually named the *shock adiabatic* or the *Hugoniot dynamic adiabate* formula, after Hugoniot [79]. It can be expressed in various ways, such as for instance:

$$\frac{P_1}{P_0} = \frac{\frac{\gamma + 1}{\gamma - 1} \frac{\rho_1}{\rho_0} - 1}{\frac{\gamma + 1}{\gamma - 1} - \frac{\rho_1}{\rho_0}} \tag{7.22}$$

$$\frac{\rho_1}{\rho_0} = \frac{1 + \frac{\gamma + 1}{\gamma - 1} \frac{P_1}{P_0}}{\frac{\gamma + 1}{\gamma - 1} + \frac{P_1}{P_0}} \tag{7.23}$$

The major result highlighted by these formulae linking the pressure and density ratios on either side of the shock is the *non-isentropic* character of the flow through

Fig. 7.7 Graphs of the
Hugoniot
adiabate—Eq. (7.25)—and
Laplace
isentropic—Eq. (7.24)—laws

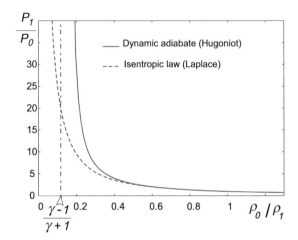

the shock.   Indeed, it should be recalled that if the phenomenon were isentropic, the pressure and density of the ideal gas would be related according to Laplace's formula, viz:

$$\frac{P_1}{P_0} = \left(\frac{\rho_0}{\rho_1}\right)^{-\gamma}. \tag{7.24}$$

In fact, it is possible to reformulate relations (7.22) or (7.23) to find, for example,

$$\frac{P_1}{P_0} = \frac{\frac{\gamma+1}{\gamma-1} - \frac{\rho_0}{\rho_1}}{\frac{\gamma+1}{\gamma-1}\frac{\rho_0}{\rho_1} - 1}. \tag{7.25}$$

This last relation, called the «*Hugoniot dynamic adiabate*» is graphically plotted in Fig. 7.7, along with Laplace's isentropic law.

The difference between the two curves can only be explained by the *non-reversible* character of the flow when crossing through a shock wave, since both relations apply to *adiabatic* evolutions. This is the reason for focusing, in the following section, on the entropy variation, which will allow us to determine the irreversibility value and confirm, in addition, that the shock corresponds to a compression, a result which cannot be solely derived from Hugoniot's relations.

Proof Hugoniot's relation, under any of the formulations (7.22), (7.23) or (7.25) can be demonstrated in different ways. The proposition is to directly derive this from the balance equations through the shock.

Thus, the energy Eq. (7.13) results in $C_p(T_0 - T_1) = \frac{1}{2}(U_1 - U_0)(U_1 + U_0)$.

Now, according to the momentum Eq. (7.12), and using the continuity equation, we can write

$$P_0 - P_1 = \rho_1 U_1^2 - \rho_0 U_0^2 = \rho_0 U_0 (U_1 - U_0).$$

By eliminating the velocity difference between these two relations we obtain:

$$C_p(T_0 - T_1) = \frac{1}{2}(U_1 + U_0)\frac{P_0 - P_1}{\rho_0 U_0} = \frac{1}{2}\left(\frac{1}{\rho_0} + \frac{1}{\rho_1}\right)(P_0 - P_1),$$

by using, once again, the continuity equation in the last equality. Taking into account the ideal gas equation of state, we know that $C_p T = \frac{\gamma}{\gamma-1}\frac{P}{\rho}$, which makes it possible to express the previous relation in the form:

$$\frac{\gamma}{\gamma - 1}\left(\frac{P_0}{\rho_0} - \frac{P_1}{\rho_1}\right) = \frac{1}{2}\left(\frac{1}{\rho_0} + \frac{1}{\rho_1}\right)(P_0 - P_1).$$

It is easy to deduce that $\dfrac{\gamma}{\gamma - 1}P_0\left(\dfrac{1}{\rho_0} - \dfrac{P_1}{P_0}\dfrac{1}{\rho_1}\right) = \dfrac{1}{2}P_1\left(\dfrac{1}{\rho_0} + \dfrac{1}{\rho_1}\right)\left(\dfrac{P_0}{P_1} - 1\right),$

then $\dfrac{\gamma}{\gamma - 1}\dfrac{P_0}{\rho_1}\left(\dfrac{\rho_1}{\rho_0} - \dfrac{P_1}{P_0}\right) = \dfrac{1}{2}\dfrac{P_1}{\rho_0}\left(1 + \dfrac{\rho_0}{\rho_1}\right)\left(\dfrac{P_0}{P_1} - 1\right).$

Thus, finally $\dfrac{\gamma}{\gamma - 1}\dfrac{P_0}{P_1}\dfrac{\rho_0}{\rho_1}\left(\dfrac{\rho_1}{\rho_0} - \dfrac{P_1}{P_0}\right) = \dfrac{1}{2}\left(1 + \dfrac{\rho_0}{\rho_1}\right)\left(\dfrac{P_0}{P_1} - 1\right),$ an expression in which only the pressure and density ratios are respectively involved. The expected result, Eq. (7.22) or Eq. (7.23), can be deduced by simply grouping the homologous terms.

HISTORICAL NOTE— Shock relations: Hugoniot and Rankine—On page 91 of his *Memoir on motion propagation in bodies and especially in perfect gases*, published in 1889, Hugoniot, [79], wrote:

«But when the density ρ experiences a sudden variation, from ρ to ρ_1, the equation $\frac{P}{\rho^m} = \frac{P_1}{\rho_1^m}$ is no longer satisfied; it is substituted by the following

$$P_1 = P\frac{(m+1)\frac{\rho_1}{\rho} - (m-1)}{m + 1 - (m-1)\frac{\rho_1}{\rho}} \cdots »$$

This relation proves that the gas behaves thermodynamically like a *barotropic fluid*. Since the crossing through the shock is further assumed to be *adiabatic*, the behaviour in question is referred to as *dynamic abiabatic*. (We are still talking about «Hugoniot's dynamic *adiabate*»). It definitively proves that, if it is legitimate to consider the shock as an adiabatic process,—the fluid *not having enough time to exchange heat with the environment* while crossing the wave—, one cannot, on the other hand, consider that the phenomenon is *reversible*. In this case, the barotropic relationship would be Laplace's formula ($P/\rho^\gamma = C^t$), still sometimes referred to as «*static adiabatic*» and reflecting an *adiabatic* **and** *reversible* behaviour. This explicitly states the nonisentropic nature of the flow passing through a shock. In 1870, Rankine [128] published an article in which he established several results on pressure wave propagation (compression and expansion), of infinitesimal and finite amplitudes. He proved that a sudden longitudinal pressure disturbance is able to steadily propagate, assuming that the heat exchange takes place between the fluid particles passing through the shock, but not with the environment. This idea is perfectly in line with the contemporary ideas on the irreversibility generation by the shock phenomenon, without contradicting its adiabatic character to the mesoscopic order. Finally, he provided the relations giving the kinetic energy and velocity variations through shock wave, as well as the propagation speed of this wave as a function of the upstream/downstream static pressure values. Although formally very different from those derived by Hugoniot and resulting from a separate analysis, these results are based on the same general philosophy—that of the *jump relations through a shock*—and are similar, according to Rayleigh [131], to Hugoniot's relations. This explains why the general jump expressions through a shock are still termed as the *Rankine-Hugoniot* relations.

7.3.7 The Entropy Variation and Stagnation Conditions

7.3.7.1 The Origin of the Entropy Variation

As we have discussed in Chap. 4, the irreversible entropy production is related to the viscosity and thermal conductivity of the fluid. While the fluid can legitimately be considered as ideal for the flow *on both sides of the shock*, this is no longer the case *within the shock wave* itself.

Experimental evidence indicates that the diffusive properties of the fluid are indeed the source of the irreversibilities. However this is in an area whose characteristic length scale is in the order of magnitude of the mean free path of the molecule agitation. Accordingly, this region is excluded from any analysis using the mesoscopic equations at the scale of the continuous medium.

The consequences of this remark are:

- legitimizing the shock schematization as a discontinuity surface on the continuous scale;
- making inoperative the determination of the entropy variation by integrating, throughout the thickness of the shock, the macroscopic dissipation expression defined in Chap. 4;
- justifying the entropy variation assessment from the upstream and downstream balance of the fluid states apart of the shock.

This set of considerations is schematically outlined in Fig. 7.8. Flow properties, such as the static pressure, are indeed subject to stiff variations, on a length scale of the same order or less than the continuous limit, where the viscosity and thermal conductivity of the fluid are actually involved.

ADDITIONAL INFORMATION—**The shock wave thickness**—We consider shock wave normal to a flow with a x direction. It is assumed that the wave front is in reality a zone of thickness δ in which the molecular effects in the fluid motion

Fig. 7.8 Schematic of the pressure variation as a *'mesoscopic jump'* through a shock

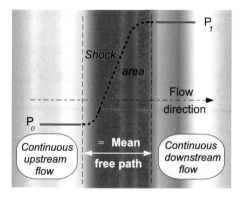

are of the same order as those of the pressure and inertial forces. Over this distance, the different flow parameters undergo *mathematically* continuous variations, so that the Mach number, for example, necessarily comes through the unit value at an abscissa x_{ref} which, measured from the upstream origin of the shock, is such that $0 < x_{ref} < \delta$. We will use the x_{ref} abscissa section as a reference for estimating the quantities within the shock area.

- From a *microscopic* point of view, the kinetic theory of gases provides that:

$$\mu_{ref} = \frac{1}{3}\rho_{ref}\upsilon_{ref}\ell_m ,$$

where υ_{ref} is a reference of the molecular agitation velocity and ℓ_m the mean free path of this same agitation motion. Taking as the reference velocity the average agitation velocity $\upsilon^2 = \frac{8}{\pi}rT \equiv \frac{8}{\pi\gamma}\mathbf{a^2}$ and applying this relation to the reference conditions for which, by convention $M_{ref} = 1$, we obtain $\upsilon_{ref} \sim \sqrt{8/\pi\gamma}\,U_{ref} \approx U_{ref}$ for the air. Thus, the *microscopic* estimate of the viscosity is

$$\mu_{ref} \approx \rho_{ref}U_{ref}\ell_m . \tag{7.26}$$

- Turning now to the mesoscopic order, it is possible to estimate the orders of magnitude of the momentum equation terms, reduced here to its sole x projection, as follows:

$$\underbrace{\rho U\frac{dU}{dx}}_{\rho_{ref}U_{ref}\frac{U_1-U_0}{\delta}} = \underbrace{-\frac{dP}{dx}}_{\frac{P_1-P_0}{\delta}} + \underbrace{\frac{4}{3}\mu\frac{d^2U}{dx^2}}_{\mu_{ref}\frac{U_1-U_0}{\delta^2}} .$$

Assuming that $P_{ref} \sim P_1 - P_0$ and $U_{ref} \sim U_1 - U_0$, the pressure and inertia terms are of an equivalent order since $U_{ref}^2 = \gamma P_{ref}/\rho_{ref}$. If the same is true of the viscosity term, we should have:

$$\mu_{ref} \approx \rho_{ref} U_{ref} \delta. \tag{7.27}$$

Hence, the comparison of both *macroscopic* and *microscopic* viscosity esti-mates proves that shock wave thickness is of the same order of magnitude as the mean free path.

A more detailed discussion of this issue can be found in Loitsyanskii's work [91], page 736, which states that the ratio of shock-thickness to the mean free path, δ/ℓ_m decreases as the Mach number increases. Thus, for instance, this ratio is of the order of 4, for $M_0 = 2$ and 2, for $M_0 = 5$.

7.3.7.2 The Entropy Variation Through a Shock

The entropy variation which an ideal gas undergoes when passing from an initial state ($_0$) to a final state ($_1$) is expressed by

$$s_1 - s_0 = C_v \ln\left[\frac{P_1}{P_0} \times \left(\frac{\rho_0}{\rho_1}\right)^{\gamma} \right]. \tag{7.28}$$

By substituting in this relation expressions (7.19) and (7.20), we directly obtain the entropy variation across the shock, in terms of the upstream Mach number, viz:

$$s_1 - s_0 = C_v \ln\left[\frac{\left(2\gamma M_0^2 + 1 - \gamma\right)}{\gamma + 1} \left(\frac{2 + (\gamma - 1) M_0^2}{(\gamma + 1) M_0^2}\right)^{\gamma} \right] \tag{7.29}$$

EXERCISE 27 The entropy variation as a function of the pressure ratio.

Several equivalent expressions of the entropy variation across a normal shock can be given.

QUESTION   Express the entropy variation of the flow crossing through a normal shock as a function of the pressure ratio P_1/P_0.

ANSWER: The requested expression is written:

$$s_1 - s_0 = C_v \ln\left[\frac{P_1}{P_0} \left(\frac{\frac{\gamma+1}{\gamma-1} + \frac{P_1}{P_0}}{1 + \frac{\gamma+1}{\gamma-1}\frac{P_1}{P_0}}\right)^{\gamma} \right].$$

7.3.7.3 Stagnation Value Variations Across a Shock

It is more suitable, in practice, to account for the irreversibilities of the flow crossing through a shock by using the '*jump variations*' of the stagnation variables. Indeed, when generating compressible gas flows in the laboratory, a conventional technique consists in storing a fluid mass of zero velocity and a given temperature in an over pressurized tank. The setting in motion is then simply carried out by the fluid expansion through a suitable conduit. With such a device, the total energy available, regardless of gravity, is defined by the *reservoir conditions*, also called the *stagnation conditions*. The flow function values referring to this state will be denoted by the index ($_i$). If the motion is *isentropic*, according to Saint-Venant's equation, the stagnation values will be retrieved at any point in the field where the velocity is locally zero. Consequently it would be preferable to speak of *isentropic* stagnation conditions to emphasize the fact that the fluid should be *isentropically* decelerated to a zero velocity.

As we have just explained, a shock wave crossing generates irreversibilities in the fluid motion, so that the *downstream* stagnation conditions are no longer, necessarily, those of the *upstream* flow. Therefore, we will establish the upstream-downstream relations for the stagnation temperature and pressure.

- **Temperature**: The energy equation directly provides

$$C_p T_0 + \frac{U_0^2}{2} \left(\equiv C_p T_{i0} \right) = C_p T_1 + \frac{U_1^2}{2} \left(\equiv C_p T_{i1} \right) . \tag{7.30}$$

We can therefore state that:

> The stagnation temperature is preserved through a shock: $T_{i0} = T_{i1}$.

- **Pressure**: The stagnation pressure variation through a shock can be determined in different ways.[6] We will proceed here by directly using the entropy variation (7.29) which has been previously established.

By applying the entropy expression of the ideal gas $s = C_p \ln T - r \ln P + C'$, to the stagnation conditions before and after the shock, we obtain:

$$s_{i1} - s_{i0} = C_p \ln \left(\frac{T_{i1}}{T_{i0}} \right) - r \left(\frac{P_{i1}}{P_{i0}} \right) .$$

However, since the only irreversibility sources in the flow are due to the shock crossing, the fluid entropy upstream of the shock (s_0) is preserved, so that $s_{i0} = s_0$ and the same applies to the fluid entropy downstream of the shock, $s_{i1} = s_1$. In addition, we have just demonstrated that the *stagnation* temperature does not change

---

[6] Note, however, that it is not possible to use the momentum equation with the stagnation condition $U_0 = 0$ or $U_1 = 0$. Indeed, for the sake of continuity—Eq. (7.11)—, the momentum equation is written in the form $P_0 + \rho_0 U_0^2 = P_1 + \rho_0 U_0 U_1$, so that the *simple condition* $U_0 = 0$, for example, implies $P_{i0} = P_{i1}$, that is, the absence of a shock.

throughout the shock. Thus, we deduce from the previous relation that:

$$s_{i1} - s_{i0} = s_1 - s_0 = -r\left(\frac{P_{i1}}{P_{i0}}\right).$$

Using Eq. (7.19), we obtain:

$$\frac{P_{i1}}{P_{i0}} = \left(\frac{2\gamma M_0^2 + 1 - \gamma}{\gamma + 1}\right)^{\frac{1}{1-\gamma}} \times \left(\frac{2 + (\gamma - 1)M_0^2}{(\gamma + 1)M_0^2}\right)^{\frac{\gamma}{1-\gamma}},$$

which we will finally put in the form:

$$\frac{P_{i1}}{P_{i0}} = \frac{\left(\dfrac{(\gamma + 1)\,M_0^2}{2 + (\gamma - 1)\,M_0^2}\right)^{\frac{\gamma}{\gamma - 1}}}{\left(\dfrac{2\gamma M_0^2 + 1 - \gamma}{\gamma + 1}\right)^{\frac{1}{\gamma - 1}}} \tag{7.31}$$

ALTERNATE PROOF— It is also possible to derive the *stagnation* pressure variation from that of the *static* pressure—Eq. (7.19)—according to the factorization:

$$\frac{P_{i1}}{P_{i0}} = \frac{P_{i1}}{P_1} \times \frac{P_1}{P_0} \times \frac{P_0}{P_{i0}}.$$

The P/P_i ratio expressions for the upstream and downstream *isentropic flows* will be demonstrated in the next chapter. They are of the form:

$$\frac{P_i}{P} = \left(1 + \frac{\gamma - 1}{2}M^2\right)^{\frac{\gamma}{\gamma - 1}},$$

where M is the Mach number of the respective flow concerned. It is then simply necessary to substitute the expressions for each of the factor products to obtain the expected result.

7.3.7.4   The Existence Requirement for a Normal Shock

We have previously discussed in Sect. 7.3, the behavioural difference between compression ($P_1/P_0 > 1$) and expansion ($P_1/P_0 < 1$) waves. We can now specify the origin of such differences.

The analytical examination of the function expressing the entropy variation (7.29) reveals that it is a monotonically increasing function of the upstream Mach number M_0. Its graph, around $M_0 = 1$ is plotted in Fig. 7.9. The only entropy variation range

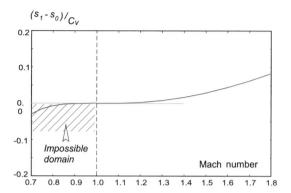

Fig. 7.9 Entropy variation near the Mach number equal to one

compatible with the second law of thermodynamics is that for which s_1 is greater than or equal to s_0. As outlined in this figure, this condition is reached for $M > 1$.

It can therefore be concluded that:

> A normal shock wave cannot occur in the subsonic flow regime.

In addition, relation (7.19) proves that if $M > 1$, then $P_1 > P_0$. Thus, the only physically acceptable pressure variation through a normal shock is that of a recompression. Therefore, there can be no *steep expansion jump*, the pressure adaptation process in this case takes the form of an expansion fan.

7.3.7.5 The Limit at a Mach Number Around the Unit

A second characteristic feature of the entropy variation results from the $s(M)$ function behaviour around $M = 1$. As plotted in the previous figure, its graph appears in fact substantially coincident with its horizontal tangent for Mach numbers ranging between 0.90 and 1.15. In other words, *infinitesimal* deviations in pressure, temperature and density generated, for example in a fluid at rest, by the displacement of a tiny solid at a velocity close to the speed of sound, correspond to applying a *quasi-isentropic* evolution to the environment. Hence, to a very high degree of approximation, slight amplitude transonic phenomena can be regarded as isentropic.

HISTORICAL NOTE—**The shock wave phenomenon**—The first *theoretical* study of finite amplitude waves was probably due to Poisson [120], in 1808. However, according to Salas [144], it was Airy who, when studying finite amplitude waves in water channels, made the major observation in a 1841 paper according to which the crests tended to overtake the troughs, so that the slope of the wave front became increasingly steep.

A few years later, in 1848, Stokes [148] mathematically confirmed this observation for a progressive sinusoidal wave, by concluding: «*It is evident that in the neighborhood of points a, c* [the compression region] *the curve becomes more and more steep as t increases, while in the neighborhood of points o, b, z* [the expansion region] *its inclination becomes more and more gentle.*» He inferred from this the possibility of a discontinuity surface being formed. However the properties of the latter, obtained by solely satisfying the continuity and momentum equations, and adopting *Boyle's relation* $a^2 = P/\rho$ for the speed of sound, contravene the energy conservation principle. The underlying reason is a lack of consideration (incomplete and erroneous) with regard to the thermodynamic aspects.

In 1860, Riemann [138] made an important contribution with respect to the mathematical inputs. Nevertheless, like Stokes, he failed to cope with the fundamental nature of shock wave in its physical dimension.

A decisive step towards a more comprehensive thermodynamic description of the phenomenon was expressed by Rankine [128], in 1870. In modern terminology, he demonstrated that the stationary shock wave could be dealt with as an adiabatic phenomenon (in the sense of the absence of heat exchange with the environment) but not reversible (production of entropy within the shock). In doing so, he removed the energy incompatibility of the previous research.

The final piece of evidence was to be put forward by Hugoniot [78, 79] in 1887 and 1889. He proved that neither Laplace's law (isentropic) nor Boyle's law (isothermal) could be applied to the flow crossing through a shock and provided the only relationship (dynamic adiabate) compatible with the irreversible adiabatic nature of the shock.

Thus, during 80 years, the physical analysis and the mathematical modelling of the shock phenomenon were essentially completed, as outlined in the review of this topic published by Rayleigh [131] in 1910. The scientific innovations in this article are indeed relatively modest and concern the entropy production across weak shocks and the estimate of shock wave thickness. On the other hand, Rayleigh confirmed that Hugoniot's contribution was «*precisely the same*» as Rankine's, fifteen years earlier, but is particularly sceptical about the very existence of the phenomenon, as modelled by Hugoniot. Thus, he noted in connection with Hugoniot's curve, which is «*however valid*» , that «*its fulfillment does not secure that the wave so defined is possible. As a matter of fact, a whole class of such waves is certainly impossible, and I would maintain, further, that a wave of the kind is never possible under the conditions, laid down by Hugoniot, of no viscosity or heat conduction.*»

EXERCISE 28 The entropy variation expansion around $M = 1$.

To quantitatively justify some properties of the small disturbance propagation, the proposition is to apply the previous results by considering these disturbances as *infinitesimal deviations* of the entropy variation in the vicinity of $M_0 = 1$.

QUESTION 1.   We take $M_0 = 1 + \epsilon$, where $\epsilon > 0$ denotes the infinitely small quantity of first order. Derive the expressions of the third order expansions in ϵ for P_1/P_0 and ρ_0/ρ_1.

QUESTION 2.   Deduce the expansion expression in ϵ^3 of the entropy variation around $M_0 = 1$.

QUESTION 3.   Give an estimate of the entropy variation for the range of acoustic pressure fluctuations $P'/P_0 \approx 10^{-5}$ to 10^{-9} in the case of the air ($\gamma = 1.4$).

SOLUTION: **1.** The answers to this first question are obtained by taking a third order expansion of the expressions for the pressure (7.19) and density (7.20) variations across the shock.

The results are written as follows:

$$P_1/P_0 = 1 + \frac{4\gamma}{\gamma + 1}\epsilon + \frac{2\gamma}{\gamma + 1}\epsilon^2, \text{ which is an exact expression to second order,}$$

and $\rho_0/\rho_1 \simeq 1 - \dfrac{4}{\gamma + 1}\epsilon + \dfrac{6}{\gamma + 1}\epsilon^2 - \dfrac{8}{\gamma + 1}\epsilon^3 + \mathcal{O}\left(\epsilon^4\right).$

2. The entropy variation is deduced from the previous results, knowing that, according to (7.28):

$$\frac{s_1 - s_0}{C_v} = \ln\left(\frac{P_1}{P_0}\right) + \gamma \ln\left(\frac{\rho_0}{\rho_1}\right).$$

To expand the logarithms, we simply have to remember that, for $u \to 0$:

$$\ln(1 + u) \simeq u - u^2/2 + u^3/3 + \mathcal{O}\left(u^4\right), \quad \text{and} \quad \ln(1 - u) \simeq -u - u^2/2 - u^3/3 \\ + \mathcal{O}\left(u^4\right).$$

Thus we obtain $\ln\left(\dfrac{P_1}{P_0}\right) \simeq \dfrac{4\gamma}{\gamma + 1}\epsilon - 2\dfrac{\gamma\,(3\gamma - 1)}{(\gamma + 1)^2}\epsilon^2 + \dfrac{8}{3}\dfrac{\gamma^2\,(5\gamma - 3)}{(\gamma + 1)^3}\epsilon^3 + \mathcal{O}\left(\epsilon^4\right),$

and $\qquad \ln\left(\dfrac{\rho_0}{\rho_1}\right) \simeq -\dfrac{4}{\gamma + 1}\epsilon + 2\dfrac{3\gamma - 1}{(\gamma + 1)^2}\epsilon^2 - \dfrac{8}{3}\dfrac{3\gamma^2 - 3\gamma + 2}{(\gamma + 1)^3}\epsilon^3 + \mathcal{O}\left(\epsilon^4\right).$

After substitution in the entropy variation expression, we note that the coefficients of the terms in ϵ and ϵ^2 cancel each other out, so that:

$$\frac{s_1 - s_0}{C_v} \simeq \frac{16}{3}\frac{\gamma\,(\gamma - 1)}{(\gamma + 1)^2}\epsilon^3 + \mathcal{O}\left(\epsilon^4\right).$$

This result quantitatively confirms the flattening of the curve over its tangent around $M = 1$, as previously demonstrated in the graph plotted in Fig. 7.9.

3. For the acoustic pressure fluctuations ranging in relative values between 10^{-9} and 10^{-5}, the expansion first order term in ϵ, namely $\epsilon \simeq (\gamma + 1)/4\gamma \, P'/P_0$ is around 0.4 times these values for the air ($\gamma = 1.4$). For the same gas, the associated entropy variations are in the order of $0.5\,\epsilon^3 \times C_v$, or again $2 \times 10^{-16}C_v$ to $2 \times 10^{-28}C_v$. They are therefore completely negligible.

7.3.8 *The Normal Shock Wave Characteristics for the Air*

We end this section on the normal shock by plotting the evolutions of the static and stagnation quantities of an air flow crossing through a normal shock wave. The curves

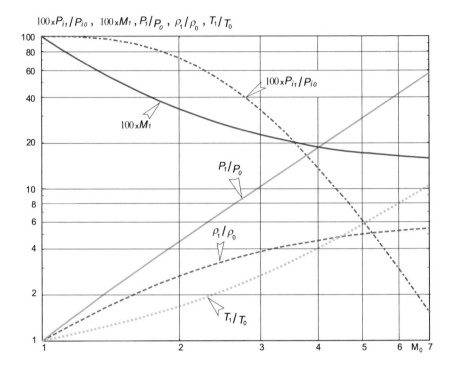

Fig. 7.10 Flow function variations through a normal shock as a function of the upstream Mach number ($\gamma = 1.4$)

plotted in Fig. 7.10 correspond to expressions (7.19), (7.20), (7.21) and (7.31), with $\gamma = 1.4$. They can provide a first assessment of the normal shock characteristics. For a more accurate calculation, tabulated values are recommended or direct reference to the previous formulae.

7.4   Oblique Shock Waves

7.4.1   The Physical Evidence

When a fixed sharp obstacle is positioned in a supersonic gas flow at infinity, the non-isentropic fluid deceleration can result in the onset of a shock wave which is inclined relative to the incident upstream flow direction. This is referred to as *oblique shock wave*.

An illustration in a two-dimensional plane geometry is outlined in Fig. 7.11, based on a hydraulic analogy. For the visualized flow, the obstacle is a dihedral with a small vertex angle.

Fig. 7.11 Hydraulic analogy of the supersonic flow pattern on a dihedral

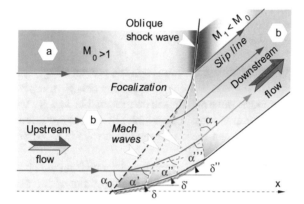

Fig. 7.12 Schematic for the supersonic flow deflection on a sharp tip

Apart from the immediate vicinity of the obstacle tip, the pressure disturbance introduced by this body takes the form of a *planar surface* inclined at an angle different from that of the dihedral.

Several parameters are likely to specify its shape, which can be locally that of a dihedral (a two-dimensional plane geometry) or cone (an axisymmetric situation), of a given angle with respect to the fluid flow direction at infinity. In order to acquire a better comprehension of this phenomenon, we will firstly qualitatively analyze the consequences of the deflection which the flow undergoes on a sharp-edged body. For this purpose, the surface of such a body is assumed to be a polygonal line consisting of a series of elementary facets, forming two by two the same angle δ, and whose succession results in a progressive increase in the flow deviation δ, δ', δ''... with respect to its initial upstream direction x at infinity.

The elementary angle δ between two adjacent facets is supposed to be *infinitely small* so that we can analyze the related flow modification in terms of *small perturbation*, as previously carried out in Sect. 7.2.1 (Fig. 7.12).

Under these conditions, on passing the first facet (an angular deviation δ), the flow is subjected to a first deflection, i.e., a direction variation of an angle α_0 with respect to its initial orientation. This angle value can be approximated by

$\sin \alpha_0 = 1/M_0$, where M_0 (> 1) is the Mach number of the upstream flow (see Additional Information below). This first disturbance, similar to a Mach wave (an isentropic compression wave of infinitesimal amplitude), leads to a downstream flow whose Mach number with respect to the velocity parallel to the new orientation is reduced, namely $M' < M_0$. This situation will continue until the second facet is reached where, *as long as the condition $M' > 1$ is still satisfied*, the flow is re-deflected due to the new slope disruption, resulting in a new wave whose angle α' is such that $\sin \alpha' = 1/M'$.

Since the Mach number has been reduced, the new deflection angle is greater than the previous angle. Thus, when the regime remains supersonic throughout the process until the final facet, the successive generated disturbances will form a converging system of Mach lines. This signifies that, at a given distance from the wall, these lines gather into a single wave. This is the oblique shock wave corresponding to a *finite amplitude* variation.

With reference to the schematic illustration in the previous Figure, the flow pattern includes two regions: (a) where the flow functions change sharply when crossing through an oblique shock wave and thus in an irreversible way, (b) where the change in direction and deceleration remain isentropic, by successive infinitesimal adaptations. The separation between these two regions results in a *contact* surface (line) which is a *slip* surface[7] (line).

Finally, as implicit in this same Figure, the distance from which the Mach waves coalesce into an oblique shock should depend on the radius of curvature of the obstacle surface. Ultimately, i.e., for the dihedral with which we began this discussion, the oblique shock wave's formation point is located in the immediate vicinity of the wedge apex. These two points will thus be merged when addressing the theoretical quantitative study of the phenomenon in the following section.[8]

Remark The above presentation of oblique shock wave formation on a discontinuous slope compression ramp can be more thoroughly replicated on a concave surface with a continuous slope variation, by using the method of characteristics. This development, outside the scope of this book, can be found in Luneau's work [92] where the case of the expansion waves on a convex wall is also addressed.

ADDITIONAL INFORMATION—**The infinitesimal oblique shock** (i)—Referring to the notations in the following Fig. 7.13, the wall angle deflection δ causes the upstream flow to deflect by a value ε with respect to its initial direction. Anticipating the subsequent result—Cf. Eq. (7.41)—, these two angles are linked by

$$\frac{\tan (\epsilon - \delta)}{\tan \epsilon} = \frac{(\gamma - 1) \, M_0^2 \sin^2 \epsilon + 2}{(\gamma + 1) \, M_0^2 \sin^2 \epsilon}$$

---

[7] On either side of such a surface (line) and at two infinitely close points, the tangential velocity can take distinct values, the normal velocity remaining zero.

[8] Viscosity effects, which tend to move the shock wave formation away from the wall, are also not taken into account in terms of this *ideal* fluid approach.

where M_0 is the upstream Mach number. For an infinitesimal deflection, $\delta \to 0$, the left hand side of this relation tends towards unity, from which it results that $M_0^2 \sin^2 \varepsilon \simeq 1$. Thus the disturbance caused by an infinitely minor directional variation can be assimilated to a Mach wave whose angle is $\alpha = \varepsilon \simeq \arcsin 1/M_0$. At the crossing of this wave, the flow is subjected to a compression. We will see— Cf. Eq. (7.38)—, that the upstream/downstream pressure values are linked by:

$$\frac{P_1}{P_0} = 1 + \frac{2\gamma}{\gamma+1} \left(M_0^2 \sin^2 \varepsilon - 1 \right).$$

Putting then $M_0^2 \sin^2 \varepsilon = 1 + u$, where u is an infinitely small positive value, this relation implies that $P_1/P_0 \simeq 1 + \frac{2\gamma}{\gamma+1} u$. Similarly, the density ratio, Eq. (7.39), proves that, under the same approximation, $\rho_0/\rho_1 \simeq 1 + \frac{2}{\gamma+1} u$. We can then estimate the entropy variation through this disturbance since:

$$\frac{s_1 - s_0}{C_v} = \ln \left[\frac{P_1}{P_0} \times \left(\frac{\rho_0}{\rho_1} \right)^\gamma \right], \quad \text{hence} \quad s_1 - s_0 \simeq 0, \quad \text{for} \quad M_0^2 \sin^2 \varepsilon = 1 + u,$$

which clearly demonstrates the isentropic nature of the compression within the limit of an infinitesimal deflection.

ADDITIONAL INFORMATION—**The infinitesimal oblique shock** (ii)—The physical mechanisms by which oblique shock wave and a Mach wave are generated are essentially of the same nature. The two wave types only differ in the level of the disturbances with which they are associated.

When the level is high, the disturbance coalescence leads to sudden changes in the flow properties and Mach number through oblique shock wave, whose angle depends in particular on the deflection imposed on the flow by the generating obstacle. The deflection angle can then be seen as a parameter related to the disturbance intensity. Thus, for evanescent deflection angles, the focus only concerns infinitesimal disturbances, resulting in a Mach wave for which this parameter should no longer be used.

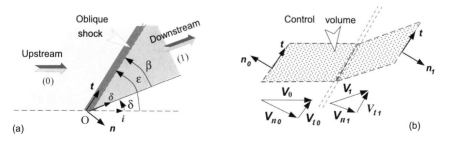

Fig. 7.13 Schematic for the oblique shock flow (**a**) and the corresponding control volume (**b**)

7.4.2 The Local Equations

Let us consider—Fig. 7.13a—a supersonic flow whose upstream velocity $\mathbf{V}_0 = V_0\mathbf{i}$ undergoes a change in direction or *deflection*, with an angle $\delta = (\mathbf{Oi}, \mathbf{O\delta})$.

After the oblique shock which is assumed to be a plane, the fluid velocity becomes $\mathbf{V}_1 = V_1\boldsymbol{\delta}$. The wave's plane inclination angle with respect to the upstream motion direction at infinity is then referred to as ϵ. The angle which this wave makes with the downstream's deflected flow direction is referred to as $\beta = \varepsilon - \delta$. Finally, we denote by $\mathbf{n}$ and $\mathbf{t}$ respectively the unit normal and tangent vectors to the shock, the orientation of the first being such that the scalar product $\mathbf{i.n}$ is positive.

As we have done for the normal shock, we consider, at the macroscopic scale, the oblique wave as a plane discontinuity surface of the flow functions. The local equations can then be easily obtained by expressing the general motion equations over an elementary control volume as outlined in Fig. 7.13b.

Under the general study assumptions (an ideal fluid in particular), these local equations are written:

Continuity:	$\rho_0 V_{n0} = \rho_1 V_{n1}$	(7.32)
Momentum / $\mathbf{n}$:	$P_0 + \rho_0 V_{n0}^2 = P_1 + \rho_1 V_{n1}^2$	(7.33)
Momentum / $\mathbf{t}$:	$V_{t0} = V_{t1}$	(7.34)
Energy :	$C_p T_0 + \dfrac{1}{2}\left(V_{n0}^2 + V_{t0}^2\right) = C_p T_1 + \dfrac{1}{2}\left(V_{n1}^2 + V_{t1}^2\right)$	
		(7.35)

7.4.3 Prandtl's Formula for the Oblique Shock

As we have demonstrated for the normal shock (Sect. 7.3.3), it is possible to deduce from the previous equations a relation between the velocity components on either side of the shock. This is Prandtl's relation which is written as follows:

$$V_{n0} V_{n1} = \mathbf{a}_c^2 - \frac{\gamma - 1}{\gamma + 1} V_t^2 \qquad (7.36)$$

where $\mathbf{a}_c$ denotes the speed of sound for $M = 1$.

Proof The energy conservation makes it possible to write:

$$C_p T + \frac{V_n^2 + V_t^2}{2} \equiv \frac{\gamma}{\gamma - 1}\frac{P}{\rho} + \frac{V_n^2 + V_t^2}{2} = C' \equiv \frac{\gamma + 1}{2(\gamma - 1)}\mathbf{a}_c^2,$$

by expressing the constant value with respect to a so-called *critical* condition (real or fictitious) such that, by definition $\|V\| = a_c$. It is then immediately deduced that:

$$\frac{P_0}{\rho_0} = \frac{\gamma+1}{\gamma}\frac{a^2_c}{2} - \frac{\gamma-1}{2\gamma}\left(V^2_{n0} + V^2_{t0}\right) \quad \text{and} \quad \frac{P_1}{\rho_1} = \frac{\gamma+1}{\gamma}\frac{a^2_c}{2} - \frac{\gamma-1}{2\gamma}\left(V^2_{n1} + V^2_{t1}\right).$$

Moreover, by applying the same procedure as for the normal shock, to the continuity and normal projection of the momentum equation, it is easy to obtain that:

$$V_{n1} - V_{n0} = \frac{P_0}{\rho_0 V_{n0}} - \frac{P_1}{\rho_1 V_{n1}}.$$

By substituting in this relation the previous expressions of P_0/ρ_0 and P_1/ρ_1 we obtain:

$$V_{n1} - V_{n0} = \frac{\gamma+1}{\gamma}\frac{a^2_c}{2}\left(\frac{1}{V_{n0}} - \frac{1}{V_{n1}}\right) - \frac{\gamma-1}{2\gamma}V^2_t\left(\frac{1}{V_{n0}} - \frac{1}{V_{n1}}\right) + \frac{\gamma-1}{2\gamma}(V_{n1} - V_{n0}),$$

an expression which only needs to be simplified to obtain the expected Prandtl relation.

Remark We can observe that the result we have just established (7.36) correctly returns Prandtl's normal shock relation, Eq. (7.16), for $V_t = 0$.

7.4.4 The Relations on Either Side of the Shock

By taking into account the tangential projection of the momentum equation, we can write the three remaining equations in the form:

$$\rho_0 V_{n0} = \rho_1 V_{n1},$$
$$P_0 + \rho_0 V^2_{n0} = P_1 + \rho_1 V^2_{n1},$$
$$C_p T_0 + \frac{V^2_{n0}}{2} = C_p T_1 + \frac{V^2_{n1}}{2}.$$

These equations are in all respects similar to those of the normal shock (Sect. 7.3.2) provided the following substitutions take place, V_{n0} to U_0, and V_{n1} to U_1. By introducing the normal Mach number components:

$$M_{n0} = \frac{V_{n0}}{a_0} \equiv \sin\epsilon \times M_0 \quad \text{and} \quad M_{n1} = \frac{V_{n1}}{a_1} \equiv \sin(\varepsilon - \delta) \times M_0, \quad (7.37)$$

we can directly write from the normal shock results:

$$\frac{P_1}{P_0} = 1 + \frac{2\gamma}{\gamma + 1}\left(M_{n0}^2 - 1\right) ,$$
(7.38)

$$\frac{\rho_1}{\rho_0} = \frac{(\gamma + 1)\, M_{n0}^2}{(\gamma - 1)\, M_{n0}^2 + 2} ,$$
(7.39)

$$\frac{T_1}{T_0} = \frac{\left(2\gamma M_{n0}^2 + 1 - \gamma\right)\left[(\gamma - 1)\, M_{n0}^2 + 2\right]}{(\gamma + 1)^2\, M_{n0}^2} .$$
(7.40)

The previous relations highlight a first difference between the *normal* and *oblique* shock. In the first case, the function variations depend on a single parameter: the upstream Mach number. On the other hand, for an oblique shock, it appears that:

> Changes in flow characteristics when crossing through an oblique shock due to a given deflection δ, depend on *two* parameters:
> – the upstream Mach number M_0;
> – the shock wave angle ε.

It is therefore necessary to carry out a more detailed study by explaining the dependencies with respect to these two parameters.

7.4.5 The Shock Angle as a Function of the Deflection and Upstream Mach Number

The first issue to be considered is the shock angle determination as a function of the deflection, for a given upstream Mach number M_0. The following relations can be deduced from the previous equations:

$$\frac{\tan(\epsilon - \delta)}{\tan \epsilon} = \frac{(\gamma - 1)\, M_0^2 \sin^2 \epsilon + 2}{(\gamma + 1)\, M_0^2 \sin^2 \epsilon} \quad \text{or} \quad \frac{1}{\tan \delta} = \left(\frac{\gamma + 1}{2}\; \frac{M_0^2}{M_0^2 \sin^2 \epsilon - 1} - 1\right)\tan \epsilon .$$
(7.41)

For a given upstream Mach number, relation (7.41) allows the shock angle to be determined as a function of the deflection angle. The generic result is graphically plotted in Fig. 7.14.

From the curve plotted in Fig. 7.14, the first following results can be pointed out:

– a cut-off value, δ_{max}, of the deflection angle appears, beyond which there is no solution with an *attached* oblique shock.
– for a deflection $\delta < \delta_{max}$, *two* shock angles are possible: for the highest, we speak of a *strong* shock wave, for the lowest, a *weak* shock wave.
– the strong shock limit with zero deflection ($\delta = 0$ and $\epsilon = \pi/2$) corresponds to the case of an attached *normal* shock wave.
– the weak shock limit at $\delta = 0$ corresponds to a Mach wave, i.e., a focalising zone for infinitesimal perturbations from a source in a supersonic flow, as we have introduced in Sect. 7.1.3. It is easily verified, using the first relation (7.41), that the

Fig. 7.14 Typical graph of the shock angle variation law as a function of the deflection, for a fixed upstream Mach number ($M_0 = 1.8, \gamma = 1.4$)

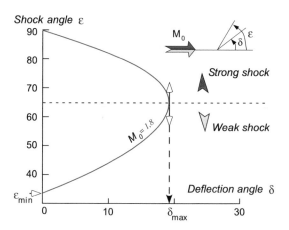

shock angle value ϵ_{min}, is such that $\sin \epsilon_{min} = 1/M_0$. This is the Mach cone angle at M_0, as defined by Eq. (7.10).

Proof To derive the first Eq. (7.41), we note that, according to the continuity equation:

$$\frac{\rho_1}{\rho_0} = \frac{V_{n0}}{V_{n1}} = \frac{V_{t1} \times \tan \epsilon}{V_{t0} \times \tan (\epsilon - \delta)} \equiv \frac{\tan \epsilon}{\tan (\epsilon - \delta)} .$$

Thus the previously demonstrated density ratio, Eq. (7.39), directly provides the first form of the expected relation. By expanding $\tan(\epsilon - \delta)$, we deduce

$$\frac{\tan\epsilon - \tan\delta}{\tan\epsilon (1 + \tan\epsilon \tan\delta)} = \frac{(\gamma - 1) M_0^2 \sin^2 \epsilon + 2}{(\gamma + 1) M_0^2 \sin^2 \epsilon},$$

which, after transformations, leads to the second expression of equations (7.41).

7.4.6 Mach Numbers on Either Side of the Shock

The relation between the Mach numbers on either side of an oblique shock is directly obtained by applying the normal shock relation, Eq. (7.16), to the Mach number's normal components. We immediately obtain:

$$M_1^2 \sin^2 (\epsilon - \delta) = \frac{(\gamma - 1) M_0^2 \sin^2 \epsilon + 2}{2\gamma M_0^2 \sin^2 \epsilon + 1 - \gamma} . \tag{7.42}$$

The shock angle ϵ being itself a function of M_0 and δ, can be eliminated from the above expression by combining it with Eq. (7.35).

The graphical solution consists of a set of curves which provides the downstream Mach number value M_1 as a function the upstream Mach number (M_0), for a given deflection angle (δ), as outlined in Fig. 7.15. From this figure, it is clear that, for a given deflection angle, the *attached* wave does not exist below a threshold value $M_{0\,lim}$ of the upstream Mach number. Beyond this value, which corresponds to the respective vertex of each curve, two situations are possible which identify two distinct flow regimes downstream of the shock wave, characterized by two different Mach number values M_1, relative to the strong and weak shock.

Finally, it can be verified that the supersonic flow regime downstream of an oblique shock is linked with the smaller of the two possible shock angle values which corresponds to the relatively smaller variations in the flow parameters through the shock.

Remark It is important to remember that, by definition, the Mach number refers to the local value of the velocity vector **module**. For this reason, the total enthalpy conservation can be written in the form:

$$\frac{\mathbf{a}^2}{\gamma - 1} + \frac{V_n^2}{2} = \frac{\gamma + 1}{\gamma - 1}\,\mathbf{a}_c^2,$$

which is wrong. On the other hand, it is correct to write

$$\left[\frac{\mathbf{a}^2}{\gamma - 1} + \frac{V_n^2}{2}\right]_0 = \left[\frac{\mathbf{a}^2}{\gamma - 1} + \frac{V_n^2}{2}\right]_1. \tag{7.43}$$

This observation also illustrates the fact that, if the tangential velocities are unchanged on either side of an oblique shock, the Mach number based on this tangential component— $M_{t0} = V_{t0}/\mathbf{a_0}$ and $M_{t1} = V_{t1}/\mathbf{a_1}$ respectively—is not preserved, due to the temperature variation and therefore the change in the speed of sound.

In order to highlight the distinction between both solutions, we extend the plotting near the vertex of a given curve. We can directly verify from Eq. (7.42) that the high ϵ values (*strong shock*) correspond to the lower branch of the curve in Fig. 7.16. The vertex location below the horizontal line $M_1 = 1$ proves that, after a *strong shock*, the flow regime is always *subsonic*. After a weak shock, the flow is essentially *supersonic*, except in a restricted area between the vertex ordinate line and the sonic line (see Fig. 7.16).

7.4.7 Summary

To summarize the overall results of the two previous sections, we return to the graphical representation of relation (7.41) plotted in the general form in Fig. 7.17.

With reference to this figure, it appears that:

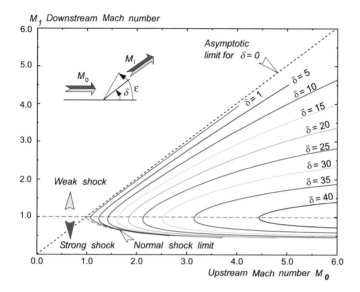

Fig. 7.15 Mach number relation for oblique shocks at a given deflection ($\gamma = 1, 4$)

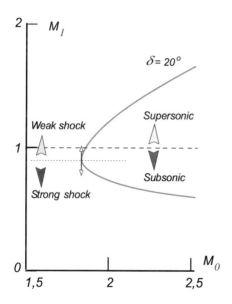

Fig. 7.16 Flow regimes associated with strong and weak shocks

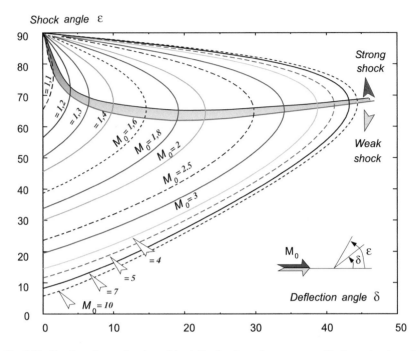

Fig. 7.17 Shock angle as a function of the deflection for a given upstream Mach number ($\gamma = 1, 4$)

- At a given upstream Mach number, an *attached* oblique shock wave can only form below a threshold value of the deflection angle. Beyond this limit, the deflection results in a sufficiently high fluid deceleration to generate a *detached* shock wave upstream of the corner apex. This is the situation schematically outlined in part (b) of Fig. 7.18.
- For a deflection $\delta < \delta_{max}$, *two* oblique wave angles are possible, associated with a *strong*, high angle shock or a *weak*, small angle shock, as shown in part (a) of Fig. 7.18.
- Downstream of a *strong* shock, the flow always reverts to the subsonic regime.
- Downstream of a *weak* shock, the flow remains supersonic, *except over a limited deflection range slightly below the maximum angle.*

To complete the presentation, an alternative plot of the general relation $f(M_0, \epsilon, \delta) = 0$—Eq. (7.41)—is provided in the next figure, where the shock angle is plotted as a function of the Mach number, for a given deflection angle.

Again, it is clear that, for a given upstream Mach number, there is a maximum deflection angle δ_{max}, below which two shock angles are possible.

ADDITIONAL INFORMATION—The Strong/Weak shock—When present, oblique shock wave can take place for two distinct shock angle values. This *theoretical*

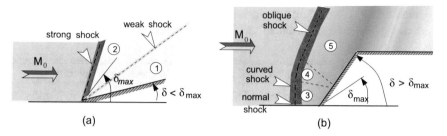

Fig. 7.18 Schematic for flow patterns around a sharp front end body for a given Mach number

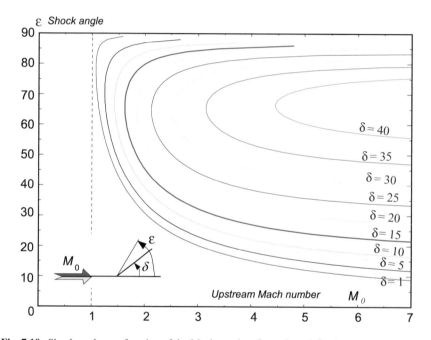

Fig. 7.19 Shock angle as a function of the Mach number, for a given deflection

result raises the question of how the flow will behave in *reality*: alternating positions between both values, selection of a preferential angle, etc. The detailed study of this issue involves stability concepts which are beyond the scope of this work and we will simply mention here the following observation facts:

– The most common configuration in practice is that of the *weak* shock, an occurrence which could be inferred from differences in the stability considerations and entropy variations with respect to the *strong* shock;

– In *internal* flows, it is possible to induce the presence of a stable strong shock by adjusting, independently of the upstream conditions, the flow counter pressure downstream of the shock.

– In *external* aerodynamics, the "*naturally*" selected flow pattern corresponds to the weak shock solution; a *strong* shock may nevertheless appear on the upstream tip of a blocked turbomachine's air intake, see Candel [25].

EXERCISE 29 The oblique shock and the supersonic stream flow velocity.

To evaluate the Mach number of a supersonic air flow, a probe is introduced parallel to the streamlines (zero incidence) with a dihedral leading edge whose apex angle is 20 degrees. Oblique plane shock wave is then formed whose angle with respect to the flow direction is measured at an angle of 42 degrees.

QUESTION   Deduce from the previous data the value of the flow Mach number.

SOLUTION: From relation (7.41) we can deduce that:

$$M_0^2 = \frac{2 \left(\tan\epsilon \times \tan\delta + 1 \right)}{\tan\epsilon \times \tan\delta \left[2 \sin^2 \epsilon \left(\dfrac{\tan\epsilon \times \tan\delta + 1}{\tan\epsilon \times \tan\delta} \right) - \gamma - 1 \right]}.$$

Considering the flow symmetry around the obstacle, the resulting deflection angle is 10 degrees. The angles ϵ and δ being known, the previous relation provides $M_0^2 = 14.597/4.136$, which finally gives $M_0 \approx 1.9$. We can observe that this value corresponds to that which can be graphically inferred from Fig. 7.19.

7.5 Practical Considerations

7.5.1 The Shock Formation in External Aerodynamics

We will have the opportunity to discuss in detail in the next chapter the conditions under which a shock wave can occur in duct flows. We will therefore only discuss here external aerodynamic configurations without entering into in-depth considerations of a more advanced level of expertise.

As we have demonstrated, the shock wave phenomenon is associated with (i) an *irreversible* evolution of (ii) a *supersonic* flow.

In practice, both conditions are fulfilled when, for example in the laboratory reference frame, a fluid moving in a wind tunnel with a supersonic Mach number reaches a fixed obstacle. Several shock configurations are then observed depending on the obstacle geometry (the shape of the airfoil leading edge, in particular) and the upstream Mach number.

• The first situation we will describe is likely to occur on a wing of a commercial aircraft in the transonic flight regime, with a subsonic upstream flow at a Mach number slightly lower than unity.

In this case, for the reasons explained in Chap. 6, we know that the fluid accelerates along the wing extrados.

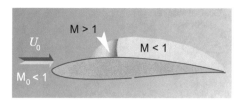

Fig. 7.20 Extrados shock in a locally supersonic flow past an airfoil

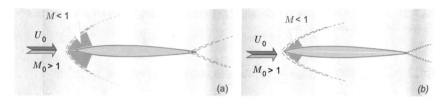

Fig. 7.21 Schematic for the shock wave systems on an airfoil in a supersonic flow

For a cruise Mach number M_0 between 0.85 and 1, in practice this acceleration can lead to the development of a *locally supersonic* flow area.

An extrados shock wave is then formed downstream of the leading edge (Fig. 7.20).

The wave positioning on the airfoil can exhibit a *strongly oscillatory* character, when its presence is coupled with an unsteady, almost periodic, vortex shedding at the trailing edge. It is the *transonic buffet phenomenon* which can cause noise pollution, mechanical fatigue of the structure and passenger discomfort.

• The second situation corresponds to the flows on fighter aircraft wings in the supersonic flight domain, where the flow at the leading edge is recompressed through a shock wave which is only very locally a normal shock. Depending on the Mach number value, this wave can be detached—Fig. 7.21a—, or attached to the profile, at a higher Mach number. At the same time, the shock waves on the extrados shift to the trailing edge[9] with an increasingly oblique direction with respect to the airfoil. The flow develops a *"fishtail shock wave"* pattern characteristic of this type of regime.

• The shock positioning also plays an important role in the in-flight operation of turbomachines which power the aircraft on a supersonic flight. Thus for a turbojet engine, the central body is at the origin of a first external fluid compression by passing through an oblique shock wave, Fig. 7.22a. In normal operation, it is a *weak* shock wave. It can become a strong shock when the air intake is blocked, Fig. 7.22b. The situation is then more like that of a scramjet (see next Chapter, Sect. 6.2.4) when the recompression of the fluid coming into the machine occurs through a normal shock wave, Fig. 7.22c.

---

[9] This location is that of the ideal fluid approach. In real fluid, the wave shifts slightly upstream of the trailing edge due to viscosity effects in the boundary layer.

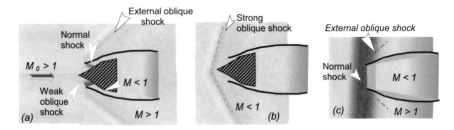

Fig. 7.22 Air intake shock waves of various jet engines: turbojet engine in normal (**a**) and blocked (**b**) operation; scramjet (**c**)

7.5.2 Wave Reflection—Shock-Boundary Layer Interaction

In practice, we are often facing the problem of a shock wave's impact on a solid wall. In an ideal fluid, the answer is easy to predict. Indeed, we have explained that when passing through an oblique shock wave, the flow is deflected towards the plane of the shock. Since at the wall, the normal velocity component is necessarily zero, it is deduced that at the impact point, the shock wave is reflected in such a way that the new flow deviation cancels out that experienced during the crossing of the incident wave, as outlined in part (a) of Fig. 7.23.

In a real fluid, the previous picture is modified by the fluid viscosity, Fig. 7.23b. As we will explain in Chap. 11, its effects, at high Reynolds numbers, are restricted to an area close to the wall called the *boundary layer*. It is in this region that the fluid motion is stopped, in a reference frame linked to a fixed obstacle. It is therefore, at least in part, the location of a subsonic regime. Hence, when the incident wave impacts the boundary layer, it causes a high pressure increase which, in subsonic conditions, can propagate upstream and cause the boundary layer to thicken. This change in thickness can, in turn, be understood as a deflection source for the external flow. The reflected wave no longer originates precisely at the impact point on the wall, but at a position shifted upstream, the flow adaptation to the fluid deceleration near the boundary layer being gradually reached through a Mach wave beam.

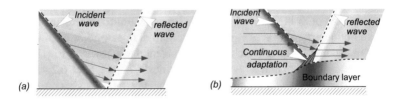

Fig. 7.23 Oblique shock wave reflection at a solid surface, **a** ideal fluid, **b** real fluid

Fig. 7.24 Schematic for the shock wave reflection with a triple point

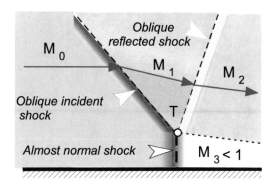

ADDITIONAL INFORMATION—**Wave reflection at a solid surface**—In terms of the ideal fluid approach, shock wave reflection on a curved solid wall can lead to phenomena different from those just described on a flat surface.

However, even on such a flat surface, the situation is not always that of the simple *normal reflection* previously explained.

Indeed, under certain conditions, the incident oblique wave may not be reflected directly on the plane of the obstacle, see Fig. 7.24. An almost normal shock element develops in the immediate vicinity of the wall on which the reflected wave takes place. This is the so-called "*Mach reflection*" configuration, characterized by the presence of a *triple point*, T in the Figure, where the incident oblique wave, the reflected oblique wave and the almost normal wave meet.

Chapter 8
One-Dimensional Steady Flows of an Ideal Compressible Gas

Abstract This chapter is concerned with steady flows of an ideal compressible gas in geometries whose cross-section is constant or slowly variable. We will begin the study by assuming furthermore that the fluid is ideal (non-viscous and non-heat-conductive). This part will lead to the flow regime analysis in a De Laval nozzle.

We will continue by examining both cases of rectilinear duct flows of a constant cross-section (*i*) with head losses due to friction effects, called *Fanno flows* and (*ii*) with heat transfer but without dissipation, or *Rayleigh flows*.

8.1 Ideal Gas Flows in Slowly Varying Cross-Section Ducts

8.1.1 The Situation Under Consideration

In this first part of the chapter (Sects. 8.2–8.5), we will study the flow properties of a weightless, inviscid and non heat-conductive fluid (ideal fluid) under the following additional assumptions:

(a) the thermodynamic behaviour of an ideal gas ($P/\rho = rT$) ;
(b) a stationary motion ($\partial/\partial t \equiv 0$) ;
(c) a unidirectional flow, which means that:

 – any flow function only depends on one space variable, the streamwise x-coordinate;
 – the streamwise variation of the flow cross section $A\,(x)$, if it exists, remains slow enough ($dA/A \ll 1$) so as to not alter the previous proposal;
 – the flow curvature, if any, should also remain small (radius of curvature (R) such as $A/R^2 \ll 1$).

© The Author(s), under exclusive license to Springer Nature Switzerland AG 2022
P. Chassaing, *Fundamentals of Fluid Mechanics*,
https://doi.org/10.1007/978-3-031-10086-4_8

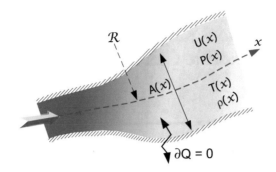

Fig. 8.1 Unidirectional flow with a slowly varying cross-section: physical domain

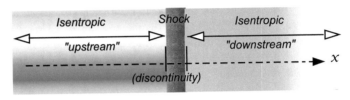

Fig. 8.2 Schematic for the regimes of a unidirectional flow with a slowly varying cross-section

This set of considerations leads to a flow configuration as outlined in Fig. 8.1. The $A(x)$ cross-section can be bounded by a solid wall. We are then dealing with an *internal duct* flow. It can also coincide with a stream tube located in an *external* flow past a solid body.

Under the previous assumptions, the evolution is necessarily *adiabatic* ($\delta Q = 0$), since the fluid can neither produce heat by *internal* dissipation, nor receive heat from the environment by *conduction*. As we will explain at a later stage (Sect. 8.1.2.1) it is also *reversible*, except in the presence of a shock. All the flows in question are therefore reduced to the same generic scheme, as outlined in Fig. 8.2. They are associated with an *isentropic* evolution, apart from the possible presence of a necessarily normal shock, (cf. Chap. 7).

8.1.2 The Isentropic Motion Equations

8.1.2.1 The Equation Model

The motion equations are obtained in a straightforward way by introducing the study assumptions in the ideal compressible fluid model (*ICF*) introduced in Chap. 4, Sect. 4.6.3. In addition to the equation of state for ideal gases, this model includes the following equations:

$$\textbf{Continuity :} \quad \rho U A = C^t \qquad (8.1)$$

$$\textbf{Momentum :} \quad \rho U \frac{dU}{dx} = -\frac{dP}{dx} \qquad (8.2)$$

$$\textbf{Enthalpy :} \quad \rho U C_p \frac{dT}{dx} = U \frac{dP}{dx} \qquad (8.3)$$

Except for the shock, the isentropic flow character validates Laplace's barotropy law and we will thus set:

$$\frac{P}{\rho^\gamma} = C^t . \qquad (8.4)$$

Remark *The isentropic character of the motion*—According to the enthalpy formulation of the energy balance, Eq. (8.3), we have $\rho\, dh = dP$, which demonstrates that the entropy variation $ds = dh - dP/\rho$ is indeed identically zero. The *differential* evolution calculation on thermodynamic equilibrium states can be extended to the *material variation* in a fluid motion, as defined in the first Chapter Sect. 1.6.1. This is the case, provided that the *local thermodynamic equilibrium* assumption can be applied at the mesoscopic scale, which is always the case here for the local flow functions, whose variation calculation throughout shock wave is, by hypothesis, excluded from the present study.

8.1.2.2 The Energy Conservation

By algebraically eliminating the pressure gradient between Eqs. (8.2) and (8.3), we immediately obtain

$$C_p dT + U dU = 0 .$$

The integration of this relation is straightforward, assuming that the fluid's specific heat capacity at constant pressure does not vary. It will then be easily verified that the result can take any of the following equivalent forms:

$$C_p T + \frac{U^2}{2} = C^t \qquad (8.5a)$$

$$\frac{\gamma r T}{\gamma - 1} + \frac{U^2}{2} = C^t \qquad (8.5b)$$

$$\frac{\gamma}{\gamma - 1}\frac{P}{\rho} + \frac{U^2}{2} = C^t \qquad (8.5c)$$

$$\frac{a^2}{\gamma - 1} + \frac{U^2}{2} = C^t \qquad (8.5d)$$

Equation (8.5c) has previously been reported in Chap. 5, Sect. 5.6.2.3, for the dynalpy conservation. It also allows retrieving Saint-Venant's relation, see Chap.

5, Sect. 5.6.2.2. By applying this equation between two local equilibrium states (1)
and (2) of the moving fluid, the relation still takes the following form:

$$U_2^2 - U_1^2 = \frac{2\gamma}{\gamma - 1}\frac{P_1}{\rho_1}\left[1 - \left(\frac{P_2}{P_1}\right)^{\frac{\gamma-1}{\gamma}}\right]. \tag{8.6}$$

Remark Relation (8.5c) can be equally directly derived from the momentum equa-
tion. Indeed, considering Laplace's formula, we have $dP/P = \gamma d\rho/\rho = K\gamma\rho^{\gamma-2}d\rho$,
by denoting $K = P/\rho^\gamma$. Thus the momentum equation becomes $UdU + K\gamma\rho^{\gamma-2}d\rho = 0$, whose integration gives the expected result.

We can also verify that for an incompressible environment, Saint-Venant's relation
returns Bernoulli's relation. Indeed for $\gamma \to \infty$, *Eq. (8.6) is written*

$$U_2^2 - U_1^2 = 2\frac{P_1}{\rho_1}\left(1 - \frac{P_2}{P_1}\right) \quad or \quad \frac{U_1^2}{2} + \frac{P_1}{\rho_1} = \frac{U_2^2}{2} + \frac{P_2}{\rho_1},$$

which is no more than Bernoulli's relation for a density fluid ρ_1.

8.2 Flow Variation Laws with the Cross-Section Area

8.2.1 The Logarithmic Differential Formulation

From the motion equations it is possible to express the logarithmic variations dU/U,
dP/P, dT/T, $d\rho/\rho$ and dM/M in the form:

Velocity:	$\dfrac{dU}{U} = \dfrac{1}{M^2 - 1}\dfrac{dA}{A}$	(8.7a)
Pressure:	$\dfrac{dP}{P} = -\gamma\dfrac{M^2}{M^2 - 1}\dfrac{dA}{A}$	(8.7b)
Temperature:	$\dfrac{dT}{T} = (1 - \gamma)\dfrac{M^2}{M^2 - 1}\dfrac{dA}{A}$	(8.7c)
Density:	$\dfrac{d\rho}{\rho} = -\dfrac{M^2}{M^2 - 1}\dfrac{dA}{A}$	(8.7d)
Mach number:	$\dfrac{dM}{M} = \dfrac{2 + (\gamma - 1)M^2}{M^2 - 1}\dfrac{dA}{A}$	(8.7e)

Proof First of all, it should be noted that, due to the isentropic nature of the evolution,
we can write:

$$\frac{dP}{P} = \gamma\frac{d\rho}{\rho}.$$

Introducing the speed of sound under the local motion conditions $\mathbf{a}^2 = \gamma P/\rho$, the previous relation directly yields $dP = \mathbf{a}^2 d\rho$. The momentum equation can be put in the form $U dU + \mathbf{a}^2 d\rho/\rho = 0$. In addition, the continuity equation provides:

$$\frac{d\rho}{\rho} + \frac{dU}{U} + \frac{dA}{A} = 0 .$$

By eliminating $d\rho/\rho$ between these last two relations, this directly leads to the first relation for dU/U. The differential expression $d\rho/\rho$ is immediately deduced from this thanks to the previous intermediate result, namely $d\rho/\rho = -M^2 dU/U$. The pressure relation results by simply using the barotropic relation $(dP/P = \gamma d\rho/\rho)$. As for the temperature, we can use the equation of state which gives:

$$\frac{dT}{T} = \frac{dP}{P} - \frac{d\rho}{\rho} .$$

Finally, returning to its definition, we can write for the Mach number:

$$\frac{dM}{M} = \frac{dU}{U} - \frac{d\mathbf{a}}{\mathbf{a}} = \frac{dU}{U} - \frac{1}{2}\frac{dT}{T} ,$$

since $\mathbf{a}^2 = \gamma r T$. After substituting the previous results, the expected differential expression is obtained.

8.2.2 Discussion and Physical Interpretation

All the previous relations demonstrate that the variation laws of the different functions are inverted according to whether the Mach number is lower or higher than one. This observation leads to the statement of the following first two theorems for isentropic one-dimensional flows:

> In a subsonic isentropic flow $(M < 1)$, the velocity and cross-section exhibit inverse variations, while the pressure, temperature[a] and density vary directly with the section area;

---

[a] As for the temperature, the conclusion on the variation direction with the section area requires that $\gamma > 1$.

Fig. 8.3 Isentropic
cross-section variation with
respect to the Mach number

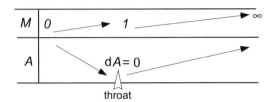

In a supersonic isentropic flow ($M > 1$), the velocity varies directly with the cross-section, while the pressure, temperature[1] and density vary inversely with the section area.

The third theorem specifies the nature of the evolutions in the vicinity of the singularity at $M = 1$ in the previous logarithmic variation expressions

Whenever an isentropic flow has sonic characteristics ($M = 1$), these are associated with a minimum cross-section or *throat*.

This statement is a direct consequence of the last relation Eq. (8.7e) which proves that the sign of dM/dA is that of $M - 1$.
We have therefore necessarily the following variation graph in Fig. 8.3.

HISTORICAL NOTE—**The sound barrier**—In external aerodynamics, the variation law singularity around $M = 1$ was at the origin of a painful page in the history of aeronautics since it was the source of a number of accidents. This is the famous epic which began in 1945 to the breaking the *sound barrier*. This expression originated in the sudden changes in the aerodynamic forces applied to an aircraft in the transonic regime. It seemed as if the atmosphere was made more compact by the relative aircraft speed by modifying the intensity and position of the lift and drag on the airfoil. This phenomenon, along with the technical and flight control difficulties implied, was not fully understood and mastered until several years after the first breaking of the sound barrier in level flight by the American Chuck Yeager, at the controls of the Bell X-1 rocket aircraft on October 14, 1947. He was assisted in his venture by Jack Ridley, second in command, who had been a student of Theodor von Kármán at Caltech. As for the aircraft itself, it was an experimental aircraft, powered by four liquid oxygen and alcohol rocket-engines and dropped at an altitude of 7,500 m by a Boeing B-29 Superfortress.

The first European jet aircraft to break the sound barrier in a slight downward flight was the French *Nord 1402A "Gerfaut 1"*, on August 3, 1954. It was a massive aircraft, 11.39 m long and with a wingspan of 7.5 m, powered by an Atar 101C engine. It was operated by one named André Turcat, who was later to become the Concorde test pilot. It was on March 3, 1955 that a Western European aircraft reached Mach 1 on a level flight. It was Dassault's *"Super Mystère B1"*.

The technical challenges of the time were first and foremost related to motorization. Indeed, the piston engine had almost reached its limits towards the end of the Second World War and, at the beginning of the 1950s, the new propulsion modes (rocket, turbojet) still did not have a high level of reliability.

The second technological challenge was the design of airfoils adapted to the supersonic flight. To acquire a better understanding of the external aerodynamic problems involved, it is sufficient to mention the following two points:

- Wing shapes which perform well on a subsonic flight become catastrophic in the transonic flight domain due to an an increase in the drag. The *knife blade* airfoils adapted to the supersonic flight, on the other hand, have a lift to drag ratio which is barely half that on a subsonic flight. When the engine fails, the plane soars like an *iron*.
- By breaking from the subsonic to the supersonic regime, the aircraft's longitudinal stability is altered. Very schematically, in the first case (subsonic), the lift force applies on the wing at a point located at the first quarter upstream of the chord. It shifts to the middle of the latter on a supersonic flight as a result of the modification of the pressure distribution law. The result is a diving moment (the aircraft tends to pitch downwards in the vertical plane) which requires appropriate control from the pilot.

8.3 The Isentropic Flow Laws

8.3.1 The Stagnation Condition Concept

We have previously mentioned, in Chap. 7, Sect. 7.3.7.3, the *stagnation condition* concept. It is easily understood if we base ourselves on the following practical considerations.

To generate a given one-dimensional duct flow with a prescribed cross-section variation law $A(x)$, a simple technique is to store the gas in a reservoir after heating and/or compressing it. This provides an energy value associated with a *zero velocity* value, hence excluding any kinetic energy form. This situation is by definition referred to as the *reservoir stagnation condition* or more briefly *stagnation condition*.

Any other function value (pressure, temperature, density) under these conditions is accordingly referred to as a stagnation or generating value, and will be identified by a symbol with the index ($_i$).

For any isentropic evolution, the stagnation values P_i, ρ_i, T_i remain constant along the motion. In other words, it is only necessary to *isentropically* decelerate the flow to a zero velocity in order to retrieve the values in question. It is for this reason that it is appropriate to speak of *isentropic* stagnation conditions.

Remark *The conservation of isentropic stagnation values*—For the reservoir condition $U_i = 0$, the energy equation gives $T_i = C^t$. As a result, $P_i/\rho_i = C^t$, which, together with Laplace's relation $P_i/\rho_i^\gamma = C^t$ leads to $P_i = C^t$ *and* $\rho_i = C^t$.

8.3.2   *Expressions Based on Stagnation Conditions*

For an isentropic flow, the local differential equations of the motion can be integrated. Taking the stagnation conditions as reference values, the different flow functions, in a dimensionless form T/T_i, P/P_i, ρ/ρ_i, $\mathbf{a}/\mathbf{a}_i$, can be expressed in terms of the sole local Mach number.

$$\frac{T_i}{T} = 1 + \frac{\gamma - 1}{2} M^2 \tag{8.8a}$$

$$\frac{P_i}{P} = \left[1 + \frac{\gamma - 1}{2} M^2 \right]^{\gamma/(\gamma-1)} \tag{8.8b}$$

$$\frac{\rho_i}{\rho} = \left[1 + \frac{\gamma - 1}{2} M^2 \right]^{1/(\gamma-1)} \tag{8.8c}$$

$$\frac{\mathbf{a}_i}{\mathbf{a}} = \left[1 + \frac{\gamma - 1}{2} M^2 \right]^{1/2} \tag{8.8d}$$

The graphical representations of the previous expressions are plotted in the following Fig. 8.4.

Proof The first integrated form of the energy equation allows us to write $C_p T + \frac{U^2}{2} = C_p T_i$, hence $\frac{T_i}{T} = 1 + \frac{\gamma r}{2C_p} \frac{U^2}{\mathbf{a}^2} = 1 + \frac{C_p \left(C_p - C_v \right)}{2C_v C_p} \frac{U^2}{\mathbf{a}^2}$, from which it is

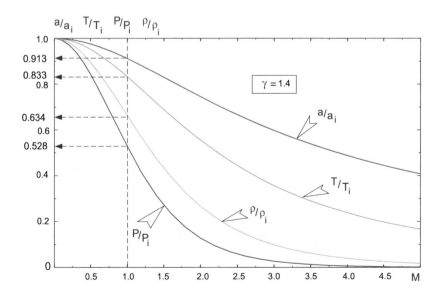

Fig. 8.4 Function variations with respect to the local Mach number in an isentropic flow

easy to deduce the first relation. Now considering the third integrated form of the energy equation, we have

$$\frac{\gamma}{\gamma-1}\frac{P}{\rho}+\frac{U^2}{2}=\frac{\gamma}{\gamma-1}\frac{P_i}{\rho_i} \quad \text{hence} \quad \frac{U^2}{2a^2}+\frac{1}{\gamma-1}=\frac{1}{\gamma-1}\frac{P_i\,\rho}{P\,\rho_i}.$$

Now, in an isentropic evolution, we know that $\rho/\rho_i = (P/P_i)^{1/\gamma}$, which, after substitution in the previous relation, leads to the expected result. The relation for ρ_i/ρ is immediately deduced from this.

EXERCISE 30 The isentropic flow from a reservoir

A reservoir is filled with dry air, considered as an ideal gas with $R = 287\ J \times Kg^{-1} \times K^{-1}$. The pressure in the reservoir is 10^6 Pa (about ten times the atmospheric pressure) and the temperature 300 K. The air discharges from the reservoir into the atmosphere through a convergent-divergent axisymmetric nozzle whose throat section is 31 mm in diameter. The exit flow Mach number is $M_s = 2$.

QUESTION 1. Calculate the fluid temperature at the nozzle throat section.
QUESTION 2. Calculate the flow velocity at the nozzle throat section.
QUESTION 3. Calculate the mass-flow rate.
QUESTION 4. Calculate the velocity and density at the nozzle exit.
QUESTION 5. Determine the nozzle exit area.
QUESTION 6. Calculate the flow pressure and temperature at the exit. Comment the results.

SOLUTION: **1.** Taking into account the Mach number exit value, the flow is necessarily sonic at the throat ($M_c = 1$). Thus, the flow is isentropic everywhere and we can calculate the temperature T_c at the throat section according to:

$$\frac{T_i}{T_c}=1+\frac{\gamma-1}{2}M_c^2=1+\frac{\gamma-1}{2}.$$

Taking $\gamma = 1.4$ we find $T_c = 300/1.2 = 250$ K, or about $-23\,°C$.

2. The throat velocity is equal to the speed of sound in this section, i.e. $U_c = \sqrt{\gamma r T_c}$. With the numerical data for use in the exercise, we find $U_c = 317$ m/s.

This result can be obtained differently without having to determine the throat temperature. Indeed, by applying the second integrated form of the energy equation at the reservoir and throat conditions, we obtain:

$$\frac{\gamma r T_c}{\gamma-1}+\frac{U_c^2}{2}=\frac{\gamma r T_i}{\gamma-1}.$$

Now, since the flow is sonic at the throat, we know that $\gamma r T_c = \mathbf{a}_c^2 \equiv U_c^{\Theta}$. Thus, the previous relation becomes:

$$\frac{\gamma + 1}{2\,(\gamma - 1)} U_c^2 = \frac{\gamma r T_i}{\gamma - 1} \qquad \text{hence} \qquad U_c = \sqrt{\frac{2\gamma r T_i}{\gamma + 1}}\,.$$

We verify that the numerical value is the same as the previous result.

3. To calculate the mass-flow rate, we will firstly determine the throat density value. For $M_c = 1$ the isentropic flow density ratio gives:

$$\frac{\rho_c}{\rho_i} = \left(\frac{2}{\gamma + 1}\right)^{1/(\gamma - 1)} = 0.634\,.$$

As for the stagnation density, it is simply, $\rho_i = P_i/rT_i$, which gives here $\rho_i = 11.60\ Kg/m^3$. The result is $\rho_c = 7.35\ Kg/m^3$.

The mass flow rate value, expressed on the throat condition basis, is $q_m = \rho_c U_c A_c$. With the relevant numerical data, we find $q_m = 1.76\ Kg/s$.

4. The flow being entirely isentropic, the relations expressing ρ_i/ρ and a_i/a are written

$$\frac{\rho_i}{\rho_s} = (2\gamma - 1)^{\frac{1}{\gamma - 1}} = 4.347 \qquad \text{and} \qquad \frac{a_i}{a_s} = (2\gamma - 1)^{1/2} = 1.342\,.$$

Thus, $\rho_s \approx 2.67\ kg/m^3$ and $a_s \approx 259$ m/s. Since the exit Mach number is 2, the flow velocity is $U_s \approx 518$ m/s.

5. The mass-flow rate conservation between the throat and outlet sections results in $\rho_c U_c A_c = \rho_s U_s A_s$. Considering the previous results, we deduce that $A_s \approx 1272\ mm^2$, which corresponds to a diameter slightly greater than 40 mm.

6. Using the isentropic flow relations for the pressure and temperature, we have:

$$\frac{P_i}{P_s} = (2\gamma - 1)^{\frac{\gamma}{\gamma - 1}} = 7.824 \qquad \text{and} \qquad \frac{T_i}{T_s} = 2\gamma - 1 = 1.8\,.$$

With the relevant numerical data, we find $P_s \approx 1.28 \times 10^5$ Pa and $T_s \approx 267$ K, or again $-106\,°C$.

The flow discharging into the atmosphere ($P_\infty \approx 10^5$ Pa) will therefore undergo a slight expansion leading to a correlative decrease in temperature. In other words, for the considered stagnation conditions, it would be possible, by slightly extending the divergent length, to accelerate the flow beyond Mach = 2 in order to isobarically discharge into the atmosphere, namely with $P_s = P_\infty$. As we will see at a later stage (cf. Sect. 8.4.1.3), the flow will then behave like a *supersonic cold jet*.

8.3.3 Expressions Based on Critical Conditions

By definition of the very stagnation conditions, it is not possible to extend relations (8.8a to 8.8d) to the section ratio when using the reservoir condition. Consequently *critical* or *sonic throat* conditions are also introduced, such as, by assumption:

$$\boxed{Critical \text{ or } sonic \text{ condition} \leftrightarrow A = A_c, \; with \; M = M_c = 1}$$

The isentropic flow relations based on the sonic throat condition are written:

$$\frac{U}{\mathbf{a_c}} = \sqrt{\frac{(\gamma+1)\,M^2}{2+(\gamma-1)\,M^2}} \qquad (8.9a)$$

$$\frac{A}{A_c} = \frac{1}{M}\left[\frac{2}{\gamma+1}\left(1+\frac{\gamma-1}{2}M^2\right)\right]^{\frac{\gamma+1}{2(\gamma-1)}} \qquad (8.9b)$$

Remark It is important to note that it is not necessary to ensure the effective presence of a real throat in the actual flow geometry in order to be able to introduce the critical condition. In fact, such a condition can be used as an ***operative reference*** associated with a fictitious sonic throat, appropriate for the sake of the calculations.

Proof To demonstrate relations (8.9a) and (8.9b) we write

$$\frac{U}{\mathbf{a_c}} = \frac{U}{\mathbf{a}}\times\frac{\mathbf{a}}{\mathbf{a_c}} \equiv M\times\frac{\mathbf{a}}{\mathbf{a_c}}.$$

Now, according to Eq. (8.5d) we know that $\dfrac{U^2}{2}+\dfrac{\mathbf{a}^2}{\gamma-1}=\dfrac{\mathbf{a}_c^2}{2}+\dfrac{\mathbf{a}_c^2}{\gamma-1}=\dfrac{\gamma+1}{\gamma-1}\mathbf{a}_c^2.$

Dividing by $\mathbf{a}_c^2$, we deduce that

$$\frac{\mathbf{a_c}}{\mathbf{a}} = \left(\frac{2+(\gamma-1)\,M^2}{\gamma+1}\right)^{1/2},$$

which directly leads to the first relation. To obtain the second equation, it is only required to write, in direct application of the continuity relation

$$\frac{A}{A_c} = \frac{\rho_c}{\rho}\times\frac{\mathbf{a_c}}{U},$$

where the ratios on the right hand side can be substituted by the previous known expressions, with $\rho_c/\rho \equiv [\rho_c/\rho_i]\times[\rho_i/\rho]$.

8.3.4 Physical Discussion

Dealing with a practical situation where the duct geometry is given and the flow generated under prescribed conditions, the main question is to determine the Mach number in any section A.

One might be inclined to conclude that the answer is provided by the direct application of the previous relation (8.9b).

In fact, it can be observed that this is not the case, due to the *non-bijective* nature of the relation $A_c/A = f(M)$.

This characteristic of relation (8.9b) appears clearly on the upper part of the following figure, where the function $A_c/A = f(M)$ is plotted. Thus, for a given single section ratio, it generally matches two distinct Mach number values, except for $M = 1$. These are associated with two different flow regimes, one subsonic and the other supersonic.

The lower part of this same Figure proves that these two regimes correspond to two distinct values of the pressure ratio, since the relation $P/P_i = f(M)$ is bijective. Thus, for the same section ratio, the flow will be subsonic if the pressure ratio is greater than 0.528 for the air, and otherwise it is supersonic.

EXERCISE 31 The subsonic flow in a diffuser

We consider a diffuser or duct whose cross-section increases in size in the flow direction. The outlet/ inlet section ratio is taken as $A_s/A_e = 1.54$. Compressible air is steadily flowing through this diffuser, with a Mach number $M_e = 0.51$.

QUESTION 1. What is the Mach number value of the flow at the diffuser outlet? (For the calculations, we will use a second-order approximation of the relation $M = g(A/A_c)$.)

QUESTION 2. Deduce the outlet/inlet static pressure ratio P_s/P_e.

SOLUTION: **1.** Let us introduce a fictitious throat corresponding to the critical (sonic) condition such that the flow is isentropic between the throat section (A_c) and the inlet section (A_e). For $M_e = 0.51$, the function $A/A_c = f(M)$ gives $A_e/A_c \approx 1.32$. We can then write that:

$$\frac{A_s}{A_c} \equiv \frac{A_s}{A_e} \times \frac{A_e}{A_c} = 1.54 \times 1.3212 \approx 2.03 \,.$$

We can deduce the Mach number value by reversing the relation $A/A_c = f(M)$. For the air ($\gamma = 1.4$), this results in:

$$\frac{A}{A_c} = \frac{1}{M}\left(\frac{1 + 0.2M^2}{1.2}\right)^3,$$

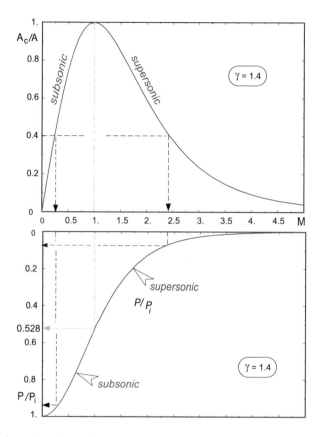

Fig. 8.5 Section ratio and expansion rate as a function of the Mach number for an isentropic flow

or, with the suggested approximation $1.2^3 \times \dfrac{A}{A_c} \times M = \left(1 + 0.2M^2\right)^3 \simeq 1 + 0.6M^2 + \mathcal{O}\left(M^4\right)$. With the numerical data for use in exercise, the Mach number is therefore obtained as a solution of the following second degree equation $0.60M^2 - 3.52M + 1 = 0$. This equation has two roots which are respectively equal to $M_1 = 0.3$ and $M_2 = 5.5$. The flow being subsonic at the diffuser inlet is necessarily subsonic at the outlet (the first isentropic flow theorem on the variation with the section), hence the solution $M_s = M_1 \approx 0.3$. It can be verified that, for this value, the fourth-order correction is indeed negligible.

2. We can write $\dfrac{P_s}{P_e} = \dfrac{P_s}{P_i} \times \dfrac{P_i}{P_e}$, when introducing the stagnation pressure P_i. Using the isentropic flow relation $P/P_i = f(M)$, we know that $P_e/P_i = f(0.51) \approx 0.84$ and $P_s/P_i = f(0.3) \approx 0.94$. Hence, we obtain that $P_s/P_e \approx 1.12$.

8.4 Examples of Ideal Compressible Fluid Flows

8.4.1 The De Laval Nozzle

8.4.1.1 Device Description

The De Laval nozzle is a converging–diverging (CD) duct, with a straight axis, as illustrated in the following Figure. In this geometry which therefore presents a minimum section or *actual throat*, various flow regimes can exist.

8.4.1.2 The Internal Flow Regimes

We will firstly examine the various flow regimes which can take place in the nozzle duct, depending upon various prescribed reservoir conditions.
To make the discussion clearer, we will assume that we proceed as follows:

- the upstream nozzle end is connected to a reservoir where the gas is stored at a stagnation temperature T_{i0} and pressure P_{i0} which are assumed to be constant.
- the downstream diffuser section opens into an almost infinite vessel, whose static pressure P_a can be can be continuously modified.

$\boxed{\text{Case n}^\circ\ 1 : P_a \equiv P_{a0} = P_{i0}}$ This first case results in the absence of flow in

the nozzle, due to the static pressure equilibrium. From this situation, let us assume that a slightly lower value than P_{i0} is imposed on the downstream static pressure. It is referred to as P'_{a0}. Hence, the fluid is set in motion by this downstream suction. However, since the pressure difference $P_{i0} - P'_{a0}$ is assumed to be low, the fluid velocity will remain predominantly subsonic everywhere. In application of the first theorem on the isentropic flow evolution (Sect. 8.2.2), the fluid will accelerate in the convergent and decelerate in the divergent, the evolution remaining isentropic everywhere.

$\boxed{\text{Case n}^\circ\ 2 : P_{a1} \leqslant P_a < P_{i0}}$ By gradually decreasing the downstream static

Fig. 8.6 The De Laval nozzle typical geometry

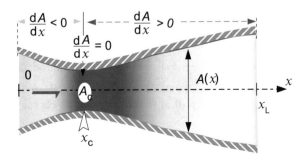

pressure, the previous situation will continue until the fluid acceleration in the convergent is sufficient to raise the flow velocity to a value which becomes exactly sonic at the throat. We will specify P_{a1} as the static pressure value for which this flow situation is reached. Thus, throughout the entire pressure range $[P_{a1}, P_{i0}]$ the mass flow rate of the fluid being sucked through the nozzle is continuously increasing. The flow regime remains *isentropic* and *subsonic* all along the nozzle, except for the marginal case ($P_a = P_{a1}$) where, by assumption, the *sonic* regime is reached at the throat.

• **The first critical regime** For $P_a = P_{a1}$, the inlet and outlet conditions at the convergent duct part are fixed since, according to the third theorem on the isentropic flow evolution (Sect. 8.2.2), the Mach number value at the throat cannot become greater than one. Consequently, the mass flow rate will not increase with a further decrease in the downstream static pressure environment below $P_a \leqslant P_{a1}$, for a fixed upstream stagnation pressure and temperature. This is a situation which corresponds to the choked flow regime, with P_{a1} denoting the first critical static pressure value. Hence, it can be stated that:

> The first critical regime in a CD nozzle is that of a fully isentropic and subsonic flow along the nozzle with a sonic condition at the throat.

Case n° 3 : $P_{a2} \leqslant P_a < P_{a1}$ As we have just explained, by further decreasing the downstream static pressure below the value P_{a1}, nothing is going to change in the flow pattern in the convergent and modifications will therefore take place in the divergent. When the applied pressure P'_{a1} is *slightly lower* than P_{a1}, the fluid acceleration carries on in the *divergent* where, according to the second isentropic flow theorem (Sect. 8.2.2), a supersonic region will form downstream of the throat. *This supersonic flow regime cannot be maintained up to the outlet section.* Indeed, if this were the case, the evolution would be *isentropic* throughout the divergent and therefore the velocity would continue to increase and the pressure to decrease. The latter would thus reach a level, at the nozzle outlet, well below P_{a1}, in contradiction with the assumed value at the beginning of this discussion. Hence the evolution can no longer be isentropic and the regime entirely supersonic along the divergent, which implies an internal adjustment to the subsonic regime. This takes the form of a standing shock[1] located in a given section of the diffuser.

• **The second critical regime**: As the pressure difference $P_{a1} - P'_{a1}$ increases, the supersonic regime domain will extend further downstream. There is therefore a situation where, for a particular value of the downstream static pressure, the shock is located at the outlet section. When this happens, we will refer to the static pressure just downstream of the shock at the nozzle exit P_{a2} as the second critical value. It can be stated that:

> The second critical regime in a CD nozzle is that of a fully isentropic flow, subsonic in the convergent and supersonic in the divergent.

---

[1] With the study assumptions, this will be a normal shock.

It is worth retaining that for the second critical regime, the static pressure at the end of the divergent, just upstream of the shock ($P_s \equiv P_{a3}$), is not equal to that of the downstream environment P_{a2}.

The adaptation to this value takes the form of a sudden recompression (normal shock) in the outlet section, see Chap. 7, Eq. (7.19). Let us finally specify that, for dry air ($\gamma = 1.4$), the ratios to the stagnation conditions of the pressure, temperature, density and sound velocity at the throat nozzle take the following numerical values at the second critical point:

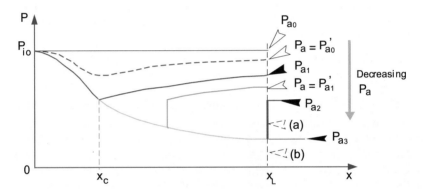

Fig. 8.7 Static pressure distribution along a convergent-divergent nozzle

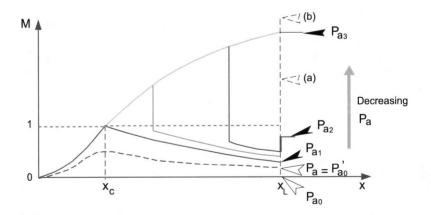

Fig. 8.8 Mach number distribution along a convergent-divergent nozzle

$$\frac{T_c}{T_i} = \frac{2}{\gamma + 1} \approx 0.833 \tag{8.10}$$

$$\frac{P_c}{P_i} = \left(\frac{2}{\gamma + 1}\right)^{\frac{\gamma}{\gamma - 1}} \approx 0.528 \tag{8.11}$$

$$\frac{\rho_c}{\rho_i} = \left(\frac{2}{\gamma + 1}\right)^{\frac{1}{\gamma - 1}} \approx 0.634 \tag{8.12}$$

$$\frac{a_c}{a_i} = \sqrt{\frac{2}{\gamma + 1}} \approx 0.913 \tag{8.13}$$

For pressure values below P_{a2}, the pressure adaptation phenomena are no longer located inside the nozzle and will be discussed in the next section.

8.4.1.3   Flow Regimes at the Nozzle Outlet

If, for the sake of the presentation clarity, we have organized the discussion by separating the phenomena inside and outside the nozzle, this should not be interpreted as distancing from the analysis mode, which remains here in line with the previous mode. We come therefore to the fourth case defined as follows.

Case n° 4 : $P_{a3} \leqslant P_a < P_{a2}$ For this entire value range, the exit pressure *at the diverging section end before any adaptation* remains below the static pressure P_{a2} of the downstream environment, as observed in zone (a) of the previous Fig. 8.8.

There will therefore be a flow recompression *past the ejection*, which results in the presence of an oblique shock wave at the nozzle lips. This external adjustment can be viewed as the *normal shock bend* of the previous case, since the recompression jump is now lower ($P'_{a2} - P_{a3} < P_{a2} - P_{a3}$).

The adaptation then takes place through the succession of an oblique shock and expansion fan, as outlined in Fig. 8.9a. Indeed, on the free jet boundary surface at a constant pressure, the compression wave reflects as an opposite sign wave (expansion), unlike that which occurs on a solid wall. Thus the first oblique shock wave is reflected in a first expansion fan. The latter, in turn, will be reflected as shock waves at the free jet boundary[2], and the process will then begin once more.

Over the pressure range (a) in Figs. 8.7 and 8.8, the pressure of the jet just leaving the nozzle exit is always below the environmental pressure. The jet is then called an *over-expanded* jet.

Case n° 5: $P_a = P_{a3}$ It will be remembered that, by convention, P_{a3} refers to the static pressure value just upstream of the normal shock wave located in the nozzle outlet section operating at the second critical point (cf. case n° 3). Thus, when the static pressure of the downstream environment is exactly equal to this value, the fluid, when discharging from the nozzle, does not experience any pressure variation.

---

[2] Under the study assumptions, this is a slip surface.

Fig. 8.9 Flow adjustment
patterns past the nozzle exit

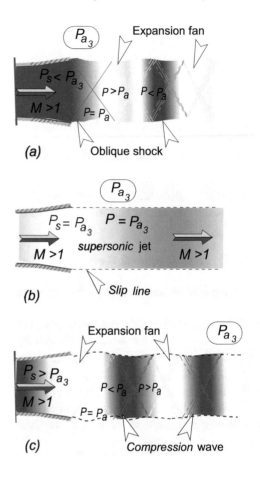

(a)

(b)

(c)

- **The third critical regime**: The flow exhausting from the nozzle is that of a *supersonic jet* whose free boundaries are slip surfaces, i.e., across which there is a discontinuity in the velocity but not in pressure. This situation, still referred to as the *design operating condition*, is outlined in Figure 8.9 (*b*).

Case n° 6: $P_a < P_{a3}$ In this case, the static pressure at the nozzle exit is higher than that of the downstream environment. The adjustment will therefore require a reduction in the flow pressure, hence the presence of a first expansion fan at the nozzle lips. The situation, which corresponds to an under-expanded jet developping in a quiescent environment, will develop further downstream, in the same way as described in the *case* n°4, without any damping, owing to the study assumptions (see diagram (*c*) in Fig. 8.9).

In addition to the schematics of Fig. 8.9, it should be noted that direct visualizations of the actual jet patterns at the exit, can be obtained by shadowgraph imaging. Some of the first to be produced are displayed on page 295 in Prandtl's work [125].

HISTORICAL NOTE—The De Laval steam-turbine—A turbine is a rotating device designed to convert the energy of a moving fluid (water, steam, air, combustion gas) in a mechanical form in order to create a torque which rotates a shaft. At the end of the 19th century, engineers concentrated their efforts on such machines, allowing, for an equivalent size, a higher power than that of a conventional reciprocating steam engine.

Thus, in the 1880s, a Swedish engineer of French origin, Carl Gustaf Patrik De Laval, designed a jet action (impulse) steam turbine [48], (see Fig. 8.10).

The rotation is driven by the impact of four jets on the wheel blade assembly in the presence of a stator vane. This device converts the thermodynamic energy of the high-pressure steam leaving the boiler into mechanical energy (kinetics) on the machine shaft.

During the turbine tuning process, De Laval discovered that the machine's efficiency was improved by designing the internal injector geometry with a first decreasing cross-section (convergent), increasing the fluid velocity to that of the speed of sound at the throat, followed by a divergent duct (diffuser), to make the jet supersonic at the exit. It is this geometry that is referred to as the *De Laval CD nozzle*.

EXERCISE 32 The flow regimes in a convergent-divergent nozzle.

A De Laval nozzle with a circular cross-section (throat diameter $D_c = 10$ mm) is supplied by a dry air source at a constant stagnation temperature and pressure. It opens into a very high capacity reservoir whose static pressure can be sustained at

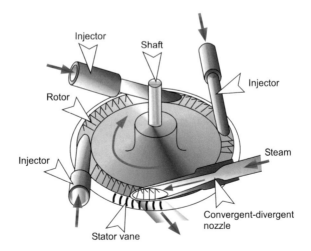

Fig. 8.10 De Laval turbine diagram and cross-sectional view of the *convergent-divergent* nozzle, adapted from Lea and Meden [87]

any constant level P_a. At the nozzle outlet, a static pressure $P_s = 0.39\,10^5\,\mathrm{N/m^2}$ and a Mach number $M_s = 2$ are measured.

We give $T_i = 400$ K, $\gamma = 1.4$ and $r = 287\ J \times Kg^{-1} \times K^{-1}$.

QUESTION 1. What are the speed of sound and flow velocity at the nozzle exit?

QUESTION 2. Determine the stagnation pressure P_i.

QUESTION 3. Calculate the fluid density at the nozzle outlet, the exit section diameter and the mass flow rate of the ejected fluid.

QUESTION 4. Under the previous nozzle operating conditions, the static pressure of the downstream environment was set to $P_a = 10^3\,\mathrm{N/m^2}$. To which exit flow regime does such a value correspond?

QUESTION 5. For which static pressure value P_a is a normal shock wave located in the nozzle exit section?

QUESTION 6. The static downstream pressure is set to the value $P_a = 1.840 \times 10^5\,\mathrm{N/m^2}$. At what value of the throat section ratio sit the normal shock recompression in the diffuser?

QUESTION 7. For which static pressure value P_a does the mass flow through the nozzle begin to change?

SOLUTION: 1. Since the flow is supersonic in the outlet section, it can be deduced that the nozzle regime is that of the second critical point. The evolution is therefore totally isentropic and the corresponding laws of Sect. 8.3.2 can be applied. Denoting the fluid temperature value in the outlet section as T_s, we know that:

$$\frac{T_i}{T_s} = 1 + \frac{\gamma - 1}{2}M_s^2.$$

We deduce $\mathbf{a}_s = \sqrt{\gamma r T_s}$ and $U_s = M_s \times \mathbf{a}_s$. With the numerical data for use in the exercise, we find $T_s = T_i/1.8 \approx 222$ K, $\mathbf{a}_s \approx 299$ m/s and $U_s \approx 598$ m/s.

2. The stagnation pressure is given by

$$\frac{P_i}{P_s} = \left[1 + \frac{\gamma - 1}{2}M_s^2\right]^{\frac{\gamma}{\gamma-1}},$$

or, with the relevant numerical data, $P_i \simeq 7.824 \times P_s \approx 3.05 \times 10^5\,\mathrm{N/m^2}$.

3. The fluid density at the outlet is directly deduced from the previous result with the equation of state. We find $\rho_s = P_s/r T_s \approx 0.612\ Kg/m^3$. The exit to throat section ratio is written:

$$\frac{A_s}{A_c} = \frac{1}{M_s}\sqrt{\left[\frac{2}{\gamma + 1}\left(1 + \frac{\gamma - 1}{2}M_s^2\right)\right]^{\frac{\gamma+1}{\gamma-1}}}.$$

With the relevant numerical data, we find $A_s/A_c = 1.6875$ from which we obtain $D_s \approx 13$ mm. The mass flow rate expressed at the outlet conditions is therefore

$$q_m = \rho_s U_s A_s \approx 4.85 \times 10^{-2} \ Kg/m^3 .$$

4. For an exit static pressure equal to $39 \times 10^3 \ N/m^2$, the flow is at an over-pressure relative to that of the downstream environment ($10^3 \ N/m^2$). The adjustment will therefore take the form of an expansion fan, in line with diagram (c) in the previous Fig. 8.7.

5. The presence of a shock at the exit section does not alter the upstream flow characteristics in the nozzle. Thus, immediately after the shock, the Mach number remains equal to 2. Using the normal shock relations (Chap. 7, Sect. 7.3.5) we directly obtain:

$$\frac{P_a}{P_s} = 1 + \frac{2\gamma}{\gamma + 1} \left(M_s^2 - 1 \right) ,$$

which gives us here $P_a = 4.5 P_s = 1.755 \times 10^5 \ N/m^2$.

6. Subject to confirmation by the outcome in the following question, it is assumed that the new static pressure value of the downstream environment past the nozzle out-let ($P_a = 1.840 \times 10^5 \ N/m^2$) corresponds to a normal shock formation inside the diffuser. To find its position, we will determine the flow Mach number M' immediately upstream of such a shock. Denoting P_{is} as the downstream stagnation pressure, we know that the value of the ratio P_{is}/P_{io} depends solely on the Mach number M'. The whole issue comes down to determining the downstream stagnation pressure P_{is}. Since beyond the shock the flow remains isentropic, this pressure remains unchanged until the downstream end of the diffuser and therefore can be determined by considering its value, in particular, in the exit section. For this purpose, we will calculate, with the relevant numerical data, the flow Mach number (M_s) in this exit section.

The rationale can therefore be symbolically summarized by the following calculation sequence of the various parameters:

$$M_s \rightarrow P_{is} \rightarrow M' \rightarrow A'/A_c.$$

(a) *Calculation of the Mach number* M_s: Since the flow is assumed to be sonic at the throat, the mass flow rate remains the same as that found previously. It can therefore be expressed that:

$$q_m = \rho_s A_s U_s \equiv \rho_s A_s M_s a_s = A_s M_s \frac{P_s}{r T_s} \sqrt{\gamma r T_s}.$$

By specifying the outlet temperature as a function of the stagnation temperature T_i, the previous relation leads to:

$$q_m = P_s A_s M_s \sqrt{\frac{\gamma}{r T_i}} \sqrt{1 + \frac{\gamma - 1}{2} M_s^2} .$$

All quantities are known except the Mach number which can be determined in this way. Numerically this yields the second degree equation $0.2x^2 + x - 0.332 = 0$ where $x = M_s^2$, whose only physically admissible root is $x \approx 0.312$, which provides $M_s \approx 0.56$.

(b) *Calculation of the pressure* P_{is}: This value is deduced from the isentropic pressure ratio:

$$\frac{P_{is}}{P_s} = \left[1 + \frac{\gamma - 1}{2} M_s^2\right]^{\frac{\gamma}{\gamma - 1}} .$$

We obtain here $P_i = 1.237 \times P_s \approx 2.28 \times 10^5 \; N/m^2$.

(c) *The stagnation pressure jump, Mach number* M' *and section ratio*: From the previous data, we can conclude that across the shock, the stagnation pressure variation is equal to $P_{is}/P_{i0} \approx 0.75$. For this value, relation (7.31) in Chapter 7, provides the upstream Mach number value, which is here $M' \approx 1.95$. The relation previously used in this exercise which expresses $A'/A_c = f(M')$ then leads to $A' \approx 1.60 A_c$. Thus the shock is located in the nozzle section whose diameter is about 12.65 *mm*. Hence it is verified retrospectively, that the initial calculation assumption (the sonic condition at the throat) is indeed satisfied.

7. The mass flow rate variation (decrease) will begin to occur for a static pressure lower than that which ensures the *subsonic* character of the flow all along the nozzle, *with a sonic condition at the throat* (first critical regime). For the section ratio A_s/A_c equal to 1.6875, the *isentropic subsonic* flow regime then corresponds to $M_s \approx 0.37$. For this Mach number value, the isentropic pressure law P_i/P provides $P \approx 0.91 P_{i0}$, hence the static pressure $P_a \approx 2.77.10^5 \; N/m^2$.

EXERCISE 33 The flow regimes in a diffuser.

We consider the diffuser configuration of Exercise 31—a divergent duct with an inlet / outlet section ratio $A_s/A_e = 1.54$—for studying the different flow regimes by using the sonic condition at a fictitious throat (first critical regime).

QUESTION 1. Determine the Mach number value of the flow at the diffuser outlet when the Mach number in the inlet section A_e is $M_e = 0.51$.

QUESTION 2. The conditions downstream of the diffuser are changed so that a normal shock wave is located in the intake diffuser section A_e. The Mach number just upstream of this shock is $M_0 = 2.53$. What is the new Mach number at the diffuser exit?

QUESTION 3. What is the maximum value which can be reached by the flow Mach number at the same diffuser exit, under supply conditions to be specified?

SOLUTION: 1. Let us imagine a critical fictitious throat such that the flow is isentropic between this throat and the inlet section. For $M_e = 0.51$, the relation expressing $A/A_c = f(M)$ gives $A_e/A_c \approx .1.32$. We deduce that:

$$\frac{A_s}{A_c} \equiv \frac{A_s}{A_e} \times \frac{A_e}{A_c} = 1.54 \times 1.3212 \approx 2.03\,.$$

For the air ($\gamma = 1.4$), relation (8.9b) is written $\dfrac{A}{A_c} = \dfrac{1}{M}\left(\dfrac{1+0,2M^2}{1,2}\right)^3$, whose non-bijective character, (cf. Fig. 8.5), leads to two Mach number values for $A_s/A_c = 2.03$. Only that of the subsonic regime is to be considered here, i.e. $M_s \approx 0.3$.

2. Relation (7.17) between the Mach numbers on either side of a normal shock leads to the Mach number just downstream of the shock $M_1 \approx 0.51$. Since the flow has become subsonic, it is isentropic in the remainder of the diffuser and the exit Mach number is therefore identical to that found in the first question.

It should be noted, however, that the mass flow rate is not necessarily the same in both cases, but that this change occurs under the same stagnation conditions upstream of the diffuser's inlet section. Consequently it is legitimate to refer to the same critical fictitious throat section for both calculations.

On the other hand, the outlet stagnation pressures are different, the equality $P_{ie} = P_{is}$ applying of course only to the first case (the isentropic *subsonic* flow all along the diffuser).

3. The maximum Mach number value at the diffuser outlet is obtained when the flow is entirely supersonic between sections A_e and A_s. Therefore we return to the isentropic stagnation pressure's equality $P_{ie} = P_{is}$ in the diffuser, but for an isentropic flow which is, this time, entirely *supersonic*. Hence, the calculation should be now carried out with a new critical fictitious throat section A'_c such that:

$$\frac{A_e}{A'_c} = \frac{1}{2.53} \times \left[\frac{1+0.2 \times (2.53)^2}{1.2}\right]^3 \approx 2.7117.$$

With this reference, the section ratio at the diffuser outlet is $A_s/A'c = A_s/A_e \times A_e/A'c \approx 4.175$. By reversing equation (8.9b) for this value we finally obtain as the Mach number in the supersonic regime $M_{s2} \approx 2.985$.

8.4.2 The Compressible Flow in a Venturi Nozzle

We have previously mentioned, in Chap. 5, Sect. 5.7.3, the Venturi effect as a means of measuring the velocity of an incompressible pipe flow.

We return here to the device operating principle *in the compressible regime*. By applying the energy equation between the two sections A and B—see Chap. 5, Fig. 5.11—we can now determine that:

$$V_B = \sqrt{\frac{\frac{2\gamma}{2-1}\frac{P_A}{\rho_A}\,(1-P_B/P_A)^{(\gamma-1)/\gamma}}{1-\sigma^2\,(P_B/P_A)^{2/\gamma}}}\,, \tag{8.14}$$

where σ stands for the contraction ratio S_B/S_A. We deduce that the mass flow rate value $q_m = \rho_B V_B S_B$ is:

$$q_m = C_d S_B \sqrt{\frac{\frac{2\gamma}{2-1} P_A \rho_A (P_B/P_A)^{2/\gamma} (1 - P_B/P_A)^{(\gamma-1)/\gamma}}{1 - \sigma^2 (P_B/P_A)^{2/\gamma}}}. \tag{8.15}$$

The C_d coefficient is a parameter introduced to take into account the overall viscosity effects, and can slightly vary from one device to another. Its value is determined, in practice, by a calibration of the instrument.

The previous relation proves that, unlike the incompressible case, the velocity value and similarly the flow rate, cannot be deduced from a simple pressure measurement. It is also necessary to determine the fluid density ρ_A, which in practice requires an additional temperature measurement.

Proof Since the flow can only be isentropic in the converging part of the Venturi tube within the limits of the two sections S_A and $S_B < S_A$, the energy equation gives:

$$\frac{\gamma}{\gamma - 1} \frac{P_A}{\rho_A} + \frac{V_A^2}{2} = \frac{\gamma}{\gamma - 1} \frac{P_B}{\rho_B} + \frac{V_B^2}{2},$$

and therefore $V_B^2 - V_A^2 = \frac{2\gamma}{\gamma - 1} \frac{P_A}{\rho_A} \left(1 - \frac{P_B}{P_A} \frac{\rho_A}{\rho_B}\right)$. By expressing the ratio ρ_A/ρ_B using Laplace's law, we obtain

$$V_B^2 - V_A^2 = \frac{2\gamma}{\gamma - 1} \frac{P_A}{\rho_A} \left[1 - \left(\frac{P_B}{P_A}\right)^{(\gamma-1)/\gamma}\right].$$

Owing to the continuity equation $V_A = \sigma V_B \rho_B/\rho_A$, we arrive, after substitution and rearrangement, at the expected result.

EXERCISE 34 The Venturi nozzle in compressible operation.

A Venturi nozzle, with a circular throat section ($\phi = 60$ mm) is arranged in a pipe with a constant circular cross-section of 100 mm in diameter. This pipeline carries compressed air. The static pressure at the Venturi inlet is $P_A = 7.5 \times 10^5 \, P_a$ and the temperature is $t_A = 37\,°C$. The static pressure value at the Venturi throat is $P_B = 3.9 \times 10^5 \, P_a$.

QUESTION   Given that the Venturi's flow rate coefficient is 0.97, determine the mass flow rate of the fluid in the pipeline.

SOLUTION: Firstly, we calculate the air density at the Venturi inlet. We know that $\rho_A = P_A/r T_A$, hence with the numerical data for use in the exercise $\rho_A = 8.43 \, kg/m^3$. Applying the flow rate relation Eq. (8.15) we directly deduce $q_m = 4.85 \, kg/s$.

8.4.3 Pitot Probing in the Supersonic Regime

As previously mentioned, the presence of a fixed solid body in a supersonic stream will lead to the flow decelerating through a shock wave system, whose geometry depends in particular on the shape of the obstacle near the stagnation point. Thus, for Pitot probes (see Chap. 5, Sect. 5.7.2), which present a blunt front-end around the total pressure measurement hole, the formation of a detached shock wave in front of the body can be observed in the supersonic regime. Locally, for the fluid lines close to the flow axis, this shock can be interpreted as a plane wave normal to the flow.

Such a device does not provide access to the *undisturbed* flow characteristics, *viz* without any perturbation due to the measuring probe. To this end, at least one piece of information is required *upstream* of the shock. We are thus led to offset the static pressure measurement, (see the upper schematic in Fig. 8.11). The static pressure point is therefore located on the periphery of an antenna, whose slender terminal end aims at reducing the fluid deceleration with a slight recompression through an oblique shock wave.

The direct flow visualization, Fig. 8.11, by differential interferometry, illustrates the detached shock wave in front of the main probe body and the presence of the oblique shock on the dart supporting the static pressure measurement. On the whole,

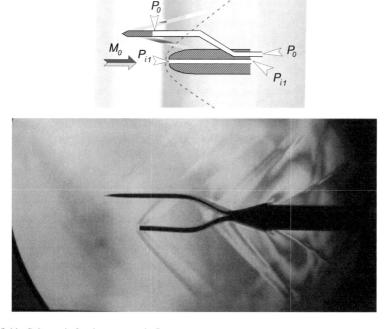

Fig. 8.11 Schematic for the supersonic flow past a Pitot tube (*top*) and visualization by differential interferometry (*bottom*), courtesy of ENSICA

the device should be designed so that there is no interference between these two wave systems upstream of the stagnation pressure's measurement point. Its geometry is therefore dependent on the Mach number range to be investigated.

With such a device, and for a suitable Mach number, we can determine the values of (*i*) the static pressure upstream of the main normal shock and (*ii*) the stagnation pressure downstream of this shock. The simple ratio of these two values directly leads to the flow Mach number by Rayleigh's formula:

$$\frac{P_{i1}}{P_0} = \left(\frac{\gamma+1}{2}\right)^{\frac{\gamma+1}{\gamma-1}} \times \frac{M_0^{\frac{2\gamma}{\gamma-1}}}{\left(\gamma M_0^2 - \frac{\gamma-1}{2}\right)^{\frac{1}{\gamma-1}}} \qquad (8.16)$$

Proof Let us start from the identity $\dfrac{P_{i1}}{P_0} = \dfrac{P_{i1}}{P_1} \times \dfrac{P_1}{P_0}$. The static pressure ratio can be expressed as a function of the *upstream* Mach number M_0, as we mentioned in Chap. 7, Sect. 7.3.5.1. The ratio P/P_i is obtained by the isentropic relation from the *downstream* number of Mach M_1. Finally, the upstream and downstream Mach numbers are linked by relation (7.17) of Chap. 7. We simply recall here these three relations which, by eliminating the downstream Mach number, lead to the expected result:

$$\frac{P_1}{P_0} = 1 + \frac{2\gamma}{\gamma+1}\left(M_0^2 - 1\right), \quad \frac{P_{i1}}{P_1} = \left(1 + \frac{\gamma-1}{2}M_1^2\right)^{\frac{\gamma}{\gamma-1}}, \quad M_1^2 = \frac{(\gamma-1)M_0^2 + 2}{2\gamma M_0^2 + 1 - \gamma}.$$

EXERCISE 35 Pitot's tube in a supersonic flow.

A Pitot tube designed for supersonic operation (see Fig. 8.11) is positioned in an air flow whose local temperature is $T = 200$ K. The measured stagnation and static pressure values are $P_{i1} = 2.2 \times 10^5$ Pa and $P_0 = 0.39 \times 10^5$ Pa respectively.

QUESTION Using the graphical representation of Rayleigh's relation, determine the flow velocity.

SOLUTION: Let us firstly calculate the flow Mach number. The pressure measurement provides $P_{i1}/P_0 = 2.20/0.39 \approx 5.6$.

Using the graphical representation in Fig. 8.12, this value corresponds roughly to a Mach number of 2. The speed of sound, under the flow conditions, is $a = \sqrt{\gamma r T} \approx 283.5$ m/s. The fluid velocity is therefore 567 m/s.

Fig. 8.12 Graphic plot of Rayleigh's formula

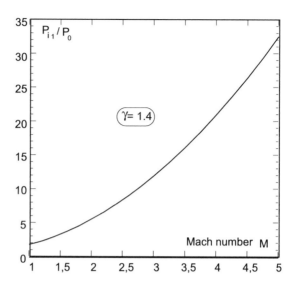

8.5 Ideal Gas Flow with Friction and Heat Transfer in a Constant Cross-Section Duct

Many practical applications involve compressed gas flows in pipelines whose cross-section does not vary in value over long distances. This is the case, for example, for gas pipelines, primary supply lines for steam or gas distribution networks, etc.

In these situations, and although the fluids do not necessarily have an ideal gas thermodynamic behaviour, the dissipative phenomena associated with the fluid intrinsic irreversibilities remain relatively negligible in the energy balance with respect to changes in:

– the momentum balance, by the existence of *parietal* friction;
– the enthalpy balance, by direct heat input *from the environment*.

Accordingly, to account for these effects, the following equation model can be adopted:

$$
\begin{array}{lll}
\textbf{Continuity:} & \dfrac{d(\rho U)}{dx} = 0 & (8.17) \\[3mm]
\textbf{Momentum:} & \rho U \dfrac{dU}{dx} = -\dfrac{dP}{dx} - \dfrac{\Lambda}{D}\dfrac{1}{2}\rho U^2 & (8.18) \\[3mm]
\textbf{Enthalpy:} & \rho C_p \dfrac{dT}{dx} = -\rho U \dfrac{dU}{dx} + \rho \dfrac{dQ}{dx} & (8.19)
\end{array}
$$

In these relations, D denotes the pipe diameter of the assumed circular cross-section, Λ is the *major head loss coefficient*, defined by equation (3.41) in Chap. 3,

Sect. 3.4.3.2, and dQ stands for the heat input per unit mass. We will continue this flow study by separately discussing the friction and heat transfer effects.

Remark *The hydraulic diameter*—If the duct does not have a circular cross-section, to use the previous equations, simply substitute the diameter D by the equivalent hydraulic diameter $D_e = 4A/C$ where A is the cross-sectional area of the duct and C is the wetted perimeter of the cross-section.

8.5.1   The Adiabatic Flow with Friction: Fanno's Flow

8.5.1.1   The Governing Equations

We consider here a flow without an external heat transfer ($dQ = 0$) but with parietal friction, whose local motion equations are written, by introducing the Mach number M:

$$Continuity: \quad \frac{d\rho}{\rho} + \frac{dU}{U} = 0, \tag{8.20}$$

$$Momentum: \quad \frac{1}{\gamma M^2}\frac{dP}{P} + \frac{dU}{U} + \frac{\Lambda}{2D}dx = 0, \tag{8.21}$$

$$Energy: \quad \frac{1}{(\gamma - 1)\,M^2}\frac{dT}{T} + \frac{dU}{U} = 0. \tag{8.22}$$

Proof By dividing the dynamic equation throughout by ρU^2, we obtain

$$\frac{dU}{U} + \frac{dP}{\gamma P}\frac{\mathbf{a}^2}{U^2} + \frac{\Lambda}{2D}dx = 0,$$

where the density is expressed as $\rho = \gamma P/\mathbf{a}^2$. Simply introducing the Mach number $M = U/\mathbf{a}$ will give us the targeted expression. The energy equation is handled in exactly the same way.

8.5.1.2   The Logarithmic Variations

From the previous relations, we can specify the logarithmic variations of all the flow functions, as well as the pressure loss coefficient, as a function of dM/M. The results are written:

Velocity:
$$\frac{dU}{U} = \frac{1}{1 + \frac{\gamma-1}{2}M^2} \frac{dM}{M} \qquad (8.23a)$$

Pressure:
$$\frac{dP}{P} = -\frac{1 + (\gamma - 1)M^2}{1 + \frac{\gamma-1}{2}M^2} \frac{dM}{M} \qquad (8.23b)$$

Temperature :
$$\frac{dT}{T} = -\frac{(\gamma - 1)M^2}{1 + \frac{\gamma-1}{2}M^2} \frac{dM}{M} \qquad (8.23c)$$

Density :
$$\frac{d\rho}{\rho} = \frac{-1}{1 + \frac{\gamma-1}{2}M^2} \frac{dM}{M} \qquad (8.23d)$$

Head loss

coefficient :
$$\frac{\Lambda dx}{2D} = \frac{(1 - M^2)}{\gamma M^2 \left(1 + \frac{\gamma-1}{2}M^2\right)} \frac{dM}{M} \qquad (8.23e)$$

Proof From the Mach number definition, we can set $U^2 = \gamma r T M^2$ which by, logarithmic differentiation, gives

$$2\frac{dU}{U} = \frac{dT}{T} + 2\frac{dM}{M}.$$

In addition, the energy equation provides $dT/T = (1 - \gamma) M^2 dU/U$. By eliminating dT/T between these two equations, we obtain the first relation expressing dU/U. The expression of dT/T, as well as that of $d\rho/\rho$, can be immediately deduced by using the continuity equation. To obtain the head loss coefficient, simply substitute all the previous expressions in the momentum equation. Thus we obtain:

$$\frac{1}{\gamma M^2}\left[\frac{-1 + (1 - \gamma)M^2}{1 + \frac{\gamma-1}{2}M^2}\right]\frac{dM}{M} + \frac{1}{1 + \frac{\gamma-1}{2}M^2}\frac{dM}{M} + \frac{\Lambda}{2D}dx = 0,$$

which simply reduces to $-\dfrac{M^2 - 1}{\gamma M^2\left(1 + \frac{\gamma-1}{2}M^2\right)}\dfrac{dM}{M} + \dfrac{\Lambda}{2D}dx = 0.$

8.5.1.3 The Entropy Variation

Using the previous results, we can specify the entropy variation in the flow as a function of the relative variation in the Mach number:

$$ds = r\,\frac{1 - M^2}{1 + \frac{\gamma-1}{2}M^2}\,\frac{dM}{M} \qquad (8.24)$$

Proof From Gibbs' expression $T\,ds = dh - dP/\rho$ we deduce, for an ideal gas $ds = C_p dT/T - r dP/P$, namely, with the previous expressions:

$$ds = -\frac{C_p\,(\gamma - 1)\,M^2}{1 + \frac{\gamma-1}{2}M^2}\,\frac{dM}{M} + r\,\frac{1 + (\gamma - 1)\,M^2}{1 + \frac{\gamma-1}{2}M^2}\,\frac{dM}{M}\,.$$

Noting that $(\gamma - 1)\,C_p = \gamma r$, we obtain, after simplifications, the expected result.

8.5.1.4   Discussion and Physical Interpretation

In order to clarify the physical interpretation of the derived results, we begin with the following three points:

1) The head loss coefficient and entropy variation expressions prove that these quantities are proportional, since indeed:

$$\frac{ds}{dx} = \frac{r\gamma M^2}{2D}\,\Lambda\,.$$

This relation means that the pressure drop along the pipe is simply another expression of the entropy variation, a result which generalizes, to the compressible case, that which will be demonstrated in Chap. 9, Sect. 6.1 for the incompressible regime.

2) As the entropy variation along the pipe can only be positive, in line with the second law of thermodynamics, it results from this expression of ds that, if $M < 1$, the flow Mach number can only increase, whereas if $M > 1$, it is necessarily decreasing.

3) The crossing through $M = 1$ is inconsistent with the study assumptions. Indeed, for this Mach number value, the head loss coefficient would be zero unless an infinite variation of dM/dx were to be considered, which makes no physical sense.

Therefore, from the previous two statements, we can conclude that:

With a given parietal friction, the compressible gas flow regime along a constant cross-section duct is maintained as unchanged:
– if the motion is subsonic, the Mach number remains smaller than one,
while continuously increasing along the duct.
– if the motion is supersonic, the Mach number remains greater than one,
while continuously decreasing along the duct.

This first set of conclusions can be visualized quite simply in the $T-s$ diagram.

Fig. 8.13 Fanno lines in the $T - s$ diagram

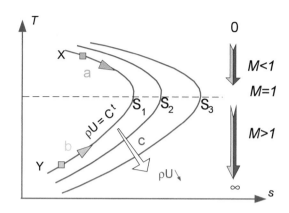

The plot for a given mass flow rate—the Fanno curve—is outlined in Fig. 8.13.

Starting from a state corresponding to a subsonic regime such as (X), the evolution can only take place along the Fanno line in the increasing entropy direction (arrow a).

For an initially supersonic regime such as (Y), it will be in the direction of the arrow b.

A second set of conclusions can be deduced by integrating the head loss coefficient equation (8.23e) between $x = 0$ $(M = M_0)$ and $x = L$ $(M = M_L)$. We obtain:

$$\frac{\Lambda L}{2D} = \frac{1}{2\gamma}\left(\frac{1}{M_0^2} - \frac{1}{M_L^2}\right) + \frac{\gamma + 1}{4\gamma}\ln\left[\left(\frac{M_0}{M_L}\right)^2 \frac{1 + \frac{\gamma - 1}{2}M_L^2}{1 + \frac{\gamma - 1}{2}M_0^2}\right].$$

Since the downstream Mach number is, in all flow cases, bounded by $M = 1$, there is an upper limit to the pipe length compatible with the study assumptions. Its value L_{max} is given by

$$\frac{L_{max}}{D} = \frac{1}{\Lambda\gamma}\left[\left(\frac{1}{M_0^2} - 1\right) + \frac{\gamma + 1}{2}\ln\left(\frac{\frac{\gamma + 1}{2}M_0^2}{1 + \frac{\gamma - 1}{2}M_0^2}\right)\right].$$

Basically, if the pipe exceeds the previous maximum length, the flow can no longer be sustained while satisfying the continuity equation $\rho U = C^t$. In the $T-s$ diagram, this means that a point such as S_1 is reached. In order to develop beyond without contravening the second law, the flow should refer to another Fanno line such that the entropy is still increasing. As can be seen in the previous Figure, this can only be achieved by switching to another line for which $\rho_2 U_2 < \rho_1 U_1$ (arrow c). In other words, the *mass-flow rate is no longer sustained*. Hence the following conclusion:

There is a duct limit length for the mass-flow rate to remain constant. For this limit, the flow is choked (sonic condition). If the pipe exceeds the choking length, a new flow is taking place whose mass-flow rate is lower.

Proof The integration of the equation expressing the head loss coefficient is obtained by first decomposing the rational fraction on the right hand side as follows:

$$\frac{\Lambda dx}{2D} = \frac{1}{\gamma}\frac{dM}{M^3} - \frac{\frac{\gamma+1}{2\gamma}}{1 + \frac{\gamma-1}{2}M^2}\frac{dM}{M}.$$

Then we deduce that $\dfrac{\Lambda L}{2D} = -\left[\dfrac{1}{2\gamma M^2} + \dfrac{\gamma+1}{4\gamma}\ln\left(\dfrac{M^2}{1 + \frac{\gamma-1}{2}M^2}\right)\right]_{M_0}^{M_L}.$

HISTORICAL NOTE—**Fanno's flow**—The motion equations, (8.20)–(8.22), as well as their general assumptions were introduced by the Italian engineer Gino Girolamo Fanno in his Master's thesis at the University of Zurich in 1904. The adiabatic frictional pipe flow has since been named after him, in recognition of his major contribution. Fanno's flow model can be applied to many industrial processes: the discharge of a pressure vessel through a short tube, the exhaust pipe of an internal combustion engine, air compressor tubing, etc. Its main limitation lies in the restriction, by hypothesis, to the absence of heat transfer with the environment. It is therefore only appropriate for low Eckert number flows where the characteristic time scale of the motion is *very low* compared to that of the heat transfer mechanisms with the surroundings.

EXERCISE **36** A compressible pipe flow with friction.

Compressed air is supplied through a circular cross-section pipe of constant diameter $D = 50$ mm. The flow Mach number at the pipe inlet is 0.2 and the constant head loss coefficient is $\Lambda = 0.02$.

QUESTION 1. Calculate the maximum value of the pipe length L_{max} for the chocking condition.

QUESTION 2. It is assumed that, for a pipe whose length is L_{max}, the flow is freely exhausting into a constant static pressure environment with $P_a = 10^5\,\text{N/m}^2$. Determine the static pressure at the pipe inlet.

SOLUTION: **1.** The maximum choking length of the pipe is expressed, with the numerical data for use in the exercise, as:

$$\frac{L_{max}}{D} = \frac{1}{0.028}\left[\left(\frac{1}{4.10^{-2}} - 1\right) + \frac{2.4}{2}\ln\left(\frac{1.2 \times 4.10^{-2}}{1 + 0.2 \times 4.10^{-2}}\right)\right],$$

which, when all calculations are made, gives $L_{max} \approx 36.3\,m$.

2. Let us observe that the logarithmic differential of the pressure is written as:

$$-\frac{dP}{P} = \frac{1+(\gamma-1)\,M^2}{1+\frac{\gamma-1}{2}M^2}\frac{dM}{M} \equiv \frac{1}{M} + \frac{1}{2}\left(\frac{\frac{\gamma-1}{2}2M}{1+\frac{\gamma-1}{2}M^2}\right),$$

whose indefinite integral is $\ln\left(\dfrac{1}{P}\right) = \ln(M) + \dfrac{1}{2}\ln\left(1 + \dfrac{\gamma-1}{2}M^2\right)$. Thus the static pressure ratio between two sections where $P = P_1$ and $P = P_2$ is equal to:

$$\frac{P_2}{P_1} = \frac{M_1}{M_2}\sqrt{\frac{1+\frac{\gamma-1}{2}M_1^2}{1+\frac{\gamma-1}{2}M_2^2}}.$$

Applying the above relation between the inlet section ($P = P_1$) and the exit section where the flow is sonic with a static pressure adapted to P_a, we obtain:

$$\frac{P_a}{P_1} = M_1\sqrt{\frac{1+\frac{\gamma-1}{2}M_1^2}{1+\frac{\gamma-1}{2}}}.$$

With the relevant numerical data, we find $P_1 \approx 5.4\,P_a = 5.4 \times 10^5\,\text{N/m}^2$.

8.5.2 The Frictionless Duct Flow with Heat Transfer, Rayleigh's Flow

8.5.2.1 The Governing Equations and Logarithmic Variations

We now come to the second type of a pipe flow of constant cross section intended to assess heat input effects in the absence of friction. The momentum and energy equations are written in this case:

$$\rho U dU + dP = 0 \quad \text{or} \quad \frac{1}{\gamma M^2}\frac{dP}{P} + \frac{dU}{U} = 0, \tag{8.25}$$

$$C_p\, dT = -U dU + dQ. \tag{8.26}$$

From the previous relations, it can be deduced that:

$$\text{Velocity :} \qquad \frac{dU}{U} = \frac{1}{1 - M^2} \frac{dQ}{C_p T} \qquad (8.27\text{a})$$

$$\text{Pressure :} \qquad \frac{dP}{P} = \frac{-\gamma M^2}{1 - M^2} \frac{dQ}{C_p T} \qquad (8.27\text{b})$$

$$\text{Temperature :} \qquad \frac{dT}{T} = \frac{1 - \gamma M^2}{1 - M^2} \frac{dQ}{C_p T} \qquad (8.27\text{c})$$

$$\text{Density :} \qquad \frac{d\rho}{\rho} = \frac{-1}{1 - M^2} \frac{dQ}{C_p T} \qquad (8.27\text{d})$$

$$\text{Mach number :} \qquad \frac{dM}{M} = \frac{1 + \gamma M^2}{2 \left(1 - M^2\right)} \frac{dQ}{C_p T} \qquad (8.27\text{e})$$

The logarithmic differential expressions, as a function of the dimensionless heat input $dQ/C_p T$, clearly reveals the Mach number influence on the variation direction of the different quantities. We will address this point in section 8.5.2.3, not without noting here that the previous expressions provide $ds = dQ/T$ for the entropy variation.

Proof The energy equation gives us:

$$\frac{dT}{T} = -U \frac{dU}{C_p T} + \frac{dQ}{C_p T} \equiv -\frac{U^2}{C_p T} \frac{dU}{U} + \frac{dQ}{C_p T} .$$

By expressing the temperature as a function of the speed of sound, $T = \mathbf{a}^2/\gamma r \equiv \mathbf{a}^2/\left(\gamma - 1\right) C_p$ we obtain:

$$\frac{dT}{T} = \left(\gamma - 1\right) M^2 \frac{dU}{U} + \frac{dQ}{C_p T} .$$

On the other hand, the equation of state allows us to write $dT/T = dP/P - d\rho/\rho$. Now, from the continuity equation $d\rho/\rho = -dU/U$ and momentum equation we know that $dP/P = -\gamma M^2 dU/U$. By substituting these expressions in the previous relation, we obtain:

$$\frac{dT}{T} = \left(1 - \gamma M^2\right) \frac{dU}{U} .$$

Eliminating dT/T directly yields the expected expression of dU/U from which all other relations are easily deduced.

8.5.2.2   The Inter-Sectional Relations

The motion equations can also be integrated between two different sections. Integrating the momentum balance equation is straightforward. Given the continuity requirement, it directly leads to the dynalpy conservation, as we have previously demonstrated in Chap. 5, Sect. 5.6.2.3. Thus we have:

$$P_1 + \rho_1 U_1^2 = P_2 + \rho_2 U_2^2 ,$$

or else

$$P_1 \left(1 + \gamma M_1^2\right) = P_2 \left(1 + \gamma M_2^2\right) .$$

Finally we obtain

$$\boxed{\frac{P_2}{P_1} = \frac{1 + \gamma M_1^2}{1 + \gamma M_2^2}} \tag{8.28}$$

On the other hand, from the Mach number definition, we have

$$\frac{M_2}{M_1} = \frac{U_2}{U_1} \times \frac{\mathbf{a_1}}{\mathbf{a_2}} \equiv \frac{U_2}{U_1} \sqrt{\frac{T_1}{T_2}} .$$

Now, due to the continuity condition

$$\frac{U_2}{U_1} = \frac{\rho_1}{\rho_2} \equiv \frac{P_1}{P_2} \times \frac{T_2}{T_1} ,$$

which, by combining with the previous expressions gives us:

$$\boxed{\frac{T_2}{T_1} = \left(\frac{M_2}{M_1} \times \frac{1 + \gamma M_1^2}{1 + \gamma M_2^2} \right)^2} \tag{8.29}$$

In this type of motion, an adiabatic stagnation temperature[3] T_a can be defined as:

$$T_a = T + \frac{U^2}{2C_p} \equiv T \left(1 + \frac{\gamma - 1}{2} M^2\right) .$$

Since the flow is not isentropic, this temperature is not preserved between the two sections. Applying this relation to two separate sections we directly obtain:

$$\boxed{\frac{T_{a2}}{T_{a1}} = \left(\frac{M_2}{M_1} \times \frac{1 + \gamma M_1^2}{1 + \gamma M_2^2} \right)^2 \left[\frac{1 + \frac{\gamma-1}{2} M_2^2}{1 + \frac{\gamma-1}{2} M_1^2} \right]} \tag{8.30}$$

---

[3] It is the temperature which the gas would attain if it were brought to rest adiabatically without performing work.

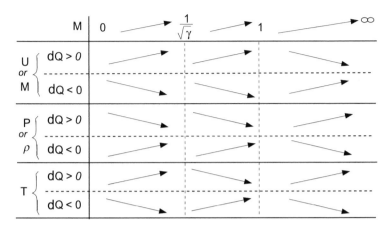

Fig. 8.14 Direction of flow function variations in a frictionless flow with heat transfer

8.5.2.3   Discussion and Physical Interpretation

Two specific Mach number values can be identified from the expressions of the logarithmic variations:

– the first is obtained for $M = 1$ and leads to infinite variations of all the parameters in the presence of a non-zero heat input;
– the second $M = 1/\sqrt{\gamma}$ is specific to the temperature evolution. It characterizes a situation where this function remains constant whatever the heat input.

From these critical values, the direction of the various flow function variations can be easily obtained and is given in the following table.

From Fig. 8.14, it appears that in the subsonic regime, any heat input results in an increase in the flow Mach number. On the other hand, in a supersonic regime, heating causes this function to decrease. Thus, in both cases, the tendency is to promote the onset of sonic conditions ($M = 1$), a situation qualified as the thermal choking condition.

A second observation can be drawn about the temperature evolution. It can be observed that any heat input (resp. output) *naturally* results in an increase (resp. decrease) of the temperature, *except* for flows whose Mach number lies between $1/\sqrt{\gamma}$ and the unit. For such flows, the *increase* in velocity (viz the fluid '*acceleration*') is combined with a *decrease* in temperature.

We shall summarize all these observations in the following two statements:

– In a Rayleigh flow, any heat supply causes the flow to evolve to a thermal choking condition $M = 1$;
– Heating (resp. cooling) of the fluid causes a temperature increase (resp. decrease) regardless of the flow's Mach number value outside the range $[1/\sqrt{\gamma}\, ; 1]$ over which the tendency is the opposite.

Fig. 8.15 Rayleigh line in the $T - s$ diagram

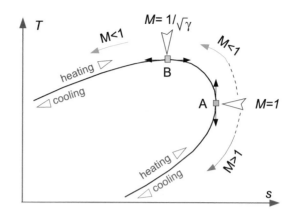

These conclusions are easily explained in the diagram $(T-s)$. Indeed, it is easily verified that the entropy variation is obviously expressed in this case by $ds = dQ/T$, so that the entropy increasing direction is the same as that of the heat input (heating). In such a diagram, the evolution is then following the Rayleigh line outlined in Fig. 8.15.

Proof From the previous results, we know that

$$\frac{dT}{T} = \frac{1 - \gamma M^2}{1 - M^2} \frac{dQ}{C_p T} \qquad \text{and} \qquad \frac{dP}{P} = \frac{-\gamma M^2}{1 - M^2} \frac{dQ}{C_p T}.$$

It is then sufficient to substitute these expressions in Gibbs' relation $ds = C_p \, dT/T - r \, dP/P$ to prove, after simplifications, that the entropy variation is indeed solely linked to the heat input, the intrinsic irreversibilities being null, according to the study assumptions.

HISTORICAL NOTE—The Ramjet—Even though the Rayleigh flow model is not appropriate for describing real situations in detail, it makes it possible to theoretically highlight some of the phenomena likely to appear in ramjet and scramjet combustion chambers.

As early as 1939, work had been undertaken in France for the development of a jet engine with no moving parts. It was the so-called "*thermo-propellant nozzle*", in the terminology of its designer René Leduc. The air is supplied to the nozzle as a result of the craft motion itself. A suitable design of the nozzle divergent allows, at a subsonic speed, the fluid to be compressed in the upstream part of the nozzle. The combustion of the kerosene injected into the flow then provides the heat input which causes the fluid to accelerate to a value further increased by the convergent shape of the final part of the nozzle (see Fig. 8.16).

Provided that, by the air supply conditions, a subsonic regime in the divergent is reached whatever the flight conditions (altitude in particular), the ramjet is

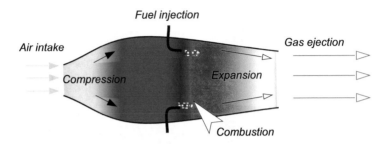

Fig. 8.16 Schematic for the ramjet operation

theoretically suitable for developing considerable thrust forces, increasing with the flight speed. Most of this thrust is obtained as a pressure form on the divergent walls, i.e., in the '*compressor*' part.

Despite its simplicity in principle, the ramjet operation proved difficult in practice due to problems with flame ignition, combustion stability, feed regularity and thermal choking. Furthermore, as it does not provide any thrust at the fixed point, this propulsion method can only be considered (i) alone, for an aircraft dropped in flight, or (ii) in combination with another engine—a turbo-ramjet engine combination—, for a self-propelled take-off aircraft.

A ramjet-powered aircraft, the Leduc 021, was built in France in 1953. Dropped from a Languedoc aircraft, it reached Mach 0.85 on its fourth flight in August 1953. Its successor was the Leduc 022, which made its first flight on December 26, 1956, operated by Jean Sarrail. In addition to the ramjet at low speeds, it used an Atar 101 D3 turbojet engine. As it suffered from excessive fuel consumption, the tests were definitively stopped on February 15, 1958. The MiG 21 and the F 104 had already taken to the air!

EXERCISE 37 The compressible gas flow in a heated duct.

In a $10cm$ diameter straight pipe an air flow is settled at a Mach number $M_0 = 0.2$ and temperature $14.7\,°C$. The supply is isentropically carried out from a reservoir whose stagnation pressure is $P_{i0} = 200\,\mathrm{kN/m^2}$.

QUESTION 1. Calculate the inlet velocity and mass flow rate.

QUESTION 2. Along a length of this pipe, immediately downstream from the inlet section L, heating elements with a total power of 150 kW are disposed. The section immediately downstream of these elements is taken as the outlet section. Calculate the fluid's adiabatic stagnation temperature at the outlet section, assuming that, due to external heat losses, the heat transfer efficiency to the moving fluid is about 92%.

(We will take $C_p = 1000\,Jkg^{-1}K^{-1}$).

QUESTION 3. Deduce the flow's Mach number value after heating, as well as the temperature and exit velocity.

SOLUTION: 1. Let us firstly calculate the flow stagnation temperature T_{i0}. Since the setting in motion is assumed to be isentropic, we have, by denoting as T_0 the inlet temperature:

$$\frac{T_{i0}}{T_0} = 1 + \frac{\gamma - 1}{2} M_0^2 .$$

For the inlet Mach number of 0.2 and temperature of 287.7 K, we find $T_{i0} = 290$ K or 17 °C. We can deduce the speed of sound under the stagnation conditions, as $a_{i0} = \sqrt{rT_{i/}} = 341$ m/s. Then, relation (8.8d)

$$\frac{a_{i0}}{a_0} = \sqrt{1 + \frac{\gamma - 1}{2} M_0^2} ,$$

provides, with the numerical data for use in the exercise, $a_0 = 340$ m/s, from which we can easily deduce the inlet flow velocity $U_0 = 68$ m/s.

To obtain the mass flow rate, it is necessary to determine the inlet density. This value (ρ_0) is deducted from that at the stagnation conditions (ρ_{i0}) by:

$$\rho_0 = \rho_{i0} / \left(1 + \frac{\gamma - 1}{2} M_0^2\right)^{1/(\gamma-1)} .$$

Knowing that $\rho_{i0} = P_{i0}/(r T_{i0}) = 2.4 \, kg/m^3$, we deduce that $\rho_0 \approx 2.35 \, kg/m^3$. The mass flow rate can then be calculated:

$$q_m = \rho_0 U_0 A_0 \approx 1.26 \, kg/s.$$

2. Due to heat losses, the total thermal power (Q_T) actually supplied to the fluid is $Q_T = 150 \times 0.92 = 138$ kW. Therefore, the heat input per unit mass of fluid is $Q = Q_T/q_m \approx 109.5$ kJ/kg. Now, by integrating the energy equation (8.28) between the input and output sections, we obtain:

$$\left[C_p T + \frac{U^2}{2}\right]_0^s = [Q]_0^s \quad \text{or} \quad C_p T_s + \frac{U_s^2}{2} = C_p T_0 + \frac{U_0^2}{2} + Q .$$

By introducing the adiabatic stagnation temperature, the previous relation directly leads to $C_p T_{is} = C_p T_{i0} + Q$. With the relevant numerical data, we obtain $T_{is} \approx 399$ K or 126° C.

3. As we have learned, the ratio of the adiabatic stagnation temperatures between two sections only depends on the flow Mach numbers in each of these sections. Thus the flow Mach number value at the outlet section can be deduced from the previous results using relation (8.30). The calculation can be processed by successive approximations or by graphic resolution. The result is $M_s \approx 0.24$.

Now, since the Mach numbers M_0 and M_s are determined, it is easy to deduce the temperature value T_S from T_0 by relation (8.29). We obtain $T_s \approx 1.375 \times T_0 \approx 396$ K or again $123\,°C$. Thus, the speed of sound at the outlet section is $\mathbf{a}_s = \sqrt{r T_s} \approx \mathbf{399}$ m/s. Accordingly the velocity after heating has increased to 96m/s.

Part III
Real Fluid Motions

Chapter 9
The Incompressible Navier-Stokes Model

Abstract This chapter opens with the formulation of the motion equations of a viscous incompressible fluid whose dimensionless expressions are further investigated in terms of a time scale analysis. The rotational and energetic properties are then discussed before presenting examples of flows which are exact solutions of these equations. The chapter concludes with physical considerations on stability and transition to turbulence.

9.1 Introduction

The chapter opens the third part devoted to viscous fluid motions. As in the rest of this work, we will assume that the medium is *monophasic* and of *homogeneous* chemical composition throughout the flow field. As we discussed in Chap. 3 Sect. 3.1.3, the *isovolume* character is then equivalent to a *constant density* evolution. In addition, we will assume that the molecular diffusive properties are expressed by constant coefficient schemes. In this chapter, the study is focused on fluid motions which can be obtained as an '*exact solution*' to the incompressible Navier-Stokes equations. Subsequently we will continue the presentation according to the classification presented in Chap. 4. Chapter 10 will thus be devoted to very low Reynolds number regimes, whereas Chaps. 11 and 12 will deal with *laminar* flows of *boundary layer* type at high Reynolds numbers.

9.2 A Return to the Model Equations

9.2.1 The Velocity-Pressure Formulation

As we have learned in Chap. 4, the thermal problem is decoupled from the dynamic problem in an isovolume evolution of a constant density fluid. The latter is governed by the following two equations for the velocity and pressure:

© The Author(s), under exclusive license to Springer Nature Switzerland AG 2022
P. Chassaing, *Fundamentals of Fluid Mechanics*,
https://doi.org/10.1007/978-3-031-10086-4_9

$$
\begin{array}{ll}
\textbf{Continuity} & \dfrac{\partial U_j}{\partial x_j} = 0 \\[3mm]
\textbf{Momentum} & \dfrac{\partial U_i}{\partial t} + U_j \dfrac{\partial U_i}{\partial x_j} = F_i - \dfrac{1}{\rho}\dfrac{\partial P}{\partial x_i} + \nu \dfrac{\partial^2 U_i}{\partial x_j \partial x_j}
\end{array}
$$

By using the vectorial identities recalled in Appendix C, we can ascertain that, under the study assumptions, the viscosity term can be simplified since:

$$
\nu \Delta \mathbf{V} \equiv \nu \left[-curl\,(curl\,\mathbf{V}) + \mathbf{grad}\,(div\mathbf{V}) \right] = -\nu\,curl\,(curl\,\mathbf{V}) .
$$

Hence the momentum equation can be put in the following equivalent form, where the physical meaning of the different terms is explained:

$$
\underbrace{\frac{\partial \mathbf{V}}{\partial t}}_{} + \underbrace{\mathbf{grad}\!\left(\frac{\|\mathbf{V}\|^2}{2}\right) + curl\,\mathbf{V} \wedge \mathbf{V}}_{} = \underbrace{\mathbf{F}}_{} \underbrace{- \frac{1}{\rho}\mathbf{grad}\,P}_{} \underbrace{- \nu\,curl\,(curl\,\mathbf{V})}_{}
$$

$$
\text{(9.1)}
$$

$$
\underbrace{\text{time variation} \qquad \text{advection}}_{} \qquad\qquad \underbrace{\text{pressure} \qquad \text{viscosity}}_{}
$$

$$
\underbrace{\text{volume} \qquad\qquad \text{surface}}_{}
$$

$$
\textit{Inertial Forces} \qquad\qquad\qquad \textit{External Forces}
$$

9.2.2 The Vorticity Equation

We can rewrite Eq. (9.1) by taking the *curl* or *rotational* of the velocity vector field $\mathbf{\Omega} = curl\,\mathbf{V}$ which is twice the vorticity vector (see Chap. 2 Sect. 2.4.1). Assuming that the external body forces derive from a potential $\mathcal{F}$, we obtain:

$$
\frac{\partial \mathbf{V}}{\partial t} + \mathbf{grad}\left(\frac{\|\mathbf{V}\|^2}{2}\right) + \mathbf{\Omega} \wedge \mathbf{V} = \mathbf{grad}\,\mathcal{F} - \frac{1}{\rho}\mathbf{grad}\,P - \nu\,curl\,\mathbf{\Omega}.
$$

By taking the *curl* throughout the equation, we deduce:

$$
\frac{\partial \mathbf{\Omega}}{\partial t} + curl\,(\mathbf{\Omega} \wedge \mathbf{V}) = -curl\left(\frac{1}{\rho}\mathbf{grad}\,P\right) - \nu\,curl\,(curl\,\mathbf{\Omega}) . \tag{9.2}
$$

Now, we have seen that $\nu\,curl\,(curl\mathbf{\Omega}) \equiv -\nu \Delta \mathbf{\Omega}$ and, since the density does not vary within the flow field:

$$curl\left(\frac{1}{\rho}\,\textbf{grad}\,P\right) = \frac{1}{\rho}\,curl\,(\textbf{grad}\,P) \equiv 0.$$

Finally, Eq. (9.2) takes the form:

$$\frac{\partial\boldsymbol{\Omega}}{\partial t} + curl\,(\boldsymbol{\Omega}\wedge\textbf{V}) = \nu\Delta\boldsymbol{\Omega} \tag{9.3}$$

By specifying the components and returning to index notations, we can ascertain that the vorticity equation can also be written in the following equivalent form:

$$\frac{\partial\Omega_i}{\partial t} + U_j\frac{\partial\Omega_i}{\partial x_j} - \Omega_j\frac{\partial U_i}{\partial x_j} = \nu\frac{\partial^2\Omega_i}{\partial x_j\partial x_j} \tag{9.4}$$

The following conclusions can therefore be deduced from this first part of the study

- The general vorticity equation of an incompressible viscous fluid flow is no longer of the classical *advection-diffusion* type, due to the presence of the additional coupling term between the rotational and velocity fields.
- The pressure explicitly disappears from the vorticity equation in an incompressible viscous fluid motion.

Remark One can very easily extend the above conclusions and Eq. (9.4) to a barotropic fluid. Indeed, we have:

$$curl\left(\frac{1}{\rho}\textbf{grad}\,P\right) = \frac{1}{\rho}\,curl\,(\textbf{grad}\,P) + \textbf{grad}\left(\frac{1}{\rho}\right)\wedge\textbf{grad}\,P,$$

or again $curl\left(\dfrac{1}{\rho}\textbf{grad}\,P\right) = \textbf{grad}\left(\dfrac{1}{\rho}\right)\wedge\textbf{grad}\,P \equiv -\dfrac{1}{\rho^2}\textbf{grad}\rho\wedge\textbf{grad}\,P.$ Now, under the barotropy assumption—$f\,(P,\rho) = 0$—the vectors $\textbf{grad}\rho$ and $\textbf{grad}\,P$ are necessarily colinear, so that their cross product is identically zero. The vorticity equation in the previous form is therefore still applicable.

9.2.3   The Velocity-Potential Equation

As previously mentioned in Chap. 5, the solenoidal character of the velocity field allows us to introduce a vector field function $\boldsymbol{\Psi}\,(M,t)$, called the velocity potential vector, such as:

$$\forall M,\ \forall t,\quad \textbf{V}\,(M,t) = curl\,[\boldsymbol{\Psi}\,(M,t)].$$

With this function, the continuity equation is identically verified, so that the calculation of the flow field is solely reduced to the determination of the vector potential, whose equation is directly deduced from Eq. (9.3). Indeed, by expressing the *curl* of the velocity field by using the function $\boldsymbol{\Psi}$, namely $\boldsymbol{\Omega} = curl\,(curl\,\boldsymbol{\Psi})$, we directly obtain:

$$\frac{\partial}{\partial t}curl\,[curl\,\boldsymbol{\Psi}] + curl[curl\,(curl\,\boldsymbol{\Psi}) \wedge curl\,\boldsymbol{\Psi}] = \nu\Delta\,[curl\,(curl\,\boldsymbol{\Psi})]\,, \quad (9.5)$$

which is a non-linear partial derivative equation, of four-order with respect to space variables.

9.2.4 The Two-Dimensional Plane Flow Case

An important simplification of the previous equations takes place in a *bidimensional plane* flow $(x,\,y)$, where, as we have seen in Chap. 5 Sect. 5.3.3, $\boldsymbol{\Psi}\,(M,t) = \psi\,(M,t)\,\mathbf{k}$ and $\boldsymbol{\Omega} = \Omega\,\mathbf{k}$, so that

$$\Omega = -\Delta\psi,$$

by denoting $\mathbf{k}$ the unit normal vector of the flow plane. As a result, the vectorial equation (9.5) is reduced to the scalar formulation:

$$\frac{\partial\,(\Delta\psi)}{\partial t} + \frac{\partial\psi}{\partial y}\frac{\partial\,(\Delta\psi)}{\partial x} - \frac{\partial\psi}{\partial x}\frac{\partial\,(\Delta\psi)}{\partial y} = \nu\Delta\,(\Delta\psi)\,. \quad (9.6)$$

On the other hand, we have

$$curl\,(\boldsymbol{\Omega} \wedge \mathbf{V}) = -\Delta\psi.curl\,(\mathbf{k} \wedge \mathbf{V}) - \mathbf{grad}\,(\Delta\psi) \wedge (\mathbf{k} \wedge \mathbf{V})\,.$$

Now, under this same assumption, it is easy to derive that $curl\,(\mathbf{k} \wedge \mathbf{V}) \equiv \mathbf{0}$ and $\mathbf{grad}\,(\Delta\psi) \wedge (\mathbf{k} \wedge \mathbf{V}) = -\mathbf{V} \odot \mathbf{grad}\,(\Delta\psi)$, so that Eq. (9.3) finally becomes:

$$\frac{\partial\Omega}{\partial t} + \mathbf{V} \odot \mathbf{grad}\Omega = \nu\Delta\Omega\,, \quad (9.7a)$$

or, in index notations

$$\frac{\partial\Omega_i}{\partial t} + U_j\frac{\partial\Omega_i}{\partial x_j} = \nu\frac{\partial^2\Omega_i}{\partial x_j\partial x_j}\,. \quad (9.7b)$$

It can therefore be concluded that:

- In any two-dimensional plane flow of a barotropic viscous fluid, the vorticity is governed by a *classical* transport equation.
- The rotational-velocity field coupling of the general formulation in Eq. (9.4), is thus specific to a *three-dimensional* evolution.

9.2.5 Poisson's Equation for the Pressure

Taking the divergence throughout Eq. (9.1), where the density and kinematic viscosity are considered as constant quantities, and assuming that body forces derive from a potential, we have:

$$div \left(\mathbf{grad} \frac{V^2}{2} \right) + div \, (\mathbf{\Omega} \wedge \mathbf{V}) = div \, (\mathbf{grad}\mathcal{F}) - \frac{1}{\rho} div \, (\mathbf{grad} P).$$

Now, it can be written $div \, (\mathbf{\Omega} \wedge \mathbf{V}) = \mathbf{V}. curl \, \mathbf{\Omega} - \mathbf{\Omega}. curl \mathbf{V} = \mathbf{V}. curl \, \mathbf{\Omega} - ||\mathbf{\Omega}||^2$. However we know that $curl\mathbf{\Omega} \equiv curl \, (curl\mathbf{V}) = \mathbf{grad} \, (div\mathbf{V}) - \Delta \mathbf{V} = -\Delta \mathbf{V}$. Putting all these results together, and remembering that $div \, (\mathbf{grad} f) \equiv \Delta f$, we finally arrive at the relation:

$$\frac{1}{\rho} \Delta P = \Delta \mathcal{F} - \Delta \left(\frac{||\mathbf{V}||^2}{2} \right) + \mathbf{V}.\Delta \mathbf{V} - ||\mathbf{\Omega}||^2 \qquad (9.8a)$$

Taking the index formulation of the momentum equation, the same process yields the equivalent expression:

$$\frac{1}{\rho} \frac{\partial^2 P}{\partial x_i \partial x_i} = \frac{\partial^2 \mathcal{F}}{\partial x_i \partial x_i} - \frac{\partial U_j}{\partial x_i} \frac{\partial U_i}{\partial x_j} \qquad (9.8b)$$

It thus appears that the pressure is governed by a Poisson equation, which radically distinguishes it from the other flow functions, which are governed by transport equations.

ADDITIONAL INFORMATION—The pressure field integration according to Green's formulation—Using Green's function for the Laplacian, we can express, for example, the pressure field from the last relation. Let us recall that, if $f \, (M)$ is a continuous function, derivable on a domain $\mathcal{D}$ as well as on its boundary $\mathcal{S}$ and with continuous derivatives, we then have:

$$f \, (M) = -\frac{1}{4\pi} \iiint\limits_{\mathcal{D}} \frac{\Delta f}{r} \, dv(M') + \frac{1}{4\pi} \iint\limits_{\mathcal{S}} \left(\frac{1}{r} \frac{\partial f}{\partial n} - f \frac{\partial \left(\frac{1}{r} \right)}{\partial n} \right) d\sigma[M'].$$

In the previous integrals, r is the distance between point M where the function f is expressed and the current point M' within the integration domain, $\mathcal{D}$ and $\mathcal{S}$ stand for the volume and bounding surface of this domain. Applied to the pressure field, this relation becomes, assuming for simplification reasons, the absence of external body forces:

$$P(M) = \frac{\rho}{4\pi} \iiint_{\mathcal{D}} \frac{\partial U_j}{\partial x_i} \frac{\partial U_i}{\partial x_j} \frac{dv(M')}{r} + \frac{1}{4\pi} \iint_{\mathcal{S}} \left(\frac{1}{r} \frac{\partial P}{\partial n} - P \frac{\partial \left(\frac{1}{r}\right)}{\partial n} \right) d\sigma(M').$$

9.3 The Time-Based Analysis of the Momentum Balance

9.3.1 The Characteristic Time Scales of the Momentum Equation

A dimensionless formulation of the momentum equation has been previously introduced in Chap. 4 Sect. 4.7.3. Here we focus on an alternative analysis aimed at basing the relative comparison of the various terms on characteristic *time* scales.

Referring to U_o, T_o, L_o, P_o and g_o as the velocity, time, length, pressure, and gravitational acceleration scale, we introduce the following dimensionless quantities:

$$U_i^* = U_i/U_o, \quad t^* = t/T_o, \quad x_i^* = x_i/L_o, \quad P^* = P/P_o, \quad F_i^* = F_i/g_o.$$

After substitution in the momentum equation, we obtain:

$$\frac{1}{T_i}\left(\frac{\partial U_i}{\partial t}\right)^* + \frac{1}{T_a}\left(U_j\frac{\partial U_i}{\partial x_j}\right)^* = \frac{F_i^*}{T_v} - \frac{1}{T_p}\left(\frac{\partial P}{\partial x_i}\right)^* + \frac{1}{T_d}\left(\frac{\partial^2 U_i}{\partial x_j \partial x_j}\right)^*.$$

This relation allows us to identify five characteristic times T_i, T_a, T_v, T_p and T_d, associated to each term of the momentum equation. Their expressions are given in the following Table.

Table 9.1 Characteristic time scales associated with the momentum equation

Symbol	Expression	Signification
T_i	T_o	Unsteadiness
T_a	L_o/U_o	Advective transport over a distance L_o
T_v	U_o/g_o	Gravitational force action
T_p	$\rho U_o L_o/P_o$	Pressure force action
T_d	L_o^2/v	Molecular diffusion over a distance L_o

The comparison of the relative orders of magnitude of the various terms in the momentum equation thus simply returns to considering *characteristic time scale ratios*. This is what will be explained in the next sections.

Remark The analysis we have just introduced on the momentum equation extends to any balance equation of any transportable property. Indeed, denoting $q(x_i, t)$ such a property expressed as an Euler variable function, its transport equation is of the form

$$\frac{\partial q}{\partial t} + U_j \frac{\partial q}{\partial x_j} = m_1 + m_2 + \cdots,$$

where the unexplained terms of the right hand side are simply denoted as m_1, m_2, etc.

Dividing the previous relation throughout by $q(x_i, t) \neq 0$ we obtain

$$\frac{1}{q}\frac{\partial q}{\partial t} + U_j \frac{1}{q}\frac{\partial q}{\partial x_j} = \frac{m_1}{q} + \frac{m_2}{q} + \cdots,$$

where it is clear that each term dimension in the equation is the *inverse of a time*. In the term-to-term comparison, the predominant contributions in any transport equation are therefore those of a higher frequency, namely a smaller time scale.

9.3.2 Characteristic Numbers as Time Scale Ratios

The ratios between pairwise time scales of Table 9.1, provide an alternative interpretation of the characteristic numbers derived from the dimensionless equation formulation explained in Chap. 4 Sect. 4.7.5. Thus, provided that *the same length scale* is chosen in the definition of the relevant time scales, we obtain, for the Euler and Reynolds number respectively:

$$E_u \equiv \frac{P_o}{\rho_o U_o^2} = \left(\frac{T_a}{T_p}\right)_{L_o}, \tag{9.9}$$

$$R_e \equiv \frac{U_o L_o}{\nu} = \left(\frac{T_d}{T_a}\right)_{L_o}. \tag{9.10}$$

Similarly, if the time scales are now defined with *the same velocity reference*, we obtain for the Froude number:

$$F_r \equiv \frac{U_o^2}{L_o g_o} = \left(\frac{T_v}{T_a}\right)_{U_o}. \tag{9.11}$$

Finally, in order to compare the characteristic time scale of the flow instationarity with that of advection, more specifically we can make use of

– the Strouhal number, in the presence of a periodic type unsteadiness whose frequency is n_o:

$$S_t \equiv \frac{n_o L_o}{U_o} = \frac{T_a}{T_i}, \tag{9.12}$$

– the Rossby number, when the flow occurs in a rotating frame whose angular velocity is Ω_o:

$$R_o \equiv \frac{U_o}{\Omega_o L_o} = \frac{T_i}{T_a}. \tag{9.13}$$

HISTORICAL NOTE—The Strouhal number—This number is named after the Czech physicist Vincent (Čeněk) Strouhal (1850–1922). With reference to the paper he published in 1878 [152], this number is used to characterize an unsteady flow of a periodic nature, such as that observed in the wake downstream of a circular cylinder, in the alternating vortex shedding regime (Bénard– von Kármán's vortex street). For $S_t \ll 1$, the motion can be considered as almost stationary.

HISTORICAL NOTE—The Rossby number—This number is named after the Swedish meteorologist Carl-Gustaf Arvid Rossby (1898–1957) as a tribute to his work. It is used, for example, in the analysis of turbomachinery flows or in geo-fluid dynamics. The Rossby number can also be understood as the ratio between inertial and rotational forces. When $R_o \gg 1$, the rotation effects can be neglected, otherwise, they should be taken into account, as is the case for a high-speed rotating machine, a large oceanic circulation, an atmospheric anticyclone, etc. For the motion study in the Earth's atmosphere, for which it was originally introduced, the Earth's angular rotation velocity, Ω_o, is substituted by the *Coriolis parameter* $f = 2\Omega_o \sin\phi$. This parameter, whose dimension is also the inverse of a time, represents in absolute value, the horizontal projection of the Coriolis force at the point of latitude ϕ.

9.3.3 The Advection/Diffusion Comparison: New Reynolds Number Interpretations

Relation (9.10), where the Reynolds number appears as the ratio between the characteristic *diffusion* and *advection* time scales, only applies if both mechanisms take place with the *same length scale*.

If the advective transport and diffusive transfer mechanisms take place at different length scales, respectively denoted by L_a and l_d, see Fig. 9.1, then the associated characteristic times are given by:

$$T_a \simeq \frac{L_a}{U_o} \quad \text{and} \quad T_d \simeq \frac{l_d^2}{\nu}.$$

Fig. 9.1 The advective transport and diffusive transfer with the associated length scales

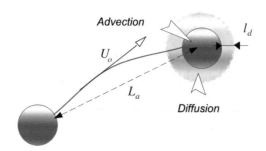

With these estimates, we can make a second comparison of the two mechanisms provided they operate *at the same time scale*. In this case, in fact, we immediately have that $\dfrac{L_a}{U_o} = \dfrac{l_d^2}{\nu}$, hence:

$$\boxed{R_e = \left(\frac{L_a}{l_d}\right)^2_T}\tag{9.14}$$

On the basis of relations (9.10) and (9.14), it can therefore be concluded that:

- The Reynolds number compares the diffusion time to the advection time, when both mechanisms occur on the *same length scale*;
- The Reynolds number compares the *square* of the advection and diffusion length scales, when these two mechanisms operate at the *same time scale*.

This provides more refined interpretations of the Reynolds number, going beyond the "mere" comparison between viscosity and inertia forces. We will have the opportunity to look at the interest of the interpretation at the *same time scale*, with regard to the boundary layer concept in Chap. 11.

ADDITIONAL INFORMATION—**The Péclet, Schmidt and Lewis numbers**—The previous *advection/diffusion* comparative analysis is obviously not only confined to the dynamic aspect regarding the momentum equation. When applied to the heat transfer equation, the characteristic (*time* or *length*) scale comparison of the pure thermodiffusion (conduction) and convection mechanisms, leads to introducing the *thermal* Péclet[1] number, equivalent to the Reynolds number, as defined by:

$$Pe_{therm} = \frac{U_o L}{a} \,,\tag{9.15}$$

where the thermal diffusivity a is substituted for the kinematic viscosity ν.

---

[1] In tribute to Jean Claude Eugène Péclet, a French physicist (1793–1857).

It is easy to verify that the *thermal* Péclet number is equal to the product of the Reynolds and Prandtl numbers, i.e., $Pe_{therm} = Re \times Pr$. It is therefore subject to the same interpretations as the Reynolds number.

When the fluid, which is still monophasic, is no longer of a homogeneous composition, the same analysis applies to the mass diffusion mechanism whose ratio to the advection introduces the *mass* Péclet number:

$$Pe_{mass} = \frac{U_o L}{\mathcal{D}} = Re \times Sc, \qquad (9.16)$$

where

$$Sc = \nu/\mathcal{D}, \qquad (9.17)$$

defines the Schmidt number,[2] as the ratio of the kinematic viscosity ν and mass diffusivity $\mathcal{D}$ of a given species in the fluid.

To complete the intercomparison of the diffusion processes solely at the molecular scale, the Lewis number[3] should be mentioned, as defined by:

$$Le = \frac{a}{\mathcal{D}} \left(\equiv \frac{a}{\nu} \times \frac{\nu}{\mathcal{D}} \right) = \frac{Sc}{Pr}. \qquad (9.18)$$

EXERCISE 38 The perfume diffusion in the air of a room.

A device generating an aroma by a liquid evaporation or solid sublimation is placed in a room whose characteristic length is $L = 5\,\mathrm{m}$. It is assumed that the olfactory principle spreads around the room by pure molecular diffusion, with a diffusivity $\mathcal{D} = 0.25\,\mathrm{cm^2/s}$.

QUESTION 1. Estimate the time T_d required to diffuse the perfume over the room length scale.

QUESTION 2. In order to shorten the process time, the flavour generator is equipped with a fan which sets in motion the air in the room at a characteristic velocity $U_o = 5\,\mathrm{cm/s}$. What is the time scale value T_a of the advective transport over the room scale?

QUESTION 3. It is assumed that the Lewis number is equal to one. In the situation defined in the previous question, calculate the Péclet number value. What is the physical signification of the obtained result?

QUESTION 4. During the time interval T_a, what is the characteristic dimension of the volume in which the fragrance's molecular diffusion takes place?

---

[2] In tribute to Ernst Heinrich Wilhelm Schmidt, a German physicist (1892–1975).
[3] In tribute to Warren Kendall Lewis, an American chemist and physicist (1882–1975).

SOLUTION: **1.** The characteristic time scale of a purely diffusive transfer over a distance L is $T_d = L^2/\mathcal{D}$. With the numerical data for use in the exercise, we find $T_d = 10^6\,\text{s} \approx 278\,h \approx 11.5$ days.

2. We have $T_a = L/U_o$, which gives us a value of $100\,\text{s}$ with the relevant data.

3. By definition of the Péclet number, we can write, when the Lewis number is equal to the unit ($a = \mathcal{D}$):

$$Pe = \frac{U_o L}{a} \equiv \frac{U_o L}{\mathcal{D}},$$

whose value here is 10^4. In the situation under consideration, the diffusion transfer and advection transport are compared at the same scale, namely the prescribed room length. We can therefore write:

$$Pe = \frac{U_o L}{\mathcal{D}} = \frac{U_o}{L} \times \frac{L^2}{\mathcal{D}} \equiv \left(\frac{T_d}{T_a}\right)_L,$$

which means that the relevant number compares the diffusion and advection time scales on the same length scale. We can verify that $T_d = 10^4\,T_a$.

4. If the diffusion process can freely proceed during a time interval $\Delta T = 10^2\,\text{s}$, the characteristic length scale δ of the space extension concerned with the diffusive transfer is given by:

$$\delta = \sqrt{\Delta T \times \mathcal{D}}.$$

With the relevant numerical numerical data, we find $\delta = 5\,\text{cm}$. This result demonstrates that, for distances much greater than δ, the diffusive transfer phenomenon can be neglected with respect to the advective transport. Conversely, if one refers to scales of a few centimeters, such a conclusion is erroneous.

9.4   The Rotational Properties of a Fluid Motion

9.4.1   The Vortical Nature of the Flow on a Solid Body

To specify the analysis and without prejudice to the conclusion's generality, we consider the case of a viscous fluid, flowing, with the velocity $\mathbf{V}_\infty$ at infinity in a given referential, on a solid body, fixed or possibly moving with a different velocity in this same referential. To simplify, we will reason in a reference frame linked to the obstacle. In such a situation, the flow field velocity should satisfy two different boundary conditions:

(a) $\mathbf{V} = \mathbf{0}$ along the solid surface;
(b) $\mathbf{V} = \mathbf{V}_\infty \neq \mathbf{0}$ at a distance from the body.

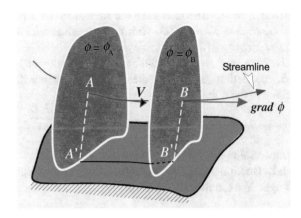

Denoting by ν the fluid kinematic viscosity and by L a characteristic length of the
obstacle, we limit the analysis here to flows whose global Reynolds number $V_\infty L/\nu$
is high. Now, let us consider a curvilinear section $\overgroup{AB}$ of any streamline, *close* to
the wall. By adopting a proof by contradiction, we will assume that the flow field is
irrotational everywhere, so that there are two equipotential surfaces $\phi = C^t (\equiv \phi_A)$
and $\phi = C^t (\equiv \phi_B)$ passing through points A and B respectively (Fig. 9.2).
The proximity of the obstacle then makes it possible to ensure the existence of a
non-empty intersection between these surfaces and the solid wall. Let A' and B' be
any two points of these intersections respectively and (Σ) a surface bounding the
contour $AA'B'B$.
Then we can write $\displaystyle\int_{\overgroup{AB}} \mathbf{V}\, \mathbf{dl} = \oint_{AA'B'B} \mathbf{V}\, \mathbf{dl} = \iint_\Sigma curl\, \mathbf{V}.\mathbf{n}\, d\sigma,$
the last equality resulting from the direct application of Stokes' theorem (Chap. 2
Sect. 2.4.3). As a result, if the velocity field were irrotational everywhere, we would
have[4]:

$$\forall A, B \quad \int_{\overgroup{AB}} \mathbf{V}\, \mathbf{dl} = 0,$$

a relation which can only be ascertained in the absence of flow, so that the motion is
necessarily rotational.

---

[4] We shall see later (Chap. 11) the reason for this restriction. Let us simply mention here that the
experimental observation reveals that at a high global Reynolds number, the irrotational character
generated by the velocity gradient at the wall does not significantly extend to the entire field to
infinity.

9.4.2 The Elementary Mechanisms of the Vorticity Dynamics

The vorticity transport equation (9.4) consists of four terms:

$$\underbrace{\frac{\partial \Omega_i}{\partial t}}_{(a)} + \underbrace{U_j \frac{\partial \Omega_i}{\partial x_j}}_{(b)} = \underbrace{\Omega_j \frac{\partial U_i}{\partial x_j}}_{(c)} + \underbrace{\nu \frac{\partial^2 \Omega_i}{\partial x_j \partial x_j}}_{(d)},$$

where one readily identifies the time variation (a), advection (b) and diffusion by molecular agitation (d) whose diffusivity is none other than the fluid kinematic viscosity ν. As previously mentioned, the specificity of this equation in a three-dimensional situation[5] is due to the presence of the *velocity-vorticity* coupling term (c), about which we will make the following two additional remarks.

- The velocity-vorticity coupling term has a *non-linear* character since the velocity and vorticity fields are not independent;
- The velocity-vorticity coupling only exists in the presence of *strain rates*. Indeed, by decomposing the velocity gradient into its symmetric and antisymmetric parts we can write:

$$\Omega_j \frac{\partial U_i}{\partial x_j} = \Omega_j \left(S_{ij} + R_{ij} \right),$$

or, taking into account the relation $R_{ij} = -\frac{1}{2} \epsilon_{ijk} \Omega_k$,

$$\Omega_j R_{ij} = -\frac{1}{2} \epsilon_{ijk} \Omega_j \Omega_k \equiv -\frac{1}{2} \epsilon_{ikj} \Omega_k \Omega_j \equiv +\frac{1}{2} \epsilon_{ijk} \Omega_j \Omega_k,$$

a chain of equalities which proves that the antisymmetric part R_{ij} is necessarily zero. Hence we have

$$\Omega_j \frac{\partial U_i}{\partial x_j} = \Omega_j S_{ij}.$$

9.4.3 The Vorticity-Velocity Interaction

Here we focus the analysis on the non-linear velocity-vorticity coupling in the $\boldsymbol{\Omega}$ transport equation. Each component of this vectorial term (c) takes the form of three additive contributions with, for example, for the first component ($i = 1$):

$$\Omega_j \frac{\partial U_1}{\partial x_j} = \Omega_1 \frac{\partial U_1}{\partial x_1} + \Omega_2 \frac{\partial U_1}{\partial x_2} + \Omega_3 \frac{\partial U_1}{\partial x_3}.$$

---

[5] This term is identically zero in a two-dimensional flow, Cf. Sect. 9.2.4.

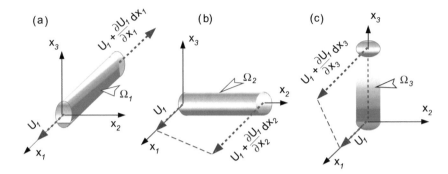

Fig. 9.3 Schematic for the velocity-vorticity interaction: **a** vortex stretching/shrinking, **b** and **c** vortex tilt

As outlined in Fig. 9.3, the interaction phenomenology associated with each of these contributions can be devided in two different mechanisms, depending on whether the curl of the velocity and strain rate components are collinear or orthogonal to each other:

- in the presence of *collinear* components—Fig. 9.3a—, the mechanism consists in the stretching/shrinking of the vortex;
- in the presence of *orthogonal* components— Fig. 9.3b, c—, the mechanism is a *vortex tilt*, in a plane normal to that defined by both the vorticity and strain rate components.

As the effects mentioned are totally characterized by the respective orientations of the vorticity and strain rates, it is easy to extend the previous remarks on the first coupling vector component to the other two components.

Thus, it can be concluded that:

> The type of the velocity-vorticity interaction phenomenology is that of:
>
> - a vortex tilt, for such terms as $\Omega_\alpha \frac{\partial U_\beta}{\partial x_\alpha}$, with $\alpha \neq \beta$ (*no α index summation*);
> - a vortex stretching/shrinking for such terms as $\Omega_\alpha \frac{\partial U_\alpha}{\partial x_\alpha}$ (*no summation*).

It should be noted that a first manifestation of the *non-linear* character of this coupling appears in the fact that tilt situations are twice as frequent as those of stretch/shrinkage.

ADDITIONAL INFORMATION—**A look back at the persistence of the vortex lines**—We have demonstrated, in Chap. 2 Sect. 2.4.6.1, the persistence of vortex lines in any flow with an acceleration potential. By assumption, this property applies to *inviscid* fluid flows. It thus originates in inertial mechanisms and can, as such, provide an example of the role played by the *velocity-vorticity* coupling term.

Fig. 9.4 Schematic for the transport of a vortex line element

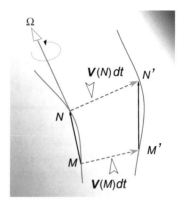

Let us first of all note that this persistence is obvious in two-dimensional plane situations. In this case, the vorticity transport equation reduces, for an *ideal fluid*, to:

$$\frac{\partial \Omega_i}{\partial t} + U_j \frac{\partial \Omega_i}{\partial x_j} \equiv \frac{d\Omega_i}{dt} = 0 .$$

Clearly the vorticity conservation, in terms of the material variation, reflects the fact that its *temporal* variation is balanced by its *convective transport* due to the vortex line displacement at the local flow velocity.

Let us now turn to the three-dimensional case and consider, at time t, a vectorial element **MN** of a vortex line. We suppose the points to be sufficiently close to take as this vector components, $dx_j = \Omega_j \, d\ell$, designating by $\mathbf{\Omega}$ the vorticity at point M and time t, and by $d\ell$ the elementary line segment length.

At time $t + dt$:

– point M has moved to M' with $\mathbf{MM'} = \mathbf{V}(M)dt$, a vector whose components are $U_i(M)dt$,

– point N has moved to N' with $\mathbf{NN'} = \mathbf{V}(N)dt$, a vector whose components are $U_i(N)dt$.

Now, we can write, to a first-order approximation:

$$U_i(N) = U_i(M) + \frac{\partial U_i}{\partial x_j} dx_j = U_i(M) + \frac{\partial U_i}{\partial x_j} \Omega_j \, d\ell ,$$

since the displacement vector **MN** is taken along the vortex line (Fig. 9.4).

From the vectorial identity, $\mathbf{MN} + \mathbf{NN'} + \mathbf{N'M'} + \mathbf{M'M} = \mathbf{0}$, we obviously derive that $\mathbf{M'N'} = \mathbf{MN} + \mathbf{NN'} - \mathbf{MM'}$.

By substituting the expressions of the right hand side vectors, we obtain for the dx'_i components of the vector $\mathbf{M'N'}$:

$$dx'_i = \Omega_i \, d\ell + \left[U_i(M) + \frac{\partial U_i}{\partial x_j} \Omega_j \, d\ell \right] dt - U_i(M)dt = \left[\Omega_i + \frac{\partial U_i}{\partial x_j} \Omega_j dt \right] d\ell .$$

Now, according to the vorticity transport equation for an *ideal fluid*, we know that $\dfrac{d\Omega_i}{dt} = \Omega_j \dfrac{\partial U_i}{\partial x_j}$. We finally deduce that:

$$dx_i' = \left[\Omega_i + \frac{d\Omega_i}{dt} dt \right] d\ell \equiv \Omega_i' d\ell \,,$$

which clearly proves that the transported element **M'N'** belongs to a vortex line. The above demonstration is inspired by Loitsyanskii's general work [91] on the mechanics of liquids and gases, taking up Helmholtz's line of reasoning [68] who was the first to put forward this result.

9.5 Energy Properties

Due to the decoupling between the dynamic and thermal problems, the energy aspects can be discussed in terms of the kinetic energy whose balance equation results in either of the following two equations:

$$
\begin{aligned}
\frac{d\left(\frac{1}{2}\rho U_i U_i\right)}{dt} &= \rho U_i F_i \ldots\ldots\ldots\ldots\ldots \quad External\ body\ force\ power \\
&= -\frac{\partial (P U_i)}{\partial x_i} \ldots\ldots\ldots\ldots \quad External\ pressure\ force\ power \\
&= \frac{\partial}{\partial x_j}\left[\mu U_i \left(\frac{\partial U_i}{\partial x_j} + \frac{\partial U_j}{\partial x_i} \right) \right] \ldots External\ viscous\ force\ power \\
&= -\frac{\mu}{2}\left(\frac{\partial U_i}{\partial x_j} + \frac{\partial U_j}{\partial x_i} \right)^2 \ldots\ldots Viscous\ dissipation
\end{aligned}
$$

$$(9.19)$$

$$
\begin{aligned}
\frac{d\left(\frac{1}{2}\rho U_i U_i\right)}{dt} &= \rho U_i F_i \ldots\ldots\ldots\ldots\ldots \quad External\ body\ force\ power \\
&= -\frac{\partial (P U_i)}{\partial x_i} \ldots\ldots\ldots\ldots External\ pressure\ force\ power \\
&= \nu \frac{\partial^2}{\partial x_j \partial x_j}\left(\frac{1}{2}\rho U_i U_i \right) \ldots Diffusion\ by\ viscosity \\
&= -\frac{\mu}{2}\left(\frac{\partial U_i}{\partial x_j} \right)^2 \ldots\ldots\ldots Pseudo{-}dissipation\ by\ viscosity
\end{aligned}
$$

$$(9.20)$$

The physical interpretation of the right hand side terms in Eq. (9.19) is straightforward, based on the general analysis results of Chap. 3 (see Table 3.1). The same is true for the first three terms of the right hand side in Eq. (9.20). With regard to the last term, we can observe that it is always negative and thus acts as a sink term in the energy balance. Therefore, it has the same effect as the *exact* dissipation listed in (9.19). However, since it is formally different, it is termed pseudo-dissipation. Returning to the first formulation, we shall remember that, in particular:

> Any *non-solidifying* motion, i.e., with a *non-zero strain rate*, of a viscous incompressible fluid is *dissipative*.

Proof By scalar multiplication of the momentum equation throughout by the velocity we directly obtain:

$$\frac{d\left(\frac{1}{2}\rho U_i U_i\right)}{dt} = \rho U_i F_i - U_i \frac{\partial P}{\partial x_i} + \mu U_i \frac{\partial^2 U_i}{\partial x_j \partial x_j} .$$

The transformation of the pressure term is straightforward for an isovolume evolution:

$$\frac{\partial (P U_i)}{\partial x_i} = U_i \frac{\partial P}{\partial x_i} + P \frac{\partial U_i}{\partial x_i} = U_i \frac{\partial P}{\partial x_i} + 0 .$$

Turning now to the viscosity terms, it is easy to ascertain that

$$\frac{\partial}{\partial x_j}\left[U_i \left(\frac{\partial U_i}{\partial x_j} + \frac{\partial U_j}{\partial x_i} \right) \right] = U_i \frac{\partial}{\partial x_j}\left(\frac{\partial U_i}{\partial x_j} + \frac{\partial U_j}{\partial x_i} \right) + \frac{\partial U_i}{\partial x_j}\left(\frac{\partial U_i}{\partial x_j} + \frac{\partial U_j}{\partial x_i} \right)$$

$$= U_i \frac{\partial^2 U_i}{\partial x_j \partial x_j} + U_i \frac{\partial}{\partial x_i}(0) + \frac{\partial U_i}{\partial x_j}\left(\frac{\partial U_i}{\partial x_j} + \frac{\partial U_j}{\partial x_i} \right) .$$

To obtain the expected result, it is simply necessary to note that, for the last term on the right hand side, we have as a result of symmetry considerations:

$$\frac{\partial U_i}{\partial x_j} \times \left(\frac{\partial U_i}{\partial x_j} + \frac{\partial U_j}{\partial x_i} \right) \equiv \frac{\partial U_j}{\partial x_i} \times \left(\frac{\partial U_j}{\partial x_i} + \frac{\partial U_i}{\partial x_j} \right) = \frac{1}{2}\left(\frac{\partial U_i}{\partial x_j} + \frac{\partial U_j}{\partial x_i} \right) \times \left(\frac{\partial U_j}{\partial x_i} + \frac{\partial U_i}{\partial x_j} \right) .$$

The second viscosity term formulation—Eq. (9.20)—is more direct since:

$$\frac{1}{2}\frac{\partial^2 (U_i U_i)}{\partial x_j \partial x_j} = \frac{\partial}{\partial x_j}\left(U_i \frac{\partial U_i}{\partial x_j} \right) = U_i \frac{\partial^2 U_i}{\partial x_j \partial x_j} + \frac{\partial U_i}{\partial x_j} \times \frac{\partial U_i}{\partial x_j} .$$

9.6   Flows as an "Exact Solution" of the Navier-Stokes Equations

Under certain specific assumptions related to the boundary and initial conditions, it is possible to analytically solve the equations governing the incompressible viscous fluid motion, Eq. (9.1). We will discuss in detail here some examples of such flows qualified as *exact solutions* of the Navier-Stokes equations. Many calculation developments will be left to the reader's discretion, as an exercise, with a deliberate emphasis here on the physical interpretations of the results. We will primarily focus on stationary motions and subsequently on unsteady flows.

9.6.1   The Plane Poiseuille Flow

This first example is that of a viscous Newtonian,[6] weightless fluid flowing between two fixed parallel infinite planes. Thus, it is assumed that the motion is steady (stationary in time), rectilinear, plane-parallel and two-dimensional. According to the terms outlined in Fig. 9.5, the x axis is taken parallel to the motion direction, the direction normal to the flow being taken as the y axis. The z axis completes the orthonormal trihedron.

We denote by U, V, W the velocity vector components in the previously defined reference frame and by P the pressure.

Due to the 2D-plane flow character, we have, for any field function $\partial/\partial z \equiv 0$ and, because of stationarity, $\partial/\partial t \equiv 0$. Under these conditions, the dynamic problem is reduced to the determination of the four scalar functions $U(x, y)$, $V(x, y)$, $W(x, y)$ and $P(x, y)$ between the two planes.

By assumption and with the adopted conventions, the parallel nature of the motion is readily expressed by:

$$V(x, y) = W(x, y) \equiv 0\,,$$

so that the continuity equation becomes $\partial U/\partial x = 0$, which means that there is no variation in the x flow direction of the longitudinal velocity component.

This property originates here from the absence of *physical* limits for this *idealized* geometry, which is infinite in the motion direction. We will see at a later stage that for a *real* flow, there are *inlet* and *outlet* conditions which make the relation $\frac{\partial U}{\partial x} = 0$, characteristic of a fully kinematically developed regime, which is only satisfied over a given value range for x.

 • *Calculation of the flow field*
The main unknown functions of the problem are ultimately reduced to $U(x)$ and $P(x, y)$ which are governed by the following equations:

---

[6] Let us recall that this shortcut expresses the fact that gravity forces are not to be taken into account in the study.

Fig. 9.5 Schematic for the
plane Poiseuille flow
geometry

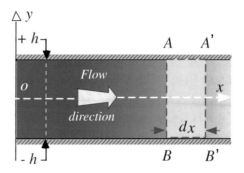

$$\begin{cases} 0 = -\dfrac{\partial P}{\partial x} + \mu \dfrac{d^2 U}{dy^2} \\ 0 = -\dfrac{\partial P}{\partial y}, \end{cases}$$

which are the Navier-Stokes equations in projections along x and y, taking into account the previous assumptions, the projection along z being identically verified. Therefore, there is no pressure variation along the vertical (y), and the momentum balance along the flow direction is reduced to the pressure and viscosity force equilibrium, according to the equation:

$$\frac{dP}{dx} = \mu \frac{d^2 U}{dy^2}, \tag{9.21}$$

where the total derivative notations are used here to account for the fact that the unknown functions solely depend upon a single argument, namely $P \equiv P(x)$ and $U \equiv U(y)$. This equation can therefore only be satisfied if both derivatives are equal to the same constant K. The pressure equation integration is straightforward and can be expressed as $P(x) = P_0 + Kx$, taking as the pressure reference $P_0 = P(0)$. The integration of the velocity differential equation leads to:

$$U(y) = \frac{K}{2\mu} y^2 + Ay + B,$$

where A and B are two constants which are to be specified by the boundary conditions. The no-slip condition at the fixed solid walls $y \pm h$ indeed produces $A = 0$ and $B = -\dfrac{K}{2\mu} h^2$. The velocity is therefore maximum on the axis where it is equal to:

$$U_{max} \equiv U(0) = -\frac{K}{2\mu} h^2.$$

The velocity profile is parabolic and is finally expressed, in a dimensionless form, as:

$$\boxed{\frac{U\,(y)}{U_{max}} = 1 - \frac{y^2}{h^2}} \tag{9.22}$$

It is then easy to ascertain that the volume flow rate per width unit is expressed by $Q_v = U_{mean} \times 2h$, where the average flow velocity is equal to $U_{mean} = \frac{2}{3}\,U_{max}$.

• **Discussion regarding the dynamic properties**
(a) It is firstly deduced from the previous results that the viscous stress tensor is reduced to:

$$\sigma_{ij}^v = \mu \begin{pmatrix} 0 & \mathrm{d}U/\mathrm{d}y & 0 \\ \mathrm{d}U/\mathrm{d}y & 0 & 0 \\ 0 & 0 & 0 \end{pmatrix}.$$

There are therefore no normal viscous stresses but solely a viscous shear which is locally expressed as:

$$\tau_{xy} = \mu \frac{\mathrm{d}U}{\mathrm{d}y} = K y\,.$$

This shear stress therefore varies linearly with the distance from the channel axis, namely, in a dimensionless form:

$$\boxed{\frac{\tau_{xy}\,(y)}{\tau_p} = \frac{y}{h}} \tag{9.23}$$

where τ_p stands for the upper wall skin friction ($\tau_p = \tau_{xy}\,(h) = Kh$).
(b) With the chosen orientation conventions, the value found for the velocity extremum is necessarily positive, which leads to the condition that K is negative. The result is a *decrease* in the *pressure* along the flow direction *without* any change in the *velocity*. No theorem of Bernoulli can therefore be invoked for this flow in which the fluid particles undergo, along their rectilinear displacement, a rotation and deformation due to the existing simple shear, see Chap. 2 Exercise 9.
(c) Equation (9.21) can be rewritten as $\dfrac{dP}{dx} = \dfrac{d\tau_{xy}}{dy}$, which by integration from 0 to h yields:

$$h\frac{dP}{dx} = \int\limits_0^h d\tau_{xy} = \tau_{xy}\,(h) - \tau_{xy}\,(0) \equiv \tau_p\,.$$

As an exercise, it can be proved that this result can also be obtained by applying Euler's theorem to a domain $ABA'B'$ as plotted in Fig. 9.5. Therefore it provides a clear *mechanical* interpretation of the pressure drop associated with the major head loss: *"The momentum conservation of the flow is the result of the exact equilibrium between the pressure and viscosity external forces throughout the entire cross-section"*, or in more general terms *"the pressure drop balances the skin friction"*.

• *Discussion regarding the energy properties*

(a) The essential energy characteristic of this steady flow is the *linear pressure drop*, which corresponds to a major or friction head loss along the flow direction. It is usual to express the pressure drop ΔP over a length L using a dimensionless head loss coefficient Λ such that:

$$\Lambda = -\frac{\Delta P}{\frac{1}{2}\,\rho\,U_{mean}^2}\,\frac{D}{L}. \tag{9.24}$$

By choosing for reference length $D = 2h$, we obtain with the previous results

$$\Lambda = \frac{24}{R_e}, \tag{9.25}$$

where $R_e = 2hU_{mean}/\nu$ is the global Reynolds number based on the channel width.

(b) By scalar multiplication of Eq. (9.21) by the local velocity $U(y)$ we obtain:

$$U\frac{dP}{dx} = \mu U\frac{d^2U}{dy^2} \equiv \mu\frac{d}{dy}\left(U\frac{dU}{dy}\right) - \mu\left(\frac{dU}{dy}\right)^2.$$

This equation is none other than that of the kinetic energy. It expresses the fact that the (reversible) power of the external pressure force balances (i) the power of the external viscosity forces (the first term of the right hand side) and (ii) the viscous dissipation (the second term). It is worth noting that, similar to the momentum and for the same reason, the kinetic energy is invariant along any streamline, i.e., at a given ordinate y.

By the integration of the previous equation throughout the channel cross section we obtain:

$$\frac{dP}{dx}\int_{-h}^{+h}U\,dy = \mu\left[U\frac{dU}{dy}\right]_{-h}^{+h} - \int_{-h}^{+h}\mu\left(\frac{dU}{dy}\right)^2 dy.$$

It is easy to verify that the term in square brackets is identically zero, so that the global energy balance over any channel cross-section is reduced to:

$$Q_v\frac{dP}{dx} = -\int_{-h}^{+h}\mu\left(\frac{dU}{dy}\right)^2 dy.$$

This last relation now allows us to give an *energetic* interpretation of the pressure drop along the flow direction, as the only external force whose power counterbalances the viscous irreversibilities throughout any cross-section.

Fig. 9.6 Velocity and
viscous friction distributions
of a Poiseuille flow in a pipe

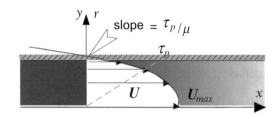

9.6.2 Poiseuille's Flow in a Pipe

Except for the geometric difference, this flow is, in all other respects, identical to
the previous flow. Thus, under the same assumptions as in Sect. 9.6.1, the viscous
weightless fluid is assumed to flow in a pipe with a circular cross-section of diameter
$d = 2a$. Hence we can easily verify, by using the cylindrical coordinates, that the
momentum equation in the flow direction is now written:

$$\frac{dP}{dx} = \mu \left(\frac{d^2U}{dr^2} + \frac{1}{r}\frac{dU}{dr} \right).$$

The streamwise pressure variation law is formally unchanged and, by integration,
the velocity profile is given by:

$$\frac{U(r)}{U_{max}} = 1 - \frac{r^2}{a^2},$$

where $U_{max} = -Ka^2/4\mu$, and the shear profile is again linear.
The spanwise velocity and viscous friction distributions are outlined in Fig. 9.6,
corresponding respectively to a parabolic and linear profile.
The head loss coefficient is now:

$$\Lambda = \frac{64}{R_e}, \tag{9.26}$$

where the global Reynolds number based on the pipe diameter is $R_e = U_{mean} \times d/\nu$.
It should be noted that, for this geometry, $U_{mean} = \frac{1}{2}U_{max}$.
From these first two examples, we can finally retain that:

> In any fully developed Poiseuille flow, the fluid's velocity, momentum and kinetic
> energy is maintained, the respective profiles being unchanged along the flow
> direction.
> In the absence of gravity forces, namely for such pressure-induced flows, the
> pressure driving action, can be explained in two complementary ways:
> – *mechanically*, by balancing the viscous friction;
> – *energetically*, by accounting for the viscous irreversible dissipation (major
> head loss).

HISTORICAL NOTE—The Hagen–Poiseuille flow—The flows we have just
   analyzed have been named after the French physician and physicist Jean-
   Léonard-Marie Poiseuille (1799–1869). Poiseuille was interested, for physiolog-
   ical reasons, in distilled water flows through glass tubes of very small diameters
   (0.65–0.015 mm). In a first work [117], reported to the Academy of Sciences in
   1842, he set forth experimentally the law according to which: «*the flow rates,
   all things being equal, are between them as the fourth power of the diameters*»,
   leading to the mathematical formula:

$$Q\,(\text{flow rate}) = k\,\frac{H\,(\text{pressure } mm\,Hg) \times D^4\,(\text{pipe diameter})}{L\,(\text{pipe length})}.$$

In this formula, k is a constant coefficient, depending upon the temperature and
the fluid nature, whose measurement was the subject of further work by Poiseuille
[118, 119].
The previous relation had been discovered earlier by Gotthilf Heinrich Lud-
wig Hagen, a German physicist and hydraulic engineer (1797–1884). It indeed
appeared in a paper, [63], published in 1839 but whose circulation among the
scientific community was undoubtedly penalized by the use of particular units (*a
Prussian ounce, a Parisian thumb*) making the conversion of the results difficult,
as Prandtl & Tietjens [159] pointed out. It is nevertheless historically justified to
associate the two names by calling this law as "Hagen–Poiseuille".

9.6.3   Couette's Axisymmetric Flow

We consider two circular coaxial cylinders, infinitely long and vertically arranged,
of respective radii R_1 and $R_2 > R_1$. The annular space is filled with a heavy and
viscous fluid, whose motion exclusively results from the uniform rotation of each
cylinder at angular velocities ω_1 and ω_2 respectively.
After a transient period, the rotational movement of the cylinders is imparted to the
entire annular fluid mass, in which a stationary motion regime is reached. This is the
only situation which is studied here.
Considering the motion as steady ($\frac{\partial}{\partial t} \equiv 0$) and axisymmetrical ($\frac{\partial}{\partial \theta} \equiv 0$), the unknown
functions of the flow field are $V_r\,(r, z)$, $V_\theta\,(r, z)$, $V_z\,(r, z)$ and $P\,(r, z)$, adopting, of
course, the cylindrical coordinates (Fig. 9.7).

• **The velocity field calculation**
As in the previous example (Sect. 9.6.1), the z-direction is assumed to be infinite
with no flow rate, so that $V_z\,(r, z) = 0$. The continuity equation then provides:

$$\frac{1}{r}\frac{\partial\,(r\,V_r)}{\partial r} = 0\,,$$

Fig. 9.7   Schematic for the
axisymmetric Couette flow

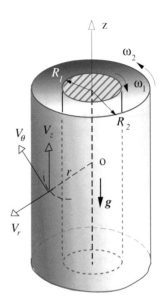

whose integration leads to $V_r = C\,(z)\,/r$. Now, for any z, we know that $V_r\,(R_1, z) = V_r\,(R_2, z) = 0$ so that the function $C\,(z)$ is necessarily null and thus also the velocity component $V_r\,(r, z)$. Hence, the main unknown functions are reduced to $V_\theta\,(r, z)$ and $P\,(r, z)$. The three projections of the Navier-Stokes equations are expressed as:

$$-\rho\frac{V_\theta^2}{r} = -\frac{\partial P}{\partial r}\,, \tag{9.27a}$$

$$0 = \mu\left(\frac{\partial^2 V_\theta}{\partial r^2} + \frac{1}{r}\frac{\partial V_\theta}{\partial r} - \frac{V_\theta}{r^2} + \frac{\partial^2 V_\theta}{\partial z^2}\right)\,, \tag{9.27b}$$

$$0 = -\rho g - \frac{\partial P}{\partial z}\,. \tag{9.27c}$$

The last of these three equations can be readily integrated in the form:

$$P\,(r, z) = -\rho g z + f\,(r)\,,$$

where the function $f\,(r)$ of the only radial distance remains to be determined. However, as of now we can conclude that, on any circular cylinder $r = C'$, the pressure obeys a *hydrostatic* distribution law along the vertical. By introducing this result into the first projection equation, we deduce that:

$$V_\theta^2 = \frac{r}{\rho}\frac{\mathrm{d}f}{\mathrm{d}r}\,,$$

which proves that the only non-zero component of the velocity vector solely depends upon r. From now on and in order to simplify the notation, we will put $V_\theta (r) \equiv V (r)$. The last equation in projection then results in:

$$\frac{\mathrm{d}^2 V}{\mathrm{d}r^2} + \frac{1}{r}\frac{\mathrm{d}V}{\mathrm{d}r} - \frac{V}{r^2} = 0 \iff \frac{\mathrm{d}}{\mathrm{d}r}\left(\frac{\mathrm{d}V}{\mathrm{d}r} + \frac{V}{r}\right) = 0 \,.$$

Therefore the velocity field is the solution of the following first-order differential equation:

$$\frac{\mathrm{d}V}{\mathrm{d}r} + \frac{V}{r} = A,$$

with, as boundary conditions $V(R_1) = \omega_1 R_1$ and $V(R_2) = \omega_2 R_2$. We can then easily demonstrate that the general integral of this equation is written under these conditions:

$$V(r) = \frac{A}{2}r + \frac{B}{r} \quad \text{with } A = 2\frac{\omega_2 R_2^2 - \omega_1 R_1^2}{R_2^2 - R_1^2} \quad \text{and} \quad B = (\omega_1 - \omega_2)\frac{R_1^2 R_2^2}{R_2^2 - R_1^2}$$

$$(9.28)$$

• **The pressure field calculation**

With the previous result, we now know that:

$$\frac{\mathrm{d}f}{\mathrm{d}r} = \frac{\rho V^2}{r} = \rho\left(\frac{A^2}{4}r + \frac{AB}{r} + \frac{B^2}{r^3}\right),$$

whose integration directly results in $f(r) = \rho\left[\frac{A^2}{8}r^2 + AB\ln(r) - \frac{B^2}{2r^2}\right] + P_0$, where P_0 stands for a constant pressure reference. The final pressure field expression is therefore:

$$P(r,z) = P_0 - \rho g z + \rho\left[\frac{A^2}{8}r^2 + AB\ln(r) - \frac{B^2}{2r^2}\right] \qquad (9.29)$$

HISTORICAL NOTE—**The Taylor–Couette flow**—The Couette flow designation for the fluid motion configuration between two coaxial rotating cylinders is somewhat reductive from a historical perspective.

Indeed, the first work on the subject probably went back to Newton (*Principia* Lib.II, Prop. 51), even if the expression of the velocity profile he set forth in 1687 is incorrect.

It is probably to Stokes [149], in 1845, that credit is due for being the first to provide, in section 8 of his paper, the exact form of the velocity variation law as $Ar + B/r$, without, however, fully explaining the constants A and B according to the boundary conditions.

This last part of the work was the task of Boussinesq [15] who, on page 628 of the report he published in 1877, provided the complete solution. He also addressed the two particular cases which are proposed hereafter as an exercise.

The reference to Couette's name pertains to the study which this scientist published in 1890 [41] on his thesis work, where the device was used as a viscometer. In this case, one of the cylinders (the inner cylinder in Couette's work) is held fixed by applying a torque opposite to that exerted by the rotating fluid (see the following exercise).

This flow configuration has also been the subject of several stability analysis, see Sect. 9.7.2.1 below. It is to Taylor [153] that we owe, in 1923, one of most comprehensive stability study, justifying the addition of his name to the appellation of this flow configuration.

EXERCISE 39 The Couette-type rotating viscometer.

It is intended to use the fluid motion configuration between coaxial cylinders as a viscosity measuring device. For that purpose, the inner cylinder is driven to rotate at a constant angular velocity ω_0 and the outer cylinder is held in a fixed position by applying a resisting torque whose intensity is denoted as $\mathbb{C}$.

QUESTION Prove that the fluid dynamic viscosity can be directly deduced from the measurement of this torque and the values of parameters to be specified.

SOLUTION: The velocity field corresponding to the problem conditions is obtained by specifying the general solution in the case where $\omega_1 = \omega_0 \neq 0$ and $\omega_2 = 0$. The result is $V(r) = Ar + B/r$, with:

$$A = \frac{-R_1^2}{R_2^2 - R_1^2}\omega_0 \quad \text{and} \quad B = \frac{R_1^2 R_2^2}{R_2^2 - R_1^2}\omega_0.$$

On any facet with a radial normal $(\mathbf{r})$, there is a tangential viscous stress (in the θ direction) of the general expression:

$$\tau_{r\theta} = \mu \left[r\frac{\partial}{\partial r}\left(\frac{V_\theta}{r}\right) + \frac{1}{r}\frac{\partial V_r}{\partial \theta} \right].$$

Under the problem conditions, this viscous friction is equal to $-2\mu B/r^2$. In particular, the fluid exerts a tangential force on the inner wall of the outer cylinder whose module, per unit length, is equal to:

$$dF = |\tau_{r\theta}(R_2)| \times R_2 d\theta = 2\mu \frac{B}{R_2}d\theta.$$

The moment relative to the cylinder center is expressed by $d\mathbb{C} = R_2 dF$, so that the resulting torque intensity is, per unit length:

$$\mathbb{C} = \int\limits_{0}^{2\pi} 2\mu B \, d\theta = 4\pi\mu \frac{R_1^2 R_2^2}{R_2^2 - R_1^2}\,\omega_0.$$

This relation demonstrates that, knowing the angular speed of rotation of the inner cylinder and the radii R_1 and R_2, the fluid dynamic viscosity measurement is reduced to that of the resisting torque which counteracts any outer cylinder rotation.

EXERCISE 40 The rotating cylinder in an infinite environment.

We consider the particular case of a Couette's axisymmetric flow where the external cylinder, supposedly motionless, has an infinite radius. Thus the motion is driven, in a viscous fluid motionless at infinity, by the rotation at a constant angular speed (ω_1), of a cylindrical rod with a circular section (radius R_1).

QUESTION 1. Give the expression of the velocity and pressure field.

QUESTION 2. Set out the relation between the velocity and pressure at the same point. What do you observe?

QUESTION 3. Justify the validity of one of Bernoulli's theorems for this type of motion (the curl of the velocity field will be calculated).

SOLUTION: **1.** For $R_2 \to \infty$ with $\omega_2 = 0$, the general expression of the velocity field solution provides that $A = 0$ and $B = \omega_1 R_1^2$. The velocity obeys in this case a hyperbolic decreasing law with respect to the axis distance, for $r \geqslant R_1$:

$$V(r) = \frac{\omega_1 R_1^2}{r}.$$

By substituting the values of the constants A and B in the general pressure field expression, we obtain:

$$P = \rho \left(-\frac{\omega_1^2 R_1^2}{2} \frac{R_1^2}{r^2} + K - gz \right).$$

2. By introducing in this relation the velocity expression, we find that:

$$P + \frac{1}{2}\rho V^2 + \rho gz = K,$$

which proves that the pressure and velocity are related through an expression similar to Bernoulli's formula.

3. From the curl operator expressions in cylindrical coordinates it is easy to verify that for this flow $\Omega_r = \Omega_\theta = \Omega_z = 0$. The application of Bernoulli's first (*strong*) theorem can therefore be considered here as valid, on the basis of the *irrotational* character of the motion. As we have discussed in Chap. 2 Sect. 2.3.2 (see Exercise 8), this irrotational motion is of a pure deformation type.

EXERCISE 41 The bulk rotating motion.

The second particular case which we examine here is that of an axisymmetric Couette flow with no internal cylinder ($R_1 = 0$). The situation is therefore that of a heavy viscous fluid, enclosed in an infinitely long cylindrical tank, which is set in motion by the tank's rotation around its supposed vertical axis.

QUESTION 1. Give the expressions of the velocity, pressure and the relation between both functions.

QUESTION 2. Show that the rotational motion of the fluid is "*block-like*" or "*solidifying*", namely without any local deformation of the fluid particles.

SOLUTION: **1.** For this particular situation we have $A = 2\omega_2$ and $B = 0$. The velocity field thus obeys a linear growth law $V = \omega_2 r$ for $0 \leqslant r \leqslant R_2$. The pressure is given by $P = \rho\left(\omega_2^2 r^2/2 - gz + K\right)$. The pressure-velocity relation therefore results in:

$$P - \frac{1}{2}\rho V^2 + \rho gz = K.$$

2. The components of the curl of the fluid velocity are $\Omega_r = \Omega_\theta = 0$, $\Omega_z = \omega_2$. Therefore the vorticity is uniform throughout the fluid. By using the deformation rate expressions in cylindrical coordinates, it is easy to ascertain that all components are zero. The fluid motion is therefore similar to that of a solid of the same shape. It is for this reason that we still speak of a *solidifying rotating bulk motion*.

9.6.4   The General Flow Between Two Parallel Plates

We resume the flow configuration between two parallel plates, common to the situations of *Poiseuille* (Sects. 9.6.1) and *Couette* (Chap. 1 Sect. 1.3.2.1). It is now assumed that the viscous weightless Newtonian fluid motion between the two parallel plates is the result of two concomitant factors[7]:

– a relative translational displacement of one of these plate on its own plane;
– a longitudinal pressure gradient parallel to the plates.

We denote by U_0 the (constant) module of the translation velocity of the upper wall with respect to the lower wall at a distant h and we introduce, for the dimensionless pressure gradient, $\Pi = \dfrac{h^2}{2\mu U_0}dP/dx$.

By writing the general motion equations in projection, it can be demonstrated, as an exercise, that in a steady state, the dimensionless pressure gradient Π is constant and the velocity profile is expressed by:

---

[7] This configuration is sometimes referred to as Poiseuille-Couette, a term that obviously has no historical roots.

Fig. 9.8 Various velocity profile types for different Π values

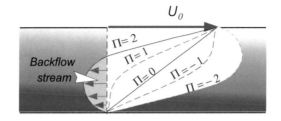

$$\frac{U\,(y)}{U_0} = \frac{y}{h} - \Pi \frac{y}{h}\left(1 - \frac{y}{h}\right).$$

The velocity profile is thus parameterized by the longitudinal pressure gradient Π, which directly leads to the following discussion (Fig. 9.8).

(a) $\Pi = 0$ The velocity profile is then linear. This is the situation of the Couette's plane flow strictly speaking, as discussed in the first chapter Sect. 1.3.2.1;

(b) $\Pi < 0$ The velocity profile is parabolic, concave for negative x. Physically, the pressure decreases in the motion direction induced by the plate displacement. The pressure gradient is therefore described as *favourable*, since it reinforces the driving viscosity effect of the upper wall translation;

(c) $0 < \Pi < 1$ The pressure forces tend to oppose the driving action of the viscosity, the pressure gradient is *unfavourable*. Accordingly, this opposite pressure effect is termed an adverse pressure gradient. The velocity profile is still parabolic, with a concavity now facing the positive x. One can easily ascertain that the slope of the velocity profile at the origin ($y = 0$) is proportional to $1 - \Pi$. Hence, for $\Pi < 1$ the parabola extremum is outside of the segment $[0, h]$;

(d) $\Pi < 1$ The difference from the previous case is that the vertex of the parabolic profile is now within the flow field. In this case a reverse or back flow takes place in a part of the fluid motion near the lower plate, in the opposite direction to that of the upper wall, as a consequence of the relative intensity of the adverse pressure gradient.

EXERCISE 42 The damping motion of the oil in a dashpot.

We consider, see Fig. 9.9, the device, named as the linear dashpot, consisting of a fixed solid cylinder, of inner radius R equipped with a mobile piston, of length l providing a radial clearance e assumed to be very small with respect to R. The chamber between the bottom of the cylinder and the piston is filled with a supposedly incompressible oil of dynamic viscosity μ whose weight will be neglected. Under the action of an axial force of module F, the piston translates at a speed V_0 expelling a fluid rate through the peripheral annulus.

QUESTION 1. In a steady state and assuming that the fluid motion between the piston and the cylinder is the same as that between two parallel plates (the clearance is assumed to be small enough to neglect curvature effects), determine the expression

Fig. 9.9 Schematic for the oil linear dashpot

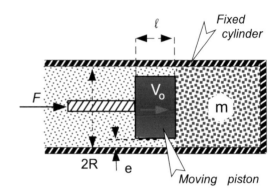

of the piston's displacement velocity as a function of the applied force, geometrical parameters and oil viscosity.

QUESTION 2. Application: $R=4\,\text{cm}$, $l=10\,\text{cm}$, $e=1\,\text{mm}$, $\mu=0.3\,\text{Pl}$, $F=1000\,\text{N}$.

SOLUTION: **1.**

The oil in the chamber is submitted to the overpressure $P \approx F/\pi R^2$. Its expulsion by the piston's displacement generates a fluid motion due to the longitudinal pressure gradient:

$$\frac{dP}{dx} = \frac{P}{l} = \frac{F}{\pi R^2 l}.$$

If we identify the situation as that of a plane motion in the clearance zone between the peripheral piston surface and the cylinder's inner wall, the velocity profile in this zone is written:

$$\frac{U}{V_0} = \frac{y}{e} - \frac{e^2 F}{2\mu\pi V_0 R^2 l} \frac{y}{e} \left(1 - \frac{y}{e}\right).$$

By integration throughout the thickness e, it is easy to deduce the expression of the fluid flow rate which, due to continuity (the oil being incompressible) is necessarily equal to $\pi R^2 V_0$. Thus we obtain the requested relation:

$$V_0 = \frac{e^3 F}{6\pi\mu\,(R+e)\,R^2 l} \approx \frac{e^3 F}{6\pi\mu R^3 l} \quad \text{for } e \ll R.$$

2. With the values of the numerical application, one finds the piston's moving speed of about $3\,\text{cm/s}$ ($2.77\,\text{cm/s}$).

9.6.5 The Unsteady Flow on a Plane in Translation, Stokes' First Problem

In the previous examples, the motion was assumed to be stationary in time. Now we will focus on a motion which is *variable over time*. It concerns a viscous, weightless fluid which extends, in a two-dimensional plane geometry all over the upper half plane $(y \geqslant 0)$. At any given time $t < 0$, this fluid is at rest. At $t = 0$ the solid wall coinciding with the plane $y = 0$ is set in motion by a rectilinear and uniform translation whose velocity components are $(U_0, 0)$.

The problem consists in deriving the function expressions of the motion which is induced, by viscosity, in the entire fluid domain as a consequence of this displacement. The pressure field will be assumed to be uniform. Since the two-dimensional flow field is infinite in the x direction, we will consider that the velocity field solely depends on the two variables y and t. The continuity equation then requires that $\partial V / \partial y = 0$ which, with the boundary condition $V(0, t) = 0$ produces $V(y, t) = 0$. The projection of the momentum equation along x is written under these conditions:

$$\frac{\partial U(y, t)}{\partial t} = \nu \frac{\partial^2 U(y, t)}{\partial y^2}.$$

The solution to this parabolic equation is found in the form $U(y, t) = U_0 f(\eta)$, with $\eta = y/g(t)$ where f and g are two unknown functions of the independent variables, η and t respectively. Under these conditions and after substitution, the previous partial differential equation results in:

$$\frac{gg'}{\nu} = -\frac{f''}{\eta f'} = \text{Constant.}$$

The length scale $g(t)$ therefore obeys a parabolic growth law over time.

By taking $g(0) = 0$ and adopting as the constant value[8] 2ν, this law can be written $g(t) = 2\sqrt{\nu t}$. The function f is then the solution of the differential equation:

$$f'' + 2\eta f' = 0,$$

with the two following boundary conditions $U(0, t) = U_0$, that is $f(0) = 1$ and $U(\infty, t) = 0$, that is $f(\infty) = 0$. The velocity field is thus expressed as:

$$U(y, t) = U_0 \left(1 - \frac{2}{\sqrt{\pi}} \int_0^{\eta} e^{-\alpha^2} \, d\alpha \right).$$

---

[8] The choice of this constant is arbitrary. Any other non-zero value is valid and will only change the expressions of $g(t)$ and η given here by the same multiplying factor.

EXERCISE 43 The Advection/Diffusion comparison in a transient motion.

Let us use the previous results to quantify, over time, the comparison between the effects of (*i*) *advection*, due to the plane translation, and (*ii*) *diffusion* of this movement by viscosity. For this purpose, we define $\delta(t)$ as the distance from the wall where the velocity, at a given time t is only equal to 1% of the parietal value U_0.

QUESTION 1. Give the variation law's expression of the distance over time $\delta\,(t)$.
(We give $\frac{1}{\sqrt{2\pi}}\int_{0}^{2.57} e^{-x^2/2}dx \approx 0.4949$).

QUESTION 2. Compare the characteristic time scales of the advective transport and diffusive momentum transfer over the same distance in the order of δ.

QUESTION 3. Discuss the results and specify the values for the following numerical application: $U_0 = 1.3\,\mathrm{m/s}$, $v = 6.35\,\mathrm{St}$ (pure glycerin).

SOLUTION: **1.** By introducing the variable change $x = \alpha\sqrt{2}$ in the general solution, we can easily ascertain that:

$$f\left(\frac{2.57}{\sqrt{2}}\right) = 1 - 0.9898 \approx 0.01.$$

Thus, we can deduce that the expected value of $\delta\,(t)$ is such that $\dfrac{\delta \times \sqrt{2}}{2\sqrt{vt}} \approx 2.57$, which finally leads to the expression $\delta\,(t) \approx 3.63\sqrt{vt}$.

2. From a physical point of view and with regard to the viscous fluid momentum balance, this type of motion leads to a competition between two mechanisms over time:

(a) a streamwise advection along x, related to the translational movement of the plate;

(b) a diffusion in the transverse direction y, due to the velocity gradient and fluid viscosity.

The advection time scale over a distance δ can be estimated as $T_a \sim \delta/U_0$ and, for the diffusion time scale, as $T_d \sim \delta^2/v$. Taking the ratio of these two quantities, we obviously find the Reynolds number based on the distance δ, i.e., $(R_e)_\delta = U_0\delta/v = T_d/T_a$.

3. In this type of motion, the previous Reynolds number is not constant over time. This result is the consequence of a *linear* growth of the advective time scale $l_a \sim U_0 t$, while the diffusive time scale grows *parabolically* ($\delta \sim \sqrt{vt}$). With the numerical data of the application, these two length scales are equal after a period of time t^* of about 5×10^{-3} s. Then, the common value is $l^* \simeq 6.4$ mm.

As plotted in Fig. 9.10, it appears that for $t < t^*$, the distance concerned by the diffusive transfer is greater than that of the advection, a conclusion which is reversed for $t > t^*$.

Fig. 9.10 Time variation of the advective and diffusive length scales

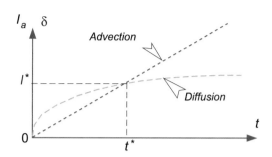

One cannot therefore conclude, in this type of motion, that there is a systematic predominance of one mechanism over the other *at any given time*.

9.7 Basic Flow Stability Concepts

9.7.1 Presenting the Problem

We will end this chapter by presenting some basic physical concepts on the *stability* issue (Sect. 9.7), *transition* and *turbulence* (Sect. 9.8) of viscous fluid flows. On this occasion, we will go slightly beyond the strict scope of this work by addressing configurations which are not exclusively related to homogeneous and monophasic fluid situations.

From a general theoretical point of view, the search for solutions to the Navier-Stokes model for incompressible fluids raises three main fundamental issues regarding:

– the existence of a solution;
– the unicity of the solution when it exists;
– the stability of the solution.

We will only deal here with the stability issue through some simple physical considerations, not without first indicating that, at present, the answers to the first two questions have not been definitely decided for three-dimensional unsteady motions.

The need for a stability study is easy to understand. Indeed, as we have seen with some examples in the previous section, an analytical solution of the Navier-Stokes equations can be obtained for some specific initial and boundary conditions. The question then naturally arises as to whether or not the experiment confirms this theoretical result. This directly appeals to the stability problem of a *theoretical flow*, namely the influence of *actual perturbations*. Such perturbations are introduced, through the initial and boundary conditions, to the motion solution obtained under *ideal mathematical* conditions.

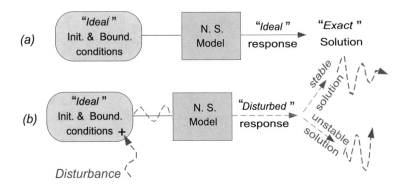

Fig. 9.11 Schematic for the flow stability in response to disturbed initial and/or boundary conditions

As outlined in Fig. 9.11, two cases are possible depending on whether the fluctuations induced in the exact solution by the external disturbances dampen (a) or, on the contrary, amplify (b).

In the first case, we will say that the solution is stable, with respect to the type of disturbance considered, whereas it becomes unstable in the second case.

From a *mechanical* point of view, i.e., by solely reasoning with regard to the momentum balance, the amplification or damping scheme of the disturbances can be interpreted as a dominance reversal between stabilizing and destabilizing forces, as we will see with the following examples.

9.7.2   Some Examples of Flow Instabilities

9.7.2.1   The Taylor-Couette Instability

The first type of instability which we present occurs in Couette's axisymmetric flow which we have previously studied (Sect. 9.6.3). It mainly takes place when, maintaining the outer cylinder stationary, the rotating speed of the inner cylinder is continuously increased from zero. The experiment has shown that, for a given value of this speed, secondary motions are generated within the rotating fluid mass. They take the form of contra-rotating toroidal cells, periodically organized along the cylinder generatrix as shown in Fig. 9.12a.

The occurrence of this space structuring pattern marks a first flow *symmetry break* whereby the two-dimensional character of the boundary conditions (invariance by translation along the generatrix) ceases to apply to the solution. The onset of this phenomenon has been experimentally correlated to a critical value of *Taylor's number*, the adimensional grouping defined as:

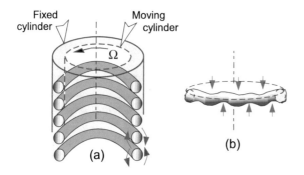

Fig. 9.12 Schematics for the Taylor-Couette instability vortex cells

$$T_a = \frac{(R_2 + R_1)\,(R_2 - R_1)^3\,\Omega^2}{2\nu^2}. \tag{9.30}$$

When assuming a low relative spacing $R_2 - R_1 \ll R_2 + R_1$, the critical Taylor number is 1712.

Beyond this value, the flow goes through a second change with the formation of an azimuthal ripple of the rollers around their diametrical symmetry plane, as outlined in diagram (b), Fig. 9.12. This phenomenon reflects a *second symmetry break*, where the rotational invariance around the cylinder axis (the axial symmetry) of the boundary conditions no longer applies to the velocity solution of the entire field.

During these first two instability phases, the stationary character of the flow is preserved. This is no longer the case if the rotation speed is further increased. For high enough values, *chaotic* motions *spontaneously* and *sporadically* originate within the fluid motion. They reveal the existence of a three-dimensional, fluctuating velocity field, indicative of a *turbulent agitation* in the fluid motion. These are called "*turbulent spots*" which first appear intermittently[9] in both space and time before extending to the entire flow field for even higher rotation speeds. The onset of turbulence spots is the ultimate phase of the flow destabilization associated with the *time invariance break*.

In this configuration, the destabilizing force is the variable centrifugal force in the fluid inter-space opposed by the stabilizing action of the viscosity.

HISTORICAL NOTE—**The Couette-Taylor instability**—The unstable character of the flow between rotating cylinders was undoubtedly revealed, for the first time, by Couette during his thesis work on the "*liquid friction*" measurement.
Indeed, in the article he published in 1890 [41], on pages 464 & 465, we can find the following description of this phenomenon:

---

[9] Intermittency is a feature which one can discover in relation to various phenomena in fluid mechanics. Consequently it is appropriate to specify here that this is a **transition intermittency**.

«In summary: 1 between 0 and 56 revolutions per minute, the friction on the inner cylinder is proportional to the outer cylinder's angular velocity, according to equation . . ., which for the present case is the simplest integral of Navier equations.
2 Between 56 and 127 revolutions, it increases very quickly, and disturbances occur;
3 From 127 to 453 revolutions, the experimental limit, it grows regularly as a second degree function of the speed.»

Thus, in Couette's arrangement, where the *inside* cylinder is fixed, the loss in flow stability caused by the increase in the *outside* cylinder rotation suddenly reveals itself.

The situation is totally different when the inner cylinder is rotated. This configuration was remarkably documented in an article published by Taylor [153] in 1923, the result of a study that had begun two years earlier. The first symmetry break, with the toric cell development, was described, as well as the various evolution patterns of these secondary motions according to their own destabilization modes. Hence it is now customary to associate these two scientists in the name of this instability.

9.7.2.2 The Rayleigh-Bénard Instability

When a weighing fluid enclosed between two horizontal parallel planes is subjected to a vertical temperature gradient, experiments have shown that, under specific conditions, it can be subjected to thermo-convective motions. If the top wall temperature is *higher* than that of the bottom plate, the heavy fluid lies below the light fluid so that the thermal stratification is stable. Indeed, a fluid particle which is displaced upwards from its original level, for example, is surrounded by a lighter fluid. It thus undergoes a downward Archimedes' thrust tending to bring it back to its original position. This is the situation in Fig. 9.13a. On the other hand, if the temperature gradient is reversed—Fig. 9.13b—, the same rationale demonstrates that the upwardly displaced particle is lighter than the surrounding environment. The result is a buoyancy force which aims at increasing the imbalance in the particle position.

In the case of an unstable stratification between the plates, experimental observations demonstrate that above a critical value of the temperature difference, a regular pattern of counter-rotating rolls with a horizontal axis, known as Bénard cells, form within the fluid. This is Rayleigh-Bénard's *thermo-convective* instability, as outlined in the following Fig. 9.14.

The dimensionless parameter which quantifies the onset of this first instability regime is the Rayleigh number defined as:

$$R_a = \frac{g\alpha \Delta T h^3}{\nu a}. \tag{9.31}$$

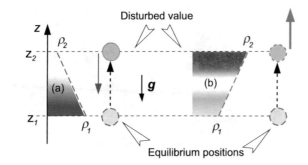

Fig. 9.13 Schematic for the Rayleigh-Bénard instability mechanism

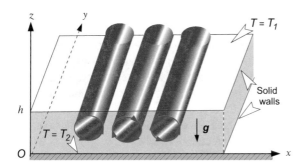

Fig. 9.14 Thermo-convection rolls of the Rayleigh-Bénard instability

In the above expression, g is the gravitational acceleration module, α the thermal expansion coefficient of the fluid, ΔT the temperature difference between the plates separated by h, ν and a the fluid kinematic viscosity and thermal diffusivity respectively. The critical value of the Rayleigh number corresponding to the onset of the previous structures is 1708.

9.7.2.3 The Bénard–Marangoni Instability

This instability mode, which was of the type studied by Bénard at the beginning of the twentieth century, is sometimes mistaken for the previous mode. In fact its mechanism is different and specific to free surface fluid situations. It occurs in a fluid film lying on a horizontal plane and subjected to a temperature difference between the solid wall and the free surface. For a critical value of the temperature difference, a *thermo-capillary* instability originates within the fluid. It results in the formation of hexagonal cells, with a vertical axis, as outlined in Fig. 9.15.

The mechanism for the instability onset is that any disturbance in the free surface temperature causes a change in the fluid interfacial (or surface) tension. If, for

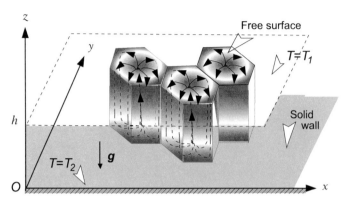

Fig. 9.15   Thermo-capillary cells of Bénard-Marangoni instability

example, the temperature locally exceeds the equilibrium temperature, the result is a surface tension gradient which drives the fluid to locally colder regions. The mass conservation can only be reached by an influx of fluid from the inner layers. Finally, a cell system is set up to provide a vertical fluid circulation which goes up at the cell center and down at its periphery.

Here again, a dimensionless grouping makes it possible to quantify the onset of this instability. This is the Marangoni number:

$$M_r = \frac{\sigma' \Delta T h}{\mu a}, \tag{9.32}$$

where $\sigma' \equiv d\sigma/dT$ is the surface tension variation rate with the temperature and ΔT the temperature difference over the fluid film thickness h.

Remark On the earth's surface, the evaporation of a solution to produce large, flawless single crystals can be disrupted by convective motions within the fluid. This explains the relevance of micro-gravity situations, where the Marangoni effect is the sole effect present in a free surface fluid.

HISTORICAL NOTE—**The thermo-convective instability**—It is probably James
    Thomson, elder brother of William (Lord Kelvin), who was the first scientist to
    report, in 1881, on thermo-convective structures in a fluid [156]. His observations
    were made at the free surface of a 10 to 12 cm thick water layer, with a tempera-
    ture slightly higher than that of the air. He then noted that after a while, tessellated
    patterns appeared at the water-air interface, the occurrence of the phenomenon
    being, it seemed, closely associated with the cooling of the surface layer. He sug-
    gested that these patterns were the evidence for the existence of vertical structures
    setting the fluid in motion throughout the layer thickness.
    In the early 1900s, Bénard [7–9] made observations of the same phenomenon,
    unaware of Thomson's earlier work. The experiment, based on a particularly
    careful protocol, involved thin fluid layers of the order of one millimeter. Due
    to the presence of a free surface, a full understanding of the phenomenon could

not be rigorously attained without analyzing the surface tension effects, which had been carried out by Marangoni [94] in 1871, whose name is therefore to be associated with this type of instability. On the other hand, Rayleigh's theoretical study [132], in 1916, although aiming at explaining some of Bénard's experiment results, addressed a fluid sheet bounded by two parallel solid planes, excluding any interfacial tension effect. Consequently it is important to distinguish Rayleigh's instability from the previous type and to keep in mind that it can be addressed by an *inviscid* analysis, where the external buoyancy forces are handled under Boussinesq's approximations.

9.7.2.4 The Kelvin–Helmholtz Instability

The three types of instabilities we have just described have a number of points in common:

- they physically correspond to a break in the balance between stabilizing and destabilizing forces;
- they appear beyond a certain threshold value of a dimensionless characteristic parameter associated with this equilibrium break;
- they result in the onset of vortex structures within the fluid itself. Therefore, they will be referred to as *internal instabilities*.

The instabilities to be discussed at present are of a different nature. In the first place, they are triggered independently of any notion of a critical threshold: the sole presence of disturbances is enough to cause them, when certain conditions are satisfied. In addition, while they also manifest themselves through an eddy structure development, they originate on particular flow surfaces, bounding fluid domains of initially distinct properties. It is for this reason that we will speak of *front instabilities*.

The Kelvin-Helmholtz instability is typical of flows whose velocity field exhibits an inflexion point. The most representative situation is that of the plane mixing layer. This region, which develops at the confluence of two parallel streams of uniform velocity but different in module, is characterized by the presence of a velocity gradient associated with a localized rotational area. Outside this zone, the flow is irrotational, so that the mixing layer can be idealized as a kinematic interface of a vortical sheet within an irrotational field—see schematic (a) in the following figure (Fig. 9.16).

The instability mechanism is as follows: Suppose a small disturbance causes a slight ripple of the layer surface—diagram (b)—and consider its convex part facing the higher velocity flow. As we have seen in Chap. 6 on the subject of the wing lift origin, this area is subject to an underpressure which tends to increase the bending amplitude. The same rationale also applies to the concave part of the layer, whose hollowing tends to increase due to overpressure effects.

Neglecting the viscosity effects, Lagrange's theorem (see Chap. 5 Sect. 5.6.2.2) requires that the vortical sheet moves with the fluid motion while preserving its vorticity intensity. Now, by a simple kinematic effect resulting from their relative

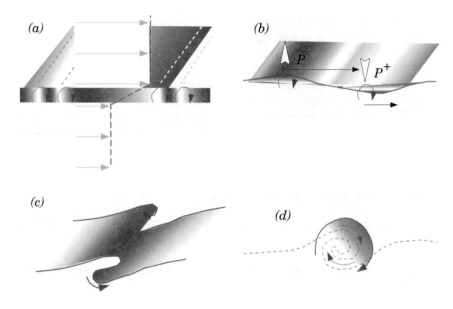

Fig. 9.16 Schematic for the Kelvin-Helmholtz instability mechanism

locations in the external streams, the crests move faster than the troughs. The distance between a crest-trough pair will thus be reduced, which will progressively lead to the vorticity concentration in vortex cores which, still according to Lagrange's theorem, remain connected to each other. From a theoretical viewpoint, the Kelvin-Helmholtz instability is governed by the inflexion point theorem which can be stated as follows:

> The presence of at least one inflexion point in the velocity profile of an ideal barotropic fluid flow is a necessary and sometimes a sufficient[10] condition for instability. The absence of an inflexion point necessarily confers stability.

HISTORICAL NOTE—**Extended Kelvin-Helmholtz type instability**—As we have presented so far, the Kelvin-Helmholtz instability is of a purely kinematic nature. In fact, in addition to the difference in the velocity modulus, the streams involved may also differ in temperature, density and the very nature of the fluids. In these cases, additional forces such as gravity, buoyancy, interfacial tension, etc can also occur.

---

[10] The first part dealing with the necessary condition of this theorem was established as early as 1879 by Rayleigh [129]. Indeed, he demonstrated that the motion of an ideal barotropic fluid could not be unstable if there was not at least one extremum in the vorticity field. The fact that this is also a sufficient condition was only proven in 1935 by Tollmien [162] in the particular case of the velocity profile of the parietal boundary layer.

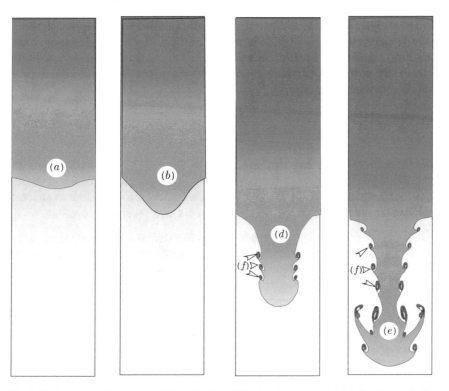

Fig. 9.17 Schematic for the Rayleigh-Taylor instability development: **a** initial disturbance, **b** linear amplification, **c** digitation, **d** mushroom-like structure formation, **f** Kelvin-Helmholtz-type secondary vortices

9.7.2.5 The Rayleigh–Taylor instability

The second type of front instability we will describe occurs at the interface separating two incompressible, non-miscible fluids of different densities, initially in hydro-static equilibrium in the gravitational field.

If the interface is *ideally planar*, an equilibrium state can be observed with a light fluid above the heavy fluid (stable situation) or, conversely, heavy fluid above light fluid. In the latter case, the equilibrium is obviously *unstable*, which means that it is broken at the slightest deviation in the interface planarity. Indeed, if due to such a disturbance, a heavy portion of fluid enters the light fluid, it will undergo an Archimedes thrust which will amplify its motion and causes it to move away from its original position. The same applies to a light parcel within the heavy fluid. We then observe an interpenetration of different density fluid masses through digitations developing from *mushroom-like structures*, see Fig. 9.17.

The analysis of this instability using a two-dimensional inviscid approach allows us to demonstrate that the slightest interface disturbance[11] (vertical deviation δ from the plane equilibrium situation in $z = 0$) grows exponentially, $\delta \propto e^{\sigma t}$, with the growth rate $\sigma = \sqrt{\alpha g \mathcal{A}_t}$, where g is the gravity acceleration, α the wave number and $\mathcal{A}_t$ the Atwood number:

$$\mathcal{A}_t = \frac{\rho_{heavy} - \rho_{light}}{\rho_{heavy} + \rho_{light}}. \qquad (9.33)$$

The influence of the surface tension and viscosity forces was analyzed by Bellman and Pennington [6], in 1954. Both properties have the effect of reducing the instability growth rate. In addition, there is a short-wave range in which the surface tension can stabilize the phenomenon (see also, for example, Guyon et al. [62], p. 68).

HISTORICAL NOTE—The Rayleigh–Taylor instability—The equilibrium nature of an incompressible heavy fluid whose density is a function of altitude was adressed by Rayleigh [130] in an article dated 1883. He developed the small perturbation linearized analysis, provided the results of the exponential amplification and derived the growth rate expression as a function of the relative density difference, now commonly referred to as the Atwood number, Eq. (9.33).

In 1950, Taylor [154] revisited Rayleigh's analysis by considering a density interface which undergoes a normal acceleration g_1 algebraically additive to that of gravity, g. He demonstrated that the new growth rate σ becomes such that $\sigma^2 = -\alpha(g + g_1)\mathcal{A}_t$ and concluded: «...when two superposed fluids of different densities are accelerated in a direction perpendicular to their interface, this surface is stable or unstable according to whether the acceleration is directed from the heavier to the lighter fluid or vice versa.»

In both cases, these are mathematical papers which do not address the physical aspects of the instability development beyond the linear amplification phase.

9.7.2.6 The Richtmyer–Meshkov Instability

The last type of frontal instability we will describe is very similar to the previous type. It can be seen as an extension of the gravitational instability situation (see Taylor's contribution in the previous Additional Information section), to the case where the interface separating two fluids of different densities undergoes a normal *impulsed* acceleration. This is known as the Richtmyer-Meshkov instability (Fig. 9.18).

If there are slight irregularities in the surface separating the two fluids, these are likely to be amplified when the acceleration is such that it tends to cause the light fluid to penetrate the heavy fluid. The development of the instability structures calls for specific mechanisms of variable density fluid flows, in particular the vorticity production by the *baroclinic torque*. These mechanisms, whose study goes far beyond

---

[11] Within the limit of *small* perturbations.

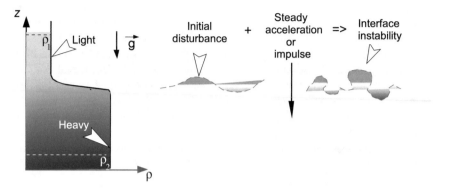

Fig. 9.18 Schematic for the Richtmyer-Meshkov Instability

the scope of the present work, are discussed in the monograph [36] by Chassaing et al.

9.8 Transition and Turbulence

9.8.1 The Flow Regime Transition and Turbulence

The examples we have just presented have demonstrated that, for the same given set of initial and boundary conditions, the motion of a real fluid can present itself under highly varied aspects. They reflect differences in nature between several evolution possibilities of the particular dynamic system which is the Newtonian fluid flow, *in response* to the disturbances it is subjected to. We will describe very schematically some of these possibilities which are usually grouped under the *flow regime* term.

(a) The first situation which can occur is when, once introduced into the flow, any kind of disturbance is unable to persist beyond a certain *space–time range*[12] from its injection point. Physically, this situation reflects the existence, within the flow itself, of stabilizing mechanisms whose action is preponderant. These confer to any experimental realization of this type of flow, a *physical determinism* character, in the sense that they ensure the *repeatability* of the experiments and measurements, apart from metrological uncertainties. When the Navier-Stokes model solution is analytically available, the agreement between the *theoretical* values and *measurements* can also be ascertained in this case. Following Favre's et al. considerations [58], it then becomes legitimate to speak in terms of *physico-mathematical determinism on a macroscopic scale*. Indeed, everything happens as if the impossibility

---

[12] One can indeed consider addressing the flow stability issue in terms of space and/or time perturbation amplification.

of any experimenter to make a physical model *perfectly reproducible* in its smallest mesoscopic details, was without consequence on the response which the flow will select. Hence, because of its very high stability level, such a regime gives the *appearance of ignoring* the discrepancies, with respect to the ideal conditions of the mathematical model, which are at the origin of the *disturbances* in any real situation.

(b) When the motion ceases to be unconditionally stable, another class of flow regime can occur which is characterized by the fact that the structures which develop in response to the existing disturbances are, to a large extent, independent of the exact nature of those disturbances.

The emblematic example of such a situation is the Bénard–Von-Kármán vortex street, where alternating vortices form in the wake of a circular cylinder for a Reynolds number between 50 and 100.

Indeed, it can be inferred from numerical simulations for example, that the stationary symmetrical roll up configuration—see Chap. 2, Figs. 2.26 and 2.28—is a solution of the unsteady two-dimensional Navier-Stokes equations over a wide Reynolds number range [20, 200]. However, as previously mentioned, such a regime is not experimentally observed beyond a critical value of the Reynolds number, of the order of 45–48. This apparent contradiction can be explained by the following additional observations:

– for a Reynolds number lower than the critical value, the simulation confirms that the symmetrical roll up configuration is stable with respect to any type of disturbance;
– for a Reynolds number above the critical value, the captive roll up configuration can only persist in the presence of *infinitesimal small* disturbances. In other words, when subjected to macroscopically significant disturbances, this regime is unstable and then directly evolves towards the periodic vortex shedding regime;
– the macroscopic properties of the new unsteady regime which is created after the destabilization of the captive roll up (vortex size, frequency, etc.) do not depend on the *disturbance characteristics* which initiated it. For instance, regardless of the nature of the disturbance causing the destabilization, the vortex shedding frequency remains the same for a given Reynolds number. Thus, except for a phase reference, which is still random because directly related to the time when the perturbation is applied, the flow, *even though it is unsteady*, still presents a high level of physico-mathematical determinism on a macroscopic scale, due to its periodic nature.

As demonstrated by Huerre and Monkewitz [77], this type of behaviour can be related to a particular instability mode. In addition to the cylinder wake previously mentioned, it is also found in *variable density* gas jets when the density ratio between the effluent and the environment is sufficiently small (a hot air or light gas jet);

(c) The third category includes flows whose destabilization mechanism may be influenced, to varying degrees, by the very features of the disturbances to which they are subjected. The typical illustrative case of this situation is that of the isovolume constant density jet, where the vortex structures which characterize the first flow destabilization stages can be controlled by the features of the disturbances introduced at a macroscopic level, see for instance Reynolds et al. [137]. Such a behaviour occurs in a wide variety of situations, in both space or time evolution;

(d) The last category is that of flows which, with regard to their response to disturbances, are positioned at the opposite end of the laminar regime. They are, in practice, the ultimate end point of a regime transition process, whose situations described in (b) and (c) are just a few examples of possible first steps.[13] These flows, referred to as *turbulent*, are those most frequently observed in nature.

9.8.2 Some Specific Features of the Turbulent Regime

The *turbulent regime* is illustrated by the fact that every flow function exhibits fluctuations which are *intrinsic to the motion*[14] and are continuously distributed over a range (spectrum) of space-time scales.[15]
It is outside the scope of this work to go into a detailed study of such a regime, whose distinctive features have made it a fully-fledged field of modern fluid mechanics. To the reader interested in an initial exploration of the matter, we highly recommend Lesieur's work [88]. For a more conventional and relatively comprehensive approach, one may refer to the monograph, admittedly a somewhat dated, but still very educational, by Hinze [72]. Finally, one will find in the works of Tennekes & Lumley [155], Pope [122], Piquet [114], and of the present author [35] more recent presentations devoted to various aspects of the subject, in particular in relation to its modelling issue.
   We will limit ourselves here to the following three general considerations, in order to give a foretaste of the fascinating complexity of a phenomenon which is so commonly and easily observed on a human scale.

– The irruption of randomness in a mesoscopic phenomenon, without changing the observation scale, is a first source of confusing questioning. Due to its sensitivity to the details of the disturbances introduced by the initial and boundary conditions,— details which are inaccessible on a macroscopic scale to any physical realization of a turbulent flow—the phenomenon seems to defy the very notion of physical determinism. Thus, two realizations which are *macroscopically identical* for the investigator, do not provide identical results, even taking into account the measurement uncertainties. To arrive at a *macroscopic determinism* degree, a *probabilistic* perspective should be adopted by considering a set of realizations representative of all the possible responses of the system which is a turbulent flow constituted under the same given initial and boundary conditions. Then, we are speaking with Favre et al. [58] of a «*statistical physico-mathematical determinism*».
– The random fluctuations which are characteristic of the turbulent regime reveal the presence of eddying motions within the flow itself at a *macroscopic scale*. How-

---

[13] The transition to turbulence can follow very different routes, depending on whether the flow is free or confined, as one might expect from the transition examples presented in the previous section.
[14] This implies that these fluctuations are present under stationary boundary conditions.
[15] The turbulent fluctuations do not reduce to a finite or countable set of discrete vibration modes, as in the case of the transition regimes (b) and (c), for example.

ever, it would a misleading exaggeration to infer from this observation that the random nature of the turbulence agitation necessarily applies to all eddies present in such a regime. In fact, some large eddy structures may retain a partially deterministic behaviour, whether or not they are reminiscent of the periodic motions of the transition regime. This is the situation of «*early turbulence*», for which, as suggested by Ha Minh et al., a semi-deterministic approach can be suggested [64, 65];

– Finally, as Lesieur noted [89], one of the main consequences of the turbulent regime in a fluid motion remains, for the engineer, the *unpredictability* of this type of situation. In fact, in as far as the knowledge of the *mean* turbulent field is no longer sufficient, the issue of this regime sensitivity to initial and boundary conditions reappears. The limit arising from the reliability of the weather forecast is a perfect illustration of this characteristic.

ADDITIONAL INFORMATION—**The analogy with the dice game**—With a view to making some of the preceding comments more readily accessible, we will take an analogy which we hope will be more easily grasped, that of the game of dice. The experiment simply consists in looking at the figure provided by the roll of a dice (*always the same*), by a player (*always the same*), on a board (*always the same*). This protocol is considered to define a *perfectly repetitive experiment* in the eyes of any observer. Nevertheless, everyone knows that it is not so for the phenomenon (rolling the dice), whose entertainment value precisely lies in the obvious randomness of the figure for each roll.

The analogy with the random nature of fluid turbulence can help to clarify some of the conceptual aspects about the origin of the phenomenon. However, we should remain cautious about the implications of such an analogy, not forgetting that the Navier-Stokes model describes the evolution of a non-linear and dissipative system, which is far from being the case of the equations governing the dice movement, considered as an indeformable solid.

Let us nevertheless further investigate the analogy with the case of a loaded dice, an object into which a massive load has been introduced, shifting the centroid center from the geometric symmetry position of the cube center. When rolling such a dice, the initial condition uncertainty is the same as for a normal dice. On the other hand, the system *sensitivity* to the inaccessible details of these conditions is changed (reduced): the evolution of the loaded dice can now be reduced, with a reliable degree of approximation, solely to that of its centroid. In other words, loading the dice means making the system evolution more "*stable*", namely less sensitive to the details of the initial conditions, and thus making this evolution quasi-deterministic, with a little skill. We have, in a way, created a "*laminar*" *dice game*!

Chapter 10
Very Low Reynolds Number Flows

Abstract The first part of the chapter analyses the specific properties of viscosity-dominated fluid motions, also known as creeping motions. In a second part, various examples of these movements are discussed in detail.

10.1 Introduction

As we have discussed in Chap. 9 Sect. 9.3, the inertial and viscosity force ratio can vary from one flow to another and, for a given flow, from one point to another in the field. The relative preponderance of either of these forces depends on the global or local value of the Reynolds number. In the case where this parameter is globally low, the flow is referred to as a *creeping motion* which corresponds to a specific model. When the fluid's viscous behaviour conforms to the Newtonian scheme, this model consists of a simplified form of the Navier-Stokes equations, known as *Stokes'* model. We will now examine this.

10.2 Stokes' Model

10.2.1 The Assumptions

By definition, the Stokes model which governs the viscous Newtonian fluid flow is based on the following two specific assumptions:
– a constant density fluid;
– negligible inertial forces, at any point and at any time, with respect to viscosity forces.

The last assumption is still expressed by the relation:

$$Re \equiv \frac{U_o L}{\nu} \ll 1, \tag{10.1}$$

where U_o is a scale representative of the fluid motion velocity, L a length scale and ν the kinematic viscosity. With the previous assumptions, the model governing the flow dynamics is reduced to the following equations:

$$\rho = C^t \text{ and } \frac{\partial U_j}{\partial x_j} = 0 ,$$

$$\frac{\partial U_i}{\partial t} = F_i - \frac{1}{\rho}\frac{\partial P}{\partial x_i} + \nu \frac{\partial^2 U_i}{\partial x_j \partial x_j}.$$

The last equation can also be written in the equivalent vector form:

$$\frac{\partial \mathbf{V}}{\partial t} = \mathbf{F} - \frac{1}{\rho}\,\mathbf{grad}\,P + \nu \mathbf{\Delta V}. \tag{10.2}$$

In reference to the general analysis developed in Chap. 9 Sect. 9.3.1, the present momentum equation involves only four time scales, characterizing respectively the instationarity (T_i), and the actions of the external body (T_F), pressure (T_P) and viscosity, (T_ν) forces. When based on a given length scale L, the latter can be estimated by

$$T_\nu = \frac{L^2}{\nu} .$$

The very low global Reynolds number assumption obviously does not allow us to deduce any information about the respective values of these different time scales. It is for this reason that we further assume that the following conditions are fulfilled:

1. $T_i \gg T_\nu$, so that the flow can be considered almost stationary;
2. $T_F \sim T_P \sim T_\nu$, which means that the three external forces are of the same order;
3. The external body forces reduce to the sole gravity forces leading to a hydrostatic pressure field $P_0(z) = \rho g z$, in the absence of motion and in a usual terrestrial referential, according to the equation:

$$\mathbf{F} = \frac{1}{\rho}\mathbf{grad}\,P_0 = \mathbf{g}.$$

Remark It should be noted that the second of the previous conditions results, in particular, in the pressure characteristic scale order of $\rho U_o^2 / R_e$.

10.2.2 The Velocity-Pressure Formulation of the Equations

Under all of the preceding conditions, the equations governing the velocity and pressure field become:

$$\begin{array}{ll} div\,\mathbf{V} = 0 & (10.3) \\ \mathbf{grad}\,(P - P_0) = \mu\,\boldsymbol{\Delta}\mathbf{V} & (10.4) \end{array}$$

10.2.3 The Pressure-Vorticity Equations

By utilizing, as we have used in Chap. 6, the vectorial identity for the Laplacian vector expression:

$$\boldsymbol{\Delta}\mathbf{V} = \mathbf{grad}\,(div\,\mathbf{V}) - curl\,(curl\,\mathbf{V}),$$

we obtain, in an isovolume situation $\boldsymbol{\Delta}\mathbf{V} = -\,curl\,\boldsymbol{\Omega}$, where we have introduced the vorticity vector $\boldsymbol{\Omega} = curl\,\mathbf{V}$. The momentum equation, known as the Stokes equation, then takes the form:

$$\mathbf{grad}\,(P - P_0) = -\mu\,curl\,\boldsymbol{\Omega}. \qquad (10.5)$$

By successively taking the divergence and curl throughout the previous equation, we deduce respectively:

$$\begin{array}{ll} \Delta P = 0 & (10.6a) \\ \boldsymbol{\Delta}\boldsymbol{\Omega} = \mathbf{0} & (10.6b) \end{array}$$

The first relation (10.6) reveals that the pressure field according to the Stokes model is governed by *Laplace*'s equation and not *Poisson*'s equation, as is the case for the Navier-Stokes model. As for the second (10.6b), its physical consequence is that there cannot be, in a *steady state*, any viscous vorticity diffusion in a Stokes flow. This result can also be ascertained by taking the curl throughout Eq. (10.2) which proves that:

$$\frac{\partial\boldsymbol{\Omega}}{\partial t} = \nu\boldsymbol{\Delta}\boldsymbol{\Omega}.$$

10.2.4 The Main Motion Properties

10.2.4.1 Linearity

Any linear combination of Stokes' stationary model solutions is a solution of this model subject to the same boundary conditions.

This result is a direct consequence of the advection term elimination which makes the Stokes model equations linear. In particular, it means that if $(\mathbf{V}_1, P_1)$ and $(\mathbf{V}_2, P_2)$ are two solutions of Eqs. (10.3) and (10.4), then the velocity–pressure field $(\mathbf{V}, P)$ is also a solution, with:

$$\mathbf{V} = \alpha_1 \mathbf{V}_1 + \alpha_2 \mathbf{V}_2 \qquad \text{and} \qquad P = \alpha_1 P_1 + \alpha_2 P_2 \,,$$

where α_1 and α_2 are any two reals.

Remark The previous property leads, as in the case of ideal fluid potential flows, to the generation of flows by "superposition". In both cases, the resulting velocity field is obtained by the vectorial addition of the elementary fields at any point in the domain, including the boundary. In a creeping motion, the same applies to the pressure, which means the algebraic addition of the elementary flow pressures. It should be remembered that the latter property is generally erroneous in a potential flow, Cf. Chap. 6 Sect. 6.2.3.

10.2.4.2   Unicity

> The flow which exists in a domain Ω, as a solution of the Stokes equations for a given set of conditions on the boundary Σ of this domain, is unique.[1]

Proof Let us consider a domain (Ω) bounded by a surface (Σ) and assume that there is a flow $(\mathbf{V}, P)$ which is a solution of the Stokes model in (Ω), satisfying the boundary condition $\mathbf{V}|_\Sigma = \mathbf{V}_0$.
Suppose there is a second flow, *i.e.* a second velocity and pressure field $(\mathbf{V}', P')$ which is a solution of the the *same* equations in the *same* domain and satisfying the *same* boundary condition $\mathbf{V}'|_\Sigma = \mathbf{V}|_\Sigma = \mathbf{V}_0$.
We will demonstrate that the velocity fields $\mathbf{V}$ and $\mathbf{V}'$ cannot be different. If they were, there would necessarily be points in the domains such that:

$$\frac{\partial V_i}{\partial x_j} \neq \frac{\partial V_i'}{\partial x_j} \,.$$

Let us then express $I \equiv \iiint_\Omega \left(\frac{\partial V_i}{\partial x_j} - \frac{\partial V_i'}{\partial x_j} \right)^2 dv \equiv \iiint_\Omega \left(\frac{\partial (V_i - V_i')}{\partial x_j} \right)^2 dv$. It can

be noted that $\left(\frac{\partial V_i}{\partial x_j} \right)^2 \equiv \frac{\partial}{\partial x_j} \left(V_i \frac{\partial V_i}{\partial x_j} \right) - V_i \frac{\partial^2 V_i}{\partial x_j \partial x_j}$, which immediately leads to transforming the previous integral as follows:

$$I \equiv \iiint_\Omega \frac{\partial}{\partial x_j} \left[(V_i - V_i') \left(\frac{\partial V_i}{\partial x_j} - \frac{\partial V_i'}{\partial x_j} \right) \right] dv - \iiint_\Omega (V_i - V_i') \left(\frac{\partial^2 V_i}{\partial x_j \partial x_j} - \frac{\partial^2 V_i'}{\partial x_j \partial x_j} \right) dv.$$

Using the divergence formula, the first integral can equally be written as:

---

[1] This property restricts the practical scope of the previous theoretical result on linearity.

$$\iiint_\Omega \frac{\partial}{\partial x_j}\left[\left(V_i - V_i'\right)\left(\frac{\partial V_i}{\partial x_j} - \frac{\partial V_i'}{\partial x_j}\right)\right]dv \equiv \iint_\Sigma \left(V_i - V_i'\right)\left(\frac{\partial V_i}{\partial x_j} - \frac{\partial V_i'}{\partial x_j}\right)n_j d\sigma\,,$$

which is identically zero, due to the equality of the velocity fields on the boundary. Introducing Stokes' Eq. (10.4), the second integral is written as follows:

$$\iiint_\Omega \left(V_i - V_i'\right)\left(\Delta V_i - \Delta V_i'\right)dv = \frac{1}{\mu}\iiint_\Omega \left(V_i - V_i'\right)\left(\frac{\partial P}{\partial x_i} - \frac{\partial P'}{\partial x_i}\right)dv$$

$$\equiv \frac{1}{\mu}\iiint_\Omega \frac{\partial}{\partial x_i}\left[\left(V_i - V_i'\right)\left(P - P'\right)\right]dv - \frac{1}{\mu}\iiint_\Omega \left(P - P'\right)\left(\frac{\partial V_i}{\partial x_i} - \frac{\partial V_i'}{\partial x_i}\right)dv\,.$$

The first integral on the right hand side is null for the same previous reason. The second is also null as a direct result of the solenoidal character of each velocity field in an isovolume situation.

If the original assumption were true, the integral I would be non-zero, which is not the case and therefore the solutions are identical.

10.2.4.3 Dissipation Minimization

> For given boundary conditions, the flow which is established in a domain Ω as the solution to the Stokes model equations is that which minimizes the viscous dissipation volume in this domain.

In other words, this theorem states that, by denoting the strain rate field associated with the Stokes solution as $2S_{ij} = \left(\partial U_i/\partial x_j + \partial U_j/\partial x_i\right)$, then:

$$D \equiv 2\mu \iiint_\Omega S_{ij}S_{ij}dv \quad \text{is } minimum \text{ within } \Omega.$$

Proof Suppose there is another flow in the same domain which satisfies the same boundary conditions. According to the previous unicity theorem, this flow is obviously *not* a solution to Stokes' equations. Let us denote the velocity and pressure of this second field as $\mathbf{V'}$ and P' respectively, with, for the corresponding deformation rate, S_{ij}'. We will begin by proving the following result:

$$I \equiv \iiint_\Omega S_{ij}\left(S_{ij}' - S_{ij}\right)dv = 0\,.$$

For this purpose, let us firstly observe that:

$$S_{ij}\left(S_{ij}' - S_{ij}\right) = \frac{1}{4}\left(\frac{\partial V_i}{\partial x_j}\frac{\partial V_i'}{\partial x_j} - \frac{\partial V_i}{\partial x_j}\frac{\partial V_i}{\partial x_j} + \frac{\partial V_j}{\partial x_i}\frac{\partial V_i'}{\partial x_j} - \frac{\partial V_j}{\partial x_i}\frac{\partial V_i}{\partial x_j} + \right.$$

$$\left. \frac{\partial V_i}{\partial x_j}\frac{\partial V_j'}{\partial x_i} - \frac{\partial V_i}{\partial x_j}\frac{\partial V_j}{\partial x_i} + \frac{\partial V_j}{\partial x_i}\frac{\partial V_j'}{\partial x_i} - \frac{\partial V_j}{\partial x_i}\frac{\partial V_j}{\partial x_i} \right),$$

an expression which, due to the nature of the dummy indices, can still be written as:

$$S_{ij}\left(S_{ij}' - S_{ij}\right) \equiv \frac{1}{2}\left(\frac{\partial V_i'}{\partial x_j} - \frac{\partial V_i}{\partial x_j}\right)\frac{\partial V_i}{\partial x_j} + \frac{1}{2}\left(\frac{\partial V_i'}{\partial x_j} - \frac{\partial V_i}{\partial x_j}\right)\frac{\partial V_j}{\partial x_i},$$

which leads to set, with obvious notations, that $2I = I_1 + I_2$.
Considering the first integral, we have:

$$I_1 = \iiint_\Omega \frac{\partial}{\partial x_j}\left[\left(V_i' - V_i\right)\frac{\partial V_i}{\partial x_j}\right]dv - \iiint_\Omega \left(V_i' - V_i\right)\frac{\partial^2 V_i}{\partial x_j \partial x_j}dv,$$

and similarly

$$I_2 = \iiint_\Omega \frac{\partial}{\partial x_j}\left[\left(V_i' - V_i\right)\frac{\partial V_j}{\partial x_i}\right]dv - \iiint_\Omega \left(V_i' - V_i\right)\frac{\partial^2 V_j}{\partial x_j \partial x_j}dv.$$

Applying the divergence formula, and taking into account that both solutions coincide on the domain boundary, the first two integrals on the right hand side in expressions I_1 and I_2 are zero. Furthermore, as a result of the isovolume condition, it is obvious that the second integral in expression I_2 is identically null. Finally, we obtain:

$$I_1 = -\iiint_\Omega \left(V_i' - V_i\right)\frac{\partial^2 V_i}{\partial x_j \partial x_j}dv.$$

Since, by assumption, the velocity field V_i verifies Stokes' equation, we have:

$$I_1 = -\frac{1}{\mu}\iiint_\Omega \left(V_i' - V_i\right)\frac{\partial\left(P - P_0\right)}{\partial x_i}dv$$

$$= -\frac{1}{\mu}\iiint_\Omega \frac{\partial}{\partial x_i}\left[\left(V_i' - V_i\right)\left(P - P_0\right)\right]dv + \frac{1}{\mu}\iiint_\Omega \left(P - P_0\right)\frac{\partial\left(V_i' - V_i\right)}{\partial x_i}dv.$$

The last integral is null due to the isovolume condition and the same for the first, since the solutions on the border coincide, which finally demonstrates that $I = 0$.
Now let us consider the dissipation within the domain Ω for the second flow, namely:

$$D' \equiv 2\mu \iiint_\Omega S_{ij}'S_{ij}'dv.$$

Noting that $S_{ij}' \equiv S_{ij} + \left(S_{ij}' - S_{ij}\right)$, we can deduce by simply squaring:

$$S'_{ij} S'_{ij} \equiv S_{ij} S_{ij} + \left(S'_{ij} - S_{ij} \right)^2 + 2 S_{ij} \left(S'_{ij} - S_{ij} \right) .$$

This identity thus makes it possible to write $D' \equiv D + 2\mu \iiint\limits_\Omega \left(S'_{ij} - S_{ij} \right)^2 dv +$

μI. However, since $I = 0$, this only leaves $D' \equiv D + 2\mu \iiint\limits_\Omega \left(S'_{ij} - S_{ij} \right)^2 dv$,

which clearly implies that $D' - D \geqslant 0$ and definitely proves the expected theorem.

10.2.4.4 The Kinematic Invariance

The last property we will mention is a direct consequence of the linearity. It can be stated as follows:

> The streamline geometry of a creeping motion is not altered by any algebraic change in the flow rate.

Indeed, for the sake of linearity, if $(\mathbf{V}, P)$ is a Stokes model solution in a domain Ω, then $(\alpha \mathbf{V}, \alpha P)$ is also a solution of the Stokes model in the same domain. Since multiplying by α does not change the velocity vector orientation, the geometry of the streamlines whose tangents are, by definition, colinear to the velocity vector, is therefore not modified by such an operation. Hence, the specific case $\alpha = -1$ corresponds to a pure and simple inversion of the movement direction along the streamlines, without altering their geometry.

The experimental evidence — The *kinematic invariance* by the flow orientation reversal is sometimes associated with the *reversibility* of the creeping motion regime. However, this terminology should not lead to confusion with the *thermodynamic* definition of the reversibility concept, since creeping motions remain irreversible in this respect.

To outline this ability to reverse of the creeping regime, we consider the following experiment inspired by G. I. Taylor's filmed sequence "Low Reynolds Number Flows". The annular space between two vertical coaxial circular cylinders is filled with a highly viscous fluid. On the free surface of this fluid, a spot of coloured dye is placed, whose evolution in time can be visualized, after setting the inner cylinder in rotation at a very low speed, so that the fluid remains in a creeping motion. By turning this cylinder in the positive trigonometric direction—path (1) in the following Figure—the spot moves and deforms, its contour changing according to the local fluid strain rate. After several turns, successive windings of the dye on itself can be observed (Fig. 10.1).

If, after N revolutions in the positive direction, the same number of revolutions *in the reverse direction*—(2) in the Figure—is carried out, the fluid particles return to their initial positions and the spot returns to the contours of the original geometry. A very slight dye dispersion appears as a result of the diffusion by the molecular

Fig. 10.1 Schematic for the kinematic reversibility of the creeping regime in an axisymmetric Couette flow

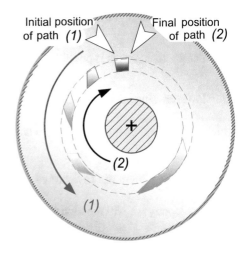

agitation, which remains the only source of the *kinematic irreversibility*. However, the mechanical power supplied to the forward motion is not retrieved on the reverse movement, which is a clear evidence of the *thermodynamically irreversible* nature of the motion.

HISTORICAL NOTE—The creeping motion—In an article dated 1851, Stokes [150] was interested in a full range of movements generated in a weighing fluid by various body oscillation, a plate, a sphere and a cylinder. To study these movements, he used the equations which he himself set out, but which he simplified by assuming that the fluid is incompressible (isovolume) and the oscillations of sufficiently small amplitude to disregard the *quadratic* advective terms: «*...the motion is supposed very small, on which account it will be allowable to neglect the terms which involve the square of the velocity»*. In doing so, Stokes introduced the class of flows with negligible inertial forces in the advective fluid transport (the non-linear terms), or creeping motions.

The key properties of this flow class were addressed in a paper by Helmholtz [69] in 1868, about the minimum dissipation rate, and by Korteweg [84] in 1883, concerning the unicity and stability issues of the creeping regime.

The reasons for studying such a flow class can be highly varied. Historically, the determination of the quantitative effects due to the air resistance on the pendulum oscillations was undoubtedly a first source of motivation, presumably dating back to Newton. In book II of the Principia, proposition XL, he examined for this purpose the oscillatory movement of a globe under the low-amplitude assumption. It is also the same type of concern which justified Stokes' interest in this situation, which is one which he dealt with in detail in his 1851 article. This creeping regime is involved in many other applications, for example, in meteorology (the size and local terminal velocity of fog droplets), in two-phase flows (idem for aerosols), in lubrication (highly viscous fluid in a confined geometry), etc.

10.3   The Validity of Stokes' Model

Here we are looking at the practical conditions under which the Stokes model can be taken into account.
As introduced in Sect. 10.2.1 of the Chapter, the creeping motion assumption reflects the preponderance of the viscosity forces over those of inertia, viz a low Reynolds number. Now, we have seen that this number could also be interpreted in terms of the scale ratio between the macroscopic advective motion and the diffusion by molecular agitation. In this respect, the creeping regime is characterized, for a given reference time, by a diffusive transfer whose length scale (ℓ) is much higher than that of the advective transport (L), according to the relation:

$$L = \left(\sqrt{Re} \right) \times \ell . \tag{10.7}$$

Remark Let us consider, as a way to illustrate, a creeping flow over a fixed solid obstacle, with a global Reynolds number of about $Re \approx 10^{-4}$ for example. The presence of this obstacle is perceived within the entire fluid by the diffusion effect of the momentum sink which the solid generates within the flow. Let us suppose that after a certain period of time, this type of information has spread by diffusive transfer up to 10 cm from the obstacle. During the same period, the fluid particles will have moved by only one millimeter, in accordance with the previous relation.

In practice, the validation of the low Reynolds number condition in the creeping regime can be obtained in three basic independent ways[2]:
– very low velocity motions;
– very small-scale movements;
– high-viscosity fluid flows.

The first situation is that of geo-material displacements. Despite the magnitude of the involved length scales, the displacement slowness is such that the global Reynolds number is still very low. Examples of such displacements include the movement of glaciers or the earth's crust.
Very small-scale length motions can occur in internal flows with minute geometries, such as channels and micro-cracks in porous media. This case may also concern the displacement of very small objects, such as bacteria or various suspensions, in a highly viscous fluid.
The motion of specific fluids such as paints, oils, bitumen, molten glass or honey seems to call for the last case. However, it should not be concluded that the Stokes

---

[2] We have studied in Chap. 9 Sect. 9.6 flows for which the inertia terms cancelled out for geometrical invariance reasons related to the boundary conditions: Poiseuille, Couette flows, etc. The equations are thus *formally* identical to those of Stokes' model, but the *physical interpretation* of the solutions is *radically different*, since their validity in terms of Reynolds' number results, for these flows, from stability considerations involving the inertia forces which do not make sense in the creeping regime.

model, as presented in Sect. 10.2.2, applies to all of these situations because many of them do involve fluids whose behaviour does not fall within Newton's scheme.

ADDITIONAL INFORMATION— **Magnitude order considerations based on characteristic scales of some typical creeping motions**—

- In tectonics, the displacement speed of the Earth's crustal plates is of the order of a few centimeters per year, *i.e.*, for an amplitude of 30 cm, a speed of about 10^{-8} m/s. With a viscosity of $\nu \approx 10^{20}\ m^2/s$, this leads to a Reynolds number of about 10^{-23} based on a 100 km length scale.
- For an object whose size is of the micrometer order and which is moving in water ($\nu \approx 10^{-6}\ m^2/s$) at a speed of 10 $\mu m/s$, the corresponding Reynolds number order is about 10^{-5}.
- Finally, for molten glass ($\nu \approx 10^{-2}\ m^2/s$) blown through an orifice of 1 cm in diameter at a speed of 1 cm/s, the Reynolds number is about 10^{-2}.

10.4   Examples of Creeping Motions

We will conclude this chapter with some examples of creeping flows. In an isovolume situation, these flows obviously belong to the motion class whose velocity field derives from a vector potential, as introduced in Chap. 5. We will therefore extend and complete here the examples which have been discussed in the above mentioned chapter.

10.4.1   The Flow in a Hele-Shaw Cell

In a Hele-Shaw flow, a highly viscous fluid moves at a low velocity (U_a) between two parallel plates spaced at a short distance from each other (a low degree of thickness $2h$). If an obstacle is placed transversely, whose characteristic dimension is L (see Fig. 10.2), it can be demonstrated that, under the creeping regime hypothesis, this configuration makes it possible to simulate, at a distance from the obstacle, the potential flow field of an incompressible ideal fluid.

By setting $\varepsilon = h/L$ and $R_e = U_a L/\nu$, it can be proved—see next Exercise—that in a steady state, when disregarding the gravity forces and under the two conditions

$$\varepsilon \ll 1 \quad \text{and} \quad R_e \ll 1,$$

the velocity vector $\mathbf{V}_0\,(x, y)$ in the cell's mid-plane ($z = 0$) verifies both equations:

Fig. 10.2 Schematic for the
Hele-Shaw flow setup

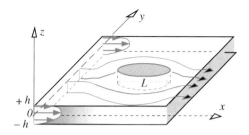

$$div\,[\mathbf{V}_0\,(x,\,y)] = 0\,,\tag{10.8a}$$

$$\mathbf{V}_0\,(x,\,y) = -\frac{h^2}{2\mu}\mathbf{grad}\,[P\,(x,\,y)]\,.\tag{10.8b}$$

The demonstration of these results is proposed as an exercise.
Let us simply point out here that the second relation expresses the irrotational character of this velocity field, owing to the identity $curl\,(\mathbf{grad}) \equiv \mathbf{0}$, a common feature of ideal fluid potential flows. Thus, in the mid-plane[3] we can state the following result, which may appear paradoxical:

> The streamlines of a viscous fluid motion in a Hele-Shaw cell ($R_e \to 0$) are the analogy of those of an inviscid fluid potential flow ($R_e \to +\infty$).

HISTORICAL NOTE—**The Hele-Shaw flow**—At the end of the XIXth century, in the context of work on the resistance exerted by a flow on an immersed solid, Hele-Shaw carried out various visualizations by injecting air or dye into the mass of a water vein. Around a finite span body, Hele-shaw [66, 67], observed the presence of a clear zone, of neat border, evidencing a condition of motion different from that of the external current, which he sought to analyze. The discovery he made then went far beyond what he expected. Let us give him the word: «*In endeavouring to investigate the markedly different condition of flow at the surface, instead of using a thick sheet of water of from three-eights to half an inch, a thin sheet of water was employed the thickness of which was not greater than that of the abovementioned border line. The result of doing this was to reveal a different state of flow in the water...and colour bands, corresponding to the stream lines of the mathematician, can be obtained.*» Stokes [151] was the one in 1898 who mathematically established that the streamlines thus highlighted were in fact equivalent to those of a potential flow of an *inviscid* fluid (see the next exercise).

---

[3] In fact, the property applies to any plane parallel to the cell sides, excluding the walls themselves, as can be seen in the following Exercise, Eq. (10.9).

EXERCISE 44 The flow field in a Hele-Shaw cell.

We consider a flow in a Hele-Shaw cell, as outlined in Fig. 10.2. The velocity components and pressure at any point (x, y, z) in the flow field are referred to as U, V, W, P. The characteristic length scale of the motion in the x and y directions is denoted by L, and by εL in the z direction, where ε represents an "*infinitely small*" positive coefficient.

QUESTION 1. Let U_a be a velocity reference such as $U \sim V \sim U_a$. Prove that, by virtue of the continuity equation, we have $W \sim \varepsilon U_a$.

QUESTION 2. Deduce, from the estimate of the different terms in the momentum equation along z, that the pressure is a function $P(x, y)$ in a weightless fluid and steady motion.

QUESTION 3. Similarly, prove that the velocity field components $U(x, y, z)$ and $V(x, y, z)$ in the (x, y) plane are linked to the pressure field $P(x, y)$ according to the equations:

$$\begin{cases} \dfrac{\partial P}{\partial x} = \mu \dfrac{\partial^2 U}{\partial x^2}, \\[2mm] \dfrac{\partial P}{\partial y} = \mu \dfrac{\partial^2 U}{\partial y^2}. \end{cases}$$

QUESTION 4. Demonstrate that the flow velocity field in the mid-plane ($z = 0$) is that of a potential, two-dimensional plane motion. Prove that this property extends to the velocity field in any plane $z = C'(\equiv z_0)$, with $-h < z_0 < +h$.

SOLUTION:

1. When denoting by U_a the reference velocity of the fluid motion in the (x, y) plane and by U_a' that of the motion along the z axis, the order of magnitude estimation of the terms in the continuity equation is as follows:

$$\frac{\partial U}{\partial x} + \frac{\partial V}{\partial y} + \frac{\partial W}{\partial z} = 0$$

$$\frac{U_a}{L} \quad \frac{U_a}{L} \quad \frac{U_a'}{\varepsilon L},$$

Hence, the non-degeneration of this equation requires that $U_a' \sim \varepsilon U_a$, which demonstrates the proposition.

2. By applying a similar process to the inertia and viscosity terms of the Navier-Stokes equations, we obtain the following estimate for the three projection directions, respectively:

$$U\frac{\partial U}{\partial x} + V\frac{\partial U}{\partial y} + W\frac{\partial U}{\partial z} = -\frac{1}{\rho}\frac{\partial P}{\partial x} + \nu\left(\frac{\partial^2 U}{\partial x^2} + \frac{\partial^2 U}{\partial y^2} + \frac{\partial^2 U}{\partial z^2}\right),$$

$$\frac{U_a^2}{L} \quad \frac{U_a^2}{L} \quad \frac{U_a^2}{\varepsilon L} \qquad \nu\left(\frac{U_a}{L^2} \quad \frac{U_a}{L^2} \quad \frac{U_a}{\varepsilon^2 L^2}\right)$$

$$U\frac{\partial V}{\partial x} + V\frac{\partial V}{\partial y} + W\frac{\partial V}{\partial z} = -\frac{1}{\rho}\frac{\partial P}{\partial y} + \nu\left(\frac{\partial^2 V}{\partial x^2} + \frac{\partial^2 V}{\partial y^2} + \frac{\partial^2 V}{\partial z^2}\right),$$

$$\frac{U_a^2}{L} \quad \frac{U_a^2}{L} \quad \frac{U_a^2}{\varepsilon L}\nu\left(\frac{U_a}{L^2} \quad \frac{U_a}{L^2} \quad \frac{U_a}{\varepsilon^2 L^2}\right)$$

$$U\frac{\partial W}{\partial x} + V\frac{\partial W}{\partial y} + W\frac{\partial W}{\partial z} = -\frac{1}{\rho}\frac{\partial P}{\partial z} + \nu\left(\frac{\partial^2 W}{\partial x^2} + \frac{\partial^2 W}{\partial y^2} + \frac{\partial^2 W}{\partial z^2}\right),$$

$$\varepsilon\frac{U_a^2}{L} \quad \varepsilon\frac{U_a^2}{L} \quad \varepsilon\frac{U_a^2}{L} \qquad \nu\left(\frac{\varepsilon U_a}{L^2} \quad \frac{\varepsilon U_a}{L^2} \quad \frac{U_a}{\varepsilon L^2}\right)$$

This leads to the following results about:

- the Laplacian's contributions, *viz* $\frac{\partial^2}{\partial z^2} \gg \frac{\partial^2}{\partial x^2}$ and $\frac{\partial^2}{\partial z^2} \gg \frac{\partial^2}{\partial y^2}$;
- the order-of-magnitude of the inertia and viscosity term:

$$[\text{Inertia}] \sim \varepsilon \times R_e \times [\text{Viscosity}] ;$$

- the components of the viscous diffusion term in all three directions:

$$[\text{Viscosity}]_x \sim [\text{Viscosity}]_y \quad \text{and} \quad [\text{Viscosity}]_x \sim \varepsilon^2 \times [\text{Viscosity}]_z .$$

Thus, retaining only the preponderant terms, Stokes' equations are reduced, in projections, to

$$0 = -\frac{\partial P}{\partial x} + \mu\frac{\partial^2 U}{\partial z^2} ,$$

$$0 = -\frac{\partial P}{\partial y} + \mu\frac{\partial^2 V}{\partial z^2} ,$$

$$0 = -\frac{\partial P}{\partial z} .$$

The pressure is therefore independent of z.

3. The integration of the first of these equations provides

$$U(x, y, z) = \frac{1}{\mu}\left(\frac{z^2}{2} + \alpha z + \beta\right)\frac{\partial P}{\partial x} ,$$

and a similar expression for $V(x, y, z)$. The integration constants are determined by prescribing the boundary conditions, *viz* $U(x, y, \pm h) = 0$, so that the solution is finally written as follows:

$$U\,(x, y, z) = \frac{1}{2\mu}\left(z^2 - h^2\right)\frac{\partial P}{\partial x}. \tag{10.9}$$

4. In the symmetry plane $z = 0$, the velocity field components are:

$$U\,(x, y, 0) \equiv U_0\,(x, y) = -\frac{1}{2\mu}h^2\frac{\partial P}{\partial x}, \tag{10.10}$$

$$V\,(x, y, 0) \equiv V_0\,(x, y) = -\frac{1}{2\mu}h^2\frac{\partial P}{\partial y}. \tag{10.11}$$

By setting $\Phi_0(x, y) = -\frac{1}{2\mu}h^2 P(x, y)$, the previous relations thus express that the velocity field $\mathbf{V}_0(x, y)$ in the mid-plane is that of a potential flow, with $\mathbf{V}_0(x, y) = \mathbf{grad}\,\Phi_0(x, y)$.

Equation (10.9) proves that this result applies to any motion in a plane $z = C'(z_0 \neq 0)$, when taking as the potential function $\Phi(x, y) = -\frac{1}{2\mu}(z_0^2 - h^2)P(x, y)$. It should be noted that in any surface parallel to the Hele-Shaw cell sides, the main velocity field can be assimilated to that of a *bicomponent* plane (x, y) motion, but variable according to z. Thus, this is not a *two-dimensional* plane motion, since $\frac{\partial}{\partial z} \neq 0$.

10.4.2 Lubrication, Viscous Film Motion and Fluid Bearing

Common experience teaches us that the relative displacement of two solids in contact with each other requires the application of a relatively large set of forces, compared to the case where a fluid film separates the two bodies. As Batchelor points out [5], this is what happens when a sheet of paper thrown flush with a smooth horizontal plane gives the impression of sliding over the plane[4]. In fact it is moving on a very thin and slightly angled air film which prevents, for a while, any direct contact with the solid wall. The explanation for this phenomenon lies in the very low degree of thickness and inclination of the fluid film which, due to the high velocity gradients, is the source of high viscosity forces. Then, Stokes' model prove that these forces can only be dynamically balanced by pressure forces. It is the latter which oppose the film to be squeezed and thus the solid parts come into contact. Such properties find many practical applications in lubrication, hydraulic slides, thrust bearings,[5] or the reduction of friction between rotating parts using fluid bearings,[6] see Fig. 10.3a.

---

[4] On page 219 of the quoted work, Batchelor describes the phenomenon in the following way «*For instance, a sheet of paper dropped onto a smooth floor will often 'float' on a film of air between it and the floor and thereby will be able to glide horizontally for some distance before coming to rest.*»

[5] The high pressures in the film of a fluid wedge are used in devices known as the *Michell thrust bearing*, which can be found on the drive shafts of ships, hydraulic turbines, etc.

[6] Air or more generally gas-lubricated bearings are also used in industrial applications to control high-speed rotating shafts in turbomachines for instance, but in this case the phenomena involved are different, since the regime is no longer that of the incompressible fluid creeping motion.

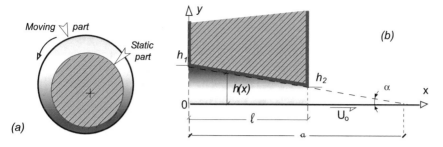

Fig. 10.3 Schematic for the fluid bearing **a** and its 2D-plane configuration **b**

We will explain the operating principle of this device by adopting a two-dimensional plane schematization.

For this purpose, let us consider the fluid motion in a domain of length l between a plate and a plane surface inclined at an angle α on it, Fig. 10.3(b). Under Stokes' approximations, the *local* value of the inertia and viscosity forces can be estimated as:

$$[\text{Inertia}] \sim \rho \frac{U_0^2}{l}, \qquad [\text{Viscosity}] \sim \mu \frac{U_o}{h'^2}, \quad \text{where} \quad h' = h_1 - h_2.$$

The characteristic Reynolds number is written, for this flow:

$$Re = \frac{U_0 l}{\nu} \times \left(\frac{h'}{l}\right)^2.$$

Assuming $R_e \ll 1$, we can then ascertain that there is, in the fluid film separating the two solid parts, a pressure field given by (see Exercise 45):

$$P(x) = P_0 + \frac{6\mu U_0 x}{\alpha^2 a (a-l)} - 6\mu \frac{x(2a-x)}{\alpha^3 a^2 (a-x)^2} Q,$$

where the expression of the volume flow rate Q is:

$$Q = \frac{a(a-l)}{2a-l} \alpha U_0.$$

From this, the value of the pressure resultant force on the moving wall can be deduced, which, per unit span, is equal to:

$$F = \frac{6\mu U_0}{\alpha^2} \left[\ln \frac{a}{a-l} - \frac{2l}{2a-l}\right]. \tag{10.12}$$

The demonstration of these results is proposed as an exercise whose solution is given hereafter.

ADDITIONAL INFORMATION—Some parametric considerations related to the slanted film flow—It is not an easy task to differentiate between the respective influences of the film thickness (h_1) and the slant angle (α) since, within the limit $\alpha \to 0$, these two quantities are linked by $h_1 = \alpha \times a$.

In fact, if we set the initial thickness h_1 and the length l of the film, then it is clear that $\alpha \to 0$ is equivalent to $l/a \to 0$. We can then deduce that:

$$\text{for } l/a \to 0, \quad \ln \frac{a}{a-l} - \frac{2l}{2a-l} \simeq \frac{1}{2}\left(\frac{l}{a}\right)^2,$$

so that the pressure force becomes $F \sim 3\mu U_o \dfrac{l^2}{h_1^2}$ for $\alpha \to 0$.

With a view to clarity the discussion, we take the following values:

$$U_0 = 10\,m/s, \ l = 10\,cm, \ \nu = 2.10^{-4}m^2/s, \ \alpha \sim \frac{h'}{l} = 2.10^{-3}rad, \ h_1 = 1\,mm.$$

Hence we obtain $R_e = 2 \times 10^{-2}$, $Q \simeq 0.4 \times 10^{-2}m^2/s$ which, with a fluid of density $10^3 kg/m^3$, leads to a pressure force of $2\,764\,N$ for a theoretical limit value ($\alpha \to 0$) of $60\,000\,N$. Let us finally point out that the dependence upon the angle α is strongly non-linear, due to the fact that if the wedge is reduced by half ($\alpha = 10^{-3}\,rad$), the force is multiplied by four, rising to $11\,056\,N$.

EXERCISE 45 The flow in a 2D-plane fluid bearing.

We refer to U, V, P as the velocity components and pressure at any point (x, y) of a bearing flow of a two-dimensional plane geometry with respect to the reference frame outlined in Fig. 10.3b. We put $h' = h_1 - h_2$ and $\varepsilon = h'/l$.

QUESTION 1. Estimate the orders of magnitude of the different terms in the continuity equation and both projections of the momentum equation.

QUESTION 2. Show that Stokes' model approximation can be expressed here by the condition: $$\varepsilon^2 \times R_e \ll 1,$$

where $R_e = U_o l/\nu$ is the Reynolds number based on the advection length.

QUESTION 3. Solve the previous equations to derive the expressions for the velocity field, pressure variation along x, volume flow rate and the components of the resultant force on the lower plate.

SOLUTION: 1. Comparing the term's order of magnitude in the continuity equation provides the first condition:

$$\frac{\partial U}{\partial x} + \frac{\partial V}{\partial y} = 0$$

$$\frac{U_o}{l} \qquad \frac{U_o'}{h'} \quad ,$$

where the estimate of the cross-wise variations is based on the length scale $h' = h_1 - h_2$. Under these conditions, the cross-wise velocity scale is given by $U_o' \sim \frac{h'}{l} U_o$ to prevent the mathematically degenerating equation.

With this result, the comparison of the inertia and viscosity terms in the Navier-Stokes equation projections along x and y is as follows:

$$U \frac{\partial U}{\partial x} + V \frac{\partial U}{\partial y} = -\frac{1}{\rho} \frac{\partial P}{\partial x} + \nu \left(\frac{\partial^2 U}{\partial x^2} + \frac{\partial^2 U}{\partial y^2} \right) ,$$

$$\frac{U_o^2}{l} \qquad \frac{U_o^2}{l} \qquad\qquad \nu \left(\frac{U_o}{l^2} \qquad \frac{U_o}{h'^2} \right)$$

$$U \frac{\partial V}{\partial x} + V \frac{\partial V}{\partial y} = -\frac{1}{\rho} \frac{\partial P}{\partial y} + \nu \left(\frac{\partial^2 V}{\partial x^2} + \frac{\partial^2 V}{\partial y^2} \right) ,$$

$$\frac{h'}{l} \frac{U_o^2}{l} \qquad \frac{h'}{l} \frac{U_o^2}{l} \qquad\qquad \nu \frac{h'}{l} \left(\frac{U_o}{l} \qquad \frac{U_o}{h'^2} \right)$$

From the previous relations, the following conclusions can be drawn:

- the cross-wise diffusion $\partial^2 / \partial y^2$ prevails over the stream-wise diffusion $\partial^2 / \partial x^2$;
- the inertia component along x is the dominant component:
  $[\text{Inertia}]_y = \varepsilon \, [\text{Inertia}]_x$;
- the viscosity component along x is the dominant component:
  $[\text{Viscosity}]_y = \varepsilon \, [\text{Viscosity}]_x$;
- the ratio between the *dominant* inertia and viscosity forces is:

$$\frac{[\text{Inertia}]_x}{[\text{Viscosity}]_x} \sim \varepsilon^2 \times Re,$$

where $Re \equiv U_o l / \nu$ stands for the usual Reynolds number and $\varepsilon = h'/l$ the slanting ratio of the fluid film.

2. Under the last equivalence, Stokes' condition, according to which the study will be carried out, is therefore expressed here as $\varepsilon^2 \times R_e \ll 1$.

By restricting the momentum equation to the dominant terms, the velocity and pressure fields are linked by:

$$\frac{\partial P}{\partial x} = \mu \frac{\partial^2 U}{\partial y^2} \quad \text{and} \quad \frac{\partial P}{\partial y} = 0 .$$

3. When prescribing the conditions on the lower and upper boundaries respectively, viz $U(x, 0) = U_o (= C')$ and $U(x, h(x)) = 0$, the integration of these equations yields:

$$U(x, y) = \left(1 - \frac{y}{h}\right) U_o - \frac{y}{h}\left(1 - \frac{y}{h}\right) \frac{h^2}{2\mu} \frac{dP}{dx},$$

where h and dP/dx are two functions of the variable x alone. Denoting by Q the volume flow rate of the fluid through the film:

$$Q = \int\limits_0^{h(x)} U(x, y)\, dy,$$

we obtain, by integration of the velocity field $\dfrac{U_o h}{2} - \dfrac{h^3}{12\mu} \dfrac{dP}{dx} = Q = C'$. The pressure field is directly derived from this:

$$P(x) = P_o + 6\mu \frac{U_o}{\alpha^2} \frac{x}{a(a-x)} - 6\mu \frac{Q}{\alpha^3} \frac{x(2a-x)}{a^2 (a-x)^2}.$$

By prescribing the same pressure condition in the inlet and outlet sections, *i.e.*, $P_o \equiv P(0) = P(l)$, we obtain the result:

$$Q = \alpha U_o \frac{a(a-l)}{2a - l}.$$

Then the pressure field is obtained as

$$P(x) = P_o + 6\mu \frac{U_o}{\alpha^2 (2a-l)} \frac{x(x-l)}{(a-x)^2},$$

which clearly proves that at any point in the film, the pressure is higher than the value outside the film. The result for the lower plate is a *thrust* opposite the upper wall which is equal to:

$$F_y = \int\limits_0^l [P(x) - P_o]\, dx = \frac{6\mu U_o}{\alpha^2} \left[\ln \frac{a}{a-l} - \frac{2l}{2a-l} \right].$$

Finally, by integrating the friction along the lower plate, the *viscous drag resultant* force is obtained:

$$F_x = \int\limits_0^l \mu \left(\frac{\partial U}{\partial y}\right)_{y=0} dx = -\frac{2\mu U_o}{\alpha} \left[\ln \frac{a}{a-l} + \frac{3l}{2a-l} \right].$$

HISTORICAL NOTE—Lubrication—In 1886, Reynolds published a study [136], aiming to explain, by the laws of hydrodynamics, the results of fluid film lubrication measurements between mechanical parts in relative movement. In fact, the appropriate equations for this problem had been given earlier by Stokes and Rayleigh. However, Reynolds pointed out the importance of the viscosity dependence on temperature in this situation and provided, on the basis of a dedicated experiment, a $\mu(T)$ relation for olive oil, by which the agreement between theory and measurement has become quite satisfactory.

10.4.3 Stokes' Flow Past a Sphere

10.4.3.1 The Problem

With this last example, we will in particular ascertain that the Stokes regime is not irrotational. The studied configuration, as outlined in the following Figure, can be defined in two different ways:

– a steady flow, with a uniform velocity V_0 at infinity, of an incompressible viscous fluid around a *fixed* sphere, of radius a;
– the uniform translational displacement, with a velocity V_0, of a sphere of radius a in a motionless fluid at infinity.

In both cases, it is assumed that the Reynolds number based on the sphere diameter ($R_e = 2aV_0/\nu$) is small enough to apply the Stokes model (Fig. 10.4).

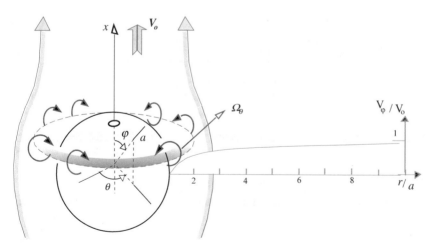

Fig. 10.4 Schematic for the creeping motion past a sphere and the velocity profile in the maximum cross section

A spherical coordinate system is adopted, whose Ox axis coincides with the flow direction at infinity. Denoting by (r, φ, θ) the coordinates in such a reference frame, the motion is axially symmetrical about the Ox axis, so that the velocity field is reduced to the two components V_r (r, φ) and V_φ (r, φ), with, in addition for any flow field function $\partial/\partial\theta \equiv 0$.

Due to the two-dimensional axisymmetrical nature of the motion, the vorticity vector is reduced to its sole component along the unit vector direction $\mathbf{e}_\theta$, and we will note

$$\mathbf{\Omega} \equiv curl\ \mathbf{V} = \Omega\,(r, \varphi)\ \mathbf{e}_\theta.$$

Thus, the entire problem can be reduced to the determination of the *scalar* function $\Omega\,(r, \varphi)$, with, as a boundary condition, $\Omega\,(\infty, \varphi) = 0$.

10.4.3.2   Stokes' Solution

The Stokes model solution for the creeping flow past a sphere consists, in the general case, of the following fields, where A, L, Q and P_0 are four constants:

– **Vorticity**:

$$\boxed{\Omega\,(r, \varphi) = A\frac{\sin \varphi}{r^2}} \tag{10.13}$$

– **Stream function**:

$$\boxed{\psi\,(r, \varphi) = \left(\frac{A}{2}r + Lr^2 + \frac{Q}{r}\right)\sin^2 \varphi} \tag{10.14}$$

– **Velocity field**:

$$\boxed{\begin{cases} V_r = \dfrac{2}{r^2}\left(\dfrac{A}{2}r + Lr^2 + \dfrac{Q}{r}\right)\cos \varphi \\[2mm] V_\varphi = -\dfrac{1}{r^2}\left(\dfrac{A}{2}r + 2Lr^2 - \dfrac{Q}{r}\right)\sin \varphi \end{cases}} \tag{10.15}$$

– **Pressure field**:

$$\boxed{P\,(r, \varphi) = P_0 + \mu A\frac{\cos \varphi}{r^2}} \tag{10.16}$$

The demonstration of these different results, given below, can usefully be carried out as an exercise. The values of the constants A, L, Q and P_0 are given in Table 10.1 with respect to the following two situations:

– a *fixed* sphere in a uniform flow of velocity V_∞ at infinity;
– a *moving* sphere by translation at a constant velocity V_0 in a rest fluid at infinity.

Proof The calculation of the vorticity—The vorticity field is obtained by solving Laplace's Eq. (10.6b) $\mathbf{\Delta\,\Omega} = \mathbf{0}$.

Table 10.1 Flows past a sphere: Stokes' solution

	Fixed sphere in a uniform flow	Moving sphere in a fluid at rest
Vorticity	$-\dfrac{3}{2}aV_\infty\dfrac{\sin\varphi}{r^2}$	$\dfrac{3}{2}aV_0\dfrac{\sin\varphi}{r^2}$
Stream function	$a^2\dfrac{V_\infty}{2}\left(-\dfrac{3}{2}\dfrac{r}{a}+\dfrac{r^2}{a^2}+\dfrac{1}{2}\dfrac{a}{r}\right)\sin^2\varphi$	$a^2\dfrac{V_0}{2}\left(\dfrac{3}{2}\dfrac{r}{a}-\dfrac{1}{2}\dfrac{a}{r}\right)\sin^2\varphi$
Velocity field	$/_r : V_\infty\left(1-\dfrac{3}{2}\dfrac{a}{r}+\dfrac{1}{2}\dfrac{a^3}{r^3}\right)\cos\varphi$	$/_r : V_0\left(\dfrac{3}{2}\dfrac{a}{r}-\dfrac{1}{2}\dfrac{a^3}{r^3}\right)\cos\varphi$
	$/_\varphi : -V_\infty\left(1-\dfrac{3}{4}\dfrac{a}{r}-\dfrac{1}{4}\dfrac{a^3}{r^3}\right)\sin\varphi$	$/_\varphi : -V_0\left(\dfrac{3}{4}\dfrac{a}{r}+\dfrac{1}{4}\dfrac{a^3}{r^3}\right)\sin\varphi$
Pressure field	$P = P_0 - \dfrac{3}{2}\mu a V_0\dfrac{\cos\varphi}{r^2}$	$P = P_0 + \dfrac{3}{2}\mu a V_0\dfrac{\cos\varphi}{r^2}$

By definition of the curl operator, we identically have $\Delta\,\mathbf{\Omega} \equiv curl\,(curl\,\mathbf{\Omega})$, so that the equation to be solved is also written:

$$curl\,(curl\,\mathbf{\Omega}) = \mathbf{0}.$$

Now, in spherical coordinates, the curl components of a vector $\mathbf{A}\left(A_r, A_\varphi, A_\theta\right)$ are expressed as:

$$\mathbf{e}_r : \frac{1}{r\sin\varphi}\left[\frac{\partial\,(A_\theta\sin\varphi)}{\partial\varphi} - \frac{\partial A_\varphi}{\partial\theta}\right],$$

$$\mathbf{e}_\varphi : \frac{1}{r}\left[\frac{1}{\sin\varphi}\frac{\partial A_r}{\partial\theta} - \frac{\partial\,(rA_\theta)}{\partial r}\right],$$

$$\mathbf{e}_\theta : \frac{1}{r}\left[\frac{\partial\,(rA_\varphi)}{\partial r} - \frac{\partial A_r}{\partial\varphi}\right].$$

Applying these relations to the vector $\mathbf{\Omega} = \Omega\,\mathbf{e}_\theta$, we obtain:

$$curl\,\mathbf{\Omega} : \begin{cases} \dfrac{1}{r\sin\varphi}\dfrac{\partial\,(\Omega\sin\varphi)}{\partial\varphi} \\[2mm] -\dfrac{1}{r}\dfrac{\partial}{\partial r}(r\Omega) \\[2mm] 0 \end{cases}$$

$$curl\,(curl\,\mathbf{\Omega}) : \begin{cases} 0 \\ 0 \\ -\dfrac{1}{r}\dfrac{\partial^2\,(r\Omega)}{\partial r^2} - \dfrac{1}{r^2}\dfrac{\partial}{\partial\varphi}\left(\dfrac{1}{\sin\varphi}\dfrac{\partial\,(\Omega\sin\varphi)}{\partial\varphi}\right) \end{cases} \tag{10.17}$$

Finally the equation to be solved for the scalar function $\Omega\,(r,\varphi)$ is written

$$r \frac{\partial^2}{\partial r^2} (r\Omega) + \frac{\partial}{\partial \varphi} \left[\frac{1}{\sin \varphi} \frac{\partial}{\partial \varphi} (\Omega \sin \varphi) \right] = 0.$$

The solution of this equation is sought in the form $\Omega (r, \varphi) = f(r) \cdot g(\varphi)$, where the functions $f(r)$ and $g(\varphi)$ depend separately on the variables r and φ respectively. By simple substitution in the equation for $\Omega (r, \varphi)$, we deduce that these functions should verify the following differential equations:

$$\frac{[r^2 f']'}{f} = C' \equiv \alpha \quad \text{and} \quad \frac{1}{g} \left[\frac{(g \sin \varphi)'}{\sin \varphi} \right]' = C' \equiv -\alpha,$$

where the prime symbol $'$ denotes the derivative with respect to the unique argument upon which the function depends. It is then easy to determine that $g(\varphi) = \sin \varphi$ is the solution of the second equation with $\alpha = 2$. The equation to be solved for the function $f(r)$ thus becomes:

$$r^2 f'' + 2rf' - 2f = 0.$$

The general solution of this differential equation is $f(r) = \dfrac{A}{r^2} + Br$, where A and B are two constants. Taking into account the boundary condition for $\Omega (r, \varphi)$ when $r \to \infty$, the constant B is necessarily zero. Thus we obtain the expected solution, Eq. (10.13).

The calculation of the stream function—The stream function of an axisymmetric flow in spherical coordinates (see Chap. 5, Table 5.1) is defined by:

$$V_r = \frac{1}{r^2 \sin \varphi} \frac{\partial \psi}{\partial \varphi},$$

$$V_\varphi = -\frac{1}{r \sin \varphi} \frac{\partial \psi}{\partial r}.$$

By substituting these expressions in the definition of the Ω_θ component, we easily obtain the equation governing the stream function:

$$\Omega_\theta \equiv \frac{1}{r} \left(\frac{\partial r V_\varphi}{\partial r} - \frac{\partial V_r}{\partial \varphi} \right) = -\frac{1}{r \sin \varphi} \frac{\partial^2 \psi}{\partial r^2} - \frac{1}{r^3} \frac{\partial}{\partial \varphi} \left(\frac{1}{\sin \varphi} \frac{\partial \psi}{\partial \varphi} \right) = A \frac{\sin \varphi}{r^2}.$$

Looking, as before, for a solution in the form of $\psi(r, \varphi) = F(r) \cdot G(\varphi)$, we obtain, after substitution:

$$r F'' + \frac{F}{r} \frac{\sin \varphi}{G} \left[\frac{G'}{\sin \varphi} \right]' = -A \frac{\sin^2 \varphi}{G}.$$

This is a second-order differential equation for the function $F(r)$, whose coefficients are either constant or a function of the argument φ. For such an equation to allow a solution, the independence of the variables r and φ requires that these coefficients be proportional, which provides two conditions for the function $G(\varphi)$:

$$\frac{\sin \varphi}{G} \left[\frac{G'}{\sin \varphi} \right]' = C_1^t \quad \text{and} \quad A \frac{\sin^2 \varphi}{G} = C_2^t.$$

These two conditions are compatible, since, for example, putting $G(\varphi) = \sin^2 \varphi$, we deduce that $C_1^t = -2$ and $C_2^t = A$. Thus, the equation to be solved for the function $F(r)$ becomes:

$$r F'' - \frac{2}{r} F = -A.$$

A particular integral of this equation is $Ar/2$ and the general integral of the homogeneous equation is $Lr^2 + Q/r$, where L and Q are two constants. By grouping these results, we obtain the expected expression, Eq. (10.14).

The calculation of the velocity field—By directly applying the definition of the stream function, we have:

$$V_r = \frac{2}{r^2} \left(\frac{A}{2} r + Lr^2 + \frac{Q}{r} \right) \cos \varphi,$$

$$V_\varphi = -\frac{1}{r^2} \left(\frac{A}{2} r + 2Lr^2 - \frac{Q}{r} \right) \sin \varphi.$$

Now, using the boundary conditions we can deduce the values of the constants A, L, Q, namely,

(a) for a uniform flow whose velocity at infinity is $V_\infty \mathbf{e}_x$ past a fixed sphere:

- along the sphere surface, $V_r(a, \varphi) = V_\varphi(a, \varphi) = 0$, $\forall \varphi$
- at infinity, $V_r(\infty, \varphi) = V_\infty \cos \varphi$ and $V_\varphi(\infty, \varphi) = -V_\infty \sin \varphi$, $\forall \varphi$

(b) for the translation of a sphere at a constant velocity $V_0 \mathbf{e}_x$, in a fluid at rest at infinity:

- along the sphere surface, $V_r(a, \varphi) = V_0 \cos \varphi$ and $V_\varphi(a, \varphi) = -V_0 \sin \varphi$, $\forall \varphi$
- at infinity, $V_r(\infty, \varphi) = V_\varphi(\infty, \varphi) = 0$, $\forall \varphi$.

With these conditions, we obtain $A = -\frac{3}{2} a V_\infty$ and $A = +\frac{3}{2} a V_0$ (resp.), $L = V_\infty/2$ and $L = 0$ (resp.), and finally $Q = a^3 V_\infty/4$ and $Q = -a^3 V_0/4$ (resp.).

The calculation of the pressure field—Assuming a weightless fluid, the pressure is deduced from the velocity by the Stokes equation

$$\mathbf{grad}\, P = \mu \mathbf{\Delta}\, \mathbf{V}.$$

Now, for an isovolume flow, we identically have $\mathbf{\Delta V} = -curl\,(curl\,\mathbf{V}) \equiv -curl\,\mathbf{\Omega}$, so that the pressure equation is written:

$$\mathbf{grad}\, P = -\mu\, curl\, \mathbf{\Omega}.$$

Then introducing the previous result for the vorticity field, we deduce:

$$
\begin{cases}
-\dfrac{1}{\mu}\dfrac{\partial P}{\partial r} \equiv \dfrac{1}{r\sin\varphi}\dfrac{\partial}{\partial\varphi}\left(\Omega\sin\varphi\right) = 2A\dfrac{\cos\varphi}{r^3}\,,\\[2mm]
-\dfrac{1}{\mu}\dfrac{1}{r}\dfrac{\partial P}{\partial\varphi} \equiv -\dfrac{1}{r}\dfrac{\partial}{\partial r}\left(r\Omega\right) = \dfrac{A\sin\varphi}{r^3}\,.
\end{cases}
$$

After integration of the first relation, we obtain $-P/\mu = -A\cos\varphi/r^2 + h\left(\varphi\right)$, which, when substituting in the second relation yields $A\dfrac{\sin\varphi}{r^3} + \dfrac{h'}{r} = A\dfrac{\sin\varphi}{r^3}$. The solution is therefore finally written as expected, namely $P = P_0 + \mu A\,\cos\varphi/r^2$, where P_0 is a constant pressure reference.

10.4.3.3   The Forces Exerted on the Sphere

As the previous results demonstrate and outlined in Fig. 10.5, a creeping flow past a sphere exerts on this obstacle both *pressure* and *viscosity* forces.

The viscous stresses are directly obtained from Newton's behaviour scheme using the previously calculated velocity field. It can thus be verified that all the viscous stress tensor components are zero except for the shear stress $\tau_{\varphi r}$ which is:

$$
\tau_{\varphi r}\left(r,\varphi\right) = \mu\left(\frac{\partial V_\varphi}{\partial r} - \frac{V_\varphi}{r} + \frac{1}{r}\frac{\partial V_r}{\partial\varphi}\right).
$$

At any point on the sphere, the pressure and viscosity forces are therefore expressed as:

$$
\mathrm{d}\mathbf{F}^P = -P\left(a,\varphi\right)\mathrm{d}s.\mathbf{n},
$$
$$
\mathrm{d}\mathbf{F}^\nu = \tau_{r\varphi}\left(a,\varphi\right)\mathrm{d}s.\mathbf{e}_\varphi.
$$

By projecting on the Ox axis and substituting the functions by their expressions, we obtain, for a sphere moving in translation at a constant velocity in a fluid at rest:

$$
\mathrm{d}F_x^P = -\left(P_0 + \frac{3}{2}\mu V_0\frac{\cos\varphi}{a}\right)\cos\varphi \times 2\pi a^2\sin\varphi\mathrm{d}\varphi,
$$
$$
\mathrm{d}F_x^\nu = -\frac{3}{2}\mu\frac{V_0}{a}\sin\varphi \times 2\pi a^2\sin\varphi\mathrm{d}\varphi,
$$

where a first integration with respect to θ was carried out on the area element, taking into account the axisymmetric geometry. By simply integrating from $\varphi = 0$ to $\varphi = \pi$, we obtain:

$$
\boxed{F_x^P = -2\pi\mu V_0 a}\qquad (10.18)\qquad\qquad\boxed{F_x^\nu = -4\pi\mu V_0 a}\qquad (10.19)
$$

which yields the total drag force:

Fig. 10.5 Schematics for the forces on a sphere in the Stokes regime

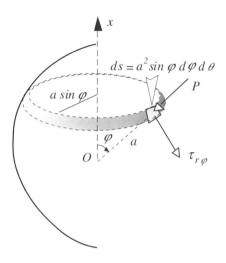

$$\boxed{F_x = F_x^P + F_x^v = -6\pi \mu V_0 a} \qquad (10.20)$$

Remark (1) The first relation proves that the pressure drag is not zero and contributes to one third of the total resistance on the sphere. Indeed, although the streamlines exhibit an "upstream-downstream" symmetry, the pressure field is not symmetrical on either side of the maximum sphere cross section, as can be seen from its expression by changing φ *into* $\pi - \varphi$. There is therefore no d'Alembert paradox for this flow, unlike the homologous configuration in an ideal (inviscid) fluid.

— (2) The previous relation, Eq. (10.20), is referred to as the Stokes formula for the drag of the sphere—Cf. historical note below. Its experimental validation has proven it could be applied beyond the *theoretical* requirement ($R_e \ll 1$) since it is still valid up to Reynolds numbers of the order of unity.

— (3) By defining the drag coefficient of the sphere as $C_x = \dfrac{|F_x|}{\frac{1}{2}\rho S V_0^2}$, we obtain, by adopting for the reference area, the maximum cross-section (πa^2):

$$C_x = \frac{24}{R_e},$$

where R_e is the Reynolds number based on the sphere diameter.

— (4) This relation only applies to a single independent sphere moving in an infinite fluid. It cannot be applied without due care to a set of spherical particles. Indeed, the disturbance induced by the presence, in a creeping motion, of a fixed obstacle is detected at a considerable large distance from it. One can easily verify this point from the results in Table 10.1 by noticing, for example, that the velocity perturbation resulting from the sphere motion in a motionless fluid will dampen as $1/r$ and thus remain sensitive at a significant distance. Thus, the change in the velocity field of the fluid at rest is still around 10% of the sphere speed, at a distance from its center of about 7.5 times its radius. The relative velocity of a second sphere positioned at this distance from the first is therefore reduced. The presence of such *long-distance* interactions in the Stokes regime thus plays a decisive role in the dynamics of suspensions, and suggests very different types of behaviour, in terms of sedimentation for example, depending on the dilution degree of the solution. In the work by Guyon, Hulin & Petit [62], one can find interesting additions on this issue, in relation to diphasic flows.

ADDITIONAL INFORMATION—The ball viscometer—The Stokes formula can be used to determine the fluid viscosity by measuring the terminal velocity of a ball with a given diameter falling through the fluid. Indeed, this velocity corresponds to a dynamic equilibrium state between the ball's weight minus the Archimedean thrust on the one hand, and on the other hand the drag of the ball. The result is:

$$V_{lim} = \frac{2}{9} \frac{\left(\rho_s - \rho_f\right) g}{\mu} a^2 ,$$

a relation which actually makes it possible to determine the dynamic viscosity μ by measuring V_{lim}. However, its validity in the experimental design should be carefully examined. In particular, any measuring device where the ball is falling through a fluid enclosed in a tube, which necessarily corresponds to a *crosswise confined* flow, can only approximately fulfill the conditions at infinity of the theoretical solution—see previous item (4). In practice, as reported by Ryhming [143], a distance to the outer walls equal to 100 times the ball's diameter reduces the confinement effect to less than 2%.

10.4.3.4   Discussion: Oseen's Correction

Stokes' solution for motions in infinite domains is not uniformly valid for $r \to \infty$. It was Oseen [110] who first brought this point to light in 1910.
We will explain this limitation from physical considerations, taking the example of the flow past a fixed sphere. The velocity field, as previously obtained in this case, can be written:

$$\mathbf{V} = \mathbf{V}_0 + \mathbf{V}' ,$$

by expliciting the disturbance velocity field $\mathbf{V}'$ generated by the presence of the sphere in the uniform flow at infinity. Its components are as follows:

$$V'_r = V_r - V_0 \cos\varphi = V_0 \left(-\frac{3}{2}\frac{a}{r} + \frac{1}{2}\frac{a^3}{r^3} \right) \cos\varphi,$$

$$V'_\varphi = V_\varphi + V_0 \sin\varphi = V_0 \left(\frac{3}{4}\frac{a}{r} + \frac{1}{4}\frac{a^3}{r^3} \right) \sin\varphi.$$

We thus find that, at a distance from the obstacle, the velocity field disturbance is mainly damping as r^{-1}, a property for which we will now examine two consequences.

(a) Taking up the *overall* momentum equation and introducing the perturbation velocity with $V' \approx \mathcal{O}(aV_0/r)$, the different contributions of the inertia and viscosity terms can be estimated as follows:

$$V_{0j}\frac{\partial V'_i}{\partial x_j} + V'_j\frac{\partial V'_i}{\partial x_j} = -\frac{1}{\rho}\frac{\partial P}{\partial x_i} + \nu\frac{\partial^2 V'_i}{\partial x_j^2},$$

$$a\frac{V_0^2}{r^2} \qquad a^2\frac{V_0^2}{r^3} \qquad\qquad \nu a\frac{V_0}{r^3}.$$

Thus, at a distance from the obstacle, the advection by the undisturbed velocity field is no longer negligible;

(b) The kinetic energy associated with the disturbance velocity field is damping as r^{-2}. For high enough r, we can therefore express the kinetic energy of an elementary volume dv in the form:

$$dE_c = \frac{\lambda(\varphi)}{r^2}dv,$$

which, with the volume element $dv = 2\pi r^2 \sin\varphi \, d\varphi \, dr$ leads formally to:

$$dE_c = 2\pi \left[\int_0^\pi \lambda(\varphi)\sin\varphi d\varphi \right] dr = \Lambda dr.$$

Thus, the value of the kinetic energy disturbance, when integrated over the whole domain ($r = a$ to $r = \infty$), diverges. Stokes' solution therefore exhibits physical inconsistencies at infinity. To overcome them, Oseen proposed to look for the perturbation velocity field as a solution of the equation:

$$V_{0j}\frac{\partial V'_i}{\partial x_j} + V'_j\frac{\partial V'_i}{\partial x_j} = -\frac{1}{\rho}\frac{\partial P}{\partial x_i} + \nu\frac{\partial^2 V'_i}{\partial x_j^2}.$$

The exact solution of this new equation cannot be derived analytically, however several approximate forms have been proposed. By means of matched asymptotic expansion methods, it can be obtained, for example, that the correction to the drag coefficient is of the form:

$$C_x = \frac{24}{R_e} \left(1 + \frac{3}{16} R_e - \frac{19}{1280} R_e^2 + \mathcal{O}\left(R_e^3\right) \right),$$

where the Stokes formula identifies the first term of the expansion.

HISTORICAL NOTE—Stokes' formula and Millikan's experiment—In his study on the pendulum movement damping by the surrounding air friction Stokes [150] published in 1851, he was led to taking an interest in the motion induced in the fluid by an oscillating sphere. He also dealt with the case of a sphere in a uniform rectilinear motion through a fluid at rest at infinity and set out, in this connection, the expression of the sphere drag (resistance). In the context of the available knowledge at that time, this result provided valuable pieces of information:

(i) the resistance was proportional to the velocity, whereas following Coulomb's experiments in 1801 [42], the resistance was taken as the function $a\,U + b\,U^2$ of the velocity U;

(ii) Stokes' formula provided a drag which is proportional to the sphere *radius* and not to the maximum cross-section area ($\propto a^2$).

Some fifty years later, Stokes' formula was to play a major role in another branch of physics. From 1909 in fact, Millikan [102] had undertaken to experimentally determine the value of the elementary electric charge (electron). The method used consisted in measuring the terminal velocity of electrically charged liquid droplets (water, alcohol, oil, mercury, etc) evolving in the gravitational field in the presence of an electrostatic field. The momentum balance, responsible for this terminal speed, allowed him to determine the charge of the electron, given that he knew all the forces involved, and in particular the air drag on the supposedly spherical droplets. The result obtained by Millikan under the conditions of the "*oil drop experiment*", and although considering a possible air slip effect on the oil drop, [103, 104] turned out to be lower than the value retained as "*exact*" thereafter. The reason took some time to emerge: it still has to do with fluid mechanics, due to the erroneous viscosity value adopted by Millikan when applying Stokes' formula!

Chapter 11
The Laminar Boundary Layer. Dynamic and Thermal Concepts

Abstract This chapter deals with high Reynolds number flows of a viscous fluid around solids, for which the significant viscosity effects are confined to a specific region called the boundary layer. After having explained the physical concept, the characteristic parameters are defined. We then derive the local and integral equations for the dynamic and thermal problems.

11.1 High Reynolds Number Flows of Real Fluids

11.1.1 The High Reynolds Number Flow Class

In the fluid motions discussed in the previous chapter, the viscosity forces were dominating. Thus the results obtained can be applied within the limit of a *global* Reynolds number R_e tending towards zero. On the other hand, in Chap. 6, we have identified the properties of flows which, for the incompressible regime, correspond to an infinite value of this same number. We have learnt on this occasion that such a situation led to paradoxes which could only be partially overcome under the ideal (inviscid) fluid assumption. The flows we are now considering are those of *real* and therefore *viscous* fluids whose global Reynolds number is *high* but *finite*. From a practical point of view, these conditions embrace a wide variety of aerodynamic and hydrodynamic situations. To set the scene, let us give a few examples.

- A motor vehicle travelling at 90 km/h in the earth's atmosphere at the sea level ($v \simeq 15 \times 10^{-6} \ MaSt$) corresponds to a Reynolds number of 2.5×10^6, based on a characteristic dimension of 1.5 m.
- For a jumbo airliner, the Reynolds number relative to the average wing chord ($9 \ m$) reaches 50×10^6 at takeoff (speed of about 300 km/h $\approx$ 83 m/s).
- The navigation of a 10 m long pleasure boat sailing at a speed of 6 m/s (approximately 21.6 km/h $\simeq$ 11.66 *knots*) is characterized by a Reynolds number ($v \simeq 10^{-6} \ MaSt$) of 60×10^6.

One could be inclined to infer from the previous numerical values that, for the flows in question, the inertia forces are far greater than the viscosity forces, by the very definition of the Reynolds number (Cf. Chap. 4 Sect. 4.7.5). In reality, such a conclusion is only partially correct, as we will explain now by first considering the isovolume situation.

11.1.2   The Viscous Effect Localization in Incompressible High Reynolds Number Flows

The Reynolds number we used in the previous discussion has been termed *global*, in the sense that it does not allow us to appreciate the inertial to viscosity force ratio at any point in the field, *i.e.*, *locally*. For this purpose, the characteristic velocity and length scales should be adapted to the local region to be analyzed. Let us take for example the viscosity forces, which, in an isovolume situation, are expressed by:

$$\tau_{ij} = 2\mu S_{ij} \equiv \mu \left(\frac{\partial U_i}{\partial x_j} + \frac{\partial U_j}{\partial x_i} \right).$$ (11.1)

If it is correct to consider here the dynamic viscosity μ as a constant for a given fluid motion, the strain rate S_{ij} varies, *a priori*, throughout the entire flow field. Thus, the key question is whether, despite a high *global* Reynolds, there are nevertheless regions within the flow field where the following condition could no longer be *locally* satisfied:

$$Inertia\ forces\ \gg\ Viscosity\ forces.$$

Obviously, for a Newtonian fluid, the answer to such a question is to be sought exclusively in those flow regions where a high velocity gradient could take place. To this end, the experiment is a useful way of readily localizing such areas. Let us consider, as a way to illustrate, the flow around a streamlined body as outlined in the following Figure.

For a streamlined body at a low incidence in an uniform velocity flow, the experiment teaches us that, even close to the obstacle, the overall advection direction is preserved, in the sense that the fluid mass transport is still driven by the motion direction imposed by the boundary condition at infinity.[1] Changes in the streamline geometry then mainly consist of a deflection in the vicinity of the obstacle, following the shape of the boundary surface itself.

When the global Reynolds number is high enough, the essential features of these changes, as highlighted, both by direct measurement or visualization, can be summarized in the following three points.

---

[1] On a high-incidence streamlined body or on a blunt body, a detachment of the fluid threads occurs on the body surface, generating a recirculation area. In such a region the fluid threads can move opposite to the flow direction at infinity. The advection therefore no longer presents a preferential direction in the entire field.

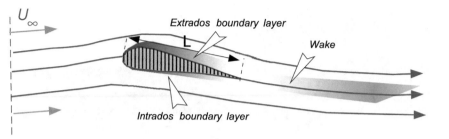

Fig. 11.1 Schematic for the high Reynolds number flow past a streamlined body

- At high global Reynolds numbers, the space velocity variation between the obstacle and the fluid at infinity does not significantly extend over the entire flow field, but is mainly confined to specific regions with steep gradients.
- These areas are located (*i*) in close proximity to the wall and (*ii*) in the wake downstream of the obstacle. They correspond to the significant viscosity influence zones.
- The more the global Reynolds number *increases*, the more the ratio of the parietal's zone thickness to the streamlined body length *decreases*.

The parietal region, where the fluid viscosity in a high Reynolds number flow has a significant influence, is called the *dynamic boundary layer*.[2]

The localization of the viscosity influence zones is visualized in the following Figure for the flow around a circular cylinder normal to the fluid motion direction at infinity.

The fluid thread visualization is obtained by differential interferometry (see Chap. 2) by slightly heating the cylinder with respect to the fluid at infinity (the temperature difference at the wall is $\Delta T \simeq 25°C$).

The first visualization (Fig. 11.2, left view) corresponds to a Reynolds number, based on the cylinder diameter, of 300. In this case, in the upstream part of the obstacle, a region can be clearly discerned, on either side of the stagnation point, whose colouring reflects the presence of a significant heat transfer. This is the thermal boundary layer, similar here to the dynamic boundary layer region. In the right-hand view of the same Figure, where the Reynolds number has now reached 4,600, this area has become indistinguishable to the naked eye due to its thinning on the wall. The influence of the Reynolds number also appears on the upper and lower mixing (or shear) layers which develop downstream of the cylinder's maximum cross-section (Fig. 11.2). They have the appearance of streaked coloured sheets, whose thickness correspondingly decreases as the Reynolds number increases. These mixing layers take place at the interface of two regions where the temperature is roughly homogeneous, namely (*i*) the domain at infinity, and (*ii*) the downstream close to the cylinder

---

[2] *"Grenzschicht"* in German, *"couche limite"* in French.

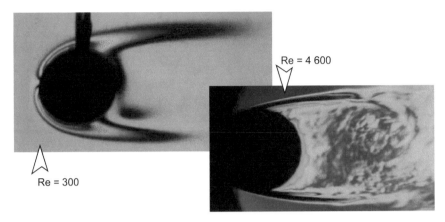

Fig. 11.2 Visualization of the near-flow area around a circular cylinder, courtesy of IMFT-TELET

where the temperature differences are reduced due to the recirculating fluid in the separation bubble behind the obstacle.

Quantitative clarifications can be made to the descriptive image which has just been put forward. They require a *local* analysis of the viscosity effects. This is precisely what we are going to undertake by adopting, in the viscosity influenced area, the appropriate interpretation of the Reynolds number as introduced in Chap. 9 Sect. 9.3.3.

11.1.3   The Advection-Diffusion Time-Equilibrium in an Isothermal Boundary Layer

Consider, as outlined in Fig. 11.3, a fluid particle moving in a boundary layer region. Its momentum is under the control of the two mechanisms of advective transport and diffusive transfer.

- Without separation from the wall, the advection takes place along streamlines whose general direction is imposed by the conditions at infinity, only slightly deviating from the external geometry of the obstacle.
  Under these conditions, a characteristic time scale for the momentum transport by advection over the distance L (see Fig. 11.3) can be taken as:

$$T_a \approx L/U_\infty.$$

- The presence of a fixed obstacle in a viscous fluid flow results, for the moving environment, in a deficit or *sink* in momentum. This momentum loss, as imposed by the wall, slows down the moving fluid when diffusing away from the obstacle, with a diffusivity which is none other than the fluid kinematic viscosity ν. If we

Fig. 11.3 Schematic for the momentum advection/diffusion in the boundary layer region

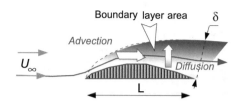

Boundary layer area δ

Advection

U_∞

Diffusion

L

denote by δ the characteristic distance, in a direction normal to the wall, of this diffusion at the trailing edge of the obstacle—namely, after a distance of the order of the length obstacle L—the time scale corresponding to the momentum transfer by diffusion over δ can be estimated as:

$$T_d \approx \delta^2/\nu.$$

When there is no crosswise limitation to the boundary layer expansion, *i.e.*, in the direction normal to the wall, we can correctly infer that both mechanisms are simultaneously acting on any fluid particle moving in the boundary layer region. In other words, the advection and diffusion mechanisms take place at the *same time scale*, which directly leads to the following result:

$$\frac{\delta}{L} \sim \sqrt{\frac{\nu}{U_\infty L}}. \tag{11.2}$$

Thus, simply based on the phenomenology we have just described, we can deduce that the relative boundary layer thickness δ/L decreases as $R_e^{-1/2}$, where $R_e = U_\infty L/\nu$ is none other than the *global* Reynolds number of the flow.

ADDITIONAL INFORMATION—**The estimation of some boundary layer thicknesses**—With the numerical values of the three examples given in Sect. 11.1.1, a straightforward application of the previous relation leads to a first *estimate* of the corresponding boundary layer thickness values. They are of the order of one millimeter, as can be seen in the following Table 11.1.

Table 11.1 Order of magnitude of some boundary layer thicknesses in the laminar regime

Machine	Global Reynolds number	$10^3 \times \delta/L$	L (m)	δ (mm)
Car	2.5×10^6	0.63	1.5	~ 0.9
Aircraft	50×10^6	0.14	9	~ 1.2
Ship	60×10^6	0.14	10	~ 1.3

Remark   The estimates that have just been given are derived from relation (11.2). It should be remembered that this was established in the presence of the viscous diffusion alone and is therefore only valid in the laminar regime. In practice, the flow regime in the boundary layer is often turbulent, which results in an increase of the diffusion mechanisms. The thicknesses for this flow regime are therefore higher than the values in Table 11.1.

11.1.4   The Convection-Diffusion Time-Equilibrium in the Thermal Boundary Layer

The discussion in the previous section also applies to the thermal diffusion when a heat transfer in addition to momentum is taking place in the boundary layer. This is the situation which occurs whenever there is a thermal imbalance between fluid and wall. Denoting, in this case, by δ_T the characteristic length scale of the diffusive heat transfer, the time scale associated with the thermal diffusion can be estimated as:

$$T_\theta = \frac{\delta_T^2}{a},$$

where a stands for the thermal diffusivity of the fluid. Since the convective time scale of the momentum transport is unchanged, we have:

$$\frac{\delta_T}{L} = \sqrt{\frac{a}{U_\infty L}}. \qquad (11.3)$$

Thus we find a relation similar to Eq. (11.2), by simply substituting for the Reynolds number the Péclet number $Pe = U_\infty L / a$.

Accordingly, the high *global* Péclet number assumption validates the *thermal boundary layer* concept for the heat transfer with a similar phenomenology as its dynamic counterpart (Cf. Sect. 11.1.2). Now, recalling that $Pe = Pr \times Re$, this condition, for fluids whose Prandtl number Pr is close to unity, simply coincides with that of a high global Reynolds number.

EXERCISE 46 The dynamic and thermal boundary layer thicknesses.

We consider the boundary layer on a fixed obstacle whose temperature is different from that of the fluid at infinity.

QUESTION Assess the ratio of the dynamic and thermal boundary layer thicknesses as a function of the Prandtl number. Discuss the cases of air and water.

SOLUTION: From the previous estimates, it is easy to deduce that:

$$\delta/\delta_T \approx P_r^{1/2}.$$

For the air ($Pr = 0.71$) we obtain $\delta \simeq 0.84\delta_T$, whereas for water ($Pr = 10$) $\delta \simeq$ $3.12\delta_T$. Thus, in this liquid, the dynamic boundary layer *encompasses* its thermal counterpart, with the situation slightly reversing in the air.

HISTORICAL NOTE— **The boundary layer concept**—It is customary to historically trace the introduction of the boundary layer concept back to Prandtl's lecture [123] in 1904, at the IIIrd International Congress of Mathematics in Heidelberg. In the corresponding paper, as it was taken up and translated into English in the NACA report [124] twenty years later, there were clear explanations related to (*i*) the formulation of the boundary layer equations, which we will see and name in the hereafter sections "*Prandtl's equations*", (*ii*) the matching with the ideal fluid solution as well as (*iii*) the flow separation phenomenon. An excellent review of Prandtl's contribution and the Göttingen School works was published by Meier [101] in 2004, in celebration of the 100th anniversary of the original paper.

Without calling into question this decisive contribution, it should however be noted that Darcy's impressive study of pipe flows [45], published in 1857, had provided an anticipatory description of the physical properties of the near-wall motion, which relates to the origin of the turbulent boundary layer concept. The contribution of this scientist being less known, we quote here some elements of his explanation for the velocity dependency of the flow resistance in pipes: « *In a vertically positioned pipe and due to the attraction of its walls, a liquid layer remains adhered to them...Obviously, the fluid cylinder which it will envelop as it passes will not leave the molecules which constitute the factitious wall motionless; it will animate them with some translation speed, some whirling motions, etc. If, therefore, on the one hand, the attraction of the walls should be considered as one of the causes for slowing down the motion, it should be realized, on the other hand, that this cause probably predominantly acts through the fluid cohesion* [viscosity] *which the outer surface of the moving cylinder has to overcome.*»

11.2 The Boundary Layer's Characteristic Parameters

11.2.1 Thicknesses

11.2.1.1 The Dynamic Boundary Layer's Standard Thickness

We shall henceforth refer to $U_E(x)$ as the *local* velocity value at the abscissa point x *on the surface* of the considered solid body in an *ideal fluid potential flow*. The standard boundary layer thickness is defined as to the distance normal to the wall where the longitudinal component of the velocity in the *viscous fluid flow*, rises to 99% of $U_E(x)$. This thickness generally varies with the abscissa along the obstacle. By conventionally writing it down as $\delta(x)$, we have, by definition:

$$U\left[x, \delta\left(x\right)\right] = 0.99 \times U_E\left(x\right). \tag{11.4}$$

11.2.1.2   The Thermal Boundary Layer's Standard Thickness

The standard thickness of the thermal boundary layer δ_T is defined similarly to that of the dynamic boundary, by substituting for the velocity profile the dimensionless temperature distribution:

$$\frac{T\left(x, y\right) - T_p\left(x\right)}{T_\infty - T_p\left(x\right)},$$

where $T\left(x, y\right)$ is the temperature at any point of the boundary layer, $T_p\left(x\right)$ the wall temperature and T_∞ the temperature of the fluid at infinity. Thus, by definition, the standard thermal boundary layer's thickness is the distance normal to the wall where the temperature difference reaches 99% of the total variation $T_\infty - T_p$. Thus we have:

$$\frac{T\left(x, \delta_T\right) - T_p\left(x\right)}{T_\infty - T_p\left(x\right)} = 0.99. \tag{11.5}$$

11.2.1.3   The Displacement Thickness

Due to the wall adherence in a viscous fluid, the decrease in velocity in the boundary layer makes the flow rate across its thickness (q_v) less than it would be, at an equal distance, in an inviscid fluid (q_p), as outlined in the following Fig. 11.4. By neglecting any variation in the inviscid fluid velocity across a distance of the order of the boundary layer thickness, this flow rate deficit can be approximated by:

$$q_p - q_v \approx \int_0^\delta \left(\rho_E U_E - \rho U\right) \mathrm{d}y.$$

By convention, it is expressed by introducing a thickness δ^* or δ_1 such that, by definition, $q_p - q_v = \rho_E U_E \delta^*$. Hence the following definition[3]:

$$\delta^*\left(x\right) = \int_0^\infty \left(1 - \frac{\rho U}{\rho_E U_E}\right) \mathrm{d}y. \tag{11.6}$$

The thickness we have just introduced is called the displacement thickness. As outlined in Fig. 11.4, this thickness physically corresponds to the distance by which the wall should be displaced outwards in order to retrieve in the potential flow over the

---

[3] The change to an infinite upper bound in the integral is to be considered in terms of the asymptotic matching limit, see Remark Sect. 11.3.5.

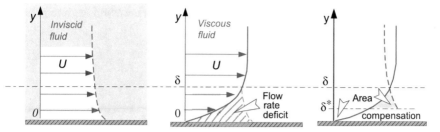

Fig. 11.4   Flow rate deficit and displacement thickness

displaced surface $\delta - \delta^*$ the same flow rate as in the viscous fluid across the entire thickness δ.

Proof   By neglecting, in an inviscid fluid, any cross-wise gradient over the thickness δ, the equality between the previous flow rates can be expressed as:

$$\int_{\delta^*}^{\delta} \rho_E U_E \mathrm{d}y = \int_{0}^{\delta} \rho U \mathrm{d}y \iff \rho_E U_E \left(\delta - \delta^*\right) = \int_{0}^{\delta} \rho U \mathrm{d}y.$$

Now it is easy to deduce from the last equality the expression for the displacement thickness δ^*. This result still simply expresses the compensation of the velocity profile areas between (*i*) the real wall for the viscous fluid flow and (*ii*) the fictive displaced wall by δ^* for the inviscid fluid potential flow, as outlined in the last diagram of the previous figure.

ADDITIONAL INFORMATION—**Taking into account the displacement thickness in wind tunneling**—To illustrate the displacement thickness concept in concrete terms, we can refer to the flow in a closed channel wind tunnel. In this case, the boundary layer's development along the walls is the source of two drawbacks in obtaining a constant velocity field in both *cross-wise* and *stream-wise* directions. The first drawback is a lack of *cross-wise* velocity uniformity in the vein. This is easily remedied by giving the cross-section dimensions high enough so that the presence of the boundary layers does not alter the velocity distribution homogeneity in the useful region at the core of the cross-section. The second is a variation of the velocity *along the tunnel axis*. It results from the longitudinal development of the boundary layers, whose thicknesses do not remain constant in the stream-wise direction. More precisely and as we will see at a later stage, the slowing area of the fluid at the wall thickens in the downstream direction. As a result, the useful core of the uniform velocity cross section reduces in size along the same direction. The flow rate conservation means that the velocity along the tunnel axis should necessarily increase in the downstream direction if the channel cross section remains constant. To overcome this second drawback, a simple solution consists in varying the wind tunnel cross-section according to a law which is exactly that given by $\delta^*(x)$ and which we will learn to calculate at a later stage.

11.2.1.4   The Momentum Thickness

The slowing down of the flow to the wall also results in a lack of momentum for the *viscous* fluid compared to the *inviscid* potential flow. To account for this, and in a similar way to the displacement thickness, a new length scale is introduced, called the momentum thickness, denoted θ, or δ_2 such that:

$$\theta(x) = \int_0^\infty \frac{\rho U}{\rho_E U_E} \left(1 - \frac{U}{U_E} \right) dy. \tag{11.7}$$

As for δ^*, this new thickness makes it possible to express the difference in momentum between the inviscid (J_p) and viscous (J_v) fluid flows, *at the same mass flow rate* ρU, as:

$$J_p - J_v \simeq \int_0^\delta \rho U U_E dy - \int_0^\delta \rho U^2 dy \equiv \rho_E U_E^2 \theta.$$

11.2.1.5   The Energy Thickness

Similarly, the (kinetic) energy thickness δ^{**} or δ_3 can be defined as:

$$\delta^{**}(x) = \int_0^\infty \frac{\rho U}{\rho_E U_E} \left(1 - \frac{U^2}{U_E^2} \right) dy. \tag{11.8}$$

Remark   In all the definitions of the previous integral thicknesses, the upper bound of integration has been taken to infinity. This is a theoretical formalism which, as already mentioned, is based on an asymptotic matching condition of the solutions for the viscous and inviscid fluid flows (see again Sect. 11.3.5, Remark #3). In practice, we frequently adopt the approximation consisting in identifying the velocity profile, beyond the standard thickness δ, with that of the inviscid fluid solution. Under this approximation the significant variation of the functions to be integrated is calculated for $0 \le y \le \delta$ and the corresponding thicknesses are evaluated, in practice, from bounded integrals over the thickness δ.

11.2.1.6   The Shape Factor

The ratio between the displacement and momentum thicknesses is referred to as the shape factor:

$$H = \delta^*/\theta. \tag{11.9}$$

This parameter takes different values depending on the laminar or turbulent nature of the flow in the boundary layer. For a boundary layer on a flat plate, its value almost halves from about 2.6 in the laminar regime to $1.3 \sim 1.4$ in the turbulent regime. The shape factor is also sensitive to the longitudinal pressure gradient. In the presence of an unfavourable gradient, it allows to characterizing the onset of the separation which corresponds to a value of H close to 4.

EXERCISE 47 The boundary layer's thicknesses for a linear velocity profile.

Calculate the dimensionless thickness values δ^*/δ, θ/δ, δ^{**}/δ along with the shape factor H for the linear velocity profile $U(x, y)/U_E(x) = y/\delta$.

SOLUTION: By direct application of the respective definition formulae, we obtain $\delta^*/\delta = 1/2$, $\theta/\delta = 1/6$, $\delta^{**}/\delta = 1/4$, which leads to a shape factor of three.

11.2.2   The Viscous Friction

11.2.2.1   The Local and Wall Friction

According to the boundary layer assumptions, the viscous stress tensor components for a 2D-plane flow are simply expressed as:

$$\tau_{ij} = \mu \begin{pmatrix} 2\dfrac{\partial U}{\partial x} & \dfrac{\partial U}{\partial y} + \dfrac{\partial V}{\partial x} \\[2mm] \dfrac{\partial U}{\partial y} + \dfrac{\partial V}{\partial x} & 2\dfrac{\partial V}{\partial y} \end{pmatrix} \approx \mu \begin{pmatrix} \sim 0 & \dfrac{\partial U}{\partial y} \\[2mm] \dfrac{\partial U}{\partial y} & \sim 0 \end{pmatrix}.$$

Hence, the viscous stress tensor reduces to the predominant component τ_{xy}, which is a shear component corresponding to the stress along the main advective direction x applied to an elementary surface whose normal direction is parallel to the y direction. At the wall ($y = 0$), the corresponding local value is that of the viscous wall friction or skin friction, namely:

$$\tau_p(x) = \mu \left(\frac{\partial U}{\partial y} \right)_{y=0}. \tag{11.10}$$

11.2.2.2   The Viscous Drag

By integration of the local skin friction along the whole body surface in contact with the boundary layer, we obtain the viscous drag. For a 2D-plane flow, the viscous drag, per span length unit, is given by:

$$D = \int_0^L \tau_p(x)\, dx.$$

11.2.2.3   The Local Friction and Drag Coefficients

The local skin friction coefficient is a dimensionless quantity defined by normalizing the viscous shear stress at the wall by the *local* dynamic pressure of the external flow:

$$C_f(x) = \frac{\tau_p(x)}{\frac{1}{2}\rho U_E^2(x)}.$$

(11.11)

Similarly, a viscous drag coefficient can be introduced, using an appropriate viscous drag normalization. Usually, the dynamic pressure of the external flow at infinity ($\frac{1}{2}\rho U_\infty^2$) and a reference area ($S$) are adopted, so that, for a 2D-plane flow

$$C_D = \frac{b \times D}{\frac{1}{2}\rho S U_\infty^2},$$

(11.12)

where D is the viscous drag per unit length in the spanwise direction and b denotes the span of the cylindrical body.

11.2.3   The Heat Transfer

11.2.3.1   The Heat Transfer Coefficient

Referring to the fluid thermal conductivity as λ, the heat flux density (the heat transfer value per unit area and time) exchanged at the wall ($y = 0$) by *conduction* is equal to:

$$\phi_p(x) = -\lambda \left(\frac{\partial T}{\partial y}\right)_{y=0}.$$

It may sometimes be more convenient, in practice, to express the same flux density using the wall temperature (T_p) and that of the fluid at infinity (T_∞). In this regard, a heat-exchange coefficient k can be introduced, such as:

$$\phi_p(x) = k\left[T_p(x) - T_\infty\right].$$

(11.13)

The coefficient k is expressed in $Wm^{-2}K^{-1}$ in the International System of Units.

11.2.3.2   The Nusselt Number

By specifying a reference length L, the heat exchange coefficient can be put into a dimensionless form, the grouping being called the Nusselt number:

$$N_u = \frac{kL}{\lambda}.$$

(11.14)

11.3 The Isovolume Boundary Layer Equations

The present section is only concerned with the dynamic problem. We will here proceed to simplify the general equations in order to establish a new flow model for the boundary layer region, as an *"intermediate"* between those of Navier-Stokes and Euler. Firstly, we will derive the local formulation, known as *Prandtl's model*, and then the integral expression or *von Kármán's equation*.
We limit ourselves to the case of a steady two-dimensional plane flow in the absence of external body forces. Under these conditions, the general Navier-Stokes equations are written in Cartesian projections:

$$\frac{\partial U}{\partial x} + \frac{\partial V}{\partial y} = 0,$$
(11.15a)

$$U\frac{\partial U}{\partial x} + V\frac{\partial U}{\partial y} = -\frac{1}{\rho}\frac{\partial P}{\partial x} + \nu\left(\frac{\partial^2 U}{\partial x^2} + \frac{\partial^2 U}{\partial y^2}\right),$$
(11.15b)

$$U\frac{\partial V}{\partial x} + V\frac{\partial V}{\partial y} = -\frac{1}{\rho}\frac{\partial P}{\partial y} + \nu\left(\frac{\partial^2 V}{\partial x^2} + \frac{\partial^2 V}{\partial y^2}\right).$$
(11.15c)

11.3.1 The Specification of the Boundary Layer Configuration

We are looking for a simplification of these equations appropriate to the specific boundary layer configuration as previously described. This simplification will therefore be based on the following so-called *boundary layer assumptions*:

- a high global Reynolds number ($R_e \gg 1$);
- a particular flow geometry due to relative thinning in the direction normal to the advection;
- a variation of the relative boundary layer thickness as a function of the global Reynolds number.

Taking the x axis as the preferential advection direction, we will mathematically express the previous assumptions by the following relations:

$$\boxed{\frac{y}{x} \sim \epsilon\,(R_e) \quad \text{and} \quad \frac{V}{U} \sim \epsilon'\,(R_e)}$$
(11.16)

In these relations, ϵ and ϵ' are both *"infinitely small"* functions of the global Reynolds number, the order of which has still to be determined. For this purpose, we will specify the Reynolds number dependency by the following power functions:

$$\boxed{\epsilon\,(R_e) = R_e^{-m}, m > 0 \quad \text{and} \quad \epsilon'\,(R_e) = R_e^{-n}, n > 0}$$
(11.17)

Accordingly, it is the values of both exponents m and n which set the respective orders of the infinitesimally small functions.

11.3.2   The Dimensionless Form of the Local Equations

We denote by L the characteristic length scale of the advection such as, by definition $x \sim L$, and by U_∞ that of velocity, such as $U \sim U_\infty$. We then introduce the following non-dimensional quantities:

$$x^* = x/L \qquad\qquad \text{and } U^* = U/U_\infty \qquad\qquad (11.18a)$$

$$y^* = y/(\epsilon L) = R_e^m \times y/L \text{ and } V^* = V/(\epsilon' U_\infty) = R_e^n \times V/U_\infty . \qquad (11.18b)$$

It should be noted that due to the scaling distortion applied to y and V normalizations, *all quantities with an asterisk* are of the *same order* of magnitude (~ 1). For the pressure, as normalized by ρU_∞^2, it will be assumed that the same applies, namely, for the dimensionless value:

$$P^* = P/\rho U_\infty^2 \sim \mathcal{O}(1) . \qquad (11.19)$$

After substituting the dimensionless functions in the Navier-Stokes equations we obtain:

$$\frac{\partial U^*}{\partial x^*} + R_e^{m-n} \times \frac{\partial V^*}{\partial y^*} = 0 , \qquad (11.20a)$$

$$U^* \frac{\partial U^*}{\partial x^*} + R_e^{m-n} \times V^* \frac{\partial U^*}{\partial y^*} = -\frac{\partial P^*}{\partial x^*} + R_e^{2m-1} \times \frac{\partial^2 U^*}{\partial y^{*2}} , \qquad (11.20b)$$

$$U^* \frac{\partial V^*}{\partial x^*} + R_e^{m-n} \times V^* \frac{\partial V^*}{\partial y^*} = -R_e^{m+n} \times \frac{\partial P^*}{\partial y^*} + R_e^{2m-1} \times \frac{\partial^2 V^*}{\partial y^{*2}} . \qquad (11.20c)$$

Proof   The derivation of the first relation is straightforward, after substitution and dividing the continuity equation throughout by U_∞/L.

As for the projection along x of the momentum equation, we have, after substitution and dividing throughout by U_∞^2/L:

$$U^* \frac{\partial U^*}{\partial x^*} + R_e^{m-n} V^* \frac{\partial U^*}{\partial y^*} = -\frac{\partial P^*}{\partial x^*} + \frac{1}{Re} \left(\frac{\partial^2 U^*}{\partial x^{*2}} + R_e^{2m} \frac{\partial^2 U^*}{\partial y^{*2}} \right) . \qquad (11.21)$$

It then appears that both Laplacian contributions are not weighted by the same factor. Consequently, and within the very high Reynolds numbers limit, it is therefore legitimate to discard the streamwise diffusion (along the x-axis) as compared to the cross-wise diffusion (along y). This simplification leads directly to the equation we were looking for.

The same process applied to the momentum equation in the y-projection provides:

$$R_e^{-n} \left(U^* \frac{\partial V^*}{\partial x^*} + R_e^{m-n} V^* \frac{\partial V^*}{\partial y^*} \right) = -R_e^m \frac{\partial P^*}{\partial y^*} + \frac{R_e^{-n}}{R_e} \left(\frac{\partial^2 V^*}{\partial x^{*2}} + R_e^{2m} \frac{\partial^2 V^*}{\partial y^{*2}} \right).$$

(11.22)

The same conclusion as before therefore also applies to the Laplacian contributions of the cross-wise velocity component. Hence, after multiplying throughout by R_e^n, we obtain Eq. (11.20c).

11.3.3 Discussion: Prandtl's Assumptions

At high Reynolds numbers, the system of Eqs. 11.20a–c can be simplified. Firstly, it should be noted that in order to discard any physically unacceptable degeneration of this system, it is necessary to impose $m = n$. Indeed, otherwise for $R_e \to \infty$, the continuity equation would lead to either $\partial U/\partial x = 0$, or $\partial V/\partial y = 0$, depending upon the sign of the difference $m - n$. Neither of these simplifications is consistent with the physical reality of the boundary layer configuration. This provides *Prandtl's first assumption*:

> In the boundary layer region, the geometric and kinematic distortions are of the same order $\dfrac{y}{x} \sim \dfrac{V}{U}$.

The *second* assumption provides the distortion factor value, based on the order of the dominant terms in the simplified formulation of the momentum equation projections.

– With $m = n$ and firstly considering the diffusion terms in Eqs. (11.21) and (11.22), it is clear that the dominant contributions are those in the y direction, owing to the factor R_e^{2m}.

– Compared to the advection terms, these dominant diffusion terms are, for each projection, in the ratio R_e^{2m-1}. Now, a positive exponent of the Reynolds number power leads, for $R_e \gg 1$, to discarding the advection with respect to the diffusion, the conclusion reversing for a negative exponent. In the first case, we retrieve the creeping motion model, inappropriate to the present situation, while the second provides the inviscid fluid model, which is also unsuitable. Thus, the only possibility to obtain a *new* model is to prescribe $2m - 1 = 0$, namely setting the order to $m = n = 1/2$.

– Finally considering the pressure term and *taking into account what has just been adopted*, all the terms (advection, pressure and diffusion) are of the *same order* in the projection of the momentum equation along x, Eq. (11.21) with $m = n = \frac{1}{2}$. On the other hand, the pressure term outweighs the other two in Eq. (11.22), because of the R_e factor applied to the pressure gradient in this equation.

In the end, *Prandtl's second assumption* can be stated as follows:

> In a boundary layer, the inertia, pressure and viscosity forces in the momentum equation can be considered of the same order, provided that the geometry of the boundary layer region is such that $\dfrac{y}{x} \sim \dfrac{1}{\sqrt{R_e}}$.

Thus, we find that the power dependence -1/2, as identified by the preliminary physical analysis, is at the root of the two fundamental boundary layer assumptions, through the dependence on the global Reynolds number, namely:

$$\boxed{\dfrac{y}{x} \sim \dfrac{V}{U} \sim \dfrac{1}{\sqrt{R_e}}} \tag{11.23}$$

In the next parts, the boundary layer's configuration specification (Sect. 11.3.1) and Prandtl's assumptions (Sect. 11.3.3) will be together referred to as *boundary layer approximations*.

11.3.4 Prandtl's Model

As a direct consequence of the previous assumptions and returning to a dimensional formulation, the Prandtl model equations of the steady, two-dimensional, plane boundary layer are written as follows:

$$\dfrac{\partial U}{\partial x} + \dfrac{\partial V}{\partial y} = 0 \tag{11.24a}$$

$$U\dfrac{\partial U}{\partial x} + V\dfrac{\partial U}{\partial y} = -\dfrac{1}{\rho}\dfrac{\partial P}{\partial x} + \nu\dfrac{\partial^2 U}{\partial y^2} \tag{11.24b}$$

$$0 = -\dfrac{\partial P}{\partial y} \tag{11.24c}$$

This model includes a second order partial differential equation with a parabolic character in space (by removal of the second order partial derivative $\partial^2 U/\partial x^2$). Therefore it corresponds to a less *"pronounced"* simplification of the Navier-Stokes model than the Euler model which is of order one. In particular, the presence of the second order derivative term enables us to specify the boundary conditions for the velocity field in a consistent way with the viscous fluid requirements, see Chap. 6 Sect. 6.1.3.

Proof With Prandtl's first assumption, the general continuity equation is *unchanged* in a boundary layer flow. According to the second, the dimensionless momentum equation along the x-axis becomes:

$$U^* \frac{\partial U^*}{\partial x^*} + V^* \frac{\partial U^*}{\partial y^*} = -\frac{\partial P^*}{\partial x^*} + \frac{\partial^2 U^*}{\partial y^{*2}} .$$

As for the equation projection along the y-axis we obtain:

$$U^* \frac{\partial V^*}{\partial x^*} + V^* \frac{\partial V^*}{\partial y^*} = -R_e \times \frac{\partial P^*}{\partial y^*} + \frac{\partial^2 V^*}{\partial y^{*2}} ,$$

leading to the prevailing of the pressure term over the other two types of forces.

11.3.5   The Viscous–Inviscid Matching

Summarizing all the elements we have just discussed, the study of a high Reynolds number flow on an obstacle revolves, in terms of the boundary layer theory, around the following four points:

- At high global Reynolds numbers and in a separation-free flow past a solid obstacle, the viscosity effects are confined within an area of relatively small thickness known as the *boundary layer* region;
- Outside the boundary layer, the fluid viscosity has a negligible effect so that the Euler model provides a satisfactory approximation for the flow solution in this part of the domain.
- The boundary layer region is characterized by a particular geometry, kinematics and dynamics on the equivalence basis:

$$\frac{y}{x} \sim \frac{V}{U} \sim \frac{1}{\sqrt{R_e}} ;$$

- Within the boundary layer, the flow is a solution of the Prandtl model, Eqs. (11.24a–c).

In concrete terms, the flow calculation can be devided into two stages dealing respectively with the *"external"* solution (the inviscid fluid potential flow outside the boundary layer) and the *"internal"* solution (the viscous fluid in terms of Prandtl's boundary layer approximations). These stages are necessarily coupled because the delimitation between the validity domains of both solutions is *a priori* unknown. The general coupled calculation procedure is directly derived from these observations and can be summarized as follows:

1. **Solving Euler's** model for the flow around the obstacle, by taking the wall as a streamline. We thus derive a first approximation of the pressure law at the wall $P_E(x)$;
2. **Solving Prandtl's** model for the boundary layer flow, using this pressure law for a first value assessment of the boundary layer thickness $\delta(x)$;
3. **Re-resolving Euler's model** for the *external* flow beyond the previous boundary $\delta(x)$ and setting a new pressure law $P_E'(x)$;
4. **Reiterate** the procedure (if necessary) from the second step onwards.

The last equation of the model ($\partial P/\partial y = 0$) allows us to validate the use of the pressure law at the wall, as obtained from the inviscid fluid flow solution, for the boundary layer calculation. Physically, the absence of any cross-wise pressure gradient in the boundary layer means that its presence does not alter the streamwise evolution of the pressure as a solution of the inviscid fluid potential flow at the outer edge of the boundary layer. In other words, denoting by $P_{v=0}(x, \delta)$ the inviscid fluid pressure distribution at the outer edge of the boundary layer, this is "*impressed*" throughout the entire layer thickness, leading to the distribution $P_E(x) \equiv P_{v=0}(x, 0)$ along the solid wall. Hence, *solely* for the *velocity field solution*, the boundary layer region can be treated as an *internal* correction zone of the *external* inviscid fluid solution based on the following equation, as a first approximation:

$$U\frac{\partial U}{\partial x} + V\frac{\partial U}{\partial y} = -\frac{1}{\rho}\frac{dP_E}{dx} + \nu\frac{\partial^2 U}{\partial y^2}. \qquad (11.25)$$

The set of conclusions relating to the calculation of the inviscid–viscous interaction is synthetically outlined in Fig. 11.5.

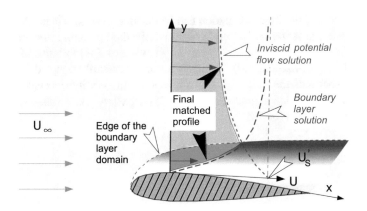

Fig. 11.5 Velocity profile on a streamlined body as the matching of the inviscid fluid and boundary layer solutions

Remark (1) It should be noted that the general procedure for the inviscid–viscous interaction, as previously outlined, is iterative in nature. In practice, it is most often applied in a truncated way to the first two steps, i.e., by considering that the approximation $P'_E(x) \simeq P_E(x)$ is sufficient. By prescribing the boundary conditions on $y = 0$ and $y = \infty$ the Prandtl model can then be solved. Hence, the values to be prescribed for the calculation of the boundary layer solution are taken as:

$$U(x, 0) = V(x, 0) = 0 \quad \text{and} \quad U(x, \infty) = U_E(x),$$

where the function $U_E(x)$ which denotes the external solution of the inviscid fluid potential velocity along the obstacle, is obtained, as a first approximation, by solving Euler's model to the body surface.

(2) By applying Bernoulli's theorem to the external potential flow, we can write along the outer free edge of the boundary layer, and again as a first approximation:

$$-\frac{1}{\rho}\frac{dP_E}{dx} = U_E \frac{dU_E}{dx},$$

so that Prantl's equation can still be written:

$$U\frac{\partial U}{\partial x} + V\frac{\partial U}{\partial y} = U_E \frac{dU_E}{dx} + \nu \frac{\partial^2 U}{\partial y^2}.$$

(3) When we confine our analysis to the first approximation, the calculation of the boundary layer solution is only possible with an asymptotic boundary condition, namely as $y \to \infty$. Therefore, the inviscid–viscous interaction can raise quite complex theoretical issues. This aspect of the problem is seldom addressed in practice, where some degree of empiricism is accepted in the determination of the outer edge of the boundary layer, with reference to the conventional thickness definition.

(4) Finally, it should be remembered that the velocity value prescribed on the outer edge for solving the boundary layer equation at a given section of abscissa x is local, namely $U_E(x)$ and not global (viz U_∞).

11.3.6 The Boundary Layer Separation

As shown in the momentum Eq. (11.25), the force balance in the boundary layer includes, in a weightless fluid, the inertia, viscosity and pressure forces. Now, according to Prandtl's theory, the latter are driven by the characteristics of the irrotational inviscid fluid motion. In an external aerodynamic situation, they therefore depend only on the shape of the obstacle and the motion conditions at infinity.
If the body takes the airfoil form in a uniform flow at infinity, we have demonstrated in Chap. 6 how the pressure resultant force can result in a lifting force on the body. As

explained at the time, this force has its main physical origin in the extrados suction (underpressure) induced by the fluid acceleration along the curvature of the airfoil's upper surface. Owing to Eq. (11.24c), we can now add:

> The existence of a boundary layer does not take part in the lift value as obtained from the ideal fluid potential theory, as long as Prandtl's approximations apply.

We will now explain how one of the fundamental assumptions of the boundary layer concept—the existence of a preferential flow direction—may fail under some conditions of the pressure field imposed by the external potential motion.

- Let us first examine what occurs, in an *inviscid* fluid flow, close to the leading edge area where the motion accelerates, as outlined in Fig. 11.6. Due to the fluid acceleration, namely $dU_E/dx > 0$, the streamwise pressure gradient along the obstacle (dP_E/dx) is negative there, according to Bernoulli's theorem. Such a gradient is termed *favourable* because it is acting *in support* of the inertial forces along the main advection direction. In concrete terms, while moving along its trajectory, a fluid particle experiences, in this situation, a downstream pressure lower than the upstream pressure. It is therefore subject to *only one drag resistance* due to the viscosity forces.

In a general way, we can therefore conclude that:

$$\boxed{\frac{dP_E}{dx} < 0 \iff Favourable\ Pression\ Gradient}$$

- Depending on the incidence and/or extrados curvature, an opposite direction of the pressure gradient can be observed further downstream on the airfoil, in a potential fluid motion. In this case, the slowing down of the fluid by viscosity can be associated with an increasing pressure $(dP_E/dx > 0)$ for a particle moving towards the trailing edge in the boundary layer. The inertia force then has to balance two resistance forces, since pressure and viscosity are now acting in the same way. For this reason, a positive pressure gradient is referred to as *unfavourable* or *adverse* to the fluid motion in the boundary layer.

If the intensity of the adverse pressure gradient is high enough, the advection direction can be reversed, as shown in the following Fig. 11.6. A backflow is then generated in the vicinity of the wall beyond the separation point where the boundary layer becomes detached from the body surface. The presence of a detachment leads to the creation of a recirculation zone in the near wake region, which invalidates the existence of a preferential motion direction.

Thus, the boundary layer theory only applies to attached flows. For the situation outlined in Fig. 11.6, this is the case as long as the convexity of the velocity profile close to the wall is directed towards the positive x direction (upstream). Since the detached region is characterized by a velocity profile whose convexity is facing the negative x direction (downstream), the boundary layer's separation point can be characterized by a zero value condition for the parietal friction, namely:

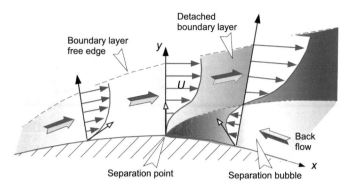

Fig. 11.6 Schematic for the boundary layer separation

$$
\text{Detachment point } x = x_D, \quad \text{so that} \quad \left[\frac{\partial U\,(x_D,\,y)}{\partial y}\right]_{y=0} = 0
$$

ADDITIONAL INFORMATION—**The wing stall phenomenon**—It is the boundary layer detachment which causes an aircraft wing to stall. Indeed, when this phenomenon occurs, the wing lift area is essentially reduced to the airfoil section upstream of the separation point. As the lift is strongly decreased, the aircraft suddenly falls to the ground. This can result in a significant loss of altitude, unless the plane is in close proximity to the ground, in which case it is a perfectly successful landing!

11.4 Integral Equations

In this section we discuss the study of the boundary layer region in terms of a *global* analysis, by considering that the flow properties are no longer described "*at each point*" in this zone, but *integrated* over the entire thickness of the boundary layer. This is our reason for speaking of the integral approach and equation (Fig. 11.7).

11.4.1 The Cross-Wise Velocity at the Boundary Layer's Free Edge

By integrating the continuity equation over the boundary layer thickness we obtain:

Fig. 11.7 Schematic for the
streamlines at the boundary
layer's free edge

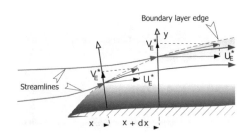

$$\int\limits_{0}^{\delta} \frac{\partial U}{\partial x} dy = - \int\limits_{0}^{\delta} \frac{\partial V}{\partial y} dy = -[V]_0^{\delta} .$$

Assuming that there is neither blowing nor suction at the wall and setting $V_E^*(x) = V(x, \delta(x))$, the previous relation leads to:

$$V_E^*(x) = - \int\limits_{0}^{\delta} \frac{\partial U}{\partial x} dy . \qquad (11.26)$$

Therefore, at the boundary layer's free edge, the velocity vector has a non-zero cross-wise component. We will see hereafter that this component does not satisfy the relation $V_E^*/U_E^* \sim d\delta/dx$, so that *the boundary layer's borderline is not a streamline*. Thus the flow undergoes a deflection at the outer edge of the boundary layer which tends to move it outwards the wall. However, this should not lead to the conclusion that the flow rate balance across the boundary layer between two sections x and $x + dx$ is negative (see Sect. 11.4.3).

11.4.2 Von Kármán's Equation

The integral equation or von Kármán's [167] can be directly deduced from Prandtl's local equations by integration with respect to the cross-wise coordinate y. It is written:

$$\boxed{\delta^* U_E \frac{dU_E}{dx} + \frac{d}{dx} \left(U_E^2 \theta \right) = \frac{\tau_p}{\rho}} \qquad (11.27)$$

Proof By integrating Prandtl's Eq. (11.25) from 0 to infinity, we obtain, after substituting the expression for the viscous shear stress $\tau = \mu \partial U / \partial y$:

$$\int\limits_{0}^{\infty} \left(U \frac{\partial U}{\partial x} + V \frac{\partial U}{\partial y} \right) dy = \int\limits_{0}^{\infty} U_E \frac{dU_E}{dx} dy + \frac{1}{\rho} \int\limits_{0}^{\infty} \frac{\partial \tau}{\partial y} dy .$$

Since the viscosity effects become, by assumption, negligible for $y = \delta$, the last integral value is none other than $-\tau_p$. Now, using integration by parts of the first integral, it can be noted that

$$\int_0^\infty V \frac{\partial U}{\partial y} dy = [UV]_0^\infty - \int_0^\infty U \frac{\partial V}{\partial y} dy \equiv - \left[U \int_0^\delta \frac{\partial U}{\partial x} dy \right]_0^\infty + \int_0^\infty U \frac{\partial U}{\partial x} dy,$$

the second equality directly resulting from the continuity equation $\partial U/\partial x + \partial V/\partial y = 0$.

Thus $\displaystyle\int_0^\infty V \frac{\partial U}{\partial y} dy = -U_E \int_0^\infty \frac{\partial U}{\partial x} dy + \int_0^\infty U \frac{\partial U}{\partial x} dy$, which by substituting in the ini-

tial equation, yields $\displaystyle -\int_0^\infty \left(2U \frac{\partial U}{\partial x} - U_E \frac{\partial U}{\partial x} - U_E \frac{dU_E}{dx} \right) dy = \tau_p/\rho$. whose left

hand side can still be written as follows:

$$\int_0^\infty \frac{d}{dx} \left(UU_E - U^2 \right) dy + \int_0^\infty (U_E - U) \frac{dU_E}{dx} dy \equiv \frac{d}{dx} \left[U_E^2 \int_0^\infty \left(\frac{U}{U_E} - \frac{U^2}{U_E^2} \right) dy \right]$$

$$+ U_E \frac{dU_E}{dx} \times \int_0^\infty \left(1 - \frac{U}{U_E} \right) dy .$$

It is then only necessary to introduce the definitions of the thicknesses δ^* and θ to obtain the expected result (Fig. 11.8).

ADDITIONAL INFORMATION—The kinetic energy integral equation—The kinetic energy transport equation can be dealt with in a similar way, as demonstrated in 1948 by Wieghardt [168]. The result is the following relation:

$$\frac{d}{dx} \left(U_E^3 \delta^{**} \right) = 2\nu \int_0^\infty \left(\frac{\partial U}{\partial y} \right)^2 dy .$$

The previous equation therefore gives access to the viscous dissipation rate in the boundary layer. The reader interested in the details of the demonstration can refer to Schlichting's work [147] in addition to the above-mentioned paper (Fig. 11.9).

EXERCISE 48 The integral equation of the skin friction coefficient.

QUESTION Deduce from the von Kármán integral relation the equation governing the local skin-friction coefficient as a function of δ^*, θ and U_E.

Fig. 11.8 The control volume for the boundary layer's global mass balance

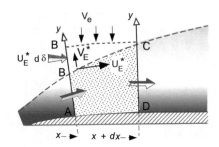

$S_{OLUTION}$: By developing the von Kármán relation, we obtain:

$$\left(\delta^* + 2\theta\right) U_E \frac{dU_E}{dx} + U_E^2 \frac{d\theta}{dx} = \frac{\tau_p}{\rho}.$$

Then dividing throughout by U_E^2, we reach the requested relation:

$$\left(\delta^* + 2\theta\right) \frac{1}{U_E} \frac{dU_E}{dx} + \frac{d\theta}{dx} = \frac{\tau_p}{\rho U_E^2} \equiv 2C_f.$$

11.4.3   The Global Mass Balance and Entrainment Velocity

Let us consider a section of the boundary layer outlined by a domain such as $ABCD$ in Fig. 11.8. The three sides AB, BC and CD are subject to a mass flux.[4] By convention, the expression of the difference in the flow rate between the AB and CD sides is based upon a velocity called the *entrainment velocity* such that, per span unit, we have:

$$V_e\left(x\right) = \frac{d}{dx} \int_0^{\delta(x)} U\left(x, y\right) dy. \qquad (11.28)$$

This entrainment velocity is not equal to the cross-wise velocity component at the boundary layer edge V_E^* which we have introduced in Sect. 11.4.1, but can be inferred from it as:

$$V_e\left(x\right) = -V_E^*\left(x\right) + U_E^*\left(x\right) \frac{d\delta}{dx}, \qquad (11.29)$$

---

[4] Therefore, this analysis only applies in the **absence** of any blowing or suction at the wall.

where $U_E^*(x)$ and $V_E^*(x)$ are the velocity vector components at the outer edge of the boundary layer,[5] *i.e.*, for $y = \delta(x)$.

Proof By the very definition of the entrainment velocity, we have:

$$
V_e \, dx = \int\limits_0^{\delta+d\delta} U\,(x+dx, y)\, dy - \int\limits_0^{\delta} U\,(x, y)\, dy
$$

$$
= \int\limits_0^{\delta} U\,(x+dx, y)\, dy - \int\limits_0^{\delta} U\,(x, y)\, dy + \int\limits_\delta^{\delta+d\delta} U\,(x+dx, y)\, dy
$$

$$
\text{or } V_e \, dx \simeq dx \int\limits_0^{\delta} \frac{\partial U}{\partial x}\,(x, y)\, dy + \int\limits_\delta^{\delta+d\delta} U\,(x, y)\, dy
$$

$$
\simeq dx \int\limits_0^{\delta} \tfrac{\partial U}{\partial x}\,(x, y)\, dy + \int\limits_\delta^{\delta+d\delta} U_E\,(x)\, dy = dx \int\limits_0^{\delta} \tfrac{\partial U}{\partial x}\,(x, y)\, dy + U_E\,(x) \times d\delta,
$$

from which the relation for V_E^* is then directly deduced.

11.4.4 The Global Momentum Balance

The same type of reasoning can be applied to the momentum balance across the same control volume $ABCD$ as outlined in Fig. 11.9. The result, of course, is none other than von Kármán's integral equation.
Indeed, the resultant momentum flux in projection along x is written:

$$
dJ = d\left[\int\limits_0^{\delta} \rho U^2 dy \right] + \rho_E U_E V_e .
$$

Regarding the resultant force along x we have:

Fig. 11.9 Control volume for the global momentum balance in a boundary layer

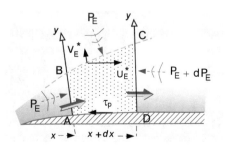

---

[5] In applications, the velocity $U_E^*(x)$ is assumed to be $U_E(x)$, which is equivalent to considering that the potential velocity profile, is uniform throughout the thickness δ.

$$dF = P_E\delta + P_E d\delta - (P_E + dP_E)(\delta + d\delta) - \tau_p dx \simeq dP_E\delta - \tau_p dx\,.$$

Euler's theorem $(dJ = dF)$ then directly leads to von Kármán's equation.

11.5   The Thermal Boundary Layer

11.5.1   The Outline of the Problem

This new section addresses some heat transfer issues in boundary layer flow situations. For an isovolume evolution, the topic loses much of its interest, as the role of temperature is reduced to that of a simple passive contaminant. This result comes directly from the decoupling of the dynamic and thermal problems, specific to this situation, as we have discussed in Chap. 4 Sect. 4.6.4. For this reason, we will temporarily disregard the incompressible fluid assumption (the isovolume evolution) in favour of that of an ideal gas thermodynamic behaviour. However, we will assume that the physical properties of the fluid are constant. The steady, two-dimensional, plane motion assumptions will be still adopted. Finally the external body forces will solely be reduced to the gravity forces. Under these conditions, the general equations of Chap. 4 Sect. 4.6.2 are written as follows:

$$\frac{\partial(\rho U)}{\partial x} + \frac{\partial(\rho V)}{\partial y} = 0 \tag{11.30}$$

$$U\frac{\partial U}{\partial x} + V\frac{\partial U}{\partial y} = g_x - \frac{1}{\rho}\frac{\partial P}{\partial x} + \nu\left[\frac{\partial^2 U}{\partial x^2} + \frac{\partial^2 U}{\partial y^2} + \frac{1}{3}\frac{\partial}{\partial x}\left(\frac{\partial U}{\partial x} + \frac{\partial V}{\partial y}\right)\right] \tag{11.31}$$

$$U\frac{\partial V}{\partial x} + V\frac{\partial V}{\partial y} = g_y - \frac{1}{\rho}\frac{\partial P}{\partial y} + \nu\left[\frac{\partial^2 V}{\partial x^2} + \frac{\partial^2 V}{\partial y^2} + \frac{1}{3}\frac{\partial}{\partial x}\left(\frac{\partial U}{\partial x} + \frac{\partial V}{\partial y}\right)\right] \tag{11.32}$$

$$\rho C_p\left(U\frac{\partial T}{\partial x} + V\frac{\partial T}{\partial y}\right) = U\frac{\partial P}{\partial x} + V\frac{\partial P}{\partial y} + \lambda\left(\frac{\partial^2 T}{\partial x^2} + \frac{\partial^2 T}{\partial y^2}\right) + \Phi_\nu, \tag{11.33}$$

where Φ_ν stands for the mechanical dissipation function equal here to:

$$\Phi_\nu = 2\mu\left[\left(\frac{\partial U}{\partial x}\right)^2 + \left(\frac{\partial V}{\partial y}\right)^2\right] + \mu\left(\frac{\partial U}{\partial y} + \frac{\partial V}{\partial x}\right)^2 - \frac{2}{3}\mu\left(\frac{\partial U}{\partial x} + \frac{\partial V}{\partial y}\right)^2\,.$$

Now, the issue we intend to analyze is whether the boundary layer's dynamic concept can be extended to the thermal situation in the presence of heat transfer, with a view to simplifying the previous equations similarly to that of Prandtl's model for the isovolume flow dynamics.

11.5.2 Basic Assumptions for the Thermal Boundary Layer

As in the case of dynamic phenomena, the thermal boundary layer concept is fundamentally related to the existence of a preferential direction of the convective heat transport with respect to the diffusive heat transfer. For the dynamic boundary layer, the high Reynolds number condition is only required in that sense. When thermal effects are present, it is necessary to involve the Prandtl number, as the ratio between the dynamic and thermal diffusivities. Thus, as we have discussed in Sect. 11.1.4, the characteristic thermal diffusion thickness is $\delta_T \propto \delta/\sqrt{Pr}$. Transposing the fundamental property of the dynamic boundary layer situation ($\delta \propto 1/\sqrt{Re}$) to the thermal case immediately results in $\delta_T \propto 1/\sqrt{Pe}$, where $Pe = Pr \times Re$ is the Péclet number. We can therefore state that:

> The thermal boundary layer approximation, as the existence of a time-scale equivalence region for the heat convection and diffusion phenomena, can apply to flow configurations such as:
> $$Pe \equiv Pr \times Re \gg 1.$$

11.5.3 The Thermal Boundary Layer Equations

Based on the isovolume case analysis,[6] we introduce the following dimensionless variables

$$x^* = \frac{x}{L}, \quad y^* = \frac{y}{L}\sqrt{Re}, \quad U^* = \frac{U}{U_0}, \quad V^* = \frac{V}{U_0},$$

$$g_x^* = \frac{g_x}{g_0}, \quad g_y^* = \frac{g_y}{g_0}, \quad P^* = \frac{P}{P_0}, \quad T^* = \frac{T}{T_0}.$$

The temperature (or temperature difference) reference is noted as T_0 and g_0 denotes the gravity acceleration module, the only body force which is considered.

By substituting in the momentum equation we obtain, given the previous results ($Re \gg 1$ for the diffusion terms):

$$U^*\frac{\partial U^*}{\partial x^*} + V^*\frac{\partial U^*}{\partial y^*} = F_r^{-1} g_x^* - E_u \frac{\partial P^*}{\partial x^*} + \frac{\partial^2 U^*}{\partial y^{*2}}, \tag{11.34a}$$

$$U^*\frac{\partial V^*}{\partial x^*} + V^*\frac{\partial V^*}{\partial y^*} = F_r^{-1}\sqrt{Re}\, g_y^* - E_u Re \frac{\partial P^*}{\partial y^*} + \frac{\partial^2 V^*}{\partial y^{*2}}. \tag{11.34b}$$

---

[6] The transposition of the isovolume case concepts requires that the density variation does not call into question the fundamental boundary layer assumptions and in particular that of the existence of the preferential advection/convection direction.

Three sets of conclusions can be drawn:

– The same geometric and kinematic scale distortion according to $\sqrt{Re}$ provides an equivalent order to the inertia and viscosity terms, as in the incompressible case. However, it should be noted that the global Reynolds number is now based on the *local* density value as:

$$Re = \rho \frac{U_0 L}{\mu} \; ;$$

– The comparison of the different terms in the momentum equation now involves the Froude and Euler numbers respectively defined by:

$$Fr = \frac{U_0^2}{g_0 L} \quad \text{and} \quad Eu = \frac{P_0}{\rho U_0^2} \; ;$$

– In the absence of gravity forces, the high Reynolds number assumption leads to the absence of the cross-wise pressure gradient in the boundary layer ($\partial P / \partial y \approx 0$). Here, the comparison of the gravity and pressure terms in the projection equation along y leads, in a weighty fluid, to the following condition:

In the presence of gravity forces, the cross-wise pressure gradient obeys a hydrostatic law in the boundary layer provided that:
$$Eu \times Fr \gg 1/\sqrt{Re} \; .$$

From now on, we will assume this condition to be fulfilled. As in the incompressible case, the pressure is then a given prescription for the boundary layer problem, which is supplied by the inviscid fluid potential solution. We will specify at a later stage how to express the streamwise pressure gradient dP_E/dx. By applying the same process to the enthalpy equation, we obtain:

$$U^* \frac{\partial T^*}{\partial x^*} + V^* \frac{\partial T^*}{\partial y^*} = Eu \, Ec \left(U^* \frac{\partial P^*}{\partial x^*} + V^* \frac{\partial P^*}{\partial y^*} \right) + Ec \left(\frac{\partial U^*}{\partial y^*} \right)^2 + Pr^{-1} \frac{\partial^2 T^*}{\partial y^{*2}} \; .$$
$$(11.35)$$

In addition to the Euler number previously mentioned, this equation introduces the Eckert and Prandtl numbers defined respectively as follows:

$$Ec = \frac{U_0^2}{C_p T_0} \quad \text{and} \quad Pr = \frac{\mu C_p}{\lambda} \; .$$

With the assumptions which have just been introduced and returning to dimensional quantities, the local equations of the dynamic and thermal laminar boundary layers are written, in a steady two-dimensional plane flow of a variable density fluid as follows:

$$\frac{\partial(\rho U)}{\partial x} + \frac{\partial(\rho V)}{\partial y} = 0$$

$$\rho\left(U\frac{\partial U}{\partial x} + V\frac{\partial U}{\partial y}\right) = \rho g_x - \frac{dP_E}{dx} + \mu\frac{\partial^2 U}{\partial y^2}$$

$$\frac{\partial P}{\partial y} = \rho g_y \qquad\qquad (11.36)$$

$$\rho C_p\left(U\frac{\partial T}{\partial x} + V\frac{\partial T}{\partial y}\right) = U\frac{dP_E}{dx} + \rho V g_y + \lambda\frac{\partial^2 T}{\partial y^2} + \mu\left(\frac{\partial U}{\partial y}\right)^2$$

Proof We begin by observing that the dominant contribution of the dissipation function Φ_ν is proportional to $\partial U/\partial y$, so that in dimensionless values, we have

$$\Phi_\nu^* = \mu\frac{U_0^2}{L^2}Re\left(\frac{\partial U^*}{\partial y^*}\right)^2.$$

Then, with regard to the enthalpic diffusion by the molecular agitation, we obtain, by only retaining the dominant contribution along y:

$$\left[\lambda\frac{\partial^2 T}{\partial x_i^2}\right]^* \approx \left[\lambda\frac{\partial^2 T}{\partial y^2}\right]^* = \lambda\frac{T_0}{L^2}Re\frac{\partial^2 T^*}{\partial y^{*2}}.$$

Returning to Eq. (11.33), we obtain, taking again dimensionless quantities:

$$\frac{\rho C_p U_0 T_0}{L}\left(U^*\frac{\partial T^*}{\partial x^*} + V^*\frac{\partial T^*}{\partial y^*}\right) = \frac{U_0 P_0}{L}\left(U^*\frac{\partial P^*}{\partial x^*} + V^*\frac{\partial P^*}{\partial y^*}\right)$$
$$+ \frac{\mu U_0}{L^2}Re\left(\frac{\partial U^*}{\partial y^*}\right)^2 + \frac{\lambda T_0}{L^2}Re\frac{\partial^2 T^*}{\partial y^{*2}}.$$

It is then only necessary to divide throughout by $\rho C_p U_0 T_0/L$ to reach the expected result.

11.5.4 Some Typical Thermal Boundary Layer Configurations

As just outlined, the dynamic and thermal boundary layer situations identifie a similar *"phenomenological reality"*. It should not be inferred, however, that these situations correspond to a *single flow type*. The thermal case is, in this respect, particularly illustrative of the variety of problems which fall within the scope of parietal flows of the boundary layer type.

Thus, the previous dynamic and thermal boundary layer model can be termed as "*generic*", in the sense that it includes, at the same level of significance, all the terms likely to be present, namely, for:

- *the momentum equation*: the inertia, body, pressure and viscosity forces;
- *the energy equation*: the convection, the pressure force power, the diffusion and the mechanical dissipation.

In practical terms, this means that when attempting to express, for example, the parietal heat transfer by means of a Nusselt number, this should be considered, *a priori*, as a function of all other significant adimensional quantities, formally:

$$N_u = f\left(E_u, F_r, R_e, P_r, E_c\right). \tag{11.37}$$

In practice, various intermediate situations exist where some of these numbers may be without major effects. They constitute as many variants of the general model which we will now explain.

As in the incompressible flow case, we will discard from this discussion the Euler number, whose value will be assumed to be of the order of unity in all circumstances. As we have discussed in the first chapter, the distinction between the compressible and incompressible flow regimes is based on the Mach number values. Although the latter does not appear explicitly in the previous analysis, it is, in fact, implicitly taken into account by the Eckert number. Indeed, by introducing the speed of sound, we can write:

$$E_c \equiv \frac{U_0^2}{C_p T_0} = \frac{U_0^2}{a_0^2} \times \frac{a_0^2}{C_p T_0} = M_0^2 \times \frac{\gamma r T_0}{C_p T_0} = \frac{C_p}{C_v}\left(C_p - C_v\right) M_0^2 = (\gamma - 1) M_0^2.$$

Thus, for $M_0 \ll 1$, the *fluid* compressibility effects, as a result of the *flow* velocity, can be disregarded. This is referred to as an *incompressible* boundary layer, which does not exclude variations in density, insofar as they occur for other reasons such as, for example, the fluid thermal expansion. On the other hand, for $M_0 \sim 1$ and beyond,[7] the simplification to be considered corresponds to the situation of the *compressible* boundary layer. We will now explain the corresponding models in more detail.

11.5.4.1   The Free Convection Boundary Layer

The *free* or *natural* convection refers to a buoyancy driven motion. Hence, the flow Mach number is very low and, in a boundary layer situation, all terms multiplied by the Eckert number in the energy equation can be discarded. Hence this equation simply reduces to:

$$\rho C_p \left(U\frac{\partial T}{\partial x} + V\frac{\partial T}{\partial y}\right) = \lambda \frac{\partial^2 T}{\partial y^2}.$$

---

[7] In practice, and as we have already pointed out, compressibility effects in air flows are actually negligible for $M_0 \lesssim 0.2$.

Under this assumption, a particularly common situation in practice is one where the variation in the fluid density, driven by the temperature difference, remains small enough to only have an effect on the momentum balance through buoyancy or Archimedean forces alone. Such a simplification, known as *Boussinesq's approximation*, extends, in isovolume boundary layer flows, to such configurations as:

- the *natural* convection, where the fluid is set in motion exclusively by the buoyancy forces:
- the *mixed* convection, which combines another origin of the fluid motion.

By denoting the isobaric thermal expansion coefficient as α (see Chap. 1 Sect. 1.3.5), the corresponding model is written as:

$$\frac{\partial U}{\partial x} + \frac{\partial V}{\partial y} = 0$$

$$\rho_0 \left(U \frac{\partial U}{\partial x} + V \frac{\partial U}{\partial y} \right) = \rho_0 g_x \alpha \left(T - T_E \right) - \frac{\mathrm{d} P_E}{\mathrm{d} x} + \mu \frac{\partial^2 U}{\partial y^2}$$

$$\frac{\partial P}{\partial y} = \rho_0 g_y$$

$$\rho_0 C_p \left(U \frac{\partial T}{\partial x} + V \frac{\partial T}{\partial y} \right) = + \lambda \frac{\partial^2 T}{\partial y^2}$$

ADDITIONAL INFORMATION—Grashof's number—The equivalence between inertia, gravity and viscosity forces in a convective boundary layer of this type corresponds to the condition $F_r \approx \Delta\rho/\rho$, or, which amounts to the same, $G_r \approx R_e^2$ with respect to the *Grashof number* $G_r = g\alpha \Delta T L^3/\nu^2$. To prove this expression, it is sufficient to take the change in density as $\Delta\rho = \alpha\rho\Delta T$. Grashof's number is then written

$$G_r = g \frac{\alpha \Delta T L^3}{\nu^2} = \frac{\Delta\rho}{\rho} \frac{gL}{U_0^2} \frac{U_0^2 L^2}{\nu^2} \equiv \frac{\Delta\rho}{\rho} R_e^2 / F_r .$$

11.5.4.2 The Forced Convection Boundary Layer in the Incompressible Regime

This case physically corresponds to low velocity flows (Mach number $\lesssim 0.2$) in the presence of temperature differences small enough to consider the temperature as a passive contaminant in the momentum balance (negligible gravity forces). In these conditions, the boundary layer equations are:

$$\frac{\partial U}{\partial x} + \frac{\partial V}{\partial y} = 0$$

$$\rho_0 \left(U \frac{\partial U}{\partial x} + V \frac{\partial U}{\partial y} \right) = -\frac{dP_E}{dx} + \mu \frac{\partial^2 U}{\partial y^2}$$

$$\rho_0 C_p \left(U \frac{\partial T}{\partial x} + V \frac{\partial T}{\partial y} \right) = U \frac{dP_E}{dx} + \lambda \frac{\partial^2 T}{\partial y^2} + \mu \left(\frac{\partial U}{\partial y} \right)^2$$

The dynamic problem is obviously that of the previous isovolume situation, and the coupling with the enthalpy equation is reached through the viscous dissipation term. For this equation, it is a *source* term which reflects a *kinetic heating effect*, as we will explain in the last chapter.

11.5.4.3 The Compressible Boundary Layer

The equations of the compressible boundary layer model are written as follows:

$$\frac{\partial \rho U}{\partial x} + \frac{\partial \rho V}{\partial y} = 0$$

$$\rho \left(U \frac{\partial U}{\partial x} + V \frac{\partial U}{\partial y} \right) = -\frac{dP_E}{dx} + \mu \frac{\partial^2 U}{\partial y^2}$$

$$\rho C_p \left(U \frac{\partial T}{\partial x} + V \frac{\partial T}{\partial y} \right) = U \frac{dP_E}{dx} + \lambda \frac{\partial^2 T}{\partial y^2} + \mu \left(\frac{\partial U}{\partial y} \right)^2$$

These equations govern the high Reynolds number flow which develops along an obstacle, of a fluid subjected to density variations driven by compressibility effects (a high Mach number) in addition to thermal effects. In this case, the gravity effects (a significant Froude number) can be disregarded.

Of course, the equation of state $P/\rho = rT$ should be added to the previous relations, since the four unknown functions U, V, T and ρ are now governed by a coupled equation set.

Remark *The external solution in an ideal fluid*—In the incompressible boundary layer, as we have explained, the longitudinal pressure gradient is linked to the external velocity field $U_E(x)$ according to Bernoulli's theorem. In a compressible flow, this gradient can still be expressed as:

$$\frac{dP_E}{dx} = -\rho_E U_E \frac{dU_E}{dx} = \rho_E C_p \frac{dT_E}{dx}.$$

The last equality simply derives from the total enthalpy conservation $C_p T + U^2/2 = C^t$.

Chapter 12
Boundary Layer Type Flows: Calculation Methods and Examples

Abstract This last chapter has a threefold objective. First of all, the classical methods for the analytical solving of the model equations derived in the previous chapter are explained in detail in order to obtain the essential dynamic and thermal features of *parietal* boundary layers. In a second step, these tools will be used to handle boundary layer evolutions on various solid bodies. Finally, in a third section, we will extend the methodology to free flows, *i.e.*, without walls, but for which the boundary layer approximations can still be applied. Such flows, referred to as boundary layer-type flows, are those of jets, wakes and mixing layers.

12.1 The Analytical Calculation Methods for the Isovolume Dynamic Boundary Layer

12.1.1 Solving the Local Equations

12.1.1.1 Resuming the Issue

Under the study assumptions—a steady two-dimensional plane flow of an incompressible, weightless fluid—the equations to be solved for the dynamic boundary layer are:

$$\frac{\partial U}{\partial x} + \frac{\partial V}{\partial y} = 0\,, \tag{12.1a}$$

$$U\frac{\partial U}{\partial x} + V\frac{\partial U}{\partial y} = U_E \frac{\mathrm{d}U_E}{\mathrm{d}x} + \nu \frac{\partial^2 U}{\partial y^2}\,. \tag{12.1b}$$

This is a system of partial differential equations, non-linear and parabolic with respect to the space variables. In the second equation, it should be recalled that the function $U_E(x)$ is a given piece of data provided by the potential flow solution. The boundary conditions prescribed for the velocity field on an impermeable wall are:

We are still speaking of free shear flows.

$$U(x,0) = V(x,0) = 0 \quad \text{and} \quad U(x,\infty) = U_E(x). \tag{12.2}$$

By introducing the stream function $\psi(x,y)$, the previous system can be reduced to a single third-order partial differential equation:

$$\frac{\partial \psi}{\partial y}\frac{\partial^2 \psi}{\partial x \partial y} - \frac{\partial \psi}{\partial x}\frac{\partial^2 \psi}{\partial y^2} = U_E \frac{dU_E}{dx} + \nu \frac{\partial^3 \psi}{\partial y^3}, \tag{12.3}$$

with, by definition $U = \dfrac{\partial \psi}{\partial y}$ and $V = -\dfrac{\partial \psi}{\partial x}$.

Nowadays, the solution of these equations can be numerically approximated for an ever-increasing number of flow configurations. This is not the type of calculation we are going to deal with here, where we confine ourselves to the development of analytical methods.

12.1.1.2 The Self-Similarity Assumption

Although simpler than the Navier-Stokes equations, Prandtl's equations are nonetheless a mathematical model for which no general method of analytical resolution is available. This remark could be redhibitory if it were a question of determining in this way any possible solution of these equations as a general rule. In fact, and as we shall discuss now, it can be justified from experimental evidence to focus, in some cases, on a particular class of solution functions which is possibly accessible by an analytical process.

In order to identify the characteristic property of this class of functions, we consider, as outlined in the following Fig. 12.1, any two sections $x = x_1$ and $x = x_2$ of the same boundary layer. We denote by $U_E(x_1)$—resp. $U_E(x_2)$—the velocity at the wall for the ideal fluid potential flow in the section $x = x_1$—resp. $x = x_2$. In each of these sections, it is assumed that the value of the longitudinal velocity component over a given thickness of the boundary layer can be recorded by direct measurement.

This thickness is defined as the distance normal to the wall where the local velocity equals a given percentage, stated here as $\sigma < 1$, of the value at the wall in an ideal fluid at the same considered section. It is obviously dependent on the boundary-layer growth and will thus be referred to as a function of x, namely[1] $g(x)$.

Thus, in the upstream section, a cross-wise velocity distribution (or profile) is obtained between $y = 0$ and $y = g(x_1)$, where $U = \sigma U_E(x_1)$. This distribution is plotted with the red square symbols in Fig. 12.1a. In the same way, the blue triangle symbols represent the profile between the wall and $y = g(x_2)$, where $U = \sigma U_E(x_2)$ in the downstream section.

---

[1] In the case where $\sigma = 0.99$, we identify $g(x) \equiv \delta(x)$, the *standard* boundary layer thickness, as defined in Chap. 11 Sect. 11.2.1.1.

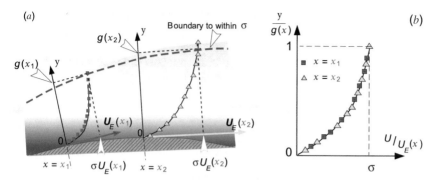

Fig. 12.1 Schematic for the experimental evidence of the self-similarity of the velocity profiles in a boundary layer

Now, by adopting a dimensionless plotting in the form $U(x, y)/U_E(x)$ as a function of $y/g(x)$, it is clear that, whatever the section, all the distributions will fit through the points $(0, 0)$ and $(\sigma, 1)$.

This observation may induce, as illustrated in the previous Figure part (b), that the velocity profiles for different sections could collapse in a single curve when plotted according to the *reduced dimensionless representation*, or in other words:

$$\frac{U(x, y)}{U_E(x)} \text{ is a function of } \eta, \text{ with } \eta = \frac{y}{g(x)}. \tag{12.4}$$

Extensive experimental observations have confirmed the relevance of this induction for many boundary layer flows. Relations (12.4) show that the dimensional profiles are geometrically deduced from one another by a self-similar or affine scaling transformation.[2] This property will be taken as an assumption related to the class of self-similar functions of Prandtl's equation solutions.

As a summary, it can be concluded that, in line with experiments:

> Some flow situations exist for which it is justified to look for a particular class of solution functions for Prandtl's equations complying with the self-similarity hypothesis, namely, such that the dimensionless velocity profile $U(x, y)/U_E(x)$ depends on the variables x and y only through the grouping $y/g(x)$, where $g(x)$ denotes any type of cross-wise length scale.

---

[2] In the field of fluid mechanics, the term "*self-similarity*" is more commonly used.

12.1.1.3   The Mathematical Formulation of the Self-Similarity Assumption

Mathematically, the self-similarity hypothesis amounts to introducing a change in both variables and functions. The first is defined as:

$$(x, y) \rightarrowtail (\xi, \eta) \quad \text{with} \quad \xi = x \text{ and } \eta = y/g(x) \equiv y/g(\xi). \tag{12.5}$$

As for the function change, it is expressed with respect to the stream function by:

$$\psi(x, y) \rightarrowtail \psi(\xi, \eta) = U_E(\xi) \times g(\xi) \times f(\eta). \tag{12.6}$$

From now on and to simplify the writing, we will only retain the symbol x to refer to the first variable of both coordinate groupings, namely (x, y) and (x, η). It will be important not to be mislead by this notation so as not to confuse the independent variables and in particular remembering that:

$$\frac{\partial \eta}{\partial x} = -\eta g'/g \quad \text{and} \quad \frac{\partial \eta}{\partial y} = 1/g. \tag{12.7}$$

Proof It is easy to verify that, for any prescribed value of ξ ($\equiv x$), it is possible to write, from the previous expression of the stream function, Eq. (12.6):

$$U = \frac{\partial \psi}{\partial y} = \frac{\partial \psi}{\partial \eta} \frac{\partial \eta}{\partial y} = \frac{1}{g} \frac{\partial \psi}{\partial \eta} = \frac{1}{g} \left(U_E g f' \right) \equiv U_E F(\eta).$$

We thus retrieve the characteristic property of the self-similarity assumption with, as the function of η in Eq. (12.4) the first derivative of the stream function, namely $F(\eta) = f'(\eta) \equiv df/d\eta$.

12.1.1.4   The Compatibility Conditions—The Solving Equation

By changing the variables and functions according to the self-similarity assumption in Eq. (12.3), we obtain:

$$\nu f''' + \left\{ g^2 \frac{dU_E}{dx} + U_E g g' \right\} f f'' + \left\{ g^2 \frac{dU_E}{dx} \right\} \left(1 - f'^2 \right) = 0,$$

where, according to Lagrange's convention, the prime notation denotes the different order derivatives of any function depending upon a *single* variable. We can now observe that the self-similarity assumption converts the initial partial differential equation into a third-order differential equation for the function $f(\eta)$. In addition, the coefficients of the derivatives of this function are either constant (ν) or a function of the second independent variable x, in the braces. Consequently, the equation will only admit solutions if these coefficients are proportional, which leads to the following conclusion:

A self-similar solution to the boundary layer equations exists if and only if the functions $U_E(x)$ and $g(x)$ fulfill the compatibility conditions:

$$U_E(x) \propto x^{\frac{\beta}{2\alpha-\beta}} \quad \text{and} \quad g(x) \propto x^{\frac{\alpha-\beta}{2\alpha-\beta}},$$

where α and β are two constants.
In this case, the reduced stream function $f(\eta)$ is a solution to the equation:

$$f''' + \alpha f f'' + \beta\left(1 - f'^2\right) = 0,$$

with the boundary conditions $f(0) = 0$, $f'(0) = 0$ and $f'(\infty) = 1$.

Proof The stream function equation—By changing the variables and functions according to the self-similarity hypothesis we obtain the following relations:

$$\frac{\partial\psi}{\partial y}\,(\equiv U) = U_E f', \quad \frac{\partial\psi}{\partial x}\,(\equiv -V) = \frac{dU_E}{dx}gf + U_E g'f + U_E gf' \times \left(-\eta g'/g\right),$$

$$\frac{\partial^2\psi}{\partial x\partial y} = U_E f' - U_E\frac{g'}{g}\eta f'', \quad \frac{\partial^2\psi}{\partial y^2} = U_E f''/g, \quad \frac{\partial^3\psi}{\partial y^3} = U_E f'''/g^2.$$

By substituting in the equation for the ψ function (12.3), we obtain:

$$U_E f'\left(\frac{dU_E}{dx}f' - U_E\frac{g'}{g}\eta f''\right) - \left(\frac{dU_E}{dx}gf + U_E g'f - U_E g\eta f''\right)U_E\frac{f''}{g} =$$

$$U_E\frac{dU_E}{dx} + \nu U_E\frac{f'''}{g^2}.$$

After simplifying and multiplying throughout by g^2 we reach the equation we are looking for:

$$\left\{g^2\frac{dU_E}{dx}\right\}\left(f'^2 - 1\right) - \left\{g^2\frac{dU_E}{dx} + U_E gg'\right\}ff'' = \nu f'''.$$

The compatibility conditions—The above equation is a differential equation for $f(\eta)$, whose coefficients are either a function of the second *independent* variable or constant. It will admit a solution only on condition that these coefficients are proportional, which immediately provides:

$$g^2\frac{dU_E}{dx} + U_E gg' = C' \equiv \alpha\nu \quad \text{and} \quad g^2\frac{dU_E}{dx} = C' \equiv \beta\nu.$$

By simply substracting throughout these relations we deduce $\alpha - \beta = U_E g g'/\nu$ and thus:

$$\frac{\beta}{\alpha - \beta} = \frac{dU_E}{U_E} \times \left(\frac{dg}{g}\right)^{-1}.$$

The functions $U_E(x)$ and $g(x)$ are thus necessarily linked by $g(x) \propto [U_E(x)]^{\frac{\alpha-\beta}{\beta}}$. By using this result, it is then easy to ascertain that the compatibility conditions are equivalent to:

$$g(x) \propto x^{\frac{\alpha-\beta}{2\alpha-\beta}} \qquad \text{and} \qquad U_E(x) \propto x^{\frac{\beta}{2\alpha-\beta}}.$$

12.1.1.5   Conclusions

The results which we have just obtained on the existence and nature of the compatibility conditions make it possible to deduce in concrete terms, from the simple examination of the external solution in an ideal fluid, that:

> When the external velocity $U_E(x)$ obeys a power law of exponent m ($\propto x^m$):
> – a self-similar solution exists in the boundary layer;
> – the corresponding potential velocity function obeys the differential equation $f''' + \alpha f f'' + \dfrac{2m\alpha}{m+1}(1 - f'^2) = 0$, where α is an arbitrary non-zero constant[3];
> – the boundary layer growth obeys the power law $\propto x^{\frac{1-m}{2}}$.

The following exercises explain how to find very simply, solely from the above considerations, the *form* of the boundary layer's growth law when the self-similarity hypothesis applies.

EXERCISE 49 The boundary layer's growth rate on a flat plate in a uniform stream.

An infinitely thin flat plate, of length L, is positioned parallel to the streamlines of a steady uniformly rectilinear flow of a viscous fluid whose velocity is U_∞. The Reynolds number is assumed to be high enough to justify taking into account the viscosity effects according to the two-dimensional plane boundary layer assumptions in the laminar regime.

QUESTION 1. Can there be a self-similar solution for this boundary layer problem?

QUESTION 2. Give the form of the boundary layer's growth law.

QUESTION 3. Which equation governs the self-similar normalized velocity profile?

---

[3] In the definition of the self-similar variable $\eta = y/g(x)$, the function $g(x)$ can be prescribed to within a multiplicative constant. The *arbitrary* choice of the α value correlatively implies the prescription of this constant, as will be explained when examining Blasius' solution in Sect. 12.2.1.1.

SOLUTION: **1.** The plate coinciding with a streamline portion does not change the velocity field of the corresponding ideal fluid potential flow. Thus, the external velocity law for the boundary layer problem is *in this case* $U_E(x) = U_\infty$. This is a particular power law x^m whose exponent is $m = 0$. There is therefore a self-similar solution to this boundary layer problem.

2. The boundary layer's growth law is given by $g(x) \propto x^{\frac{1-m}{2}}$, namely here $g(x) \propto \delta(x) \propto \sqrt{x}$. The streamwise variation of the boundary layer thickness is thus parabolic. It is clear that with this simple reasoning, it is not possible to *calculate* the value of the proportionality coefficient. This can only be obtained after having completely resolved the velocity field (see Sect. 12.2.1.1) for the affinity function which is written here as:

$$\frac{U}{U_\infty} = f'\left(\frac{y}{a\sqrt{x}}\right) \equiv f'(\eta),$$

by setting $\eta = y/a\sqrt{x}$, where a is an arbitrary constant.

3. The equation governing the function $f(\eta)$ then necessarily takes the form:

$$f''' + \alpha f f'' = 0,$$

which is none other than the particular form of the general expression for the present problem, where $\beta = 0$. Let us finally specify that the coefficients a and α are interdependent, as we will explain in the full calculation of this flow.

EXERCISE 50 The boundary layer on a flat plate in a sink flow.

We consider an infinitely thin flat plate of length L, whose trace coincides with the segment $[0, L]$ along the Ox-axis of the XOY reference frame, see Fig. 12.2-(b). It is assumed to be immersed in a flow which, outside the boundary layer, is that generated in an ideal fluid, by a sink located at the origin of the reference frame.

QUESTION 1. Is there a longitudinal pressure gradient along the plate?

QUESTION 2. Is there a self-similar solution for the boundary layer problem on this obstacle?

QUESTION 3. Give the form of the boundary layer's expansion law along the plate.

SOLUTION: **1.** Locating the sink at the plane origin, see Fig. 12.2b, we know—Cf. Chap. 6 Sect. 6.3.1.2—that the ideal fluid potential flow is characterized by the complex potential function $f(Z) = -D/2\pi \, Log(Z)$, where $Z \equiv X + iY \neq 0$ denotes any point in the plane (X, Y) and D is a positive real. The corresponding complex velocity is expressed as $w(Z) = -\frac{D}{2\pi}Z^{-1}$. Since the streamlines are straight lines passing through the origin, the presence of the plate does not change the flow configuration of the ideal fluid. Adopting, for the boundary layer study, the coordinates (x, y) relating to the reference frame linked to the plate, we then have:

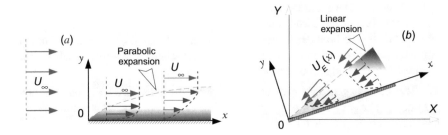

Fig. 12.2 Examples of self-similar boundary layer solutions on a flat plate: **a** uniform flow, **b** sink flow

$$U_E\,(x) = -\frac{D}{2\pi}x^{-1}.$$

By applying Bernoulli's potential flow theorem, we deduce the value of the pressure gradient along the obstacle:

$$\frac{\mathrm{d}P_E\,(x)}{\mathrm{d}x} = -\rho U_E\frac{\mathrm{d}U_E}{\mathrm{d}x} = \rho\frac{D^2}{4\pi^2}x^{-3}.$$

There is therefore a *favourable* pressure gradient along the plate.

2. The external velocity solution being a power-law with $m = -1$, there is therefore a self-similar solution for the boundary layer problem in this situation.

3. With the previous exponent value, we deduce that this boundary layer has a linear expansion $\delta\,(x) \propto x$.

12.1.2   The Integral Method Based Calculation

12.1.2.1   The General Principle of the Integral Method

As demonstrated by the existence of compatibility conditions, the resolution of the local boundary layer equations under the self-similarity assumption is not possible in all cases.

We will therefore now present a method for calculating the boundary layer properties which can be applied with or without this assumption. It is based on the use of the von Kármán integral equation as derived in Chap. 11 Sect. 11.4.2. At first glance, the problem may seem ill-posed since this single equation involves *three* unknown functions $\delta^*\,(x)$, $\theta\,(x)$ and $\tau_p\,(x)$. In fact, the three unknown in question are not independent but implicitly interlinked through the dimensionless velocity profile U/U_E.

The general principle of the method is to make this dependence *explicit*, by presupposing an *analytical form* of this velocity profile. Hence the method which consists of the following four steps can be qualified as an *approximate resolution*:

1. Specifying a dimensionless form of the velocity profile $f(y/\delta) = U/U_E$;
2. Prescribing the boundary conditions in the expression of f;
3. Determining the expressions of $\delta^*(x)$, $\theta(x)$ and τ_p as a function of $\delta(x)$ for the velocity profile obtained in the previous step;
4. Substituting in the integral equation and resolving for *a single* main unknown function.

12.1.2.2   The Kármán–Pohlhausen Implementation

Among the various approximation formulae for the adimensional boundary layer's velocity profile, we will more particularly address the polynomial expression as originally put forward by Pohlhausen [116] in 1921. The presentation will follow the previous four steps.

1– *Specifying a dimensionless profile form:*
According to Pohlhausen, the dimensionless velocity distribution is expressed as a fourth-degree polynomial:

$$\frac{U(x,y)}{U_E(x)} = a(x) + b(x)\,\eta + c(x)\,\eta^2 + d(x)\,\eta^3 + e(x)\,\eta^4, \qquad (12.8)$$

when puting $\eta = y/\delta(x)$. The boundary layer's thickness law $\delta(x)$ is obviously unknown at this stage.
It should be noted that the above expression does not *necessarily* satisfy the self-similarity requirement, since the coefficients, from $a(x)$ to $e(x)$ are not, *a priori* constant but depend on the x variable.

2– *Prescribing the boundary conditions:*
With this velocity profile, the following five boundary conditions can be prescribed:
– at the wall:

$$U(x,0) = 0, \qquad (12.9a)$$

$$V(x,0) = 0, \qquad (12.9b)$$

– at the free boundary[4]:

$$U(x,\delta) = U_E(x), \qquad (12.9c)$$

$$(\partial U/\partial y)_{y=\delta} = 0, \qquad (12.9d)$$

$$\left(\partial^2 U/\partial y^2\right)_{y=\delta} = 0. \qquad (12.9e)$$

---

[4] In this *approximate* resolution method, the matching to the external potential solution takes place at a finite distance $y = \delta(x)$ and no longer asymptotically as $y \to \infty$.

All of the above conditions can be directly introduced into the polynomial expression (12.8), with the exception of the second, which refers to the cross-wise velocity component V. To make it possible to pass on this condition to the U profile, simply apply Prandtl's equation at the wall. Indeed for $y \to 0$, we have:

$$\underbrace{U \frac{\partial U}{\partial x}}_{\downarrow \atop 0} + \underbrace{V \frac{\partial U}{\partial x}}_{\downarrow \atop 0} = U_E \frac{dU_E}{dx} + \nu \underbrace{\frac{\partial^2 U}{\partial y^2}}_{\left(\frac{\partial^2 U}{\partial y^2}\right)_{y=0}}$$

Thus, the no-slip condition at the wall results in the relation:

$$\left(\frac{\partial^2 U}{\partial y^2}\right)_{y=0} = -\frac{U_E}{\nu} \frac{dU_E}{dx} . \tag{12.9f}$$

By requiring the function $f(\eta)$ to satisfy the set of conditions (12.9a), (12.9b), (12.9c), (12.9d) and (12.9f), it can be deduced that the polynomial velocity profile is necessarily of the form:

$$\boxed{\frac{U(x, y)}{U_E(x)} = F(\eta) + \Lambda(x) \times G(\eta)} \tag{12.10}$$

with :

$$F(\eta) = 1 - (1 + \eta)(1 - \eta)^3 \equiv 2\eta - 2\eta^3 + \eta^4 \tag{12.11a}$$

$$G(\eta) = \frac{1}{6}\eta(1 - \eta)^3 = \frac{1}{6}\left(\eta - 3\eta^2 + 3\eta^3 - \eta^4\right) \tag{12.11b}$$

$$\Lambda(x) = \frac{\delta^2(x)}{\nu} \frac{dU_E}{dx} \tag{12.11c}$$

This leaves only one coefficient as a function of x, denoted $\Lambda(x)$ and called the *first boundary layer shape factor*, in the sense that it determines the profile form as a linear combination of two dimensionless polynomial functions depending solely upon the self-similar variable, namely $F(\eta)$ and $G(\eta)$.

The variation range of the shape factor Λ is bounded on $[-12; +12]$ for the following two physical reasons:

- the expected velocity profile is supposed to be that of a boundary layer without separation. This condition leads to prescribing $\Lambda \geqslant -12$;
- in a steady flow, the cross-wise variation of the streamwise velocity component is monotonous, with $U \leqslant U_E$ at any point of the boundary layer. This condition then yields $\Lambda \leqslant +12$.

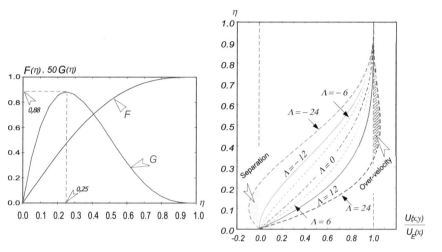

Fig. 12.3 Pohlhausen's polynomial distribution: $F(\eta)$ and $G(\eta)$ functions (left), velocity profile $F(\eta) + \Lambda(x)\,G(\eta)$ (right)

Thus, the dimensionless Pohlhausen velocity profiles constitute a family of one-parameter curves, as exemplified in the Fig. 12.3.

Proof The prescription of the boundary conditions—The five boundary conditions result respectively in:

$$0 = a\,, \tag{12.12a}$$

$$-\frac{U_E}{\nu}\frac{\mathrm{d}U_E}{\mathrm{d}x} = 2c\,\frac{U_E}{\delta^2}\,, \tag{12.12b}$$

$$1 = a + b + c + d + e\,, \tag{12.12c}$$

$$0 = b + 2c + 3d + 4e\,, \tag{12.12d}$$

$$0 = 2c + 6d + 12e\,. \tag{12.12e}$$

By introducing the shape factor Λ, it is easy to deduce from these relations that $b = 2 + \Lambda/6$, $c = -\Lambda/2$, $d = -2 + \Lambda/2$ and $e = 1 - \Lambda/6$. After substitution, we directly obtain expression (12.10).

The non over-velocity condition—Let us put $Y(\eta) = 1 - U/U_E$. The physically acceptable variation range of this function is the [0,1] interval. Now, the extrema of $Y(\eta)$ are obtained for

$$\frac{\mathrm{d}Y}{\mathrm{d}\eta} = (1 - \eta)^2 \left[-2 - \frac{\Lambda}{6} + \left(-4 + \frac{2\Lambda}{3} \right) \eta \right] = 0.$$

The first double root ($\eta_1 = \eta_2 = 1$) corresponds to $y = \delta$, a point where $U = U_E$ which is a valid value. Another root (η_3) is obtained for:

$$\eta_3 = \frac{2 + \Lambda/6}{2\Lambda/3 - 4},$$

which has to be set outside the domain. One can easily ascertain that $\eta_3 \geqslant 1$ for $\Lambda \leqslant 12$.

The non-separation condition—As we have explained in Chap. 11 Sect. 11.3.6, the separation point can be typified by the condition $(\partial U/\partial y)_{y=0} = 0$. Now:

$$\left(\frac{\partial U}{\partial y}\right)_{y=0} = U_E\,(2 + \Lambda/6),$$

hence the critical value $\Lambda = -12$. As outlined in Fig. 12.3 the velocity profile shows, below this limit, an inverted concavity close to the wall which characterizes the existence of a reverse flow. This is inconsistent with the attached boundary layer assumptions, therefore we necessarily have $\Lambda \geqslant -12$.

3 – *Determining the characteristic parameter expressions:*

By direct application of the definitions, we obtain:

$$\frac{\delta^*(x)}{\delta(x)} = \int_0^1 (1 - f(\eta))\,\mathrm{d}\eta = \frac{3}{10} - \frac{\Lambda}{120}, \tag{12.13a}$$

$$\frac{\theta(x)}{\delta(x)} = \int_0^1 (1 - f(\eta))\,f(\eta)\,\mathrm{d}\eta = \frac{37}{315} - \frac{\Lambda}{945} - \frac{\Lambda^2}{9072}, \tag{12.13b}$$

$$\tau_p(x) = \mu\frac{U_E(x)}{\delta(x)}\left(2 + \frac{\Lambda}{6}\right). \tag{12.13c}$$

4– *Substituting in the integral equation and solving the equation:*

Using the von Kármán equation in the form

$$\frac{\mathrm{d}\theta}{\mathrm{d}x} + (2\theta + \delta^*)\frac{1}{U_E}\frac{\mathrm{d}U_E}{\mathrm{d}x} = \frac{\tau_p}{\rho U_E^2},$$

and multiplying throughout by $U_E^2\theta/\nu$ we obtain:

$$\frac{1}{2}U_E\frac{\mathrm{d}\left(\theta^2/\nu\right)}{\mathrm{d}x} + \left(2 + \frac{\delta^*}{\theta}\right)\frac{\theta^2}{\nu}\frac{\mathrm{d}U_E}{\mathrm{d}x} = \frac{\tau_p}{\mu U_E}.$$

This expression suggests introducing the new parameter:

$$K(x) = \frac{\theta^2(x)}{\nu}\frac{\mathrm{d}U_E(x)}{\mathrm{d}x}, \tag{12.14}$$

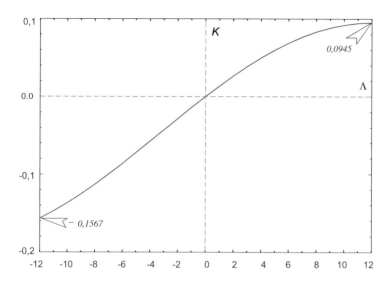

Fig. 12.4   Shape parameter $K(x)$ and $\Lambda(x)$ dependence

known as the second shape factor; It is deduced from the first by the relation:

$$\frac{K(x)}{\Lambda(x)} \equiv \frac{\theta^2(x)}{\delta^2(x)} = \left(37/315 - \Lambda/945 - \Lambda^2/9072\right)^2 . \tag{12.15}$$

In Fig. 12.4, we can notice that, within the valid range of the Λ variation, the correspondence $K(\Lambda)$ is bijective, so that it is completely justified to formally set out that:

$$\frac{\delta^*}{\theta} \equiv \frac{3/10 - \Lambda/120}{37/315 - \Lambda/945 - \Lambda^2/9072} = g_1(K) ,$$

$$\frac{\tau_p \theta}{\mu U_E} \equiv (2 + \Lambda/6) \left(37/315 - \Lambda/945 - \Lambda^2/9072\right) = g_2(K) .$$

The equation to be solved is finally written:

$$\frac{\mathrm{d}}{\mathrm{d}x} \left(\frac{K}{\mathrm{d}U_E/\mathrm{d}x}\right) = -\frac{4K + 2g_1(K) - g_2(K)}{U_E} . \tag{12.16}$$

This is a first-order differential equation which can only be numerically solved in the general case. Once the shape factor K is determined, the value of Λ can be deduced from relation (12.15) and then all the other characteristic parameters from relations (12.13a–c).

ADDITIONAL INFORMATION—An analytical approximate solution—As demonstrated by Pohlhausen [116] in 1921, we can proceed to an approximate analytical integration of Eq. (12.16). Indeed, from the previous expressions of the polynomials $g_1(K)$ and $g_2(K)$, we can verify that the function of K on the right hand side of Eq. (12.16) can be roughly approximated by the linear law[5]:

$$-4K - 2g_1(K) + g_2(K) \simeq 0.470 - 6K .$$

Using this approximation, the equation to be solved becomes

$$\frac{\mathrm{d}}{\mathrm{d}x}\left(\frac{K}{\mathrm{d}U_E/\mathrm{d}x}\right) = \frac{0.47 - 6K}{U_E} ,$$

or again $U_E \dfrac{\mathrm{d}}{\mathrm{d}x}\left(\dfrac{\theta^2}{\nu}\right) + 6\dfrac{\theta^2}{\nu}\dfrac{\mathrm{d}U_E}{\mathrm{d}x} = 0.47$. After multiplying throughout by U_E^5 we obtain:

$$U_E^6 \frac{\mathrm{d}}{\mathrm{d}x}\left(\frac{\theta^2}{\nu}\right) + 6U_E^5 \frac{\theta^2}{\nu}\frac{\mathrm{d}U_E}{\mathrm{d}x} \equiv \frac{\mathrm{d}}{\mathrm{d}x}\left(U_E^6\frac{\theta^2}{\nu}\right) = 0.47 U_E^5 .$$

Therefore, it is finally possible to integrate the equation in the following form:

$$\theta^2(x) = \frac{0.47\nu}{U_E^6(x)} \int_0^x U_E^5(\alpha)\,\mathrm{d}\alpha . \tag{12.17}$$

Thus, with this method, simply knowing the external velocity law is sufficient to determine by a straightforward integration the approximate values of the boundary layer's characteristic properties.

12.2 Calculation Examples Related to the Isovolume Dynamic Boundary Layer

12.2.1 Flows as Exact Solutions of the Local Equations

We intend here to use some special cases to outline the implementation of the method for solving the local boundary layer equations according to the self-similarity assumption. The results obtained will be referred to as "*exact solutions*" as opposed to those based on the integral calculation method (Sect. 12.2.2 hereafter). We will start with the simplest configuration, which historically was the first application of Prandtl's theory. This is the boundary layer on a flat plate parallel to a uniform flow at infinity, as investigated by Blasius in his thesis in 1907 [11] and published in a paper in 1908 [12].

---

[5] For $-0.08 \leqslant K \leqslant +0.08$, the discrepancy with the exact law is less than 10%.

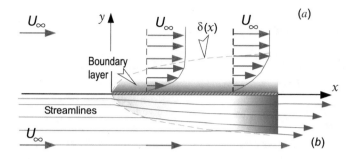

Fig. 12.5 Schematic for the boundary layer configuration along a flat plate at zero-incidence in a uniform stream at infinity: **a** velocity profiles; **b** streamline deflections

12.2.1.1   Blasius Solution for the Boundary Layer Over a Flat Plate

A flat, infinitely thin plate of length L and span b is positioned parallel to a uniform flow of a viscous ($\nu \neq 0$) weightless fluid of velocity U_∞ at infinity. The motion is assumed to be permanent and the aspect ratio b/L such that it allows a two-dimensional plane approach. Finally the global Reynolds number $U_\infty L/\nu$ is taken high enough to use the boundary layer approximations.

As outlined in Fig. 12.5a, the viscous nature of the fluid causes a cross-wise distribution (profile) of the streamwise velocity component, associated with the streamline deflection *(b)*.

In an ideal fluid, with the obstacle merging with a streamline, the flow field remains uniform at all points, with the value U_∞ to which a constant pressure field obviously corresponds. According to Bernoulli's theorem along the plate, we deduce that $\mathrm{d}P_E/\mathrm{d}x = 0$.

In accordance with Prandtl's model and since the longitudinal pressure gradient is zero, the velocity correction to the potential flow solution, within the boundary layer is governed by the equations:

$$\frac{\partial U}{\partial x} + \frac{\partial V}{\partial y} = 0 \quad \text{and} \quad U\frac{\partial U}{\partial x} + V\frac{\partial U}{\partial y} = \nu\frac{\partial^2 U}{\partial y^2} \, ,$$

with, as boundary conditions $U(x, 0) = V(x, 0) = 0$ and $U(x, \infty) = U_\infty \left(\equiv C' \right)$. We know that a self-similar solution exists in this case which corresponds to a stream function of the form

$$\psi = U_\infty g(x) f(\eta) \, ,$$

where the variable η is defined as $\eta = y/g(x)$. By noting that:

$$\mathrm{d}\eta = -y\frac{g'(x)}{g^2(x)}\mathrm{d}x + \frac{\mathrm{d}y}{g(x)} \, ,$$

we immediately infer that $\partial\eta/\partial x = -\eta g'/g$ and $\partial\eta/\partial y = 1/g$.
The following results are then readily obtained:

$$U = \frac{\partial\psi}{\partial y} = \frac{\partial\psi}{\partial\eta}\frac{\partial\eta}{\partial y} = U_\infty f', \qquad -V = \frac{\partial\psi}{\partial x} = \frac{\partial\psi}{\partial\eta}\frac{\partial\eta}{\partial x} = U_\infty g'\left(\eta f' - f\right),$$

along with $\dfrac{\partial U}{\partial x} = -U_\infty \dfrac{g'}{g}\eta f''$, $\qquad \dfrac{\partial U}{\partial y} = U_\infty \dfrac{f''}{g}$, $\qquad \dfrac{\partial^2 U}{\partial y^2} = U_\infty \dfrac{f'''}{g^2}$.

The substitution of these expressions in the momentum equation—that of continuity being identically verified by using the stream function—leads, after simplifications, to the relation $\dfrac{U_\infty}{\nu}gg' = -\dfrac{f'''}{ff''}$ which, due to the separation of the variables, can only be satisfied with:

$$\frac{U_\infty}{\nu}gg' = C^t \equiv K \qquad and \qquad -\frac{f'''}{ff''} = C^t \equiv K\,.$$

The choice of the constant K ($\neq 0$) remains arbitrary and following Blasius' proposal, we will take $K = 1/2$. The change of variables is then defined as $\eta = y\sqrt{U_\infty/\nu x}$ and the differential equation for solving the velocity field becomes:

$$2f''' + ff'' = 0\,. \tag{12.18}$$

With the boundary conditions $f(0) = f'(0)$ and $f'(\infty) = 1$, this equation can be numerically solved with a high rate of accuracy. A graphical representation of the solution as obtained by Howarth [75] in 1938 is plotted in Fig. 12.6.

Quantitatively, Blasius' solution provides the following results:

$$\boxed{\frac{\delta(x)}{x} = \frac{4.92}{\sqrt{R_x}}} \tag{12.19} \qquad \boxed{\frac{\delta^*(x)}{x} = \frac{1.72}{\sqrt{R_x}}} \tag{12.20} \qquad \boxed{\frac{\theta(x)}{x} = \frac{0.664}{\sqrt{R_x}}} \tag{12.21}$$

where $R_x = U_\infty x/\nu$ denotes the *local* Reynolds number based on the abscissa of any point along the plate.

The local parietal friction is given by

$$\tau_p(x) = f''(0)\,\mu U_\infty\sqrt{\frac{U_\infty}{\nu x}} \approx 0.332\mu U_\infty\sqrt{\frac{U_\infty}{\nu x}}\,,$$

from which the local skin friction coefficient is deduced as:

$$\boxed{C_f(x) = \frac{\tau_p(x)}{\frac{1}{2}\rho U_\infty^2} = \frac{0.664}{\sqrt{R_x}}} \tag{12.22}$$

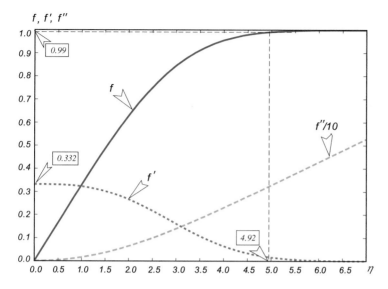

Fig. 12.6 Plot of the Blasius solution function and its first two derivatives

ADDITIONAL INFORMATION—**The viscous solution limit for** $R_e \to \infty$—In Chap. 4 the ideal fluid concept was introduced based on thermodynamic considerations. Its dynamic reduction to a Newtonian fluid is that of an inviscid medium. Thus, by its very definition, any *ideal* fluid motion corresponds to an *infinite* Reynolds number value.

Since Prandtl's model inherently addresses *high* Reynolds number flows of *viscous* Newtonian fluids, one can legitimately wonder what the solution of such a model becomes within the infinitely high R_e limit. More explicitly, the question is whether, by taking the $\nu \to 0$ limit in the viscous solution, one obtains the solution of the inviscid model ($\nu = 0$).

As the comprehensive examination and resolution of this issue is beyond the scope of this course, we will limit ourselves to the presentation of some specific aspects, with reference to Blasius' flat-plate boundary layer solution.

According to Eq. (12.19), when $R_e \to \infty$, the boundary layer thickness tends towards zero, reflecting the collapse of the viscosity influence area on the obstacle wall. Thus, for $\nu \to 0$ in the viscous solution, the boundary layer thickness becomes evanescent. However, the boundary conditions remain unchanged, so that the velocity field is still a solution of the Blasius equation:

$$2f''' + ff'' = 0\,,$$

with $f(0) = 0$, $f'(0) = 0$ and $f'(\infty) = 1$.

As a result, the velocity variation near the wall becomes *increasingly steep*. By comparison, the *non-viscous* equation reduces to

Fig. 12.7 Comparison of
the solutions for $R_e \to \infty$
(viscous model) and
$R_e = \infty$ (ideal fluid) of the
flow on a flat plate

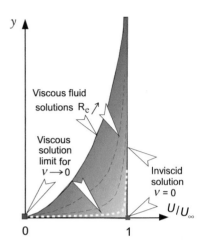

$$ff'' = 0,$$

where the third-order derivative term due to the viscosity has now disappeared.

The *non-viscous solution* of the previous non-trivial equation ($f''_{(\nu=0)} = 0$) is readily written as $f_{(\nu=0)}(\eta) = a\eta + b$. $f_{(\nu=0)}(\eta) = \eta$. It only allows two boundary conditions to be prescribed and is therefore ultimately reduced to $f_{(\nu=0)}(\eta) = \eta$. It clearly corresponds to a uniform velocity profile ($f'_{(\nu=0)}(\eta) = C^t$), with in particular, a *non-zero velocity at the wall*. Thus, as outlined in Fig. 12.7, it can be stated that:

> Within the limit $\nu \to 0$ ($R_e \to \infty$), the viscous solution of the boundary layer velocity cannot tend towards the ideal (inviscid) fluid solution $\nu = 0$ ($R_e = \infty$) up to the solid wall.

12.2.1.2　Boundary Layer Flows Over a Flat Plate with a Streamwise Pressure Gradient

It is important to understand that, for the same obstacle, the boundary layer development depends on the longitudinal pressure gradient. To demonstrate this, we propose to return, as an exercise, to the example previously examined in Sect. 12.1.1.5, Exercise 50.

EXERCISE 51 The boundary layer calculation on a flat plate in a sink/source flow.

We consider a flat plate in a flow generated by a sink located at one of its extremity, as outlined in the previous Fig. 12.2. The Ox axis is taken along the plate and the analysis will be restricted to the $y \geqslant 0$ portion of the domain.

QUESTION 1. Write the local equations which govern the velocity field of the boundary layer flow which develops on the plate in the (x, y) reference frame.

QUESTION 2. By introducing the self-similarity assumption, show that, with a suitable choice of the self-similar variable η, the solving equation for the stream function can be expressed in the following differential form:

$$f'^2 - f''' - 1 = 0.$$

QUESTION 3. After multiplying throughout by f'' the previous equation, show that, by prescribing the boundary conditions $f'(\infty) = 1$ and $f''(\infty) = 0$, this equation can be integrated in the form:

$$f''^2 - \frac{2}{3}\left(f' + 2\right)\left(f' - 1\right)^2 = 0.$$

QUESTION 4. Knowing that $\displaystyle\int \frac{dx}{(1-x)\sqrt{x+2}} = \frac{2}{\sqrt{3}} atanh\left(\sqrt{\frac{2+x}{3}}\right)$ integrate the previous relation in the form of $\eta\left(f'\right)$. After reversal, deduce the expression for the velocity profile $f'(\eta)$. (We give $atanh(\sqrt{2/3}) \approx 1.146$).

QUESTION 5. What happens if the sink is replaced with a source of the same flow rate?

SOLUTION: 1. Taking as the complex potential function for the sink flow of an ideal fluid $f(z) = -U_0 L_0 \log(z)$, with $U_0 L_0 > 0$, the external velocity law along the plate is written as $U_E(x) = -U_0 L_0 / x$. The symbols U_0 and L_0 denote two parameters whose dimensions are that of a velocity and length respectively. With this result, Prandtl's equation becomes:

$$U\frac{\partial U}{\partial x} + V\frac{\partial U}{\partial y} = -\frac{U_0^2 L_0^2}{x^3} + \nu\frac{\partial^2 U}{\partial y^2}.$$

2. The self-similarity assumption results in this case in a stream function of the general expression:

$$\psi = -\frac{U_0 L_0}{x} g(x) f(\eta),$$

from which we can derive $U = -\dfrac{U_0 L_0}{x} f'$, $V = \dfrac{U_0 L_0}{x^2} gf - \dfrac{U_0 L_0}{x} g'f - \dfrac{U_0 L_0}{x} g'\eta f'$,
$\dfrac{\partial U}{\partial x} = \dfrac{U_0 L_0}{x^2} f' + \dfrac{U_0 L_0}{x}\dfrac{g'}{g}\eta f''$, $\dfrac{\partial U}{\partial y} = -\dfrac{U_0 L_0}{x}\dfrac{f''}{g}$, $\dfrac{\partial^2 U}{\partial y^2} = -\dfrac{U_0 L_0}{x}\dfrac{f'''}{g^2}$.

By substitution in Prandtl's equation and after simplifications we obtain:

$$f'^2 + \left(x\frac{g'}{g} - 1\right)ff'' = 1 + \frac{\nu}{U_0 L_0}\frac{x^2}{g^2}f'''.$$

This differential equation with respect to the function $f(\eta)$, whose coefficients are either constant or a function of the second independent variable, has a solution only if the latter are also constant. Taking for example $\alpha^2 = \nu x^2 / (U_0 L_0 \, g^2)$, the boundary layer's expansion law is expressed as:

$$g(x) = \frac{1}{\alpha} \sqrt{\frac{\nu}{U_0 L_0}} \, x \,.$$

The *linear* law already highlighted by the simple application of the compatibility conditions can thus be rediscovered at this point and confirmed from an analytical demonstration. It can be further ascertained that $xg' = g$, so that the differential equation simply reduces to:

$$f'^2 - \alpha^2 f''' - 1 = 0 \,.$$

By choosing to define the self-similar variable as $g(x) = \sqrt{\nu / U_0 L_0} \, x$, namely, by adopting $\alpha = 1$, we obtain the form of the expected solving equation.

3. Multiplying throughout by f'' we obtain $f'' f''' - f'^2 f'' + f'' = 0$, whose integration is straightforward

$$\frac{f''^2}{2} - \frac{f'^3}{3} + f' = K \,.$$

To determine the integration constant, we use the two conditions at infinity which give $K = 4/3$, hence finally:

$$f''^2 - \frac{2}{3} \left(f'^3 - 3f' + 2 \right) \equiv f''^2 - \frac{2}{3} \left(f' + 2 \right) \left(f' - 1 \right)^2 = 0 \,.$$

4. Since the function f' has a physical meaning only if it has positive values, the previous relation can be expressed in the form:

$$\frac{df'}{d\eta} = \sqrt{\frac{2}{3} \left(f' + 2 \right) \left(f' - 1 \right)^2} \quad \text{or} \quad \eta = \sqrt{\frac{3}{2}} \int_0^{f'} \frac{dx}{(1 - x) \sqrt{x + 2}} \,,$$

since, necessarily, $f' < 1$ and $f'(0) = 0$. Using the recalled result, the right hand side integral can be explained to give:

$$\eta = \sqrt{2} \left[atanh \sqrt{\frac{2 + f'}{3}} - 1.146 \right] \,.$$

Finally, after reversal, the velocity profile is written:

$$U(x, y) = -\frac{U_0}{x} f'(\eta) = -\frac{U_0}{x} \left[3th^2 \left(\frac{\eta}{\sqrt{2}} + 1.146 \right) - 2 \right] \,.$$

This expression enables us to deduce the conventional boundary layer thickness which is equal in this case to:

$$\frac{\delta}{x} \approx 3.39 \sqrt{\frac{\nu}{U_0 L_0}} \, .$$

With regard to the wall friction, we have by definition:

$$\tau_p \left(x \right) = \mu \left(\frac{\partial U}{\partial y} \right)_{y=0} = -\mu \frac{U_0 L_0}{x^2} \sqrt{\frac{U_0 L_0}{\nu}} \, f'' \left(0 \right) .$$

The value of $f'' \left(0 \right)$ is directly obtained from the solving equation, where $f'' \left(0 \right) = 0$. Thus we have $f'' \left(0 \right) = 2 / \sqrt{3}$, which finally leads to the local friction coefficient:

$$C_f \left(x \right) \equiv \frac{\tau_p \left(x \right)}{\frac{1}{2} \rho U_E^2 \left(x \right)} \approx \frac{3.30}{\sqrt{R_x}} \, ,$$

where the local Reynolds number is defined as $R_x = x U_E \left(x \right) / \nu$.

5. When substituting a source to the sink, the external velocity law becomes $U_E \left(x \right) = U_0 L_0 / x$. There is therefore, *in an ideal fluid*, a simple inversion of the motion direction along the streamlines: if, by convention, the fluid particles move towards the sink in the negative x direction, they move in the positive x-direction with a source. However, in both cases, the longitudinal pressure gradient $(d P_E \propto d \left(U_E^2 \right))$ is the same. Hence, for a sink, this gradient is favourable to the motion, whereas in the presence of a source it becomes unfavourable, viz opposite to the flow direction. In a *viscous* fluid, the development of the boundary layer may therefore give rise to the occurrence of the detachment phenomenon. This remark suggests why, in variable cross section ducts, it is better to set fluids in motion by suction from downstream, in order to reduce the risk of separation.

12.2.1.3   The Falkner–Skan Boundary Layer Flow Over a Wedge

We consider in an ideal fluid (see Fig. 12.8), the flow whose complex potential function in the (X, Y) plane writes $f \left(Z \right) = K Z^{\frac{2}{2-\beta}}$, where K and β are two non-zero real numbers. As we have demonstrated in Chap. 6, this function represents various flows in and around corners. For $\beta \geqslant 0$, the flow takes place within an angular sector $\widehat{X O x}$ whose $O x$ side is identified by the value $\beta \pi / 2$, as outlined in the following Figure. When materializing this side, the given function can be interpreted as that of the flow on the upper face of a two-dimensional wedge whose half-angle is $\beta \pi / 2$. The general expression for the corresponding complex velocity is $w \left(Z \right) = \frac{2K}{2-\beta} Z^{\frac{\beta}{2-\beta}}$.

Fig. 12.8 Schematic for the boundary layer flow over a wedge

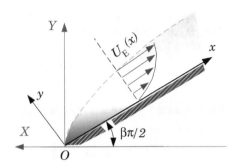

Along the Ox axis, where the argument of any complex number is $arg\,(Z) = \pi - \beta\pi/2$, the velocity vector is colinear with the dihedral side, so that in the local reference frame (x, y) the external velocity can be simply expressed as:

$$(x) = kx^m \,,$$

with $k = 2K/2 - \beta$ and $m = \beta/2 - \beta$.

The boundary layer calculation in this case is therefore based on the following Prandtl equation:

$$U\frac{\partial U}{\partial x} + V\frac{\partial U}{\partial y} = mk^2 x^{2m-1} + \nu\frac{\partial^2 U}{\partial y^2} \,.$$

The stream function, according to the self-similarity assumption, is written as $\psi = kx^m g\,(x)\,f\,(\eta)$.

Hence $U = kx^m f'$, $V = -kmx^{m-1}gf - kx^m \left(fg' - g'\eta f'\right)$ and

$$\frac{\partial U}{\partial y} = kx^m \frac{f''}{g} \,, \quad \frac{\partial^2 U}{\partial y^2} = kx^m \frac{f'''}{g^2} \,.$$

After substitution and simplification we obtain:

$$f'^2 - 1 - ff'' - \left[\frac{x}{m}\frac{g'}{g}\right]ff'' = \left[\frac{\nu}{km}\frac{x^{1-m}}{g^2}\right]f''' \,.$$

Now, the separation of the variables requires that:

$$\frac{x}{m}\frac{g'}{g} = C^t\,(\equiv \alpha) \quad \text{and} \quad \frac{\nu}{mk}\frac{x^{1-m}}{g^2} = C^t\,(\equiv \gamma) \,.$$

Based on the work published in 1931 by Falkner and Skan [57], we will adopt $\gamma = (m + 1)/m$. It is then easy to deduce that $\alpha = (1 - m)/2m$, so that the solving equation finally returns to:

$$f'^2 - 1 - \frac{m+1}{2m}ff'' = \frac{m+1}{m}f''' \quad \text{or} \quad 2f''' + ff'' - \beta\left(f'^2 - 1\right) = 0 \,.$$

The associated change of variables is defined by:

$$\eta = \sqrt{\frac{(m+1)\,k}{\nu}}\; x^{\frac{m-1}{2}}.$$

The resulting equation can be solved numerically, the solution being parameterized by the wedge angle value. Thus, in particular for $\beta = m = 0$, we retrieve the Blasius variables and solving equation for the boundary layer on a flat plate with a zero pressure gradient, Eq. (12.18).

12.2.2 Examples of Boundary Layer Calculations Using the Integral Method

12.2.2.1 The Flat Plate in a Uniform Flow

Here we reuse the flow pattern on a flat plate in a uniform flow. The objective is twofold:
– providing a *straightforward* example of the integral method implementation;
– assessing the method by comparing its results with those of the Blasius solution (Sect. 12.2.1.1).
Two options are suggested for use as exercises.

EXERCISE 52 The boundary layer's polynomial solution on a flat plate.

To calculate the properties of the boundary layer developing on a flat plate in a uniform velocity flow U_∞, we propose to apply the Kármán–Pohlhausen integral method, as previously discussed (Sect. 12.1.2.2). Accordingly, the following polynomial approximation of degree four is adopted to express the dimensionless velocity profile:

$$\frac{U(x,y)}{U_E(x)} = a + b\eta + c\eta^2 + d\eta^3 + e\eta^4,$$

where the a, b, c, d, e coefficients can be, *a priori*, a function of the x abscissa.

QUESTION 1. What is the general polynomial expression reduced to in the particular case of this boundary layer on a flat plate with a zero pressure gradient?

QUESTION 2. Show that in this case the quantities δ^*/δ, θ/δ and $\tau_p\,\delta$ are constants whose values will be expressed.

QUESTION 3. Deduce, by integrating von Kármán's equation, the variation law of the boundary layer thickness $\delta\,(x)$.

QUESTION 4. Give the expressions of $\delta\,(x)\,/x$, $\delta^*\,(x)\,/x$, $\theta\,(x)\,/x$ as well as the wall friction coefficient $\tau_p\,(x)\,/\rho U_\infty^2$ solely based on the *local* Reynolds number $R_x = U_\infty x/\nu$.

SOLUTION: **1.** For this obstacle in such a flow, the external solution ($U_E\,(x) = U_\infty$) proves that there is no longitudinal pressure gradient, so that the shape factors are identically zero. Result (12.10) therefore directly yields the answer:

$$\frac{U\,(x,y)}{U_\infty} = F\,(\eta) = 2\eta - 2\eta^3 + \eta^4 .$$

2. In direct application of expressions (12.13a, b, c) we find in this case $\delta^*/\delta = 3/10$, $\theta/\delta = 37/315$, $\tau_p\delta = 2\mu U_\infty$ which are indeed constants.

3. The von Kármán equation here reduces to $\dfrac{d\theta}{dx} = \dfrac{\tau_p}{\rho U_\infty^2}$, which, after substitution of the previous expressions, yields $\delta d\delta = \dfrac{630}{37}\dfrac{\nu}{U_\infty}dx$.

By prescribing the condition $\delta\,(0) = 0$, the integration of this differential equation finally leads to

$$\delta\,(x) = \sqrt{\frac{1260}{37}}\sqrt{\frac{\nu x}{U_\infty}} \approx 5.84\sqrt{\frac{\nu x}{U_\infty}} .$$

4. By dividing the above expression throughout by x, we immediately deduce $\delta\,(x)/x \approx 5.84/\sqrt{R_x}$, when introducing the local Reynolds number. Then we can easily obtain that:

$$\delta^*\,(x)/x \approx 1.75/\sqrt{R_x}, \quad \theta\,(x)/x \approx 0.69/\sqrt{R_x} .$$

Finally, regarding the skin friction, we have:

$$\frac{\tau_p}{\rho U_\infty^2} \approx \frac{2}{5.84}\frac{\nu}{U_\infty}\sqrt{\frac{U_\infty}{\nu x}} =\approx 0.34/\sqrt{R_x} .$$

EXERCISE 53 The boundary layer's sinusoidal solution on a flat plate.

The previous exercise is reused, this time choosing, as the general form of the dimensionless velocity profile, the expression:

$$\frac{U\,(x,y)}{U_\infty} = a\cos\,(\alpha\eta) + b\sin\,(\beta\eta) ,$$

where $\eta = y/\delta\,(x)$ and a, b, α, β are four parameters, *a priori* function of the sole x variable.

QUESTION 1. Give the values of $a,\ b,\ \alpha,\ \beta$ which are compatible with the flow boundary conditions.

QUESTION 2. Give the expressions of δ^*/δ, θ/δ and $\tau_p\,\delta$. Confirm whether these are still constants for this flow.

QUESTION 3. Determine the variation law of the boundary layer thickness $\delta\,(x)$.

QUESTION 4. Compare the dimensionless values of $\delta\,(x)\,/x$, $\delta^*\,(x)\,/x$, $\theta\,(x)\,/x$ and $\tau_p\,(x)\,/\rho U_\infty^2$ with those obtained in the previous exercise.

QUESTION 5. The numerical application: calculate the thicknesses at the end of a a two-meter long plate in a flow whose velocity is $U_\infty = 20$ m/s of a fluid whose kinematic viscosity is $\nu = 1.4\ 10^{-5}\ m^2/s$.

SOLUTION: 1. A priori, the proposed profile expression is able to fulfill four conditions since it involves four coefficients. We will therefore prescribe:

$$U\,(x,0) = 0, \quad V\,(x,0) = 0, \quad U\,(x,\delta) = U_\infty, \quad \text{and} \quad (\partial U/\partial y)_{y=\delta} = 0\,.$$

The first of these conditions necessarily leads to $a = 0$, so that there are only two remaining coefficients to be determined. The last two conditions then give $b\sin\beta = 1$ and $b\beta\cos\beta = 0$. The only non-trivial solution to this system is $\beta = \pi/2$ and $b = 1$. Finally, the velocity profile of the prescribed form compatible with the three allowable boundary conditions is written:

$$\frac{U\,(x,y)}{U_\infty} \equiv f\,(\eta) = \sin\left(\frac{\pi}{2}\eta\right)\,.$$

2. With this expression, we obtain:

$$\frac{\delta^*}{\delta} \equiv \int_0^1 \left[1 - \sin\left(\frac{\pi}{2}\eta\right)\right]\,d\eta = \frac{\pi-2}{\pi}\,,\quad \frac{\theta}{\delta} \equiv \int_0^1 \left[1 - \sin\left(\frac{\pi}{2}\eta\right)\right]\sin\left(\frac{\pi}{2}\eta\right)\,d\eta = \frac{4-\pi}{2\pi}$$

and similarly for the skin friction $\tau_p = \mu\left(\dfrac{\partial U}{\partial y}\right)_{y=0} = \mu\dfrac{U_\infty}{\delta}f'\,(0) = \dfrac{\pi}{2}\mu\dfrac{U_\infty}{\delta}\,.$

3. By substituting in the integral equation we have $\delta d\delta = \dfrac{\pi^2}{4-\pi}\dfrac{\nu}{U_\infty}dx$, which with the same integration condition—$\delta\,(0) = 0$—gives:

$$\frac{\delta\,(x)}{x} = \pi\sqrt{\frac{2}{4-\pi}}\frac{1}{\sqrt{R_x}} \approx \frac{4.80}{\sqrt{R_x}}\,.$$

4. The comparison of the results obtained with both approximate formulations (polynomial and sinus) is provided in the following Table 12.1, along with the "exact values" of the Blasius solution. One might be surprised that the sinus profile, which satisfies one less boundary condition than the polynomial relation, still gives quite acceptable results. In reality, although the condition $V = 0$ cannot be directly prescribed, it is nevertheless fulfilled by the sinus profile. Indeed, taking into account Prandtl's equation for $y = 0$, this condition still results in $(\partial^2 U/\partial y^2)_{y=0} = 0$. Now, it is clear that $\partial^2 U/\partial y^2 \propto \sin\left(\frac{\pi}{2}\eta\right)$ which is actually zero for $\eta = 0$.

Table 12.1 Thickness, skin friction and shape factor comparison in the boundary layer on a flat plate in a uniform flow, according to different prediction methods

Parameter profile	$\frac{\delta(x)}{x} \times \sqrt{R_x}$	$\frac{\delta^*(x)}{x} \times \sqrt{R_x}$	$\frac{\theta(x)}{x} \times \sqrt{R_x}$	$\frac{\tau_p(x)}{\rho U_\infty^2} \times \sqrt{R_x}$	H
$\sin\left(\frac{\pi}{2}\eta\right)$	4.80	1.74	0.66	0.33	2.64
$2\eta - 2\eta^3 + \eta^4$	5.84	1.75	0.69	0.33	2.54
Blasius	4.92	1.72	0.66	0.33	2.59

Since this observation is by no means general, it should be kept in mind that the approximated methods can give quite acceptable results for some characteristics, without there necessarily being entire concordance at any point with the velocity profile solution of the *local* equations.

5. With the given numerical values, we find $\delta(L) \simeq 5.7$ mm, $\delta^*(L) \simeq 2.1$ mm, $\theta(L) = 0.8$ mm.

12.2.2.2   The Boundary Layer on a Circular Cylinder

This new application of the integral method is intended to illustrate a situation where the resolution with the self-similarity assumption is not possible. As outlined in the following Figure, the configuration is that of a uniform flow of velocity U_∞ at infinity around an infinitely long cylinder of a circular cross-section (radius a) (Fig. 12.9).

(a) *The external solution in an ideal fluid*: As we have learned in Chap. 6 Sect. 6.3.2.3, the complex potential of the ideal fluid flow is expressed by $f(Z) = U_\infty \left(Z + \frac{a^2}{Z}\right)$ for $\|Z\| \geq a$, where Z denotes the affix of a point in the (X, Y) reference frame. The complex velocity is written $W(Z) = U_\infty \left(1 - a^2/Z^2\right)$ whose restriction at any point on the cylinder ($Z = ae^{i\varphi}$) is:

$$W(Z)|_{circle} = U_\infty \left(1 - e^{-2i\varphi}\right).$$

By locating any point on the circle $r = a$ by the polar angle α measured from the upstream stagnation point and counted positively in the reverse trigonometric direction, it is easy to determine that, in the local reference frame ($x = a\alpha$, y), the variation law of the velocity module on the circle is expressed by:

$$|U_E(x)| \equiv |U_E(\alpha a)| = 2U_\infty |\sin\alpha|. \tag{12.23}$$

As expected, we can find that the velocity on the circle, in an ideal fluid without circulation, is zero for $\alpha = 0$ (upstream stagnation point), $\alpha = \pi$ (downstream stagnation

Fig. 12.9 Schematic for the boundary layer on a circular cylinder

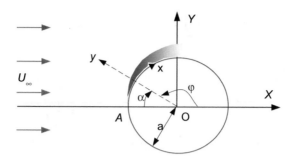

point) and is maximum for $\alpha = \pm\pi/2$. Hence expression (12.22) is not in line with the compatibility requirement $U_E(\alpha) \propto \alpha^n$, so that there is no solution complying with the self-similarity assumption.

Proof The angular correspondence characterizing the change in the coordinate system (X, Y) to (x, y) at any point on the circle is simply defined by $\varphi = \pi - \alpha$. The complex velocity on the circle is therefore $W(Z)|_{circle} = U_\infty \left(1 - e^{2i\alpha}\right)$. We immediately deduce that:

$$\|W\|_{circle} = U_\infty \sqrt{(1 - \cos 2\alpha)^2 + \sin^2 2\alpha} = U_\infty \sqrt{2(1 - \cos 2\alpha)} = 2U_\infty |\sin \alpha|.$$

In the local reference frame *i.e.*, for $x = a\alpha$, this expression is none more than the external velocity law $U_E(x)$ for the boundary layer calculation. By applying Bernoulli's theorem between the upstream stagnation point and any point on the obstacle surface, we have $P(M) = P(A) - 2\rho U_\infty^2 \sin^2 \alpha$, which leads to the longitudinal pressure gradient:

$$\frac{1}{\rho} \frac{dP_E}{dx} \equiv \frac{1}{\rho a} \frac{dP(M)}{d\alpha} = -4 \frac{U_\infty^2}{a} \sin \alpha \cos \alpha.$$

(b) The boundary layer calculation: We propose to carry out the boundary layer correction according to the Kármán–Pohlhausen procedure and adopting a linearized approximation of the relation between the shape factors (see Sect. 12.1.2.2, Additional Information). First of all, let us explain these two shape parameters, limiting ourselves to the boundary layer which develops along the upper half of the cylinder $(0 \leqslant \alpha \leqslant \pi)$.

By definition $\Lambda(\alpha) = \dfrac{\delta^2(\alpha)}{\nu} \dfrac{dU_E(\alpha)}{a d\alpha} = 2\dfrac{U_\infty}{a\nu}\delta^2 \cos \alpha$, and similarly:

$K(\alpha) = \dfrac{\theta^2(\alpha)}{\nu} \dfrac{dU_E(\alpha)}{a d\alpha} = 2\dfrac{U_\infty}{a\nu}\theta^2 \cos \alpha$.

Under the linear approximation mentioned above, the momentum thickness is determined as:

$$\theta^2(x) = \frac{0,47\nu}{U_E^6(x)} \int\limits_0^x U_E^5(\sigma)\,d\sigma, \text{ namely here } \theta^2(\alpha) = \frac{0,47\nu}{2U_\infty \sin^6 \alpha} \int\limits_0^\alpha \sin^5\sigma\, a d\sigma,$$

or after calculation:

$$\theta^2(\alpha) = \frac{0,47a\nu}{2U_\infty \sin^6 \alpha} \left(\frac{8}{15} - \cos\alpha + \frac{2}{3}\cos^3\alpha - \frac{1}{5}\cos^5\alpha \right), \qquad (12.24)$$

which gives us, for the second shape factor:

$$K(\alpha) = \frac{0,47\cos\alpha}{\sin^6 \alpha} \left(\frac{8}{15} - \cos\alpha + \frac{2}{3}\cos^3\alpha - \frac{1}{5}\cos^5\alpha \right). \qquad (12.25)$$

Once the $K(\alpha)$ law is determined, it should be remembered that than for Λ is directly deduced and accordingly the boundary layer thickness and velocity profile.

Proof Knowing that $\int \sin^n \varphi d\varphi = -\frac{1}{n}\sin^{n-1}\varphi\cos\varphi + \frac{n-1}{n}\int \sin^{n-2}\varphi d\varphi$, we obtain

$$\int\limits_0^\alpha \sin^5 \varphi d\varphi = \left[-\frac{1}{5}\sin^4\varphi\cos\varphi + \frac{4}{5}\left(-\frac{1}{3}\sin^2\varphi\cos\varphi - \frac{2}{3}\cos\varphi \right) \right]_0^\alpha$$

$$= \left[-\frac{1}{5}\sin^4\varphi\cos\varphi - \frac{4}{15}\sin^2\varphi\cos\varphi - \frac{8}{15}\cos\varphi \right]_0^\alpha$$

$$= \frac{8}{15} - \frac{8}{15}\cos\alpha - \frac{4}{15}\sin^2\alpha\cos\alpha - \frac{1}{5}\sin^4\alpha\cos\alpha,$$

from which we can easily deduce the expected result.

(c) The separation point localization: The graph of the function $K(\alpha)$ corresponding to Eq. (12.25) is plotted in Fig. 12.10.

As we have discussed in Chap. 11, the detachment of the fluid threads from the body surface, which illustrates the boundary layer separation, theoretically occurs for $\Lambda = -12$. By virtue of the K and Λ relation, namely $K = \Lambda \times (37/315 - \Lambda/945 - \Lambda^2/9072)$, the second factor value at the *separation point* is thus $K = -0.1567$. This result means that a detachment does occur for $\alpha \simeq 1.88$ radians, or about 108°. This value is consistent with that quoted by Schlichting [147], resulting from a calculation by a different method. However, they are both higher than the experimental data where a *laminar* boundary layer detachment is observed slightly upstream of the projected frontal cross-section ($\alpha < 90^o$). It should be noted that the experimental comparison is biased by the *plane flow* hypothesis, which is clearly not justified here because of the curvature effects of the circular profile.

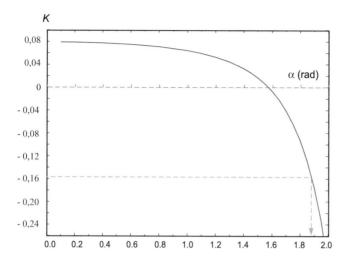

Fig. 12.10 Second shape factor for the laminar boundary layer on a circular cylinder

ADDITIONAL INFORMATION—**Flow Separation and Turbulence**—A fairly widespread confusion among the general public is to confuse the detachment and
turbulence phenomena, or, which conversely amounts to the same thing, non-
separation and laminarity. Thus, for example, in some practices involving the
wind action on a sail, it can be argued that a *laminar* flow on the profile should be
preserved in order to avoid the *separation* of the fluid threads. Such a formulation
is not correct, from the strict point of view related to fluid mechanics. We will
briefly shed some light on this issue, a more in-depth discussion of the underlying
concepts relates to a deeper knowledge of turbulence (see for example Chassaing
[35]).

In the present study, the boundary layer flow is assumed to develop in the
laminar regime. In real situations of course, there are many cases where the
flow regime is turbulent. The transition between these two regimes essentially
depends on the balance between stabilizing (viscosity) and destabilizing (inertia)
forces, as discussed in Chap. 9 Sect. 9.7. The detachment for its part, is due to the
competition, close to the wall, between inertial forces (advection) and pressure
forces in the main flow direction. For the purposes of this discussion, only the
following two points will be retained, in a schematic manner:

– the laminar / turbulent transition and boundary layer detachment (separation)
  are two distinct phenomena;
– in specific flow configurations, they can be put into an interacting situation.

It is in this interaction that we should look for the explanation of the sudden drag
reduction of a cylinder or a sphere which is observed for a *critical* Reynolds
number R_c. We are still speaking of the '*drag crisis*' in this regard.

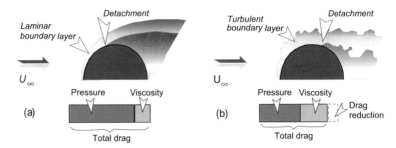

Fig. 12.11   Influence of the transition to turbulence on the detachment pattern and its impact in terms of drag: **a** the sub-critical regime and laminar boundary layer, **b** the super-critical regime and turbulent boundary layer

- In the sub-critical regime, such as, by definition $R < R_c$, the boundary layer which develops on the front part of the cylinder is laminar and detaches upstream of the maximum frontal coss-section. The result is a moderate *viscous drag*, but a high *pressure drag* due to the extensive expansion of the detachment area.
- In the supercritical regime, $(R > R_c)$, the experiment has revealed that the boundary layer is turbulent, a property which makes it more resistant to wall separation. In fact, the separation settles at a backward position which is located slightly beyond the maximum frontal cross-section of the cylinder. This reduction in the detached wake area results in a decrease in the pressure drag, which is greater, in absolute value, than the increase in the viscous friction due to the turbulent nature of the flow in the attached boundary layer. Thus, in total, a slight increase in the Reynolds number beyond R_c leads to a significant decrease in the resulting drag (pressure + viscosity) of the obstacle.

These considerations are outlined by schematics (a) and (b) in Fig. 12.11.

In practice, the measured value of the *critical* Reynolds number for an infinite span cylinder is between 4×10^5 and 5×10^5, and 2×10^5 and 4×10^5, for a sphere, this range reflecting the influence of experimental factors on the transition, such as the surface roughness or the turbulence intensity of the external flow (see next Historical Note).

HISTORICAL NOTE—**Drag crisis—Critical Reynolds number—Transition to turbulence**—The first wind tunnel tests date back to the very beginning of the 20th century, including the work of the Orville and Wilbur Wright brothers in 1901 in the USA and Gustave Eiffel in 1909 in France. They were aimed at determining the aerodynamic forces (mainly lift and drag) on wing profiles, as well as the resistance (drag) of various bluff bodies, including the sphere.

In those early days, turbulence in wind tunnels was ignored, and its aerodynamic effects on streamlined and bluff bodies still unknown (see previous Additional Information). This explains the great disparity in the values of the drag coefficient (C_x) of the sphere deduced, at the same global Reynolds number, from measurements carried out in different wind tunnels.[6] The reason lies in the same phenomenon as we have just described concerning the cylinder.

Once it was understood that the reduction in the sphere resistance at the critical regime was in connection with the onset of turbulence in the boundary layers, the Reynolds number determination of the drag crisis served as a measure of the turbulence level in the facility, as Prandtl [125] points out. On page 299 of Rebuffet's work [134], we find a compilation of these levels for some French wind tunnels, including that of the IMFT.[7] At a later date, new and more precise parameters have been introduced to characterize the turbulent agitation level in wind tunnels, notably the turbulence intensity. We owe it to Dryden *et al.*, [50] for experimentally determining, in 1938, the interconnection between the sphere's critical Reynolds number and the turbulence intensity.

12.3   The Thermal Boundary Layer on a Flat Plate

As highlighted in Chap. 11, the parietal heat transfer in boundary layer configurations, can give rise to different situations depending on the nature of the dominant terms in the equations. These situations range from *natural convection* for a low-velocity, isovolume-evolving fluid, to *forced convection*, with kinetic heating for high-speed, compressible fluid flows. It is outside the scope of our work here to deal with all cases. We will limit ourselves to a few examples to illustrate the analogies and differences between dynamic and thermal phenomena in the boundary layer on a flat plate, with a zero-pressure gradient.

In all situations, we will consider a steady, two-dimensional plane motion and assume the physical properties of the fluid to be constant.

12.3.1   *Free Convection on a Heated Vertical Flat Plate*

A flat plate of length L is vertically arranged in a weighing fluid, initially at rest and at a uniform temperature $T = T_\infty$ throughout the space. We intend to determine some properties of the heat transfer which takes place, in a steady state, between the fluid and the plate, when the latter is maintained at a constant temperature $T = T_p > T_\infty$.

Near the wall, the hot and therefore lighter fluid is set in motion by the Archimedes' thrust alone, giving rise to a purely convective heat transfer (Fig. 12.12).

The temperature difference will be assumed such that, for the fluid under consideration, the condition $G_r \approx R_e^2$ is satisfied, in order to validate the convective boundary layer's model (see Chap. 11, Sect. 11.5.4.1).

---

[5] According to Rebuffet [134], Eiffel had measured a $C_x = 0.176$, whereas the Göttingen tests led to the value $C_x = 0.44$.

[7] Institut de Mécanique des Fluides de Toulouse.

Fig. 12.12 Natural
convective boundary layer on
a vertical plate

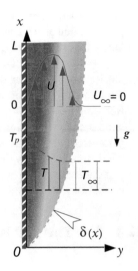

Under the general study assumptions, the external pressure gradient along the
plate is zero so that the momentum and enthalpy equations can be written with the
Boussinesq approximation as follows:

$$U\frac{\partial U}{\partial x} + V\frac{\partial U}{\partial y} = \alpha g\,(T - T_\infty) + \nu\frac{\partial^2 U}{\partial y^2}\,,\tag{12.26}$$

$$U\frac{\partial T}{\partial x} + V\frac{\partial T}{\partial y} = a\frac{\partial^2 T}{\partial y^2}\,.\tag{12.27}$$

12.3.1.1 The Velocity and Temperature Fields

It can be proved that a self-similar solution can be found for the previous equations
along with the continuity equation. Indeed, by setting:

$$\psi = h\,(x)\,f\,(\eta)\,,\qquad \frac{T - T_\infty}{T_p - T_\infty} = \Theta\,(\eta)\qquad \text{with}\ \eta = \frac{y}{l\,(x)}\,,$$

all compatibility conditions can be fulfilled with $l\,(x) \propto x^{1/4}$ and $h\,(x) \propto x^{3/4}$. The
boundary layer's growth rate is therefore no longer parabolic in this case.
Based on these results and according to Pohlhausen, we will define:

$$f\,(\eta) = \frac{\psi}{\nu G_{r_x}^{1/4}}\qquad \text{and}\qquad \eta = G_{r_x}^{1/4}\frac{y}{x}\,,$$

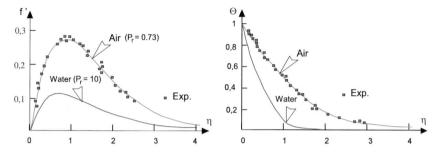

Fig. 12.13   Dimensionless velocity and temperature profiles: Solution from Pohlhausen and experiment, adapted from Grimson [59]

where $Gr_x = \alpha g x^3 \left(T_p - T_\infty\right)/\nu^2$ is the *local* Grashof number, *i.e.*, based on the abscissa along the plate. Under these conditions, the velocity field components are defined as

$$U = \frac{\nu}{x} Gr_x^{1/2} f'(\eta), \qquad V = \frac{\nu}{4x} Gr_x^{1/4} \left[\eta f'(\eta) - f(\eta)\right],$$

and the equations for solving the functions $f(\eta)$ and $\Theta(\eta)$ are written as follows:

$$2f''' + \frac{3}{2} f f'' - f'^2 + 2\Theta = 0, \tag{12.28}$$

$$2\Theta'' + \frac{3}{2} P_r f \Theta' = 0. \tag{12.29}$$

The solution of these equations should satisfy the boundary conditions $f(0) = f'(0) = 0$, $\Theta(0) = 1$, $f'(\infty) = 0$ and $\Theta(\infty) = 1$. It is plotted in Fig. 12.13 adapted from Grimson [59], for two Prandtl number values, air and water. One can also observe the relevant agreement with the experiment.

Proof Denoting by $l(x)$ a length scale function of the single coordinate x, we introduce the self-similarity variable $\eta = y/l(x)$. We then look for the velocity field from the stream function $\psi = h(x) \times f(\eta)$ and the temperature field as the reduced function:

$$\frac{T(x, y) - T_\infty}{T_p - T_\infty} = \Theta(\eta).$$

We deduce from this $U = h f'/l$, $V = -h' f + h l' \eta f'/l$,
$\dfrac{\partial U}{\partial x} = h' f'/l - h l' f'/l^2 - h l' \eta f''/l^2$, $\dfrac{\partial U}{\partial y} = h f''/l^2$ and $\dfrac{\partial^2 U}{\partial y^2} = h f'''/l^3$.

Substituting in the momentum equation and after simplification we obtain:

$$\left[l h' - h l'\right] f'^2 - \left[l h'\right] f f'' = \nu f''' + \alpha g \left(T_p - T_\infty\right) \left[l^3/h\right] \Theta.$$

Since the terms in square brackets are functions of the single variable x, this differential equation requires the following compatibility conditions:

$$l^3/h = A, \quad lh' = B, \quad hl' = C,$$

where A, B, C are three constants. It can easily be deduced that $l(x) \propto x^{1/4}$ and $h(x) \propto x^{3/4}$. It remains to ascertain that these conditions also lead to a self-similar solution for the energy equation. After substitution by the dimensionless quantities, the latter becomes:

$$[hl'] f\Theta' + a\Theta'' = 0.$$

It therefore requires no other condition than $hl' = C$, a relation which has already been prescribed.

12.3.1.2   The Heat Transfer

The local heat transfer, per unit of time and area, is here expressed as:

$$\phi_p(x) \equiv -\lambda \left(\frac{\partial T}{\partial y}\right)_{y=0} = -\lambda (T_p - T_\infty) \frac{G_{r_x}^{1/4}}{x} \Theta'(0).$$

It can also be associated with the local Nusselt number

$$Nu_x \equiv \frac{x\phi_p(x)}{\lambda(T_p - T_\infty)} = -G_{r_x}^{1/4}\Theta'(0).$$

For the air ($P_r = 0.73$), we obtain $\Theta'(0) \simeq -0.36$, so that the Nusselt's number is locally equal to

$$Nu_x \simeq 0.36\, G_{r_x}^{1/4}. \tag{12.30}$$

ADDITIONAL INFORMATION—**Thermal convection on a flat plate in any fluid: the Prandtl number's influence**—For a fluid other than the air, the previous expression (12.30) is no longer appropriate. Referring to Q as the total heat flux on one side of the plate whose area is $b \times L$, we define the global transfer coefficient k by the relation $Q = k(T_p - T_\infty) bL$. We deduce, in a dimensionless form, the expression for the *global* Nusselt number:

$$\overline{Nu} = \frac{kL}{\lambda} = \frac{Q}{\lambda(T_p - T_\infty)b}.$$

The previous result—Eq. (12.30)—suggests looking for an expression of this Nusselt number *as a function* of Grashof and *Prandtl* numbers, in the form

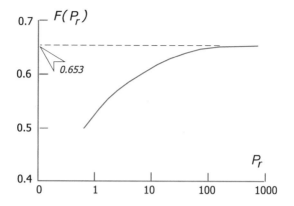

Fig. 12.14 Nusselt's law coefficient, Eq. (12.31), as a function of the Prandtl number in a convective boundary layer on a vertical flat plate, adapted from Grimson [59]

$$\overline{N}u = C \times \left(Pr.Gr_{L}\right)^{1/4}, \qquad (12.31)$$

where Gr_{L} is the Grashof number based on the plate length L.
The dimensionless coefficient C is assumed to be constant.
In fact, according to Grimson [59], the experiment reveals that this coefficient is slightly dependent on the Prandtl number. More precisely, taking $C = F\,(Pr)$, the measurements (Cf. Fig. 12.14) have revealed that the significant variation range of the F function corresponds to Prandtl numbers between 1 and 100, this function becoming substantially constant (≈ 0.653) for $200 < P_{r} < 1000$.

EXERCISE 54 Heat convection on a flat plate in an oil bath.

A flat plate, 10 cm high and 50 cm wide, is vertically immersed in an oil bath at a regulated temperature of $80^{o}C$. Both sides of the plate are heated to the constant temperature of $T_{p} = 200^{o}C$.

QUESTION 1.   Express the total heat flux Q on one side of the plate as a function of the Grashof number Gr_{L} based on the plate length and the value at the wall of the dimensionless temperature gradient $\Theta'(0)$.

QUESTION 2.   We set $\overline{N}u = F\,(P_{r}) \times \left(P_{r}\,Gr_{L}\right)^{1/4}$ where $\overline{N}u$ stands for the global Nusselt number $(Q/\left[\lambda b\left(T_{p} - T_{\infty}\right)\right])$. Give the expression of F as a function of the Prandtl number and the parietal value of the dimensionless temperature gradient $\Theta'(0)$.

QUESTION 3.   We give, for the physical properties of the oil:
– kinematic viscosity: $\nu = 0.5 \times 10^{-4}\,m^{2}/s$,
– thermal diffusivity: $0.5 \times 10^{-7}\,m^{2}/s$,
– thermal conductivity: $\lambda = 0.14\,W m^{-1} K^{-1}$,
– thermal expansion coefficient: $\alpha = 0.8 \times 10^{-3}\,K^{-1}$.
Using the graphical representation of the function $F\,(P_{r})$ in Fig. 12.14, calculate the value of the total heat flux released by the plate.

QUESTION 4. Proceed with the previous question, with all conditions unchanged except for the layout of the plate, which is now placed vertically along its width, namely 50 cm high and 10 cm wide (*It is assumed that two-dimensional plane flow assumptions are still valid.*).
What conclusion do you draw from the comparison of the results in the two situations?

SOLUTION: **1.** The heat flux on one side of the plate is:

$$Q = b \int_0^L \phi_p(x)\,dx = -\lambda b \left(T_p - T_\infty\right) \Theta'(0) \int_0^L \frac{Gr_x^{1/4}}{x}\,dx = -\frac{4}{3}\lambda b \left(T_p - T_\infty\right) \Theta'(0)\, Gr_L^{1/4}.$$

2. By definition, we have $\overline{Nu} = \dfrac{Q}{\lambda b \left(T_p - T_\infty\right)} = -\dfrac{4}{3}\Theta'(0)\, Gr_L^{1/4}$ which, by simple identification with the proposed form, yields $F(Pr) = -\dfrac{4}{3}\Theta'(0)\, Pr^{-1/4}$.

3. Let us begin by calculating the oil's Prandtl number. With the numerical data for use in the exercise we obtain $Pr = 1000$. It can be deduced (see the graph in Fig. 12.14) that $F(1000) = 0.653$.
Hence in this case, we have $\overline{Nu} = 0.653 \times (Pr\,Gr_L)^{1/4}$. Now we have to calculate the Grashof number value, which is:

$$Gr_L = \frac{\alpha g L^3 \left(T_p - T_\infty\right)}{\nu^2} = \frac{0.8 \times 9.81 \times (0.1)^3 \times 120}{\left(0.5 \times 10^{-4}\right)^2} \approx 3.8 \times 10^5.$$

We deduce that $\overline{Nu} = 0.653 \times \left(3.8 \times 10^8\right)^{1/4} \approx 91$. Thus, for both sides of the plate we obtain:

$$Q = 2\lambda b \overline{Nu}\left(T_p - T_\infty\right) = 2 \times 0.14 \times 10^{-3} \times 0.5 \times 91 \times 120,$$

or finally $Q \simeq 1.5\,kW$.

4. Simply changing the orientation of the plate does not change the Prandtl number of the fluid.
With regard to the Grashof number, we can take, for the first arrangement $Gr_1 = \alpha h^3$ and for the second $Gr_2 = \alpha b^3$, where α is a constant, h ($= 10\,cm$) and b ($= 50\,cm$) standing for the width and length of the plate respectively.
Accordingly we can set, β denoting another constant, $\overline{Nu}_1 = \beta h^{3/4}$ and $\overline{Nu}_2 = \beta b^{3/4}$. This results in heat fluxes respectively equal to $Q_1 = \gamma b h^{3/4}$ and $Q_2 = \gamma h b^{3/4}$, either by eliminating the constant γ, $Q_1 = (b/h)^{1/4}\,Q_2$. With the relevant numerical data, we obtain $Q_1 \approx 1.5 Q_2$.
This example thus makes it clear that, all other things being equal, the *heat transfer by natural convection on a flat rectangular plate depends not only on the plate area but also on its arrangement (aspect ratio) in the flow field.*

12.3.2 Forced Convection in Low Speed Flows

12.3.2.1 Introduction to the Issue

We are now interested in the heat transfer between a plate and a fluid in a uniform motion parallel to the plate at infinity. Contrary to the previous case, the influence of buoyancy forces will now be taken as negligible, which means in particular that the fluid is set in motion by the action of forces which are not directly related to the heat exchange, such as the pressure forces. This is referred to as forced convection.

In the first situation which we will consider here, we will assume that the dissipation by viscosity, resulting from the stopping of the fluid on the plate, releases a negligible heat volume, which is the case for moderate values of the external flow velocity. The complementary case (a high velocity external flow) will be addressed in the next Sect. 12.3.3.

Under the hypotheses which characterize this thermal boundary layer configuration, the solution of the problem, for an adiabatic wall condition (a zero parietal heat flux), is trivial and is obtained with a plate temperature equal to that of the fluid at infinity. For this reason, only the case of an isothermal wall (a constant imposed plate temperature) will be discussed.

12.3.2.2 The Governing Equations

The situation we are considering therefore verifies the following hypotheses:

H1: an isovolume evolution;

H2: negligible buoyancy forces, *i.e.* $G_r/R_\ell^2 \ll 1$, so that any free convection transfer can be discarded;

H3: a Prandtl number close to one (the same order for the momentum and heat diffusivities);

H4: an Eckert number much lower than the Reynolds number, so that the kinetic heating by dissipation can be disregarded ($E_C/R_e \ll 1$).

The local equations are then written:

$$\frac{\partial U}{\partial x} + \frac{\partial V}{\partial y} = 0, \tag{12.32}$$

$$U\frac{\partial U}{\partial x} + V\frac{\partial U}{\partial y} = \nu\frac{\partial^2 U}{\partial y^2}, \tag{12.33}$$

$$U\frac{\partial T}{\partial x} + V\frac{\partial T}{\partial y} = \frac{\lambda}{\rho C_p}\frac{\partial^2 T}{\partial y^2}. \tag{12.34}$$

The boundary conditions associated with the previous model are:

- for $y = 0$: $U = V = 0$, $T = C^t \left(\equiv T_p \right)$ or $\partial T / \partial y = 0$,   (12.35)
- for $y = \infty$: $U = U_\infty$, $T = T_\infty$.   (12.36)

It should be noted that for the temperature value at the wall, either one of the two conditions (12.35) can be prescribed. The first is that of an isothermal plate ($T(y = 0) = T_p$), the second of an adiabatic wall (a zero parietal heat flux). Due to the decoupling of the dynamic and thermal problems, the set of the first two equations can be solved independently of the last relation. We recognize, Cf. Sect. 12.2.1.1, the equations of the Blasius problem which has the self-similar solution:

$$U (x, y) = U_\infty f' (\eta), \quad V (x, y) = \frac{1}{2} \sqrt{\frac{\nu U_\infty}{x}} \left[\eta f' (\eta) - f (\eta) \right],$$

with $\eta = y \sqrt{U_\infty / \nu x}$, the function $f (\eta)$ being, for its part, the solution of the differential Eq. (12.17) under the boundary conditions $f (0) = f' (0) = 0$ and $f' (\infty) = 1$. By extending the self-similarity assumption to the temperature field[8] we propose to look for the solution of the thermal problem in the form of:

$$\frac{T (x, y) - T_\infty}{T_p - T_\infty} = h (\eta) .$$ (12.37)

It is then easily deduced from the enthalpy equation that the function $h (\eta)$ is governed by the following differential equation:

$$2h'' + P_r f h' = 0 ,$$ (12.38)

where $P_r = \nu / a = \mu C_p / \lambda$ denotes the Prandtl number of the fluid.

Proof The expressions of the various temperature field derivatives as deduced from the previous self-similar relation are:

$$\frac{\partial T}{\partial x} = \frac{\partial T}{\partial \eta} \frac{\partial \eta}{\partial x} = -\frac{h'}{2} \frac{y}{x} \sqrt{\frac{U_\infty}{\nu x}}, \quad \frac{\partial T}{\partial y} = h' \sqrt{\frac{U_\infty}{\nu x}} \quad \text{and} \quad \frac{\partial^2 T}{\partial y^2} = h'' \frac{U_\infty}{\nu x} .$$

By substituting in the energy equation, we obtain, after a first simplification:

$$-\frac{U_\infty}{2x} \eta f' h' + \frac{U_\infty}{2x} \left(\eta f' - f \right) h' = \frac{U_\infty}{x} \frac{\lambda}{\mu C_p} h'' .$$

Introducing the Prandtl number ($P_r = \mu C_p / \lambda$) and reducing the left hand side, we reach the expected result.

---

[8] This approach is *natural*, due to of the formal analogy of the momentum and enthalpy equations.

12.3.2.3   The Isothermal Wall Solution

The general solution of the enthalpy equation can still be put in the form:

$$\frac{h''}{h'} = -P_r \frac{f}{2} \, .$$

Now, since the function f (η) is the solution of Blasius' differential equation ($2 f''' + ff'' = 0$), we still have:

$$\frac{h''}{h'} = P_r \frac{f'''}{f''} \, , \tag{12.39}$$

whose integration is straightforward and leads to:

$$\boxed{h\left(\eta\right) = 1 - \frac{\int\limits_0^{\eta} \left[f''\left(u\right)\right]^{P_r} \mathrm{d}u}{\int\limits_0^{\infty} \left[f''\left(u\right)\right]^{P_r} \mathrm{d}u}} \tag{12.40}$$

Proof   A first integration of relation (12.39) leads to $h' = A \left[f''\right]^{P_r}$, where A stands for an integration constant. We thus deduce that:

$$h = A \int\limits_0^{\eta} \left[f''\left(u\right)\right]^{P_r} \mathrm{d}u + B \, .$$

Now, for $\eta = 0$, we have for an *isothermal wall* $h\left(0\right) = 1$, so that $B = 1$. In addition, for $\eta \to \infty$, $h\left(\eta\right) \to 0$, so that the integration constant A is given by the relation:

$$A \int\limits_0^{\infty} \left[f''\left(u\right)\right]^{P_r} \mathrm{d}u = -1 \, .$$

Remark   The set of results which we have just established clearly shows that, for a fluid whose Prandtl number would be exactly equal to the unit, there would be a perfect coincidence of the dynamic and thermal properties, in terms of the function distributions, thicknesses and parietal quantities (friction and heat flux).

EXERCISE 55   Heat transfer in low speed forced convection.

We consider a flat plate maintained at a constant temperature ($T = T_p$) parallel to a uniform flow of distinct temperature ($T = T_\infty$). It is assumed that the heat transfer between the plate and the fluid takes place in a laminar boundary layer in the low velocity forced convection regime.

QUESTION 1. Express the heat flux at any point on the plate surface as a function of the dimensionless temperature profile $h(\eta) = \frac{T-T_\infty}{T_p-T_\infty}$, where $\eta = y\sqrt{U_\infty/\nu x}$.

QUESTION 2. By adopting the following Pohlhausen interpolation $h'(0) = -0.332 P_r^{1/3}$, give the expression of the local dimensionless heat transfer coefficient (referred to the abscissa x along the plate), solely as a function of the local Prandtl and Reynolds numbers.

SOLUTION: **1.** The heat flux at the wall is obtained by:

$$\phi_p(x) = -\lambda\left(\frac{\partial T}{\partial y}\right)_{y=0} \equiv -\lambda\left(T_p - T_\infty\right) h'(0) \sqrt{\frac{U_\infty}{\nu x}}.$$

Defining the local heat transfer coefficient $\alpha(x)$ as $\phi_p(x) = \alpha(x)\left(T_p - T_\infty\right)$ we obtain:

$$\alpha(x) = -\lambda h'(0) \sqrt{\frac{U_\infty}{\nu x}}.$$

We note that the heat transfer is not constant along the plate and that the heat transfer coefficient varies according to a $1/\sqrt{x}$ law, as does the wall friction.

2. The local dimensionless coefficient or Nusselt number based on the abscissa along the plate is expressed, by definition, as:

$$Nu_x = \frac{\alpha(x) \times x}{\lambda},$$

or, with the Pohlhausen interpolation $Nu_x = 0.332 P_r^{1/3} R_{e_x}^{1/2}$.

12.3.3 Forced Convection in High Speed Isovolume Flows

12.3.3.1 The Outline of the Problem

We are now interested in the heat transfer by forced convection in the boundary layer configuration on a flat plate when the external flow velocity is high enough to generate, by stopping the fluid on the obstacle, a bulk heat input due to the viscous dissipation. However, it is assumed that the velocity change is limited to such values that the fluid compressibility is of no significance. The opposite case will be discussed in the next section. Unlike the forced convection situation without dissipative heating discussed in the previous section, the problem can now be divided into two variants, i.e., with (i) an isothermal boundary condition (a plate at a constant temperature) or (ii) adiabatic (a zero flux at the wall). We will therefore consider these two situations separately in later sections.

12.3.3.2   The Governing Equations

The boundary layer's heat transfer situation on a flat plate which we are studying fulfils the same hypotheses as stated in the previous section (Sect. 12.3.2.2), with the exception of the last hypothesis. Since the Eckert number is no longer negligible compared to the Reynolds number, the enthalpy balance equation contains a *source* term due to the viscous dissipation.
The corresponding equation expressing this balance is therefore written as:

$$ U \frac{\partial T}{\partial x} + V \frac{\partial T}{\partial y} = \frac{\lambda}{\rho C_p} \frac{\partial^2 T}{\partial y^2} + \frac{\mu}{\rho C_p} \left(\frac{\partial U}{\partial y} \right)^2 . \qquad (12.41) $$

As in the previous case, we are looking for a dimensionless temperature profile according to the self-similarity assumption. Putting $T = T(\eta)$, Eq. (12.41) is written:

$$ 2T'' + P_r f T' = -2 \frac{U_\infty^2}{C_p} P_r f''^2 . \qquad (12.42) $$

Without prejudice to the validity of the physical conclusions, but in order to simplify the mathematical expressions, we will discuss, in the next section, the solution properties of this equation for the single case $P_r = 1$ where it becomes:

$$ 2T'' + f T' = -2 \frac{U_\infty^2}{C_p} f''^2 . \qquad (12.43) $$

12.3.3.3   The General Solution with Unit Prandtl Number

It is known that the solution of the previous second-order differential equation is the sum of the integral of the corresponding homogeneous equation and a particular integral of the entire equation. The homogeneous equation is the same as in the previous case, Eq. (12.38). The corresponding solution is taken in the form:

$$ \theta_1(\eta) \equiv \frac{T(\eta) - T_\infty}{T_p - T_\infty} = A \int_0^\eta f''(u) \, \mathrm{d}u + B = A f'(\eta) + B , \qquad (12.44) $$

in order to comply with the same boundary conditions as the low speed isothermal case. As for the particular integral of the entire equation, the right hand side suggests looking for an expression of the form:

$$ \theta_2(\eta) = \frac{2C_p}{U_\infty^2} [T(\eta) - T_\infty] . \qquad (12.45) $$

Thus the differential equation in θ_2 becomes $2\theta_2'' + f\theta_2' = -4f''^2$. One can then easily verify that a particular integral of this equation is $\theta_2 = 1 - f'^2$. The general solution is therefore:

$$T(\eta) - T_\infty = (T_p - T_\infty)\,\theta_1(\eta) + \frac{U_\infty^2}{2C_p}\theta_2(\eta)\,, \qquad (12.46)$$

or else, by explaining the values of θ_1 and θ_2

$$T(\eta) - T_\infty = (T_p - T_\infty)\left[Af'(\eta) + B\right] + \frac{U_\infty^2}{2C_p}\left[1 - f'^2(\eta)\right]\,, \qquad (12.47)$$

where both A and B constants are still to be determined on the basis of the boundary conditions. However, for $\eta \to \infty$, we know that $T(\infty) = T_\infty$ and $f'(\infty) = 1$, so that $A = -B$. The generic form of the temperature profile is finally:

$$\boxed{T(\eta) - T_\infty = A\left(T_p - T_\infty\right)\left[1 - f'(\eta)\right] + \frac{U_\infty^2}{2C_p}\left[1 - f'^2(\eta)\right]} \qquad (12.48)$$

The remaining constant is to be determined according to the boundary condition at the wall and therefore depends on whether the wall is isothermal or adiabatic. It is for this reason that we are now going to address these two cases separately.

Proof Let us introduce the value $\theta_2 = 1 - f'^2$ into the equation to solve. We obtain $-4\left(f''^2 + f'f'''\right) - 2ff'f'' = -4f''^2$, or again, after reduction, $2f'\left(2f'''+ ff''\right) = 0$. This relation is identically verified because the function is, by property, a solution of the Blasius equation $2f''' + ff'' = 0$.

12.3.3.4 The Flow Past an Adiabatic Wall. The Recovery Temperature

For an adiabatic boundary condition between the plate and the flow (an athermal wall), the heat flux at the wall is zero, which results in:

$$\left(\frac{\partial T}{\partial y}\right)_{y=0} = 0\,.$$

With expression (12.48), this condition leads to:

$$0 = -A\left(T_p - T_\infty\right)f''(0) - \frac{U_\infty^2}{C_p}f'(0)f''(0)\,.$$

Since $f'(0) = 0$, the previous relation requires $A = 0$. The temperature profile on an adiabatic wall is therefore reduced to:

$$T(\eta) - T_\infty = \frac{U_\infty^2}{2C_p}\left[1 - f'^2(\eta)\right] \qquad (12.49)$$

The solution which has just been determined proves that the plate takes a *constant* temperature *different* from that of the fluid at infinity. This value, which is referred to as the *adiabatic wall temperature* or the *recovery temperature*, here is equal to:

$$T_f = T_\infty + \frac{U_\infty^2}{2C_p}. \qquad (12.50)$$

It is therefore always *higher* than that of the fluid at infinity.

ADDITIONAL INFORMATION— **Physical interpretation**—The viscous dissipation on the right hand side of the enthalpy balance Eq. (12.43) is always a *source* term. In the absence of any compressibility effect, it is therefore the only cause for the fluid kinetic heating, by stopping the flow on the obstacle. Thus, near the wall (a strong transverse velocity gradient area), the fluid takes the temperature $T_f > T_\infty$. As a result, if the plate were maintained at the temperature $T_p = T_\infty$, a heat volume would be transferred from the fluid to the plate which would no longer be athermal. In order to prevent such a heat flux, the temperature gradient normal to the wall should be cancelled. This can only be carried out by warming the plate to the same temperature as that of the fluid particles with which it is in contact.

12.3.3.5   The Temperature Field for an Isothermal Wall

In this case, the boundary condition to be prescribed for the temperature field at the wall is written as follows $T(0) = C^t\left(\equiv T_p\right)$, which results, for the solution of the temperature field $\theta_2(\eta)$, in:

$$T_p - T_\infty = A\left(T_p - T_\infty\right)\left(1 - f'(0)\right) + \frac{U_\infty^2}{2C_p}\left(1 - f'^2(0)\right).$$

We deduce that $A = 1 - \dfrac{U_\infty^2}{2C_p\left(T_p - T_\infty\right)} \equiv 1 - E_{c\Delta}$, by referring to the Eckert number $E_{c\Delta}$ based on the temperature difference $\Delta = T_p - T_\infty$. The temperature profile then takes the form:

$$\frac{T(\eta) - T_\infty}{T_p - T_\infty} = 1 - (1 - E_{c\Delta})f'(\eta) - E_{c\Delta}f'^2(\eta) \qquad (12.51)$$

We deduce, for the heat flux at the wall:

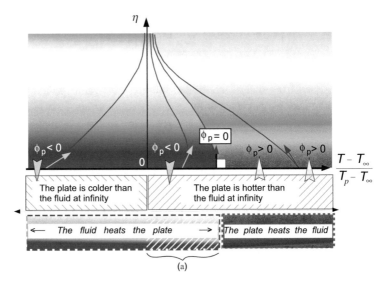

Fig. 12.15 Heat transfer with kinetic heating on an isothermal flat plate

$$\phi_p\left(x\right) = -\lambda\left(\frac{\partial T}{\partial y}\right)_{y=0} = \lambda\sqrt{\frac{U_\infty}{\nu x}}\left(T_p - T_\infty\right)\left(1 - E_{c\Delta}\right)f''\left(0\right).$$

By introducing the recovery temperature Eq. (12.50), this result can be rewritten as follows:

$$\phi_p\left(x\right) = \lambda\sqrt{\frac{U_\infty}{\nu x}}\left(T_p - T_f\right)f''\left(0\right). \tag{12.52}$$

As a direct consequence of the results which have just been demonstrated, the various following cases can be distinguished.

- For $T_p < T_\infty$, the plate is *colder* than the fluid at infinity and the heat flux is from the fluid to the plate ("*the fluid heats the plate*").
- For $T_\infty < T_p < T_f$, the plate is *hotter* than the fluid at infinity but the heat flux is still directed from the fluid to the plate, Cf. range (a) in Fig. 12.15. This situation, where a fluid *colder than the plate at infinity* is nevertheless "*heating the plate*", is therefore not paradoxical.
- For $T_p = T_f$, the plate, although *hotter* than the fluid at infinity does not transfer any heat flux to the fluid. This is the adiabatic situation.
- For $T_p > T_f$, the plate is *hotter* than the fluid at infinity and the heat flux is from the plate to the fluid. Now "*the plate actually heats the fluid*".

The temperature profiles for these different cases are displayed in Fig. 12.15.
   In summary, we can retain that:

> With kinetic heating and without compressibility effects, the plate heats the fluid when its wall temperature is higher than the recovery temperature and not the fluid temperature at infinity.

Remark It can be noted that the influence of the viscous dissipation in the expression of the temperature field (12.51) at the origin of the terms where the Eckert number $E_{c\Delta}$ stands. Thus, in the absence of kinetic heating, the solution reduces to:

$$\frac{T(\eta) - T_\infty}{T_p - T_\infty} = 1 - f'(\eta) .$$

It can be observed that this expression is indeed the one to which relation (12.40) leads, when one makes $P_r = 1$.

ADDITIONAL INFORMATION—**The case of a Prandtl number different from one**—The equations can only be analytically integrated if $P_r = 1$. For practical applications where this condition is generally not verified, the following elements can be taken into account:

(i) *The temperature distribution*: Instead of solution (12.51) for the isothermal wall, the following general form of the temperature field should be substituted:

$$\frac{T(\eta) - T_\infty}{T_p - T_\infty} = [1 - E_{c\Delta}a(P_r)] H_1(\eta, P_r) + E_{c\Delta} H_2(\eta, P_r) ,$$

which is an expression parameterized according to the Prandtl number.

(ii) *The heat flux at the wall*: The value of the local Nusselt number corresponding to the previous distribution is written as follows:

$$N_{u_x} = 0.332 P_r^{1/3} R_{e_x}^{1/2} [1 - E_{c\Delta}b(P_r)] .$$

For a graphic representation of the functions $a(P_r)$ and $b(P_r)$, one can refer to the work of Schlichting [147], page 316.

(iii) *The recovery temperature*: For $P_r \neq 1$, the expression of the recovery temperature also depends on the Prandtl number in the form:

$$T_f = T_\infty + r(P_r) \frac{U_\infty^2}{2C_p} ,$$

where a satisfactory approximation of the coefficient r, due to Pohlhausen, is simply $r(P_r) \simeq \sqrt{P_r}$. The physical interpretation of the coefficient in question is revealed when comparing the real and ideal fluid situations. In the latter case, as explained in Chap. 8, the enthalpy conservation equation results in:

$$C_p T + \frac{U^2}{2} = C^t .$$

This relation makes it possible to introduce, at any point where the velocity is zero, a *necessarily adiabatic* stagnation temperature T_a such that:

$$T_a = T_\infty + \frac{U_\infty^2}{2C_p} \, . \tag{12.53}$$

With a real fluid, the kinetic heating by stopping the flow is indeed responsible for a "*thermal buffer*" formation, which also causes, in the presence of an athermal wall, the rise in temperature of the latter. However, the thermal conductivity of the fluid makes it impossible to speak of an adiabatic stagnation condition, due to *internal* heat transfers in the fluid. Consequently, except for the singular case where $P_r = 1$, the recovery temperature, in a real fluid, remains lower than the adiabatic stagnation temperature, in an ideal fluid, all other things being equal. Therefore the coefficient r is called the temperature recovery factor. It accounts for the enthalpy rate recovered by stopping a real fluid, in reference to an equivalent adiabatic situation for an ideal fluid.

EXERCISE 56 Kinetic heating on a flat plate.

A flat plate of constant temperature T_p is positioned parallel to a uniform flow of an incompressible, viscous and thermally conductive fluid, whose velocity and temperature at infinity are respectively equal to $U_\infty = 100\,\mathrm{m/s}$ and $T_\infty = 300\,\mathrm{K}$. One will take for the heat capacity at constant pressure $C_p = 1000\,\mathrm{JKg^{-1}K^{-1}}$ and for the Prandtl number $P_r = 1$.

QUESTION 1. The plate is heated to a constant temperature *higher* by 4 degrees than that of the fluid at infinity ($T_p = 304\,\mathrm{K}$). Does the fluid heat up on contact with the plate?

QUESTION 2. What is the value of the Eckert number based on the temperature difference in an athermal situation?

SOLUTION: **1.** As we have learned, the direction of the heat flux between the plate and the fluid depends on the sign of the temperature difference $T_p - T_f$ and not on that $T_p - T_\infty$. To answer the question, the value of the recovery temperature T_f should firstly be calculated. Knowing that $T_f = T_\infty + U_\infty^2/2C_p$, Eq. (12.50), we obtain here $T_f = 305\,\mathrm{K}$. The plate therefore does not heat up the fluid, this situation only occurs here for a temperature difference between the wall and the fluid at infinity greater than 5 degrees.

2. In an adiabatic situation, we have just found that $\Delta T = 5$ degrees. The corresponding Eckert number is $E_{c\Delta} = U_\infty^2/2C_p\Delta T$, whose value is equal to one, with the numerical data for use in the exercise.

12.3.4 Some Basics on the Compressible Boundary Layer at $P_r = 1$

This last section is devoted to the compressible boundary layer. Such a topic alone would deserve a comprehensive development which is not possible in the context of this course. We will therefore limit ourselves to the presentation of some elementary properties in the case of a fluid whose Prandtl number is unity.

12.3.4.1 Busemann's Theorem

This result first used by Busemann [23] in 1935 can be stated as follows:

> • In a two-dimensional, steady, plane, compressible boundary layer, the temperature field can be deduced from that of the longitudinal velocity when the Prandtl number is equal to unity and either of the following two conditions is verified:
>
> – a zero longitudinal pressure gradient;
> – any obstacle with an athermal wall.
>
> • The general relation $T\ (U)$ is expressed in the form $T = -\dfrac{U^2}{2C_p} + \alpha U + \beta$,
> are two constants determined by the boundary conditions.

Proof Let us take again the enthalpy equation of Chap. 11, Sect. 11.5.4.3 regarding this compressible flow:

$$\rho C_p \left(U \frac{\partial T}{\partial x} + V \frac{\partial T}{\partial y} \right) = U \frac{dP_E}{dx} + \lambda \frac{\partial^2 T}{\partial y^2} + \mu \left(\frac{\partial U}{\partial y} \right)^2 .$$

We are looking for the assumptions under which it admits a solution of the form $T = T\ (U)$, where U is the longitudinal component of the velocity vector. We have:

$$\frac{\partial T}{\partial x} = T' \frac{\partial U}{\partial x}, \quad \frac{\partial T}{\partial y} = T' \frac{\partial U}{\partial y}, \quad \frac{\partial^2 T}{\partial y^2} = T' \frac{\partial^2 U}{\partial y^2} + T'' \left(\frac{\partial U}{\partial y} \right)^2 .$$

Thus the energy equation becomes:

$$\rho C_p \left(U \frac{\partial U}{\partial x} + V \frac{\partial U}{\partial y} \right) T' = U \frac{dP_E}{dx} + \lambda T' \frac{\partial^2 U}{\partial y^2} + \left(\lambda T'' + \mu \right) \left(\frac{\partial U}{\partial y} \right)^2 .$$

Now, according to the momentum equation, we know that, for this same flow:

$$\rho \left(U \frac{\partial U}{\partial x} + V \frac{\partial U}{\partial y} \right) = -\frac{dP_E}{dx} + \mu \frac{\partial^2 U}{\partial y^2} .$$

By a linear combination it is deduced that:

$$0 = \left(U + C_p T'\right) \frac{dP_E}{dx} + \left(\lambda - \mu C_p\right) T' \frac{\partial^2 U}{\partial y^2} + \left(\lambda T'' + \mu\right) \left(\frac{\partial U}{\partial y}\right)^2 . \qquad (12.54)$$

Under the condition $P_r = 1$, we have $\lambda = \mu C_p$ so that the equation becomes:

$$\left(U + C_p T'\right) \frac{dP_E}{dx} = -\left(\lambda T'' + \mu\right) \left(\frac{\partial U}{\partial y}\right)^2 .$$

It obviously reduces to $\lambda T'' + \mu = 0$ in the absence of a longitudinal external pressure gradient. Otherwise $(dP_E/dx \neq 0)$, the relation imposes that $U + C_p T' = 0$. It proves that the temperature field is a function of the velocity with necessarily for $y = 0$, $T'(0) = 0$—athermal wall—since $U(0) = 0$.

12.3.4.2   The Recovery Temperature

In the presence of an athermal wall, the condition $T'(0) = 0$ leads to $\alpha = 0$, so that Busemann's relation becomes $T = -U^2/2C_p + \beta$. By requiring the conditions of linking to the external solution—$T(\infty) = T_E(x)$, $U(\infty) = U_E(x)$, we obtain:

$$T(x, y) - T_E(x) = \frac{1}{2C_p} \left[U_E^2(x) - U^2(x, y)\right]. \qquad (12.55)$$

This equation proves that the wall of the obstacle $(U = 0)$ takes the recovery temperature $T_f(x)$ whose value is:

$$T_f(x) = T_E(x) + \frac{U_E^2(x)}{2C_p} . \qquad (12.56)$$

By introducing the local external Mach number $M_E = U_E/a_E$ and with $a_E^2 = (\gamma - 1) C_p T_E$ we still obtain:

$$T_f(x) = T_E(x) \left[1 + \frac{\gamma - 1}{2} M_E^2\right] \qquad (P_r = 1) . \qquad (12.57)$$

ADDITIONAL INFORMATION—**Prandtl number different from unity**—As in the incompressible case, the recovery temperature obtained with $P_r = 1$ and the adiabatic stagnation temperature are identical. For all other values of the Prandtl number, a *temperature recovery factor* should be applied. The recovery temperature is then expressed as:

$$T_f\,(x) = T_E\,(x)\left[1 + r\,\frac{\gamma - 1}{2}\,M_E^2\right] \quad (P_r \neq 1)\,. \tag{12.58}$$

A range of experimental work on flat plates and cones at various incidences has revealed that the temperature recovery factor could be correctly estimated, for the air, as in the case of the incompressible regime by $r \simeq \sqrt{P_r}$.

EXERCISE 57 The cruising temperature at the nose of the *Concorde* aircraft.

In the supersonic cruising regime, *Concorde* Aircraft was flying at a Mach number $M_\infty = 2$, at an altitude where the air temperature is close to $T_\infty = -50^oC$.

  QUESTION   Knowing that the recovery temperature on a substantially conical surface can be estimated by $T_f = T_\infty\left(1 + r\,(\gamma - 1)\,M_\infty^2/2\right)$, assess the value of the recovery temperature at the nose of the aircraft. We will take as the recovery factor value $r = 0.85$).

SOLUTION: With the numerical data for use in the exercise, we obtain $T_f \approx 102^oC$. The measurements give a slightly higher result ($\sim 115^oC$).

EXERCISE 58 Kinetic heating in a compressible boundary layer.

A flat plate is positioned at a zero incidence in a uniform flow of air ($\gamma = 1.4$ and $P_r = 0.72$) whose temperature is $T_\infty = 0^oC$. Two values of the Mach number are considered, respectively equal to $M_\infty = 1$ and $M_\infty = 3$.

  QUESTION 1.   What is the adiabatic stagnation temperature for each Mach number?

  QUESTION 2.   Calculate the kinetic heating of the assumed athermal plate in each case.

SOLUTION: **1.** The adiabatic stagnation temperature corresponds to the isentropic stagnation enthalpy in an ideal fluid, namely $T_a = T_\infty\left(1 + (\gamma - 1)\,M_\infty^2/2\right)$. With the numerical data for use in the exercise we obtain $T_a = 1.2T_\infty \approx 327.5$ K ($M_\infty = 1$) and $T_a = 2.8T_\infty \approx 764.5$ K ($M_\infty = 3$).

2. In an athermal situation, the kinetic heating of the plate is equal to that obtained in an adiabatic stagnation situation, to within the recovery factor. Indeed, we have:

$$\frac{T_f - T_\infty}{T_\infty} = r\,\frac{\gamma - 1}{2}\,M_\infty^2 \equiv r\,\frac{T_a - T_\infty}{T_\infty}\,.$$

With the relevant numerical data, we obtain $\Delta T = 46.5^o$ and $\Delta T = 417.5^o$ respectively.

12.3.4.3   Heat Transfer in a Boundary Layer on an Isothermal Flat Plate

In the case of an isothermal flat wall ($T(0) = T_p = C'$), positioned without incidence in a uniform flow of Mach number M_∞ and temperature T_∞ at infinity, Busemann's theorem becomes:

$$\left(1 - \frac{T_p}{T_\infty}\right) \frac{U}{U_\infty} + \frac{\gamma - 1}{2} M_\infty^2 \frac{U}{U_\infty} \left(1 - \frac{U}{U_\infty}\right). \tag{12.59}$$

From this, the parietal heat flux is expressed as:

$$\phi_p = -\lambda \left(\frac{\partial T}{\partial y}\right)_{y=0} = -\lambda T'(0) \left(\frac{\partial U}{\partial y}\right)_{y=0},$$

or, after calculation:

$$\phi_p = -\lambda \frac{T_\infty}{U_\infty} \left[\frac{T_\infty - T_p}{T_\infty} + \frac{\gamma - 1}{2} M_\infty^2\right] \left(\frac{\partial U}{\partial y}\right)_{y=0}.$$

The sign of the heat flux which conditions the heat transfer direction is that of the term in square brackets. It can therefore be concluded that for $T_p - T_\infty < U_\infty^2/2C_p$, or what amounts to the same $(T_p - T_\infty)/T_\infty < (\gamma - 1) M_\infty^2/2$, the heat flux is *negative*, which means that the fluid cools on contact with the plate, the heat transfer being directed from the solid to the fluid. The conclusion is the opposite for $T_p - T_\infty > U_\infty^2/2C_p$, as outlined in the following Figure (Fig. 12.16).

Fig. 12.16 Schematic for heat transfer in a compressible boundary layer on an isothermal flat plate for $Pr = 1$

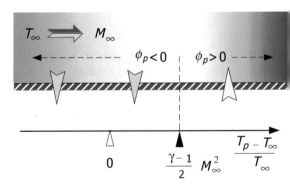

12.4 The Two-Dimensional Thin Shear Layers and Free Shear Flows

12.4.1 The Thin Shear Layer Approximation in Free Flows

As explained in Chap. 11 Sect. 11.3.1, the two-dimensional boundary layer assumptions for a two-dimensional boundary layer are based on:

– a sufficiently high global Reynolds number of the flow;
– a distinctive geometry, with preferential directions for the advection and diffusion;
– the development of both advective transport and diffusive transfer on the same time scale.

If these conditions are indeed satisfied along the wall of any streamlined body correctly positioned in the flow[9] they can also be in other configurations which do not necessarily involve the presence of a solid obstacle. We then speak of the *thin shear layer approximation* for *free flows*, a category which includes jets, wakes and mixing layers.

(a) **The jet**—An *ideal* jet flow consists of the injection of a momentum flow rate into an infinite quiescent atmosphere from a point source. This last condition is never carried out in practice, where the jet exhausts from a *finite section* nozzle or convergent, the momentum flux emission being necessarily associated in this case with that of a mass flow rate.[10] Consequently the thin shear layer approximation only applies, in this case, beyond a certain distance downstream of the emission, see the left Fig. 12.17. In this slender zone, we can then consider that the advective transport is coupled with the laminar jet expansion due to the diffusive action of the *molecular agitation* during the same time scale. In the case of a turbulent jet—which will not be studied here—this equilibrium condition remains unchanged by applying it, this time, to the diffusion by the *turbulent agitation*. The only modification results from the jet's diffusion being driven by *continuous motions*, much more intense than those occuring at the molecular level. This explains the much higher expansion rate of the turbulent jet.

(b) **The wake**—Any fixed obstacle disposed in a moving fluid behaves, towards the flow, like a '*sink in momentum*'. In this sense, the wake configuration can be thought of as the opposite of the previous configuration. Thus, at a sufficient distance downstream of the obstacle, see the central diagram in Fig. 12.17, the molecular diffusion gradually tends to make up for the velocity deficit in a wake zone where the thin shear layer approximation becomes applicable. As regards the influence of the *laminar* or *turbulent* nature of the regime on the expansion law of the wake, the same remarks as those previously introduced for the jet, also apply here.

By this is meant a situation where there is no detachment of the fluid threads from the body in question.
[10] With the exception of the *synthetic jet*, see Additional Information.

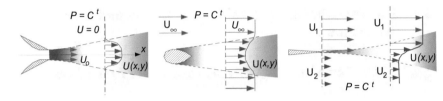

Fig. 12.17   Schematics for the jet, wake and mixing layer flows

(c) The mixing layer—*Ideally*, the mixing layer is the region which develops between two parallel and uniform flows of different modulus velocities.[11] It is therefore associated with a *step* in the momentum module. In practice, the generation of such a step between parallel currents requires the presence of a solid wall which applies the same zero velocity condition to both flows, with, possibly, the development of boundary layers for each of them, see the diagram on the right in Fig. 12.17. Without being infinite, the transverse velocity gradient nevertheless remains high enough to lead, by diffusion, to the spreading of the mixing layer, as the downstream advective flow develops (Fig. 12.18).

ADDITIONAL INFORMATION—**The synthetic Jet Actuator**—Active flow control is a means for advanced performance optimisation in a variety of applications.

In external aerodynamics, for example, it has been documented that, by appropriate boundary layer manipulation, it is possible to delay or prevent the boundary layer detachment. The control techniques involve different *actuators*, among which *synthetic jets.*

A synthetic jet is obtained by means of a device composed of a cavity with an orifice, which encloses a variable volume of air driven by the deformation of an oscillating diaphragm. The diaphragm vibrations (at a frequency ranging between 100 and 1000 Hz) generate a succession of toric vortices right in front of the orifice, which constitute the *synthetic jet*. In contrast to the *usual jet*, there is no mass flow rate ejection during an oscillation period, the cavity expelling and drawing in the same volume of air over each half period.

Thus, the fluid motion in a *synthetic jet* is very different from that in the *usual jet* where a mass flow rate comes out of a nozzle. In particular, for the synthetic jet, the velocity results solely from vortex induction mechanisms of Biot & Savart type (Cf. Chap. 2, Sect. 2.4.8), mechanisms which may be responsible for the specific behaviour of the vortices themselves, as described, for example, in Guyon's *et al.* work [62].

---

[11] At a later stage, we will consider the *space evolving* mixing zone. There is also a *time evolving* mixing layer configuration which is not addressed here.

Fig. 12.18 Schematic diagram of a synthetic jet type actuator

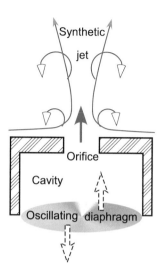

12.4.2   The Plane Free Jet

12.4.2.1   The Local Motion Equations

According to what has been previously mentioned, the exit conditions of an *ideal plane jet* would be an injection of a momentum flow rate through an *infinitely thin slot*. In practice, for a x-directed jet, we simply assume a high aspect ratio between the height (according to y) and the width (according to z) of the rectangular outlet section, in order to validate the two-dimensional flow character ($\partial/\partial z \equiv 0$). The jet is supposed to develop in an infinite environment of constant static pressure P_a. It then turns out from Prandtl's third equation ($\partial P/\partial y = 0$) that, in the jet flow region where the boundary layer approximations apply, the static pressure is also constant and equal to P_a. Therefore, in a steady state and for a weightless fluid, the local motion equations of the jet according to the thin shear layer approximation are written:

$$\frac{\partial U}{\partial x} + \frac{\partial V}{\partial y} = 0 , \tag{12.60}$$

$$U\frac{\partial U}{\partial x} + V\frac{\partial U}{\partial y} = \nu\frac{\partial^2 U}{\partial y^2} . \tag{12.61}$$

Regarding the boundary conditions and taking by reason of symmetry the $y = 0$ coordinate as the jet axis, we have, $U(x,0) = U_{axis}(x)$, $(\partial U/\partial y)_{y=0} = 0$ and $V(x,0) = 0$. In addition, given that the atmosphere is assumed to be free of any colinear motion with the jet axis (co-flow), we derive the following condition $U(x,\infty) = 0$.

By multiplying the momentum equation throughout by U, we obtain the kinetic energy balance equation, which can be put in the form:

$$U\frac{\partial}{\partial x}\left(\frac{U^2}{2}\right) + V\frac{\partial}{\partial y}\left(\frac{U^2}{2}\right) = \nu\frac{\partial}{\partial y}\left(U\frac{\partial U}{\partial y}\right) - \nu\left(\frac{\partial U}{\partial y}\right)^2, \qquad (12.62)$$

where one identifies the first term of the right hand side as the power of the external viscosity forces and the second as the viscous dissipation.

12.4.2.2   The Integral Properties

By integrating with respect to y the local equations, the three following so-called *integral* properties can be derived:

In a free isovolume plane jet under steady-state conditions:
- the *mass* flow rate increases in the forward direction;
- the *momentum* flow rate is constant in the forward direction;
- the *kinetic energy* flow rate decreases in the forward direction.

Proof We begin with the second proposition. By integration of the momentum equation, we obtain

$$\int_0^\infty U\frac{\partial U}{\partial x}\mathrm{d}y + \int_0^\infty V\frac{\partial U}{\partial y}\mathrm{d}y = \nu\left[\frac{\partial U}{\partial y}\right]_0^\infty.$$

Owing to the prescribed boundary conditions, the right hand side of this relation is zero (no shear in the surrounding fluid in the y-direction at infinity and maximum velocity on the jet axis by symmetry). An integration by parts of the second integral on the left hand side yields, after taking into account the continuity equation:

$$\int_0^\infty V\frac{\partial U}{\partial y}\mathrm{d}y = [UV]_0^\infty - \int_0^\infty U\frac{\partial V}{\partial y}\mathrm{d}y = \int_0^\infty U\frac{\partial U}{\partial x}\mathrm{d}y.$$

Using this result, the integral momentum balance equation is written:

$$2\int_0^\infty U\frac{\partial U}{\partial x}\mathrm{d}y \equiv \frac{\mathrm{d}}{\mathrm{d}x}\int_0^\infty U^2\mathrm{d}y = 0, \qquad (12.63)$$

which demonstrates the proposition.
For the energy Eq. (12.62), we obtain in the same way:

$$\int_0^\infty \frac{U}{2}\frac{\partial U^2}{\partial x}dy + \int_0^\infty \frac{V}{2}\frac{\partial U^2}{\partial y}dy = \nu\left[U\frac{\partial U}{\partial y}\right]_0^\infty - \nu\int_0^\infty \left(\frac{\partial U}{\partial y}\right)^2 dy,$$

where the contribution of the $\nu\left[U\frac{\partial U}{\partial y}\right]_0^\infty$ term is zero owing to the boundary conditions. Applying to the second term of the left hand side a process similar to that applied to the momentum equation (integration by parts and substitution by the continuity equation), the final result is:

$$\int_0^\infty \left[U\frac{\partial}{\partial x}\left(\frac{U^2}{2}\right) + \frac{U^2}{2}\frac{\partial U}{\partial x}\right]dy \equiv \frac{d}{dx}\int_0^\infty U\frac{U^2}{2}dy = -\nu\int_0^\infty \left(\frac{\partial U}{\partial y}\right)^2 dy,$$

(12.64)

where the right hand side expression clearly demonstrates the negative value of the result.

Finally, the continuity equation directly leads to the following result:

$$\int_0^\infty \frac{\partial U}{\partial x}dy = -\int_0^\infty \frac{\partial V}{\partial y}dy \equiv -[V]_0^\infty = -V_E(x),$$

by denoting by $V_E.(x)$ the entrainment velocity at the jet edge, as discussed in Chap. 11, Sect. 11.4.1. It only remains to demonstrate that $V_E(x) < 0$ or else:

$$\int_0^\infty \frac{\partial U}{\partial x}dy > 0, \quad \text{knowing that} \quad \int_0^\infty U\frac{\partial U}{\partial x}dy = 0 \quad \text{and} \quad \int_0^\infty U^2\frac{\partial U}{\partial x}dy < 0.$$

Now, for any given $x \equiv x_0$, the $U(x_0, y)$ function is monotonically decreasing[12], and conditions (12.63) and (12.64) necessarily lead to variations in the integrating factors of the respective integrals such as outlined in Fig. 12.19a and b.

It is then clear that, by dividing the values of the function $[U\partial U/\partial x](y)$ by those of $U(y)$, we obtain a graph such as (c) for the term $\frac{\partial U}{\partial y}$ and thus a positive area under the curve.

12.4.2.3   The Self-Similar Solution

By extension of the self-similar assumption introduced for solving the boundary layer equations on an obstacle (Sect. 12.1.1.2), we are looking for a solution of the plane jet motion equations based on a stream function of the form:

---

[2] This *intuitive* result will be demonstrated at a later stage.

Fig. 12.19 Distribution of
the axial velocity and mass
flow **a**, momentum **b** and
kinetic energy **c** integrants

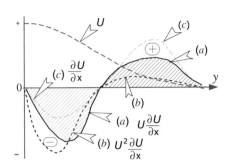

$$\psi = U_{axis}\,(x)\,g\,(x)\,f\,(\eta) \quad \text{with} \quad \eta = y/g\,(x)\,, \tag{12.65}$$

where $g\,(x)$ denotes a length scale typical of the lateral jet expansion. Under this assumption, the following results can be demonstrated.

- The center-line velocity decreases according to the law:
  $U_{axis}\,(x) = 3\,(J\nu/2x)^{1/3}$, where ρJ is the momentum flux per unit span and ν the fluid kinematic viscosity;
- The jet expansion law is of the form $g\,(x) \propto x^{2/3}$;
- The velocity distribution is as follows:

$$U\,(x,y) = \frac{1}{2}\left(\frac{3J^2}{4\nu x}\right)^{1/3}\left[1 - th^2\left(\frac{y}{2}\left(\frac{J}{6\nu^2 x^2}\right)^{1/3}\right)\right]$$

Proof The streamwise evolution laws—By directly using the calculations explained in detail in Sect. 12.1.1.4, we can immediately write, by introducing the self-similarity relations in the momentum equation:

$$g^2 U'_{axis}\left(f'^2 - ff''\right) - U_{axis}gg'ff'' = \nu f'''\,.$$

This equation then leads to the compatibility conditions:

$$U_{axis}\,(x) \propto x^{\frac{\beta}{2\alpha-\beta}} \quad \text{and} \quad g\,(x) \propto x^{\frac{\alpha-\beta}{2\alpha-\beta}}\,,$$

by which it becomes $\dfrac{\beta}{2\alpha - \beta}f'^2 - \dfrac{\alpha}{2\alpha - \beta}ff'' = \nu f'''$.

Now taking into account the momentum flux conservation, we can write

$$\int_0^\infty U^2 dy = \int_0^\infty U^2_{axis}\,(x)\,f'^2\,(\eta)\,g\,(x)\,d\eta = C'\,, \text{ so that } U^2_{axis}\,(x) \times g\,(x) = C'.$$

This last relation proves that $\beta = -\alpha$. Consequently, the streamwise variation laws of the center-line velocity and jet thickness, according to the self similarity assumption, are necessarily of the form $U_{axis}\,(x) \propto x^{-1/3}$ and $g\,(x) \propto x^{2/3}$.

The axial velocity profile—In line with the previous results, we are of course setting $U_{axis} = ax^{-1/3}$ and $g(x) = x^{2/3}$ where a is a positive constant. The solving equation of the function $f(\eta)$ now becomes $f'^2 + ff'' + \frac{3\nu}{a} f''' = 0$. After a first integration we obtain $ff' + \frac{3\nu}{a} f'' = 0$, the integration constant being zero since $f(0) = f'(0) = 0$. A second integration results in:

$$\frac{f^2}{2} + \frac{3\nu}{a} f' = \frac{3\nu}{a} ,$$

the value of the constant notably resulting from the condition $f'(0) = 1$. Taking $\gamma^2 = a/6\nu$, the previous differential equation takes the form:

$$d\eta = \frac{df}{1 - \gamma^2 f} ,$$

whose integration leads to $\eta = \int_0^f \frac{d\upsilon}{1 - \gamma^2 \upsilon} = \frac{atanh(\gamma f)}{\gamma} .$

Finally the function f is written $f(\eta) = \frac{1}{\gamma} tanh(\gamma \eta) .$

The dimensionless velocity profile is deduced as $\frac{U}{U_{axis}} = f'(\eta) = 1 - tanh^2(\gamma \eta) .$

The constant γ can now be expressed from the momentum flux value per unit of jet span. Indeed, we have

$$\rho J = 2 \int_0^\infty \rho U^2 dy = 2\rho a^2 \int_0^\infty f'^2(\eta) \, d\eta = \frac{2\rho a^2}{\gamma} \int_0^\infty \left[1 - tanh^2(u)\right]^2 du = \frac{4\rho a^2}{3\gamma} ,$$

from which we can deduce $2a = \left(3J^2/4\nu\right)^{1/3}$, which finally leads to the expected results.

12.4.3   The Plane Wake

12.4.3.1   The Local Motion Equations

Under the general assumptions of a steady, two-dimensional plane motion of an incompressible weightless fluid, we now consider the wake region developing downstream of a fixed obstacle in a flow of uniform velocity U_∞ at infinity. This region is characterized by a *velocity deficit* which is denoted by

$$U_D(x) = U_\infty - U(x, y) , \qquad (12.66)$$

where $U(x, y)$ stands for the local longitudinal component of the wake velocity vector. We are interested in the region at a sufficient distance from the obstacle so that:

(a) the structural specificities of the near wake can be discarded,
(b) the velocity deficit is small enough to allow a linear approximation of the equations.

By introducing relation (12.66) in Prandtl's equation, we obtain:

$$(U_\infty - U_D) \frac{\partial (U_\infty - U_D)}{\partial x} + V \frac{\partial (U_\infty - U_D)}{\partial y} = U_\infty \frac{dU_\infty}{dx} + \nu \frac{\partial^2 (U_\infty - U_D)}{\partial y^2},$$

which, at first order in U_D is simply reduced to:

$$U_\infty \frac{\partial U_D}{\partial x} = +\nu \frac{\partial^2 U_D}{\partial y^2}. \tag{12.67}$$

12.4.3.2   The Drag Expression

In a very general way, *i.e.*, without taking into consideration the detail of the obstacle's geometry, we can derive the following result:

> The drag of a body in a steady, two-dimensional plane flow of a weightless fluid whose velocity is uniform at infinity is, per span unit, equal to
>
> $$\boxed{D = \rho U_\infty Q_D} \tag{12.68}$$
>
> where Q_D is the flow rate deficit in the distant wake.

Proof This result is demonstrated by applying Euler's momentum theorem to the fluid within a control domain bounded by a $\{ABCD\}$ surface surrounding the obstacle at a sufficient distance and the obstacle contour $\{\Sigma\}$, see Fig. 12.20.

Let us first express the momentum flux through this domain. Assuming the absence of any blowing or suction effect at the obstacle wall, we have, by projecting along x and per span unit:

$$M = \int_{ABCD} \rho U \, (\mathbf{V.n}) \, dl = -\int_{AB} \rho U_\infty^2 dy + \int_{CD} \rho U^2 (x, y) \, dy + \int_{BC+DA} \rho U_\infty V_E dx.$$

In addition, the mass flow rate conservation allows the integral of the entrainment velocity to be expressed as:

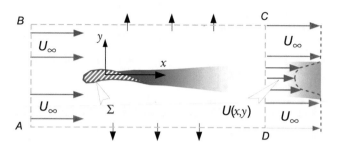

Fig. 12.20 Control domain for the calculation of the drag around an obstacle

$$\int\limits_{BC+DA} \rho V_E \mathrm{d}x = \int\limits_{AB} \rho U_\infty \mathrm{d}y - \int\limits_{CD} \rho U\,(x,y)\,\mathrm{d}y \equiv \int\limits_{-h}^{+h} \rho U_D \mathrm{d}y\,.$$

Thus the momentum flux is finally expressed as follows:

$$\frac{M}{\rho} = \int\limits_{-h}^{+h} \left[-U_\infty^2 + U^2 + U_\infty\,(U_\infty - U) \right] \mathrm{d}y = \int\limits_{-h}^{+h} U\,(U - U_\infty)\,\mathrm{d}y$$

$$\equiv -\int\limits_{-h}^{+h} U_D\,(U_\infty - U_D)\,\mathrm{d}y\,.$$

Taking the section at a sufficient distance from the obstacle to assume a linear approximation, we obtain for $h \to \infty$:

$$M = -\rho U_\infty \int\limits_{-\infty}^{+\infty} U_D \mathrm{d}y = -\rho U_\infty Q_D \quad \text{with} \quad Q_D \equiv \int\limits_{-\infty}^{+\infty} U_D \mathrm{d}y\,, \qquad (12.69)$$

where Q_D denotes the flow rate deficit in the far wake.

As for the resultant force along the flow direction at infinity, it is none other than the opposite of the obstacle drag (D). Indeed, along any closed contour $\{ABCD\}$ the resultant pressure force is zero (owing to the constant pressure) and the viscous stresses are negligible according to the boundary layer approximations. Euler's theorem projection along x finally results in the relation $-\rho U_\infty Q_D = -D$.

12.4.3.3 The Self-Similar Solution

In a two-dimensional symmetrical plane wake at a sufficient distance from the obstacle, it can be demonstrated that a self-similar region exists where:

- the wake expansion is parabolic as a function of downstream distance;
- the longitudinal velocity component is expressed as

$$
U\left(x,y\right)=U_{\infty}\left[1-\frac{D}{2\rho U_{\infty}^2 L}\sqrt{\frac{Re_L}{\pi}}\sqrt{\frac{L}{x}}\,e^{-\frac{U_{\infty}y^2}{4\nu x}}\right],
$$

where L is a characteristic length scale, D the drag of the obstacle and U_∞ the fluid velocity at infinity.

Proof Let us define, according to the self-similarity assumption $U_D\left(x,y\right)=U_{ref}\left(x\right)\times f\left(\eta\right)$ where η still denotes the normalized dimensionless transverse coordinate $(\eta=y/g\left(x\right))$. It is clear that:

$$
\frac{\partial U}{\partial x}=U'_{ref}f-U_{ref}\frac{g'}{g}\eta f'\quad\frac{\partial U}{\partial y}=U_{ref}\frac{f'}{g}\quad\frac{\partial^2 U}{\partial y^2}=U_{ref}\frac{f''}{g^2}\,.
$$

Thus, by substituting in the wake momentum equation we obtain:

$$
\left[\frac{U'_{ref}}{U_{ref}}g^2\right]f-\left[gg'\right]\eta f'=\frac{\nu}{U_\infty}f''\,,
$$

from which we derive the following compatibility conditions:

$$
\frac{U'_{ref}}{U_{ref}}g^2=C^t\;(\equiv\alpha)\quad\text{and}\quad gg'=C^t\;(\equiv\beta)\,.
$$

We immediately deduce that $g=\sqrt{2\beta x}$, by choosing the abscissa origin such that $g\left(0\right)=0$. As a result, it follows that $U_{ref}\left(x\right)\propto x^{\alpha/2\beta}$. An additional relation between the constants α and β exists because the obstacle drag is necessarily independent of x. Using the integral expression previously determined for the drag, Eq. (12.69) we deduce that:

$$
\rho U_\infty U_{ref}\left(x\right)\times g\left(x\right)\times\int_{-\infty}^{+\infty}f\left(\eta\right)\mathrm{d}\eta=C^t\,,
$$

which finally leads to $\alpha=-\beta$. As a result $U_{ref}\left(x\right)\propto x^{-1/2}$. Hence, the wake expansion in the self similar zone is parabolic ($g\left(x\right)\propto\sqrt{x}$), the reference velocity obeying a decay law in $1/\sqrt{x}$. Finally, the velocity deficit profile is the solution of the differential equation:

$$
f+\eta f'=-\frac{\nu}{\beta U_\infty}f''.
$$

By choosing $\beta = 1/2$ in order to find the Blasius variable change ($\eta = y\sqrt{U_\infty/\nu x}$, see Sect. 12.2.1.1), this differential equation is put in the form $(\eta f)' = -2f''$ whose integration directly results in $\eta f = -2f'$, the integration constant being zero by reason of symmetry ($f'(0) = 0$). A new integration finally leads to the solution $f(\eta) = Ae^{-\eta^2/4}$. It is easy to observe that the constant A can be expressed in terms of the drag, since:

$$D = A\rho U_\infty U_{ref} g \int_{-\infty}^{+\infty} e^{-\eta^2/4} \, d\eta \, .$$

Knowing that the value of the integral is none other than $2\sqrt{\pi}$, we obtain, by taking the reference velocity in the dimensionless form $U_{ref} = U_\infty \sqrt{L/x}$, where L is an arbitrary length scale:

$$A = \frac{D}{2\rho U_\infty^2 L} \sqrt{\frac{Re_L}{\pi}} \, . \tag{12.70}$$

The expected solution can then be easily deduced from the results which have just been demonstrated.

Remark One might be surprised that the solution could be provided independently of the obstacle geometry. This statement also applies to the jet flow where the velocity field can be determined without the need for a detailed specification of the emission conditions. In both cases, the explanation is the same: the analytical resolution only applies to an idealized configuration where the characteristic flow width (jet or wake) is zero at the origin—$g(0) = 0$. Such a condition is never reached in practice, so that any experimental verification of the self similarity laws can only be carried out within a translation of the flow origin along the x-axis. It is for this reason that one introduces experimentally the concept of fictitious origin or pole of the flow (x_0). The expansion law, for example, is then rewritten in the generic form $g(x) \propto (x - x_0)^m$ where only the exponent m remains constant for a similar flow configuration.

EXERCISE 59 The drag of a flat plate.

A flat plate of length L is positioned parallel to a uniform flow of velocity U_∞ at infinity.

QUESTION 1. Using the Blasius solution results, derive the expression of the plate drag per unit span as a function of the Reynolds number $U_\infty L/\nu$.

QUESTION 2. By applying Euler's theorem to a fluid domain to be specified, determine the expression of this same drag as a function of the momentum thickness. Then retrieve the result of the first question.

QUESTION 3. Express the longitudinal velocity field in the wake zone downstream of the boundary layers beyond the plate.

SOLUTION: **1.** Relation (12.22) provides

$$\tau_p\,(x) = 0.332\frac{\rho U_\infty^2}{\sqrt{R_x}} = 0.332\rho U_\infty^2\sqrt{\frac{\nu}{U_\infty}}\frac{1}{\sqrt{x}}\,.$$

For this obstacle, the entire drag is reduced to the viscous drag and is therefore equal per span unit:

$$D = 2\int_0^L \tau_p\,(x)\,\mathrm{d}x = 0.664\,\rho U_\infty^2\sqrt{\frac{\nu}{U_\infty}}\left[2\sqrt{x}\right]_0^L\,,$$

or finally $\dfrac{D}{2\rho U_\infty^2} = 0.664\dfrac{L}{\sqrt{R_{e_L}}}$, where $R_{e_L} = U_\infty L/\nu$ is the Reynolds number based on the plate length.

2. The half-domain to be used for applying Euler's theorem to this particular case is the rectangle whose length coincides with one face of the plate and height $\delta\,(L)$, the boundary layer thickness at the plate end. By using the exact general expression, one can write without re-demonstrating:

$$D = \int_{-\delta(L)}^{+\delta(L)} \rho U\,(U_\infty - U)\,\mathrm{d}y \quad \text{or} \quad \frac{D}{\rho U_\infty^2} = \int_{-\delta(L)}^{+\delta(L)} \frac{U}{U_\infty}\left(1 - \frac{U}{U_\infty}\right)\mathrm{d}y = 2\theta\,(L)\,,$$

where θ is the momentum thickness defined in Chap. 11, Sect. 11.2.1.4. Now for the boundary layer on a flat plate with a zero pressure gradient, the Blasius solution Eq. (12.19) provides:

$$\theta\,(L) = 0.664\sqrt{\frac{\nu L}{U_\infty}}\,.$$

The expression obtained in the previous question can equally be found in this way.

3. The velocity field in the wake is directly obtained by specifying the constant A of the general solution to the present flow situation. Using relation (12.68), it simply results that $A = 0.664/\sqrt{\pi} \approx 0.375$. The solution is finally written:

$$U\,(x, y) = U_\infty\left[1 - 0.375\sqrt{\frac{L}{x}}\,e^{-\frac{U_\infty y^2}{4\nu x}}\right]\,.$$

Appendix A
Thermodynamics

A.1 General Background

Our scope is limited to recalling the basic thermodynamic concepts which are used in the main body of the present work. These are therefore concepts and results of the *equilibrium thermodynamics* related to energy transformations between various equilibrium states of the system.[1]

Such an equilibrium assumption could appear to be in contradiction with the application to fluid flows, which is the subject of this work. In this case indeed, the medium is *out of equilibrium*, its properties (velocity, pressure, temperature, etc.) being non-uniformly distributed in space and unsteadily in time. Let us recall, however, that all fluid motions considered here are under the *local thermodynamic equilibrium*[2] assumption which allows the transposition of the entire corpus of classical thermodynamics.

A.2 The State Variables and Functions

For the purposes of this course, we limit ourselves exclusively to *single-phase* gas or liquid media, chemically *non-reactive* and *homogeneous*, i.e., formed from a single chemical species or a mixture of invariable composition. The following reminders therefore only pertain to this particular situation. For a more general and extensive presentation, the works of Bruhat [21], Roy [140] and Diu et al. [49] can be referred to, among many others.

In this regard, the transformations of matter, which are only involved through diffusive mass transfers, are not concerned.

See first chapter, Sect. 1.2.3.5.

© The Editor(s) (if applicable) and The Author(s), under exclusive license to Springer Nature Switzerland AG 2022
P. Chassaing, *Fundamentals of Fluid Mechanics*,
https://doi.org/10.1007/978-3-031-10086-4

In terms of the equilibrium thermodynamics for this fluid medium, we will consider as state variables[3]:
- the density,
- the pressure,
- the temperature.

With regard to the state functions, we will consider the following *specific* quantities, i.e., defined per unit of mass:
- the internal energy,
- the enthalpy,
- the entropy.

A.3 The Equation of State

Using Huerre's terminology[4] [76], we define:

$$- \text{ the } \textit{canonical} \text{ equation of state }, \quad e = e\,(\rho, s)\,, \tag{A.1}$$

$$- \text{ the } \textit{thermal} \text{ equation of state }, \quad e = e\,(\rho, T)\,. \tag{A.2}$$

A.4 The Pressure, Temperature and Characteristic Coefficients

The partial derivatives of the state equations make it possible to express the following state variables[5, 6]:

$$- \text{ the } \textit{thermodynamic} \text{ pressure }, \quad P = \rho^2 \left(\frac{\partial e}{\partial \rho}\right)_s\,; \tag{A.3}$$

$$- \text{ the } \textit{temperature}, \quad T = \left(\frac{\partial e}{\partial s}\right)_\rho\,; \tag{A.4}$$

---

[3] In classical fluid mechanics, these *thermodynamic state variables* are to be considered as *functions of the flow field coordinates and time.*

[4] The thermal state equation according to this author is called the calorific state equation by Ryhming, [143].

[5] The subscript means that the corresponding quantity is taken as a constant.

[6] This qualifier makes it possible to distinguish it from the mechanical pressure, as defined in Chap. 3, Sect. 3.2.2.3.

The following characteristic coefficients are introduced in a similar way:

- the *isothermal* compressibility coefficient, $\quad \chi_T = \dfrac{1}{\rho} \left(\dfrac{\partial \rho}{\partial P} \right)_T ;$ $\quad$ (A.5)

- the *isentropic* compressibility coefficient , $\quad \chi_T = \dfrac{1}{\rho} \left(\dfrac{\partial \rho}{\partial P} \right)_s ;$ $\quad$ (A.6)

- the *isobaric* thermal expansion coefficient, $\quad \alpha = -\dfrac{1}{\rho} \left(\dfrac{\partial \rho}{\partial T} \right)_P ;$ $\quad$ (A.7)

- the heat capacities at constant *pressure* and *volume*

$$C_p = T \left(\frac{\partial s}{\partial T} \right)_P , \qquad\qquad C_v = T \left(\frac{\partial s}{\partial T} \right)_\rho . \qquad (A.8)$$

A.5 The First Law

The first law introduces a function which solely depends on the state of the system and accounts for the energy exchanges between the system and the environment during a transformation between equilibrium states.

The exchanges in question are in the form of heat (with the elementary specific value δQ_{ext}) and work of macroscopic forces acting from the system surroundings (with the elementary specific value δW_{ext}).

The first law is then mathematically expressed by:

$$d(e + e_c) = \delta Q_{ext} + \delta W_{ext} , \qquad (A.9)$$

where the symbol ' d ' on the left hand side refers to the differential of the *point functions* under consideration,[7] while the symbol ' δ ' stands for an *elementary variation* which depend on the path followed by the system during any given transformation.

---

[7] Mathematically this allows to simply writing $\displaystyle\int_A^B de(e + e_c) = e(B) + e_c(B) - e(A) - e_c(A).$

A.6   The Second Law

If the first law can be considered as a principle of *quantitative* order, the second is rather of the *qualitative* type. Indeed, this law makes it possible to draw a distinction, in the entire energy exchanges with the surroundings, between the *reversible* and *irreversible* contributions. Under the same exchange conditions as those previously mentioned, it allows stating the following two proposition:

1. There is a state function, called the specific entropy, whose differential is such that:

$$T\,\mathrm{d}s = \delta Q_{ext} + \delta f\,;\qquad\qquad\qquad (A.10)$$

2. The δf contribution which quantifies the *entropy production*[8] by the *irreversibilities* occurring during the transformation satisfies the condition:

$$\delta f \geq 0\,.\qquad\qquad\qquad (A.11)$$

When $\delta f = 0$, the transformation is termed *reversible* and the entropy variation (increase or decrease) is directly derived from the heat exchange with the environment. If a *reversible* transformation is also *adiabatic* ($\delta Q_{ext} = 0$), then the entropy of the system is invariant. We speak in this case of an *isentropic* transformation ($\mathrm{d}s = 0$).   To summarize, the previous formulation—Eq. (A.10)—of the second law of thermodynamics states that the *entropy production* for an *irreversible* transformation is greater than that which would be observed for a *reversible* transformation of the same heat exchange with the surroundings and between the same initial and final equilibrium states.

A.7   The Irreversibility Sources

A.7.1   The Extrinsic Irreversibility

This first type of irreversibility, referred to as *extrinsic* can be easily explained in the context of the classical thermodynamics of a zero kinetic energy system. Using relation (A.14) hereafter, the first law yields[9]:

$$\mathrm{d}e \equiv T\,\mathrm{d}s - P\,\mathrm{d}\upsilon = \delta Q_{ext} + \delta W_{ext}\,.$$

By substituting the entropy variation from the second law, Eq. (A.10), on the left hand side, we obtain:

---

[8] This contribution is here homogeneous to an energy. The specific entropy s, corresponds to an energy per unit of temperature. It is thus expressed in $J \times K^{-1}$ units.

[9] To simplify, it is assumed that the heat exchange is isentropic.

$$\delta f = \delta W_{ext} - (-P \mathrm{d}v) .$$

This result expresses the fact that the amount of irreversibilities is due to the difference between the work exchanged by the system with the surroundings for a *real* transformation and a *reversible* transformation between the same equilibrium states of the system.[10]

In rational mechanics, we are interested, first of all, in *simple* or *idealized* objects, which are the '*material*' point or the '*indeformable, perfectly rigid*' solid. In this case, the irreversible character of such object motions can only result from the interaction with the surroundings, through the action of non-conservative contact forces. Even if it still exists in the mechanics of deformable media—and therefore in fluid mechanics—, this mode of irreversibility, which only involves external interactions, is no longer the only relevant mode for these environments.

A.7.2   The Intrinsic Irreversibility

Irreversibilities of a different nature indeed manifest themselves in a *system of discrete material points* not rigidly linked to each other.[11] In this case, the kinetic energy theorem of rational mechanics teaches us that the kinetic energy variation of the system is equal to the sum of the work of all forces acting on the system elements, both from *outside* (W_{ext}) and *inside* (W_{int}). By expressing this result in specific values, i.e., per system mass unit, we will write, for an elementary transformation:

$$\mathrm{d}e_c = \delta W_{ext} + \delta W_{int} . \qquad (\mathrm{A}.12)$$

We then deduce from the first law, Eq. (A.9), that $\mathrm{d}e = \delta Q_{ext} - \delta W_{int}$.
Making use of relation (A.14) hereafter, we can still write that:
$Q_{ext} + \delta W_{int} = T \mathrm{d}s - P \mathrm{d}v$, or $\delta Q_{ext} + \delta W_{int} = \delta Q_{ext} + \delta f - P \mathrm{d}v$, by explaining the entropy variation expression according to the second law. Finally, we obtain[12]

$$\delta f = -\delta W_{int} + P \mathrm{d}v .$$

Considering, for easier interpretation, that the transformation is carried out without variation of the elementary volume,[13] we can notice that

---

[10] It should be recalled that for a *reversible* transformation, the work of compression is carried out under a pressure equilibrium condition between the system and the environment, which is no longer the case with a *real* transformation.

[11] It is also the case of deformable solids *not perfectly* elastic and all *real* fluids.

[12] Again, we assume, that the heat exchange is isentropic.

[13] We speak, in classical thermodynamics, of an *isochoric* process. In fluid mechanics, it is termed as an *isovolume* or *solenoidal* evolution, in the sense of a divergence-free velocity field.

$$\delta f = -\delta W_{int} \,. \tag{A.13}$$

Thus, in this case, the work of the *internal* forces of the system is at the origin of an irreversible entropy generation. This second type of irreversibility is qualified as *intrinsic* and is of the type we will systematically retrieve in real *fluid* motions.

A.8   The Differential Relation to the Internal Energy and Entropy Specific Values

For a *reversible* transformation, the volume of heat exchanged with the environment is expressed by $\delta Q_{ext} = T ds$, where T is the temperature of the system, still equal to the environmental temperature. As for the work exchanged with the environment, it is equal to $\delta W_{ext} = -P dv = \frac{P}{\rho^2} d\rho$, where the system pressure P is again equal to the external pressure. Applying the first principle to this *reversible* transformation leads to:

$$de = T ds - P dv = T ds + \frac{P}{\rho^2} d\rho \,. \tag{A.14}$$

Since all differentials relate to state variables, the previous expression applies to any transformation where the variables undergo the same variations, whether these result from A *reversible* or an *irreversible* evolution.[14] Relation (A.14) can be linked to a more general expression involving the free energy (free enthalpy) $G = e + P v - T s$, introduced by Gibbs. It is for this reason that it is referred to, in shorthand, as the *Gibbs relation* in the body of this work.

A.9   The Ideal Gas Assumption

The ideal gas concept refers to a non-chemically reacting medium, in a thermodynamic equilibrium and which fulfills the following conditions[15]:

$$- \textit{equation of state:} \quad \frac{P}{\rho} = RT \equiv \frac{\mathcal{R}}{\mathcal{M}} T \quad \text{and } C_p - C_v = R \,, \tag{A.15}$$

[14] When the transformation is *irreversible*, the heat exchange is determined by $\delta Q_{ext} = T ds - \delta f$ according to the second law (A.10). But in this case, the work of the *external* pressure force $\delta W_{ext} = -P_{ext} dv$ is higher than that of a *reversible* transformation, where $P_{ext} \equiv P$, the pressure of the system. Indeed, it should be increased by the work due to the intrinsic irreversibilities of the system, δW_{int}. Equality (A.13) clearly proves that expression (A.14) is still valid.

[15] Some of the properties listed here also refer to the thermally perfect gas.

where $\mathcal{R} = 8,314472\,\mathrm{J\,K^{-1}\,mol^{-1}}$ is the molar (universal) gas constant, $\mathcal{M}$ the gas molar mass. In the International System of Units (SI), the value of R is expressed as $\mathrm{J\,K^{-1}\,kg^{-1}}$. For the air, $R = 287\,\mathrm{J\,K^{-1}\,kg^{-1}}$.

$$- Joule's\ first\ law: \qquad \mathrm{d}e = C_v\mathrm{d}T\,, \qquad (A.16)$$

(the internal energy of a perfect gas only depends on its temperature);

$$- Joule's\ second\ law: \qquad \mathrm{d}h = C_p\mathrm{d}T\,, \qquad (A.17)$$

(the enthalpy of a perfect gas only depends on its temperature);

A.10 The Incompressibility Assumption

In a considerable number of hydraulic applications, the constant fluid density can be considered a highly acceptable assumption. With the terminological reservations explained in Chap. 3, Sect. 3.1.3, the equation of state to be adopted now is that of the *incompressible fluid*. In terms of thermodynamics, the density no longer acts as a *state variable*, so that the canonical state equation reduces to:

$$e = e\,(s)\,. \qquad (A.18)$$

Thus, the temperature remains defined by relation (A.4), while relation (A.3) no longer allows defining the pressure in the themodynamic manner. This quantity is thus to be taken in a purely mechanical context, as the trace of the stress tensor. The fluid thermodynamic behaviour is therefore entirely characterized, in this case, by the caloric state equation, in the following simplified form:

$$\mathrm{d}e = C\mathrm{d}T \quad \text{with} \quad C = C_v = C_p = T\frac{\mathrm{d}s}{\mathrm{d}T}\,. \qquad (A.19)$$

Appendix B
The Tenso-Vectorial Notations and Operators

t is assumed that the reader is familiar, with a sufficient degree of knowledge, n the field of vector and tensor algebra and calculus. The present appendix is only concerned with scalar, vector and tensor manipulations, with no mathematical developments of the subject.

A right-handed Cartesian coordinate system is adopted as the frame referential of he usual three-dimensional Euclidean space ($\mathbb{R}^3$).

B.1   General Conventions

• *Index formalism* The unit vectors of the Cartesian coordinate system are denoted as $\mathbf{e}_i$, where $i = 1, 2, 3$. Any vector $\mathbf{A}$ can be expressed as a linear combination of his *basis*

$$\mathbf{A} = A_1\mathbf{e}_1 + A_2\mathbf{e}_2 + A_3\mathbf{e}_3 \equiv \sum_{i=1}^{i=3} A_i\mathbf{e}_i.$$

The *three* scalars A_i, with $i = 1, 2, 3$ are called the vector *components* related to the basis, *viz*, $A_i = \mathbf{A} \cdot \mathbf{e}_i$ where the dot symbol denotes the *scalar product* or the *dot product* and using Einstein summation convention (see hereafter);

Similarly, the *nine* components of a second-order tensor $\overline{\overline{A}}$ are denoted as A_{ij}, with $i = 1, 2, 3$ and $j = 1, 2, 3$. Such a tensor can be expressed, in $\mathbb{R}^3$, as a linear combination of the nine *unit dyads*:

$$\overline{\overline{A}} = \underbrace{A_{11}\mathbf{e}_1\mathbf{e}_1 + A_{12}\mathbf{e}_1\mathbf{e}_2 + \ldots A_{33}\mathbf{e}_3\mathbf{e}_3}_{nine\ terms} \equiv \sum_{i=1}^{i=3}\sum_{j=1}^{j=3} A_{ij}\mathbf{e}_i\mathbf{e}_j$$

Einstein summation convention Any index appearing twice in any multiplicative grouping implies a summation on that index. It is thus a '*dummy*' index since it can e replaced by any symbol i, j, k etc., which reflects a contraction of one unit in the rder of the expression under consideration.

© The Editor(s) (if applicable) and The Author(s), under exclusive license
© Springer Nature Switzerland AG 2022
. Chassaing, *Fundamentals of Fluid Mechanics*,
https://doi.org/10.1007/978-3-031-10086-4

For example $A_i B_i$ means $A_1 B_1 + A_2 B_2 + A_3 B_3 \equiv \sum_{i=1}^{i=3} A_i B_i$.

This expression is thus the scalar product of the two vectors **A** and **B**.

The dummy index (i) has only an operative function, so the scalar product in question may just as well be written $A_j B_j$.

On the other hand, $A_i B_j$ denotes a second-order tensor, the indices i and j being different are called '*free*', which means that they have to run through the 1, 2 and 3 values.

Finally, the dummy-index summation convention will be extended to index repetition in a partial derivation operator. Thus, for instance:

$$\frac{\partial A_i}{\partial x_j} B_j \left(\equiv \sum_{j=1}^{j=3} \frac{\partial A_i}{\partial x_j} B_j \right) = \frac{\partial A_i}{\partial x_1} B_1 + \frac{\partial A_i}{\partial x_2} B_2 + \frac{\partial A_i}{\partial x_3} B_3 ,$$

which stands for the (right) contracted product of the vector **A** derivatives, namely the tensor $\overline{\overline{A}}$ ($A_{ij} = \partial A_i / \partial x_j$, $i = 1, 2, 3$ and $j = 1, 2, 3$) by the vector **B**(B_l, $l = 1, 2, 3$). Accordingly, the result is a *vector*, as demonstrated by the presence of one single *free* index, i in this case.

B.2   Some Usual Symbols

• *The Kronecker delta* δ_{ij}, with $i = 1, 2, 3$ and $j = 1, 2, 3$ denotes the second-order unit tensor.

$$\overline{\overline{I}} = \begin{pmatrix} 1, 0, 0 \\ 0, 1, 0 \\ 0, 0, 1 \end{pmatrix} \text{ or } \delta_{ij} = \begin{cases} 1, & for \ i = j , \\ 0, & for \ i \neq j . \end{cases}$$

• *The 3-dimensional Levi-Civita symbol* or *permutation symbol* ϵ_{ijk}, where each index takes values 1, 2, 3 denotes the pseudo-tensor whose components are

$\epsilon_{ijk} = +1, \ for \ i, j, k \ in \ the$ direct *permutation of* 1, 2, 3 ;
$\epsilon_{ijk} = -1, \ for \ i, j, k \ in \ the$ reverse *permutation of* 1, 2, 3 ;
$\epsilon_{ijk} = 0, \ in \ all \ other \ cases.$

B.3   The Main Operators

A scalar, vector and tensor field is a real-valued, vectorial and tensorial function which respectively returns a scalar (i.e., a real number), a vector (i.e., three scalar components) and a nth-order tensor (i.e., n scalar components) for any point of a given space domain.

• *The gradient of a scalar field*—Assuming that the scalar field $f(x, y, z) \equiv f(x_i)$ is a *continuously differentiable* function of the coordinates, the *vector* whose components are $\partial f/\partial x$, $\partial f/\partial y$ and $\partial f/\partial z$ is called the *gradient*. From this definition, it results that:

$$df = \frac{\partial f}{\partial x}dx + \frac{\partial f}{\partial y}dy + \frac{\partial f}{\partial z}dz \equiv \mathbf{grad}\, f \cdot d\mathbf{M}, \tag{B.1}$$

where $d\mathbf{M}\,(dx, dy, dz)$ is the differential displacement vector and $\cdot$ denotes the scalar product. Equation (B.1) can be equally written as follows :

$$df = d\mathbf{M} \cdot \nabla f, \tag{B.2}$$

where the *differential operator nabla* (∇) in Cartesian coordinates (O; $\mathbf{e}_1$, $\mathbf{e}_2$, $\mathbf{e}_3$) is defined as

$$\nabla \equiv \frac{\partial}{\partial x_1}\mathbf{e}_1 + \frac{\partial}{\partial x_2}\mathbf{e}_2 + \frac{\partial}{\partial x_3}\mathbf{e}_3. \tag{B.3}$$

• **The gradient of a vector field**—Assuming a *continuously differentiable* vector field $\mathbf{A}$, the gradient of the vector field is the second-order tensor $\overline{\overline{grad}}\,\mathbf{A}$ whose components A_{ij} in Cartesian coordinates are:

$$A_{ij} = \begin{pmatrix} \dfrac{\partial A_1}{\partial x_1} & \dfrac{\partial A_1}{\partial x_2} & \dfrac{\partial A_1}{\partial x_3} \\[2ex] \dfrac{\partial A_2}{\partial x_1} & \dfrac{\partial A_2}{\partial x_2} & \dfrac{\partial A_2}{\partial x_3} \\[2ex] \dfrac{\partial A_3}{\partial x_1} & \dfrac{\partial A_3}{\partial x_2} & \dfrac{\partial A_3}{\partial x_3} \end{pmatrix} \tag{B.4}$$

where $A_i(x_1, x_2, x_3), i = 1, 2, 3$ are the vector components in Cartesian coordinates.

• **The divergence of a vector, tensor field**—In Cartesian coordinates (x_1, x_2, x_3), the *divergence* of a vector field $\mathbf{A}$ is defined as the *scalar*:

$$div\,\mathbf{A}\ (or\ \nabla \cdot \mathbf{A}) = \frac{\partial A_1}{\partial x_1} + \frac{\partial A_2}{\partial x_2} + \frac{\partial A_3}{\partial x_3} = \frac{\partial A_i}{\partial x_i},$$

where the summation convention with respect to the derivative index is used in the last expression. Similarly, the divergence of a second-order tensor $\overline{\overline{A}}$ is a *vector* such as, for instance

$$div^L\,\overline{\overline{A}}\ (or\ \nabla \cdot \overline{\overline{A}}) = \left(\frac{\partial A_{11}}{\partial x_1} + \frac{\partial A_{21}}{\partial x_2} + \frac{\partial A_{31}}{\partial x_3} \right)\mathbf{e}_1 +$$

$$\left(\frac{\partial A_{12}}{\partial x_1} + \frac{\partial A_{22}}{\partial x_2} + \frac{\partial A_{32}}{\partial x_3} \right)\mathbf{e}_2 + \left(\frac{\partial A_{13}}{\partial x_1} + \frac{\partial A_{23}}{\partial x_2} + \frac{\partial A_{33}}{\partial x_3} \right)\mathbf{e}_3$$

$$= \frac{\partial A_{ij}}{\partial x_i}\mathbf{e}_j.$$

Table B.1 Exprtessions of some common operators in index notation

Function	Operator	Result
$\phi\,(x_1, x_2, x_3)$	$\mathbf{grad}\ \phi\ :\quad \dfrac{\partial \phi}{\partial x_i}$	Vector
	$\Delta\phi = \dfrac{\partial^2 \phi}{\partial x_j \partial x_j}$	Scalar
$\mathbf{A}\,(x_1, x_2, x_3)$	$\mathrm{div}\mathbf{A} = \dfrac{\partial A_i}{\partial x_i}$	Scalar
	$curl\mathbf{A}\ :\quad \varepsilon_{ijk}\dfrac{\partial A_k}{\partial x_j}$	Vector
	$\Delta\mathbf{A}\ :\quad \dfrac{\partial^2 A_i}{\partial x_j \partial x_j}$	Vector
	$\overline{\overline{grad}}\mathbf{A}\ :\quad \dfrac{\partial A_i}{\partial x_j}$	2-order Tensor
$\overline{\overline{T}}\,(x_1, x_2, x_3)$	$\mathrm{div}\overline{\overline{T}} = \dfrac{\partial T_{ij}}{\partial x_j}$	Vector
	$\overline{\overline{grad}}\,\overline{\overline{T}}\ :\quad \dfrac{\partial T_{ij}}{\partial x_k}$	3-order Tensor

Hence the divergence operation corresponds to a contraction which decreases the tensor order by one unit. When the derivative applies to the first (left) index of the second-order tensor, it is referred to as a *left*-divergence. Obviously, the contraction of a second-order tensor can also be applied to the second index, introducing the *right*-divergence

$$div^R\,\overline{\overline{A}} = \frac{\partial A_{ij}}{\partial x_j}\,\mathbf{e}_i.$$

In general $div^R\,\overline{\overline{A}} \neq div^L\,\overline{\overline{A}}$, unless $\overline{\overline{A}}$ is a *symmetrical* second-order tensor *viz*, $A_{ij} = A_{ji}$.

B.4   Some Expressions of Tenso-Vectorial Operators in Index Notation

In Table B.1 $\phi\,(x_1, x_2, x_3)$ denotes a scalar function of the space coordinates, $\mathbf{A}\,(x_1, x_2, x_3)$ and $\mathbf{B}\,(x_1, x_2, x_3)$ vector functions, $\overline{\overline{T}}\,(x_1, x_2, x_3)$ a second-order tensor function.[16]

---

[16] For the applications which will be those of this course, the second-order tensors will be symmetrical, so that the distinction between the divergence on the right and left index is not relevant.

Appendix C
Vector Calculus Identities

In the subsequent relations, f, p and q are scalar functions of the coordinates of any point M, $\mathbf{A}$ and $\mathbf{B}$ vector functions of the same coordinates. It can be demonstrated that the following relations are verified as identities.[17]

$$\mathbf{grad}\ (p+q) = \mathbf{grad}\ (p) + \mathbf{grad}\ (q) \tag{C.1}$$

$$\mathbf{grad}\ (p.q) = p.\mathbf{grad}\ (q) + q.\mathbf{grad}\ (p) \tag{C.2}$$

$$\mathbf{grad}\ [f\ (p)] = f'(p).\mathbf{grad}\ (p) \tag{C.3}$$

$$div\ (\mathbf{A+B}) = div\ \mathbf{A} + div\ \mathbf{B} \tag{C.4}$$

$$\mathbf{grad}(\mathbf{A.B}) = \overline{\overline{grad}}\mathbf{A.B} + \mathbf{A} \wedge curl\mathbf{B} + \overline{\overline{grad}}\mathbf{B.A} + \mathbf{B} \wedge curl\mathbf{A} \tag{C.5}$$

$$div\ (p\mathbf{A}) = p\ div\ \mathbf{A} + \mathbf{A}.\mathbf{grad}\ (p) \tag{C.6}$$

$$div\ (\mathbf{A} \wedge \mathbf{B}) = \mathbf{B}.\,curl\mathbf{A} - \mathbf{A}.curl\mathbf{B} \tag{C.7}$$

$$div\ (\mathbf{grad}\ p) = \Delta p \qquad (Laplacian) \tag{C.8}$$

$$div\ (curl\mathbf{A}) = 0 \tag{C.9}$$

$$div\ (\Delta\mathbf{A}) = \Delta\ (div\mathbf{A}) \tag{C.10}$$

$$curl\ (\mathbf{A+B}) = \ curl\ (\mathbf{A}) + curl\ (\mathbf{B}) \tag{C.11}$$

$$curl\ (p\mathbf{A}) = p\ curl\ \mathbf{A} + \mathbf{grad}\ p \wedge \mathbf{A} \tag{C.12}$$

---

[17] We confine ourselves here to some identities which may prove useful in fluid mechanics. For further detailed information, the reader will profitably refer to Coirier's work [39] and for a complete form to the aide-memoire [40] by the same author.

© The Editor(s) (if applicable) and The Author(s), under exclusive license   537
to Springer Nature Switzerland AG 2022
P. Chassaing, *Fundamentals of Fluid Mechanics*,
https://doi.org/10.1007/978-3-031-10086-4

$$curl\,(curl\mathbf{A}) = \mathbf{grad}\,(div\mathbf{A}) - \Delta\mathbf{A} \qquad\qquad (C.13)$$

$$curl\,(\mathbf{grad}\,p) = \mathbf{0} \qquad\qquad (C.14)$$

$$curl\,(\mathbf{A} \wedge \mathbf{B}) = \overline{\overline{grad}}\mathbf{A}.\mathbf{B} + \mathbf{A}\,\,div\mathbf{B} - \overline{\overline{grad}}\mathbf{B}.\mathbf{A} - \mathbf{B}\,\,div\mathbf{A} \qquad\qquad (C.15)$$

Appendix D
Vector Operator Expressions in Projection

D.1 Definitions and Notations

For practical applications, it is sometimes appropriate that the motion be described with a coordinate system suited to the problem geometry. We will limit ourselves here to the common reference frames, whose defining elements are outlined in the next figure (Fig. D.1).

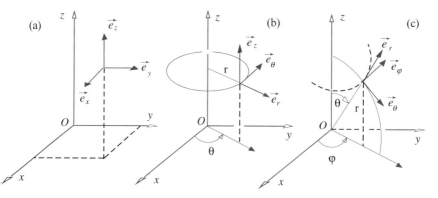

Fig. D.1 Definition elements of **a** the cartesian, **b** cylindrical and **c** spherical coordinate systems

© The Editor(s) (if applicable) and The Author(s), under exclusive license
to Springer Nature Switzerland AG 2022
P. Chassaing, *Fundamentals of Fluid Mechanics*,
https://doi.org/10.1007/978-3-031-10086-4

The correspondence between the x, y, z Cartesian coordinates and the two other reference frame's coordinates is expressed in the next Table.

Cylindrical system	Spherical system
$x = r \cos \theta$	$x = r \sin \theta \cos \varphi$
$y = r \sin \theta$	$y = r \sin \theta \sin \varphi$
$z = z$	$z = r \cos \theta$

Finally, the length and volume elements in the different reference frames are specified in the table below.

System	Rectangular Cartesian	Cylindrical	Spherical
Unit vector	$\mathbf{e}_1 \to x$	$\mathbf{e}_1 \to r$	$\mathbf{e}_1 \to r$
	$\mathbf{e}_2 \to y$	$\mathbf{e}_2 \to \theta$	$\mathbf{e}_2 \to \theta$
	$\mathbf{e}_3 \to z$	$\mathbf{e}_3 \to z$	$\mathbf{e}_3 \to \varphi$
Metric	$dx^2 + dy^2 + dz^2$	$dr^2 + r^2 d\theta^2 + dz^2$	$dr^2 + r^2 d\theta^2 + r^2 \sin^2\theta \, d\varphi^2$
Volume element	$dx \times dy \times dz$	$dr \times r d\theta \times dz$	$dr \times r d\theta \times r \sin \theta \, d\varphi$

D.2 The Expression of the Various Operators

In the following $f(M)$ stands for a scalar function of the space coordinates and $\mathbf{A}(M)$ denotes a vector function whose components in the three referentials are respectively:

$$\mathbf{A}\left(A_x, A_y, A_z\right), \qquad \mathbf{A}\left(A_r, A_\theta, A_z\right), \qquad \mathbf{A}\left(A_r, A_\theta, A_\varphi\right).$$

D.2.1 The Divergence of a Vector

$$div\mathbf{A} = \frac{\partial A_x}{\partial x} + \frac{\partial A_y}{\partial y} + \frac{\partial A_z}{\partial z}$$

$$div\mathbf{A} = \frac{1}{r}\frac{\partial (r A_r)}{\partial r} + \frac{1}{r}\frac{\partial A_\theta}{\partial \theta} + \frac{\partial A_z}{\partial z}$$

$$div\mathbf{A} = \frac{1}{r^2}\frac{\partial \left(r^2 A_r\right)}{\partial r} + \frac{1}{r \sin \theta}\frac{\partial (\sin \theta A_\theta)}{\partial \theta} + \frac{1}{r \sin \theta}\frac{\partial A_\varphi}{\partial \varphi}$$

D.2.2 The Laplacian of a Scalar Field

$$\Delta f = \frac{\partial^2 f}{\partial x^2} + \frac{\partial^2 f}{\partial y^2} + \frac{\partial^2 f}{\partial z^2}$$

$$\Delta f = \frac{1}{r} \frac{\partial}{\partial r} \left(r \frac{\partial f}{\partial r} \right) + \frac{1}{r^2} \frac{\partial^2 f}{\partial \theta^2} + \frac{\partial^2 f}{\partial z^2}$$

$$\Delta f = \frac{1}{r^2} \frac{\partial}{\partial r} \left(r^2 \frac{\partial f}{\partial r} \right) + \frac{1}{r^2 \sin \theta} \frac{\partial}{\partial \theta} \left(\sin \theta \frac{\partial f}{\partial \theta} \right) + \frac{1}{r^2 \sin^2 \theta} \frac{\partial^2 f}{\partial \varphi^2}$$

D.2.3 The Gradient of a Scalar Function

$$\mathbf{e}_x : \frac{\partial f}{\partial x} \qquad \mathbf{e}_r : \frac{\partial f}{\partial r} \qquad \mathbf{e}_r : \frac{\partial f}{\partial r}$$

$$\mathbf{e}_y : \frac{\partial f}{\partial y} \qquad \mathbf{e}_\theta : \frac{1}{r} \frac{\partial f}{\partial \theta} \qquad \mathbf{e}_\theta : \frac{1}{r} \frac{\partial f}{\partial \theta}$$

$$\mathbf{e}_z : \frac{\partial f}{\partial z} \qquad \mathbf{e}_z : \frac{\partial f}{\partial z} \qquad \mathbf{e}_\varphi : \frac{1}{r \sin \theta} \frac{\partial f}{\partial \varphi}$$

D.2.4 The Curl of a Vector Field

$$\mathbf{e}_x : \frac{\partial A_z}{\partial y} - \frac{\partial A_y}{\partial z} \qquad \mathbf{e}_r : \frac{1}{r} \frac{\partial A_z}{\partial \theta} - \frac{\partial A_\theta}{\partial z} \qquad \mathbf{e}_r : \frac{1}{r \sin \theta} \left[\frac{\partial \left(A_\varphi \sin \theta \right)}{\partial \theta} - \frac{\partial A_\theta}{\partial \varphi} \right]$$

$$\mathbf{e}_y : \frac{\partial A_x}{\partial z} - \frac{\partial A_z}{\partial x} \qquad \mathbf{e}_\theta : \frac{\partial A_r}{\partial z} - \frac{\partial A_z}{\partial r} \qquad \mathbf{e}_\theta : \frac{1}{r \sin \theta} \frac{\partial A_r}{\partial \varphi} - \frac{1}{r} \frac{\partial \left(r A_\varphi \right)}{\partial r}$$

$$\mathbf{e}_z : \frac{\partial A_y}{\partial x} - \frac{\partial A_x}{\partial y} \qquad \mathbf{e}_z : \frac{1}{r} \left[\frac{\partial \left(r A_\theta \right)}{\partial r} - \frac{\partial A_r}{\partial \theta} \right] \qquad \mathbf{e}_\varphi : \frac{1}{r} \left[\frac{\partial \left(r A_\theta \right)}{\partial r} - \frac{\partial A_r}{\partial \theta} \right]$$

D.2.5 The Divergence of a Symmetrical Second-Order Tensor

The following formulae apply to a symmetrical second-order tensor $\overline{\overline{T}}$, $(T_{ij} = T_{ji})$.

$$\mathbf{e}_x : \frac{\partial T_{11}}{\partial x} + \frac{\partial T_{12}}{\partial y} + \frac{\partial T_{13}}{\partial z} \qquad \mathbf{e}_r : \frac{\partial T_{11}}{\partial r} + \frac{1}{r} \frac{\partial T_{12}}{\partial \theta} + \frac{\partial T_{13}}{\partial z} + \frac{T_{11} - T_{22}}{r}$$

$$\mathbf{e}_y : \frac{\partial T_{21}}{\partial x} + \frac{\partial T_{22}}{\partial y} + \frac{\partial T_{23}}{\partial z} \qquad \mathbf{e}_\theta : \frac{\partial T_{21}}{\partial r} + \frac{1}{r} \frac{\partial T_{22}}{\partial \theta} + \frac{1}{r} \frac{\partial T_{23}}{\partial z} + \frac{2 T_{12}}{r}$$

$$\mathbf{e}_z : \frac{\partial T_{31}}{\partial x} + \frac{\partial T_{32}}{\partial y} + \frac{\partial T_{33}}{\partial z} \qquad \mathbf{e}_z : \frac{\partial T_{31}}{\partial r} + \frac{1}{r} \frac{\partial T_{32}}{\partial \theta} + \frac{\partial T_{33}}{\partial z} + \frac{T_{31}}{r}$$

$$\mathbf{e}_r \;:\; \frac{\partial T_{11}}{\partial r} + \frac{1}{r}\frac{\partial T_{12}}{\partial \theta} + \frac{1}{r\sin\theta}\frac{\partial T_{13}}{\partial \varphi} + \frac{1}{r}\left(2T_{11} - T_{22} - T_{33} + \frac{T_{12}}{\tan\theta}\right)$$

$$\mathbf{e}_\theta \;:\; \frac{\partial T_{21}}{\partial r} + \frac{1}{r}\frac{\partial T_{22}}{\partial \theta} + \frac{1}{r\sin\theta}\frac{\partial T_{23}}{\partial \varphi} + \frac{1}{r}\left(3T_{21} + \frac{T_{22} - T_{33}}{\tan\theta}\right)$$

$$\mathbf{e}_\varphi \;:\; \frac{\partial T_{31}}{\partial r} + \frac{1}{r}\frac{\partial T_{32}}{\partial \theta} + \frac{1}{r\sin\theta}\frac{\partial T_{33}}{\partial \varphi} + \frac{1}{r}\left(3T_{31} + 2\frac{T_{23}}{\tan\theta}\right)$$

D.2.6   The Gradient of Vector Field

The $\overline{\overline{grad}}\mathbf{A}$ tensor components are written $\overline{\overline{grad}}\mathbf{A} = \begin{pmatrix} A_{11} & A_{12} & A_{13} \\ A_{21} & A_{22} & A_{23} \\ A_{31} & A_{32} & A_{33} \end{pmatrix}$.

The expressions of these components can be explained in the three referentials as follows:

Cartesian:
$$A_{11} = \frac{\partial A_1}{\partial x} \qquad A_{12} = \frac{\partial A_1}{\partial y} \qquad A_{13} = \frac{\partial A_1}{\partial z}$$
$$A_{21} = \frac{\partial A_2}{\partial x} \qquad A_{22} = \frac{\partial A_2}{\partial y} \qquad A_{23} = \frac{\partial A_2}{\partial z}$$
$$A_{31} = \frac{\partial A_3}{\partial x} \qquad A_{32} = \frac{\partial A_3}{\partial y} \qquad A_{33} = \frac{\partial A_3}{\partial z}$$

Cylindrical:
$$A_{11} = \frac{\partial A_1}{\partial r} \qquad A_{12} = \frac{1}{r}\frac{\partial A_1}{\partial \theta} - \frac{A_2}{r} \qquad A_{13} = \frac{\partial A_1}{\partial z}$$
$$A_{21} = \frac{\partial A_2}{\partial r} \qquad A_{22} = \frac{1}{r}\frac{\partial A_2}{\partial \theta} + \frac{A_1}{r} \qquad A_{23} = \frac{\partial A_2}{\partial z}$$
$$A_{31} = \frac{\partial A_3}{\partial r} \qquad A_{32} = \frac{1}{r}\frac{\partial A_3}{\partial \theta} \qquad A_{33} = \frac{\partial A_3}{\partial z}$$

Spherical:
$$A_{11} = \frac{\partial A_1}{\partial r} \quad A_{12} = \frac{1}{r}\frac{\partial A_1}{\partial \theta} - \frac{A_2}{r} \quad A_{13} = \frac{1}{r\sin\theta}\frac{\partial A_1}{\partial \phi} - \frac{A_3}{r}$$
$$A_{21} = \frac{\partial A_2}{\partial r} \quad A_{22} = \frac{1}{r}\frac{\partial A_2}{\partial \theta} + \frac{A_1}{r} \quad A_{23} = \frac{1}{r\sin\theta}\frac{\partial A_2}{\partial \varphi} - \frac{A_3}{r\tan\theta}$$
$$A_{31} = \frac{\partial A_3}{\partial r} \quad A_{32} = \frac{1}{r}\frac{\partial A_3}{\partial \theta} \quad A_{33} = \frac{1}{r\sin\theta}\frac{\partial A_3}{\partial \varphi} + \frac{A_1}{r} + \frac{A_2}{r\tan\theta}$$

D.2.7 *The Material Derivative of a Scalar Function*

For the following formulae, $f(M, t)$ is a scalar function of the Euler variables, denoted as (x, y, z, t), (r, θ, z, t) and (r, θ, φ, t) in the three referentials respectively. The velocity vector components are denoted as (V_x, V_y, V_z), $(V_r, V_\theta, V_z,)$ and $(V_r, V_\theta, V_\varphi)$ in each coordinate system respectively.

With these notations, the material derivative of the scalar function f is expressed respectively by:

$$\frac{df}{dt} = \frac{\partial f}{\partial t} + V_x \frac{\partial f}{\partial x} + V_y \frac{\partial f}{\partial y} + V_z \frac{\partial f}{\partial z}$$

$$\frac{df}{dt} = \frac{\partial f}{\partial t} + V_r \frac{\partial f}{\partial r} + \frac{V_\theta}{r} \frac{\partial f}{\partial \theta} + V_z \frac{\partial f}{\partial z}$$

$$\frac{df}{dt} = \frac{\partial f}{\partial t} + V_r \frac{\partial f}{\partial r} + \frac{V_\theta}{r} \frac{\partial f}{\partial \theta} + \frac{V_\varphi}{r \sin\theta} \frac{\partial f}{\partial \varphi}$$

Appendix E
The Projections of the Navier-Stokes Equations in Various Coordinate Systems

E.1 The Cartesian Coordinates

The velocity vector components in this referential are referred to as U, V and W.

The continuity equation:
$$\frac{\partial \rho}{\partial t} + \frac{\partial (\rho U)}{\partial x} + \frac{\partial (\rho V)}{\partial y} + \frac{\partial (\rho W)}{\partial z} = 0$$

The strain rate tensor:

$$S_{xx} = \frac{\partial U}{\partial x} \qquad S_{xy} = S_{yx} = \frac{1}{2}\left(\frac{\partial U}{\partial y} + \frac{\partial V}{\partial x}\right)$$

$$S_{yy} = \frac{\partial V}{\partial y} \qquad S_{xy} = S_{yx} = \frac{1}{2}\left(\frac{\partial U}{\partial y} + \frac{\partial V}{\partial x}\right)$$

$$S_{zz} = \frac{\partial W}{\partial z} \qquad S_{xz} = S_{zx} = \frac{1}{2}\left(\frac{\partial U}{\partial z} + \frac{\partial W}{\partial x}\right)$$

The Navier-Stokes equations:

$$\frac{\partial U}{\partial t} + U\frac{\partial U}{\partial x} + V\frac{\partial U}{\partial y} + W\frac{\partial U}{\partial z} = F_x - \frac{1}{\rho}\frac{\partial P}{\partial x} + \nu\left(\frac{\partial^2 U}{\partial x^2} + \frac{\partial^2 U}{\partial y^2} + \frac{\partial^2 U}{\partial z^2}\right)$$

$$\frac{\partial V}{\partial t} + U\frac{\partial V}{\partial x} + V\frac{\partial V}{\partial y} + W\frac{\partial V}{\partial z} = F_y - \frac{1}{\rho}\frac{\partial P}{\partial y} + \nu\left(\frac{\partial^2 V}{\partial x^2} + \frac{\partial^2 V}{\partial y^2} + \frac{\partial^2 V}{\partial z^2}\right)$$

$$\frac{\partial W}{\partial t} + U\frac{\partial W}{\partial x} + V\frac{\partial W}{\partial y} + W\frac{\partial W}{\partial z} = F_z - \frac{1}{\rho}\frac{\partial P}{\partial z} + \nu\left(\frac{\partial^2 W}{\partial x^2} + \frac{\partial^2 W}{\partial y^2} + \frac{\partial^2 W}{\partial z^2}\right)$$

© The Editor(s) (if applicable) and The Author(s), under exclusive license
© Springer Nature Switzerland AG 2022
Chassaing, *Fundamentals of Fluid Mechanics*,
https://doi.org/10.1007/978-3-031-10086-4

E.2   The Cylindrical Coordinates

The velocity vector components in this referential are referred to here as V_r, V_θ and V_z.

The continuity equation:
$$\frac{\partial \rho}{\partial t} + \frac{1}{r}\frac{\partial (\rho\, r\, V_r)}{\partial r} + \frac{1}{r}\frac{\partial (\rho V_\theta)}{\partial \theta} + \frac{\partial (\rho V_z)}{\partial z} = 0$$

The strain rate tensor:

$$S_{rr} = \frac{\partial V_r}{\partial r} \qquad\qquad S_{r\theta} = S_{\theta r} = \frac{1}{2}\left(\frac{1}{r}\frac{\partial V_r}{\partial \theta} + \frac{\partial V_\theta}{\partial r} - \frac{V_\theta}{r}\right)$$

$$S_{\theta\theta} = \frac{1}{r}\frac{\partial V_\theta}{\partial \theta} \qquad\qquad S_{\theta z} = S_{z\theta} = \frac{1}{2}\left(\frac{\partial V_\theta}{\partial z} + \frac{1}{r}\frac{\partial V_z}{\partial \theta}\right)$$

$$S_{zz} = \frac{\partial V_z}{\partial z} \qquad\qquad S_{rz} = S_{zr} = \frac{1}{2}\left(\frac{\partial V_r}{\partial z} + \frac{\partial V_z}{\partial r}\right)$$

The Navier-Stokes equations:

$$\frac{\partial V_r}{\partial t} + V_r\frac{\partial V_r}{\partial x} + \frac{V_\theta}{r}\frac{\partial V_r}{\partial \theta} + V_z\frac{\partial V_r}{\partial z} - \frac{V_\theta^2}{r} = F_r - \frac{1}{\rho}\frac{\partial P}{\partial r} +$$
$$\nu\left(\frac{\partial^2 V_r}{\partial r^2} + \frac{1}{r^2}\frac{\partial^2 V_r}{\partial \theta^2} + \frac{\partial^2 V_r}{\partial z^2} + \frac{1}{r}\frac{\partial V_r}{\partial r} - \frac{2}{r^2}\frac{\partial V_\theta}{\partial \theta} - \frac{V_r}{r^2}\right) \quad \text{(E.1)}$$

$$\frac{\partial V_\theta}{\partial t} + V_r\frac{\partial V_\theta}{\partial x} + \frac{V_\theta}{r}\frac{\partial V_\theta}{\partial \theta} + V_z\frac{\partial V_\theta}{\partial z} + \frac{V_r V_\theta}{r} = F_\theta - \frac{1}{\rho r}\frac{\partial P}{\partial \theta} +$$
$$\nu\left(\frac{\partial^2 V_\theta}{\partial r^2} + \frac{1}{r^2}\frac{\partial^2 V_\theta}{\partial \theta^2} + \frac{\partial^2 V_\theta}{\partial z^2} + \frac{1}{r}\frac{\partial V_\theta}{\partial r} + \frac{2}{r^2}\frac{\partial V_r}{\partial \theta} - \frac{V_\theta}{r^2}\right) \quad \text{(E.2)}$$

$$\frac{\partial V_z}{\partial t} + V_r\frac{\partial V_z}{\partial x} + \frac{V_\theta}{r}\frac{\partial V_z}{\partial \theta} + V_z\frac{\partial V_z}{\partial z} = F_z - \frac{1}{\rho}\frac{\partial P}{\partial z} +$$
$$\nu\left(\frac{\partial^2 V_z}{\partial r^2} + \frac{1}{r^2}\frac{\partial^2 V_z}{\partial \theta^2} + \frac{\partial^2 V_z}{\partial z^2} + \frac{1}{r}\frac{\partial V_z}{\partial r}\right) \quad \text{(E.3)}$$

E.3   The Spherical Coordinates

The velocity vector components in this referential are referred to here as V_r, V_θ and V_φ.

The continuity equation:

$$\frac{\partial \rho}{\partial t} + \frac{1}{r^2}\frac{\partial \left(\rho\, r^2 V_r\right)}{\partial r} + \frac{1}{r\sin\theta}\frac{\partial\left(\rho\sin\theta\, V_\theta\right)}{\partial\theta} + \frac{1}{r\sin\theta}\frac{\partial\left(\rho V_\varphi\right)}{\partial\varphi} = 0$$

The strain rate tensor:

$$S_{rr} = \frac{\partial V_r}{\partial r}$$

$$S_{\theta\theta} = \frac{1}{r}\frac{\partial V_\theta}{\partial\theta}$$

$$S_{\varphi\varphi} = \frac{1}{r\sin\theta}\frac{\partial V_\varphi}{\partial\varphi} + \frac{V_r}{r} + \frac{V_\theta}{r\tan\theta}$$

$$S_{r\theta} = S_{\theta r} = \frac{1}{2}\left(\frac{1}{r}\frac{\partial V_r}{\partial\theta} + \frac{\partial V_\theta}{\partial r} - \frac{V_\theta}{r}\right)$$

$$S_{\theta\varphi} = S_{\varphi\theta} = \frac{1}{2}\left(\frac{1}{r\sin\theta}\frac{\partial V_\theta}{\partial\varphi} + \frac{1}{r}\frac{\partial V_\varphi}{\partial\theta} - \frac{V_\varphi}{r\tan\theta}\right)$$

$$S_{\varphi r} = S_{r\varphi} = \frac{1}{2}\left(\frac{1}{r\sin\theta}\frac{\partial V_r}{\partial\varphi} + \frac{\partial V_\varphi}{\partial r} - \frac{V_\varphi}{r}\right)$$

The Navier-Stokes equations:

$$\frac{\partial V_r}{\partial t} + V_r\frac{\partial V_r}{\partial r} + \frac{V_\theta}{r}\frac{\partial V_r}{\partial\theta} + \frac{V_\varphi}{r\sin\theta}\frac{\partial V_r}{\partial\varphi} - \frac{V_\varphi^2 + V_\theta^2}{r} = F_r - \frac{1}{\rho}\frac{\partial P}{\partial r} +$$
$$v\left(\Delta V_r - 2\frac{V_r}{r^2} - 2\frac{V_\theta}{r^2\tan\theta} - \frac{2}{r^2}\frac{\partial V_\theta}{\partial\theta} - \frac{2}{r^2\sin\theta}\frac{\partial v_\varphi}{\partial\varphi}\right)$$

$$(E.4)$$

$$\frac{\partial V_\theta}{\partial t} + V_r\frac{\partial V_\theta}{\partial r} + \frac{V_\theta}{r}\frac{\partial V_\theta}{\partial\theta} + \frac{V_\varphi}{r\sin\theta}\frac{\partial V_\theta}{\partial\varphi} - \frac{V_\varphi^2}{r\tan\theta} + \frac{V_\theta V_r}{r} = F_\theta - \frac{1}{\rho r}\frac{\partial P}{\partial\theta} +$$
$$v\left(\Delta V_\theta - \frac{V_\theta}{r^2\sin^2\theta} + \frac{2}{r^2}\frac{\partial V_r}{\partial\theta} - 2\frac{\cos\theta}{r^2\sin^2\theta}\frac{\partial V_\varphi}{\partial\varphi}\right)$$

$$(E.5)$$

$$\frac{\partial V_\varphi}{\partial t} + V_r\frac{\partial V_\varphi}{\partial r} + \frac{V_\theta}{r}\frac{\partial V_\varphi}{\partial\theta} + \frac{V_\varphi}{r\sin\theta}\frac{\partial V_\varphi}{\partial\varphi} + \frac{V_\varphi V_\theta}{r\tan\theta} + \frac{V_\varphi V_r}{r} = F_\varphi - \frac{1}{\rho r\sin\theta}\frac{\partial P}{\partial\varphi} +$$
$$v\left(\Delta V_\varphi - \frac{V_\varphi}{r^2\sin^2\theta} + \frac{2}{r^2\sin\theta}\frac{\partial V_r}{\partial\varphi} + 2\frac{\cos\theta}{r^2\sin^2\theta}\frac{\partial V_\theta}{\partial\varphi}\right)$$

$$(E.6)$$

List of Exercises

© The Editor(s) (if applicable) and The Author(s), under exclusive license
to Springer Nature Switzerland AG 2022
P. Chassaing, *Fundamentals of Fluid Mechanics*,
https://doi.org/10.1007/978-3-031-10086-4

References

1. A.J. Barré De Saint-Venant, Note sur la définition de la pression dans les corps fluides ou solides en repos ou en mouvement. Bull. Soc. Phil. Paris 134–138 (1843)
2. A.J. Barré De Saint-Venant, Mémoire sur la théorie de la résistance des fluides. Solution du paradoxe proposé à ce sujet par d'Alembert aux géomètres. Comparaison de la théorie aux expériences. C. R. Acad. Sci. **24**, 243–246 (1847)
3. A.J. Barré De Saint-Venant, Sur l'hydrodynamique des cours d'eau. C. R. Acad. Sci. Paris **74**, 570–577 (1872)
4. Bass, *Cours de Mathématiques - Tome II*, 3ème éd. (Masson & Cie, Paris, 1964)
5. G.K. Batchelor, *An Introduction to Fluid Dynamics* (Cambridge University Press, 2000)
6. R. Bellman, R.H. Pennington, Effects of surface tension and viscosity on Taylor instability. Quart. Appl. Math. **12**, 151–162 (1954)
7. H. Bénard, Les tourbillons cellulaires dans une nappe liquide - Première partie: description générale des phénomènes. Rev. Gén. Sci. Pures & Appl. **11**, 1261–1271 (1900)
8. H. Bénard, Les tourbillons cellulaires dans une nappe liquide - Deuxième partie?: Procédés mécaniques et optiques d'examen - Lois numériques des phénomènes. Rev. Gén. Sci. Pures & Appl. **11**, 1309–1328 (1900)
9. H. Bénard, Les tourbillons cellulaires dans une nappe liquide transportant de la chaleur par convection en régime permanent. Ann. Chim. Phys. **7**(23), 62–144 (1901)
10. D. Bernoulli, *Hydrodynamica - De viribus et motibus fluidorum commentarii* (Opus Academium - Sumptibus Johannis Reinholdi Dulseckeri, 1738)
11. H. Blasius, *Grenzschichten in flüssigkeiten mit kleiner reibung* (Ph.D. Georgia Augusta, Göttingen, 1907)
12. H. Blasius, Grenzschichten in Flüssigkeiten mit kleiner Reibung. Zeit. Math. Phys. **56**(1), 1–37 (1908)
13. L. Boltzmann, *Vorlesungen über Gastheorie - I. Teil: Theorie der gase mit einatomigen molekülen, deren dimensionen gegen die mittlere weglänge verschwinden* (Verlag von Johann Ambrosius Barth, Leipzig, 1896)
14. J.-C. Borda, Mémoire sur l'écoulement des fluides par les orifices des vases. Hist. & Mém Acad. Roy. Sci. 143–150 (Hist.) & 579–607 (Mém.) (1766)
15. J. Boussinesq, Essai sur la théorie des eaux courantes. Mém. Acad. Sci. Paris **23**, 1–680 (1877)
16. J. Boussinesq, *Théorie analytique de la chaleur mise en harmonie avec la thermodynamique et avec la théorie mécanique de la lumière. Tome 1er. Problèmes généraux* (Gauthier-Villars, Paris, 1902)
17. J. Boussinesq, *Théorie analytique de la chaleur mise en harmonie avec la thermodynamique et avec la théorie mécanique de la lumière*. Tome 2nd (Gauthier-Villars, Paris, 1903)

© The Editor(s) (if applicable) and The Author(s), under exclusive license to Springer Nature Switzerland AG 2022
P. Chassaing, *Fundamentals of Fluid Mechanics*,
https://doi.org/10.1007/978-3-031-10086-4

18. J. Bouttes, *Mécanique des Fluides* (Edition Marketing Paris, 1988)
19. M. Braza, A. Sevrain, *Développement d'un logiciel de résolution des équations de Navier-Stokes sur IBM 3090 VF* (1989)
20. G.O. Brown, *Henry Darcy's Perfection of the Pitot Tube*, ASCE Conference on "Henry P. G. Darcy and Other Pioneers in Hydraulics" (Philadelphia, 2003)
21. G. Bruhat, *Thermodynamique* (Masson & Cie, 1962)
22. E. Buckingham, On Physically Similar Systems; Illustrations of the Use of Dimensional Equations. Phys. Rev. **4**(4), 345–376 (1914)
23. A. Busemann, Gasstromung mit laminarer Grenzschicht entlang einer Platte. ZAMM **15**, 23–25 (1935)
24. C. Camichel, *Rapport sur le tube de Pitot appliqué à la mesure de la vitesse des liquides*, Assoc. Sci. Hydrology, Commission on hydraulics measurements, Rep. no. 7 (1939)
25. S. Candel, *Mécanique des fluides* (Dunod Université, Bordas. Paris, 1990)
26. S. Carnot, Réflexions sur la puissance motrice du feu et sur les machines propres à développer cette puissance. Ann. Sci. E.N.S. (réimpression), 2ème Sér. **1**, 393–457 (1824)
27. A.-L. Cauchy, Théorie de la propagation des ondes à la surface d'un fluide pesant d'une profondeur indéfinie. Mém. Acad. Roy. Sci., in Œuvres complètes **1**(ser. 1), 5–318 (1827)
28. A.-L. Cauchy, Mémoire sur les dilatations, les condensations et les rotations produites par un changement de forme dans un système de points matériels. Exercices de physique et d'analyse mathématique, t. **2**, 302–330 (1841)
29. A.-L. Cauchy, Sur les fonctions de variables imaginaires. C. R. Acad. Sci. **32**, 160 (1851)
30. J.-B. Cazalbou, T. Lili, *Calcul d'écoulement instationnaire à recirculation de fluide visqueux*, Fac. Sci. Tunis, Report Dept. de Physique (1985)
31. J.-B. Cazalbou, *Calcul d'écoulement de fluide parfait incompressible par une méthode de singularités*, Report ENSICA-DMF (1995)
32. S.A. Chaplygin, Sur la pression d'un écoulement plan parallèle sur un solide immergé (théorie de l'aéroplane) (in Russian). Math. Coll. Moscow **28**, 120–126 (1910)
33. S. Chapmann, The kinetic theory of simple and composite monatomic gases: viscosity, thermal conduction, and diffusion. Proc. Roy. Soc. Lond., Ser. A **93**(646), 1–20 (1916)
34. P. Chassaing, J.-P. Clet, *Pratique d'une méthode de visualisation dans l'étude de l'écoulement et des propriétés rhéologiques d'un fluide non-newtonien* (Tech. Rep. D, I-ENSEEIHT, 1969)
35. P. Chassaing, *Turbulence en mécanique des fluides - Analyse du phénomène en vue de sa modélisation à l'usage de l'ingénieur*, Cépaduès-Ed. (Collection Polytech) (2000)
36. P. Chassaing, R.A. Antonia, F. Anselmet, L. Joly, S. Sarkar, *Variable Density Fluid Turbulence* (Kluwer Academic Publishers, 2002)
37. B. Chebbi, S. Tavoularis, Pitot-static tube response at very low Reynolds numbers. Phys. Fluids A **3**(3), 481–483 (1991)
38. R. Clausius, Sur diverses formes facilement applicables, qu'on peut donner aux équations fondamentales de la théorie mécanique de la chaleur. (French Translation H.-F. Bessard). J. Math. Pures & App., Paris **10**, 361–400 (1865)
39. J. Coirier, *Mécanique des milieux continus - Cours et exercices corrigés* (Dunod, Paris, 2001)
40. J. Coirier, *Mécanique des milieux continus - Aide mémoire* (Dunod, Paris, 2001)
41. M. Couette, Etudes sur le frottement des liquides. Ann. Chim. Phys. **21**, 433–510 (1890)
42. C.-A. Coulomb, Expériences destinées à déterminer la cohérence des fluides et les lois de leur résistance dans les mouvements très lents. Mémoires de Coulomb Paris Gauthier-Villars, Paris, tome **1**, 333–357 (1820)
43. L. Crocco, Eine neue Stromfunktion fur die Erforschung der Bewegung der Gase mit Rotation. ZAMM **17**(1), 1–17 (1937)
44. J. Le Rond D'Alembert, Paradoxe proposé aux géomètres sur la résistance des fluides. Opuscules **5**(34), 132–138 (1768)
45. H. Darcy, Recherches expérimentales relatives au mouvement de l'eau dans les tuyaux. Mém. Acad. Sci. **15**, 1–268 (1857)
46. H. Darcy, Note relative à quelques modifications à introduire dans le tube de Pitot. Ann. Ponts & Chaussées **15**(Ser 3), 351–359 (1858)

47. R. Dautray, J.-L. Lions, *Analyse mathématique et calcul numérique pour les sciences et les techniques -Tome 1: Modèles physiques, INSTN-CEA* (Coll. Ens. Math, Elsevier Masson, 1987)
48. C.G.P. De Laval, *Steam turbine*, U.S. Patent No. 522,066 (1894)
49. B. Diu, C. Guthmann, D. Lederer, B. Roulet, *Eléments de physique statistique* (Hermann, 2001)
50. H.L. Dryden, G.B. Schubauer, W.C. Mock Jr., H.K. Skramstad, *Measurements of intensity and scale of wind-tunnel turbulence and their relation to the critical Reynolds number of spheres*, NACA - Report No. 581 (1938)
51. P. Duhem, Théorie thermodynamique de la viscosité, du frottement et des faux équilibres chimiques. Mém. Soc. Sci. Phys. & Nat. Bordeaux **2** 5ème Sér., 1–210 (1896)
52. P. Duhem, Recherches sur l'hydrodynamique. Ann. Fac. Sci. Toulouse **3** 2ème Sér., 315–317 (1901)
53. P. Dupin, M. Teissié-Solier, M. *Les tourbillons alternés et les régimes d'écoulement d'un fluide autour d'un obstacle*, Inst. Elec. & Appl. Mec., Toulouse Univ. Gauthier-Villars (1928)
54. D. Enskog, *Kinetische Theorie der Vorgänge in mässig verdünnten Gasen* (Uppsala University, Ph.D, 1917)
55. L. Euler, Recherches physiques sur la nature des moindres parties de la matière. Opuscula varii argumenti **1**, 287–300 (1746)
56. L. Euler, Principes généraux du mouvement des fluides. Mém. Acad. Sci. Berlin **11**, 274–315 (1757)
57. V.M. Falkner, W. Skan, Solutions of the Boundary-layer Equations. Lond. Ed. Dublin Phil. Mag. **12**(Ser. 7), 865–896 (1931)
58. A. Favre, H. Guitton, J. Guitton, A. Lichnerowicz, E. Wolff, *De la causalité à la finalité. A propos de la turbulence* (Maloine S. A., 1988)
59. J. Grimson, *Advanced Fluid Dynamics and Heat Transfer* (McGRAW-HILL Book Company, 1971)
60. Groupe français de rhéologie, *Dictionnaire de Rhéologie* (SEBTP Paris, 1988)
61. A. Guilbaud, *L'hydrodynamique dans l'œuvre de D'Alembert 1766–1783: histoire et analyse détaillée des concepts pour l'édition critique et commentée de ses Œuvres complètes et leur édition électronique*, Ph.D. Université Claude Bernard Lyon 1 (2007)
62. E. Guyon, J.-P. Hulin, L. Petit, *Hydrodynamique physique*, 3ème édition (EDP Sciences, 2012)
63. G.H.L. Hagen, Über die Bewegung des Wassers in engen cylindrischen Röhren. Ann. Phys. Chem. (Poggendorff) **2**(46), 423–442 (1839)
64. Ses capacités et ses limitations, H. Ha Minh, La modélisation statistique de la turbulence. C. R. Acad. Sci. **327**, 343–358 (1999)
65. H. Ha Minh, A. Kourta, Semi-deterministic turbulence modelling for flows dominated by strong organized structures, in *9th Symposium on Turbulent Shear Flows* (1993)
66. H.S. Hele-Shaw, The flow of water. Nature **58**, 34–36 (1898)
67. H.S. Hele-Shaw, Stream-line motion of a viscous film - (I) Experimental investigation of the motion of a thin film of viscous fluid. Br. Assoc. Adv. Sci. Rep. **68** (Bristol), 136–142 (1898)
68. H. Helmholtz, Über Integrale der Hydrodynamischen Gleichungen, welche den Wirbelbewegungen entsprechen. J. Ang. Math., Berlin **55**, 25–55 (1858)
69. H. Helmholtz, Zur Theorie der stationären Ströme in reibenden Flüssigkeiten. Beiträge Verhand. natur.-med. Vereins Heidelberg **5**, 1–9 (1871)
70. C. Herschel, The Venturi water meter: an instrument making use of a new method of gauging water; Applicable to the cases of very large tubes, and of a small value only, of the liquid to be gauged. Trans. Am. Soc. Civil Eng. **17**, 228–259 (1888)
71. J. Hilsenrath, *Tables of Thermodynamic and Transport Properties of Air* (Carbon Dioxide, Carbon Monoxide, Hydrogen, Nitrogen, Oxygen and Steam (Pergamon Press, Argon, 1959)
72. J.O. Hinze, *Turbulence - An Introduction to Its Mechanism and Theory* (McGraw-Hill Book Company, 1959)
73. G.-A. Hirn, Sur les principaux phénomènes que présentent les frottements médiats et sur les diverses manières de déterminer la valeur mécanique des matières employées au graissage des machines. Bull. Soc. Indus. Mulhouse **26**, 188–238 (1855)

554 References

74. G.-A. Hirn, Sur les lois de la production du calorique par les frottements médiats. Bull. Soc. Indus. Mulhouse **26**, 238–277 (1855)
75. L. Howarth, On the Solution of the Laminar Boundary Layer Equations. Proc. Roy. Soc. Lond. A **164**, 547–579 (1938)
76. P. Huerre, *Mécanique des fluides - Tome I Cours* (École Polytechnique, 1998)
77. P. Huerre, P.A. Monkewitz, Absolute and convective instabilities in free shear layers. J. Fluid Mech. **159**, 151–168 (1985)
78. H. Hugoniot, Mémoire sur la propagation du mouvement dans un fluide indéfini (Première Partie). J. Math. Pures & Appl. **3**(Sér. 4), 477–492 (1887)
79. H. Hugoniot, Mémoire sur la propagation du mouvement dans les corps et spécialement dans les gaz parfaits (Deuxième partie). J. École Polytechnique **58**, 1–125 (1889)
80. I.E. Idel'cik, *Mémento des pertes de charges - Coefficients de pertes de charge singulières et de pertes de charge par frottement* (Eyrolles, 1986)
81. N. Joukowsky, Über die Konturen der Tragflächen der Drachenflieger. ZFM **22**, 81–284 (1910)
82. D. La Verne Katz, *Handbook of Natural Gas Engineering* (McGraw-Hill, 1959)
83. V.J. Katz, The history of Stokes' theorem. Math. Mag. **52**(3), 146–156 (1979)
84. D.J. Korteweg, On a general theorem of the stability of the motion of a viscous fluid. Lond. Ed. Dublin Phil. Mag. **16**(6), 112–118 (1883)
85. J.-L. Lagrange, Mémoire sur la théorie du mouvement des fluides. Nouv. Mém. Acad. Roy. Sci. Berlin 695–748 (1781)
86. H. Lamb, *Hydrodynamics*, 4th edn. (Cambridge University Press, 1916)
87. E.S. Lea, E. Meden, The De Laval steam-turbine. Proc. Inst. Mech. Eng. **67**(1), 697–714 (1904)
88. M. Lesieur, *La turbulence* (Presses Universitaires de Grenoble, 1994)
89. M. Lesieur (ed.), *Turbulence et déterminisme* (Presses Universitaires de Grenoble, 1998)
90. P.E. Liley, Survey of recent work on the viscosity, thermal conductivity, and diffusion of gases and liquefied gases below 500 K, in *Progress in International Research on Thermodynamic and Transport Properties* (1962), pp. 10–13
91. L.G. Loitsyanskii, *Mechanics of Liquids and Gases* (Pergamon Press, 1966)
92. J.-L. Luneau, *Dynamique des fluides compressibles*, Cépaduès-Editions (1975)
93. G. Magnus, Ueber die Abweichung der Geschosse, und Ueber eine abfallende Erscheinung bei rotirenden Körpern. Ann. Phys. Chem. **88**, 1–29 (1853)
94. C. Marangoni, Über die Ausbreitung der Tropfen einer Flüssigkeit auf der Oberfläche einer anderen. Ann. Phy. Chem. **143**, 337–354 (1871)
95. S.J. Marcy, Evaluating the second coefficient of viscosity from sound dispersion or absorption data. AIAA J. **28**, 171–173 (1990)
96. J. Mathieu, *Mécanique des Fluides* (École Centrale de Lyon, 1970)
97. J.C. Maxwell, Illustrations of the dynamical theory of gases. Part I. On the motions and collisions of perfectly elastic spheres. Lond. Ed. Dub. Phil. Mag. **19**(124), 19–32 (1860)
98. J.C. Maxwell, On the viscosity or internal friction of air and other gases. Phil. Trans. Roy. Soc. Lond. **156**, 249–268 (1866)
99. J.C. Maxwell, On the dynamical theory of gases. Trans. Roy. Soc. Lond. **157**, 49–88 (1866)
100. J.R. Mayer, Sur la transformation de la force vive en chaleur, et réciproquement. C. R. Acad. Sci. Paris **27**, 385–387 (1848)
101. G.E.A. Meier, Prandtl's boundary layer concept and the work in Göttingen. A historical view on Prandtl's scientific life, in *Proceedings of the IUTAM Symposium on "One Hundred Years of Boundary Layer Research" DLR-Göttingen*, Germany (2004), pp. 1–18
102. R.A. Millikan, A new modification of the cloud method of determining the elementary electrical charge and the most probable value of that charge. Phil. Mag. **19**(110), 209–228 (1910)
103. R.A. Millikan, The isolation of an ion, a precision measurement of its charge, and the correction of Stokes's law. Phys. Rev. **32**(4), 349–397 (1911)
104. R.A. Millikan, On the elementary electrical charge and the Avogadro constant. Phys. Rev. **2**(2), 109–143 (1913)

105. C.L.M.H. Navier, Mémoire sur les lois du mouvement des fluides. Mém. Acad. Roy. Sci. **6**, 389–440 (1823)
106. I. Newton, *Philosophiae Naturalis Principia Mathematica*, 1st edn. (Londini, 1687)
107. I. Newton, *Philosophiae Naturalis Principia Mathematica*, Editio Secunda (Cantabrigiae, 1713)
108. I. Newton, *Philosophiae Naturalis Principia Mathematica*, 3rd edn. (G. & J. INNYS, Londini, 1726)
109. A. Oberbeck, Über die Wärmeleitung der Flüssigkeiten bei Berücksichtigung der Strömungeninfolge vor Temperaturedifferenzen. Ann. Phys. Chem., Neue Folge **7**, 271–292 (1879)
110. C.W. Oseen, Über die stokes'sche formel und über eine verwandte aufgabe in der hydrodynamik. Ark. Mat. Astr. Fys. **6**(29), 1–20 (1910)
111. R.L. Panton, *Incompressible Flow* (Wiley, 1984)
112. V.C. Patel, Calibration of the Preston tube and limitations on its use in pressure gradients. J. Fluid Mech. **33**(1), 185–208 (1965)
113. F. Pavie, *Simulation numérique d'écoulements décollés compressibles à des vitesses élevées: aspects physiques et numériques* (Ph.D, INPT (ENSICA), 1990)
114. J. Piquet, *Turbulent flows - Models and Physics*, Rev. 2nd Print. (Springer, 2001)
115. H. Pitot, Description d'une machine pour mesurer la vitesse des eaux courantes et le sillage des vaisseaux. Mém. Acad. Roy. Sci. 363–376 (1732)
116. K. Pohlhausen, Zur näherungsweisen Integration der Differentialgleichung der laminaren Grenzschicht. ZAMM **1**(4), 252–290 (1921)
117. J.M. Poiseuille, Recherches expérimentales sur le mouvement des liquides dans les tubes de très-petits diamètres. Ann. Chim. Phys. **7**, 50–74 (1843)
118. J.M. Poiseuille, Recherches expérimentales sur le mouvement des liquides dans les tubes de très-petits diamètres. Ann. Chim. Phys. **21**, 76–110 (1846)
119. J.M. Poiseuille, Recherches expérimentales sur le mouvement des liquides dans les tubes de très-petits diamètres. Mém. Acad. Roy. Sci. **9**, 433–543 (1846)
120. S.D. Poisson, Mémoire sur la théorie du son. J. École Polytechnique **7**, 319–392 (1808)
121. S.D. Poisson, Mémoire sur les équations générales de l'équilibre et du mouvement des corps solides élastiques et des fluides. J. École Polytechnique **13**(20), 1–174 (1831)
122. S.B. Pope, *Turbulent Flows* (Cambridge University Press, 2000)
123. L. Prandtl, *Über Flüssigkeitsbewegung bei sehr kleiner Reibung*, IIIrd Int. Kong. Math. (Heidelberg, 1904)
124. L. Prandtl, *Motions of fluids with very little viscocity*, NACA TM-452 (1928)
125. L. Prandtl, *Guide à travers la mécanique des Fluides* (Dunod, 1952)
126. J.H. Preston, The determination of turbulent skin friction by means of pitot tubes. Aeronaut. J. **58**(518), 109–121 (1954)
127. W.J.M. Rankine, On plane water-lines in two dimensions. Phil. Trans. Roy. Soc. Lond. **154**, 369–391 (1864)
128. W.J.M. Rankine, On the thermodynamic theory of waves of finite longitudinal disturbances. Phil. Trans. Roy. Soc. Lond. **160**, 277–287 (1870)
129. J.W. Rayleigh, On the stability, or instability, of certain fluid motions. Proc. Lond. Math. Soc. **11**(1), 57–72 (1880)
130. J.W. Rayleigh, Investigation of the character of the equilibrium of an incompressible heavy fluid of variable density. Proc. Lond. Math. Soc. **14**, 170–177 (1883)
131. J.W. Rayleigh, Aerial plane waves of finite amplitude. Proc. Roy. Soc. Lond., Ser. A **84**, 247–284 (1910)
132. J.W. Rayleigh, On convection currents in a horizontal layer of fluid when the higher temperature is on the under side. Lond. Ed. Phil. Mag., Sixth Ser. **32**(192), 529–546 (1916)
133. J.W. Rayleigh, Dynamics of revolving fluids. Proc. Roy. Soc. Lond., Ser. A **93**, 148–154 (1916)
134. P. Rebuffet, *Aérodynamique expérimentale* (Librairie polytechnique, Ch. Béranger, 1962)

135. O. Reynolds, An experimental investigation of the circumstances which determine whether the motion of water shall be direct or sinuous, and of the law of resistance in parallel channels. Phil. Trans. Roy. Soc. Lond. **174**, 935–982 (1883)
136. O. Reynolds, On the theory of lubrication and its application to Mr. Beauchamp tower's experiments, including an experimental determination of the viscosity of olive oil. Phil. Trans. Roy. Soc. Lond. **177**, 157–234 (1886)
137. W.C. Reynolds, D.E. Parekh, P.J.D. Juvet, M.J.D. Lee, Bifurcating and blooming Jets. Ann. Rev. Fluid Mech. **35**, 295–315 (2003)
138. B. Riemann, Ueber die Fortpflanzung ebener Luftwellen von endlicher Schwingungsweite. Abhand. König. Gesell. Wissen. Göttingen **8**, 43–65 (transcribed by D.R. Wilkins) (1860)
139. B. Robins, *New Principles of Gunnery: Containing the Determination of the Force of Gunpowder, and an Investigation of the Difference in the Resisting Power of the Air to Swift and Slow Motions* (London J. Nourse, 1742)
140. M. Roy, *Thermodynamique macrosopique - Notions fondamentales* (Dunod Paris, 1964)
141. M. Roy, *Mécanique - II Milieux continus* (Dunod Paris, 1966)
142. B. Rumford, An inquiry concerning the source of the heat which is excited by friction. Phil. Trans. Roy. Soc. Lond. **88**, 80–102 (1798)
143. I.L. Ryhming, *Dynamique des fluides* (EPFL Press, 2004)
144. M.D. Salas, *The Curious Events Leading to the Theory of Shock Waves*, Invited lecture, 17th Shock Interaction Symp., Rome (2006)
145. B. Sepulvado, *Prandtl number for a typical loading of MY-T-OIL*, Personal technical paper (2009)
146. A. Sevrain, J. Meyer, J.-C. Pons, H. Boisson, Etude des structures bidimensionnelles de la zone proche d'un jet plan, in *Visualisation et Traitement d'images en Mécanique des Fluides*, ed. by M. Stanislas, J.C. Monnier (Lille, 1990), pp. 72–78
147. H. Schlichting, *Boundary-Layer Theory* (McGRAW-HILL Book Company, 1979)
148. G.G. Stokes, On a difficulty in the theory of sound. Phil. Mag. **33**, 349 (1848)
149. G.G. Stokes, On the theories of the internal friction of fluids in motion, and of the equilibrium and motion of elastic solids. Trans. Cam. Phil. Soc. **8**, 287–319 (1849)
150. G.G. Stokes, On the effect of the internal friction of fluids on the motion of pendulums. Trans. Cam. Phil. Soc. **9**(2), 8–106 (1851)
151. G.G. Stokes, Mathematical proof of the identity of the stream lines obtained by means of a viscous film with those of a perfect fluid moving in two dimensions. British Assoc. Adv. Sci. Report **68**(Bristol), 143–145 (1898)
152. V. Strouhal, Ueber eine besondere Art der Tonerregung. Ann. Phys. Chem. Leipzig **5**(10), 216–251 (1878)
153. G.I. Taylor, Stability of a viscous liquid contained between two rotating cylinders. Phil. Trans. Roy. Soc. Lond., Ser. A **223**, 289–343 (1923)
154. G.I. Taylor, The instability of liquid surfaces when accelerated in a direction perpendicular to their planes. Proc. Roy. Soc. Lond. **201**, 192–196 (1950)
155. H. Tennekes, J.L. Lumley, *A First Course in Turbulence* (MIT Press, 1972)
156. J. Thomson, On a Changing Tesselated Structure in certain Liquids [Due to Convective Circulation]. Proc. Phil. Soc. Glasgow **13**, 464–468 (1881–2)
157. W. Thomson, On vortex motion. Trans. Roy. Soc. Ed. **25**, 217–260 (1869)
158. O. Thual, *Introduction à la mécanique des milieux continus déformables* (Cepadues-Editions Toulouse, 1997)
159. O.G. Tietjens, *Applied Hydro- and Aeromechanics Based on Lectures of L. Prandtl* (McGRAW-HILL Book Company, 1934)
160. L. Tisza, Supersonic absorption and stokes' viscosity relation. Phy. Rev. **61**, 531–536 (1942)
161. G.A. Tokaty, *A History and Philosophy of Fluid Mechanics* (Dover Publications Inc., New-York, 1971)
162. W. Tollmien, Ein allgemeines Kriterium der instabilität laminarer Geschwindigkeitsverteilungen. Nach. Gesell. Wissen. Göttingen (Math.) **1**(5), 79–114 (1935)
163. A. Vaschy, Sur les lois de similitude en physique. J. Univ. Elec. 3ème Sér. **18**, 243–246 (1892)

164. J.B. Venturi, *Recherches expérimentales sur le principe de la communication latérale du mouvement dans les fluides, appliqué à l'explication de différens phénomènes hydrauliques* (Houel & Ducros, Paris, 1797)

165. H. Villat, As luck would have it - A few mathematical reflections. Ann. Rev. Fluid Mech. **4**, 1–6 (1972)

166. S. Viscardy, *Viscosity from Newton to Nonequilibrium Statistical Mechanics*, arXiv e-print (arXiv:cond-mat/0601210) (2006), pp. 1–51

167. Th. von Kármán, Über laminare und turbulent Reibung. ZAMM **21**(4), 233–252 (1921)

168. K. Wieghardt, Über einen Energiesatz zur Berechnung laminarer Grenzschichten. Ingenieur-Archiv **16**, 231–242 (1948)

169. RKh. Zeytounian, A historical survey of some mathematical aspects of Newtonian fluid flows. Appl. Mech. Rev. **54**, 525–562 (2001)

170. H. Ziegler, *An Introduction to Thermodynamics* (North-Holland, 1983)

Index

Printed in the United States
by Baker & Taylor Publisher Services